Turfgrass Insects of the United States and Canada

TURFGRASS INSECTS

OF THE **UNITED STATES**
AND **CANADA**

SECOND EDITION

Patricia J. Vittum

Associate Professor, Department of Entomology
University of Massachusetts, Amherst

Michael G. Villani

Professor, Department of Entomology
New York State Agricultural Experiment Station, Geneva
A Division of the College of Agriculture, Cornell University

Haruo Tashiro

Professor Emeritus, Department of Entomology
New York State Agricultural Experiment Station, Geneva
A Division of the College of Agriculture, Cornell University

Comstock Publishing Associates A DIVISION OF

Cornell University Press | ITHACA AND LONDON

First published 1999 by Cornell University Press
Printed in the United States of America
Color plates printed in Hong Kong

Library of Congress Cataloging-in-Publication Data

Vittum, Patricia J., 1951–
Turfgrass insects of the United States and Canada / Patricia J. Vittum, Michael G. Villani, Haruo Tashiro. — 2nd ed.
p.   cm.
Rev. ed. of: Turfgrass insects of the United States and Canada / Haruo Tashiro. 1987.
Includes bibliographical references.
ISBN 0-8014-3508-0
1. Turfgrasses—Diseases and pests—United States.   2. Turfgrasses—Diseases and pests—Canada.
3. Insect pests—United States.   4. Insect pests—Canada.   I. Villani, Michael G.   II. Tashiro, Haruo,
1917–   .   III. Tashiro, Haruo, 1917–   . Turfgrass insects of the United States and Canada.   IV. Title.
SB608. T87V58   1999
635.9′64297—dc21   99-23831

Cornell University Press strives to use environmentally responsible suppliers and materials to the fullest extent possible in the publishing of its books. Such materials include vegetable-based, low-VOC inks and acid-free papers that are recycled, totally chlorine-free, or partly composed of nonwood fibers. Books that bear the logo of the FSC (Forest Stewardship Council) use paper taken from forests that have been inspected and certified as meeting the highest standards for environmental and social responsibility. For further information, visit our website at www.cornellpress.cornell.edu.

Cloth printing   10 9 8 7 6 5 4 3 2 1

Contents

Preface to Second Edition   ix

Acknowledgments to Second Edition   xi

Preface to First Edition   xv

1   Turfgrass in the Modern Environment   1
Grass Structure   1
Grass Identification   2
Turfgrass Climatic Adaptations   3
Major Turfgrasses in the United States and Canada   5
Drought Dormancy and Its Relationship to Insect Damage   8
Dichondra Lawns   10
Economic Impact of Turfgrass Culture   10

2   Insects and Near Relatives   12
Phylum Arthropoda   12
Form and Function of Insects and Mites   12
Types of Mouthparts and Turf-Feeding Damage   17
Orders of Turfgrass-Damaging Insects and Mites   20

3   Insects and Mites: Turf Association   21
Habitats of Turfgrass Insects   21
Seasonal Presence of Injurious Stages   27

4   Acarine Pests   30
Bermudagrass Mite   30
Zoysiagrass Mite   34
Buffalograss Mite   34
Winter Grain Mite   35
Clover Mite   39
Banks Grass Mite   40

5   Orthopteran Pests: Family Gryllotalpidae   42
Southern and Tawny Mole Crickets   42

6   Hemipteran Pests: Suborder Heteroptera   54
Chinch Bug Taxonomy   54
Hairy Chinch Bug   55

Southern Chinch Bug   63
Buffalograss Chinch Bug   68
Common Chinch Bug   70

7   Hemipteran Pests: Suborder Homoptera   71
Greenbug   71
Two-lined Spittlebug   79
Rhodesgrass Mealybug   82
Ground Pearl   84

8   Lepidopteran Pests: Family Pyralidae   87
Temperate-Region Sod Webworms   87
Tropical-Region Sod Webworms   100

9   Lepidopteran Pests: Family Noctuidae   106
Cutworms and Armyworms   106
Black Cutworm   113
Variegated Cutworm   115
Armyworm   116
Bronzed Cutworm   117
Fall Armyworm   118
Yellow-striped Armyworm   120
Lawn Armyworm   121

10   Lepidopteran Pests: Family Hesperiidae   124
Fiery Skipper   124

11   Coleopteran Pests: Family Scarabaeidae   129
Overview   129

12   Scarabaeid Pests: Subfamily Aphodinae   141
Black Turfgrass Ataenius   141
Aphodius spp.   149

13   Scarabaeid Pests: Subfamily Cetoniinae   152
Green June Beetle   152

14   Scarabaeid Pests: Subfamily Dynastinae   158
Masked Chafers   158

15   Scarabaeid Pests: Subfamily Melolonthinae   167
Asiatic Garden Beetle   167
European Chafer   172
May or June Beetle   183

16   Scarabaeid Pests: Subfamily Rutelinae   196
Japanese Beetle   196
Oriental Beetle   209

17   Coleopteran Pests: Family Chrysomelidae   215
Dichondra Flea Beetle   215

18   Coleopteran Pests: Family Curculionidae   219
Billbug Taxonomy   219
Bluegrass Billbug   219

Hunting Billbug　226
Phoenician Billbug, Denver Billbug, and Other Billbug Species　229
Annual Bluegrass Weevil　230

19　Dipteran Pests: Families Tipulidae and Chloropidae　243
European Crane Fly　243
Frit Fly　249

20　Hymenopteran Pests: Family Formicidae　255
Ants—General Information　255
Red Imported Fire Ant　258
Harvester Ant　261
Lasius neoniger　264

21　Hymenopteran Pests: Families Sphecidae and Vespidae　266
Bees and Wasps　266

22　Minor Insect Pests　270
Northern Mole Cricket　270
Short-tailed Cricket　271
Grasshoppers　271
Periodical Cicadas　272
Buffalograss Mealybugs　273
Leafhoppers　273
Leafbugs and Fleahoppers　274
Bermudagrass Scale　275
Turfgrass Scale　275
Cottony Grass Scale　276
Lucerne Moth　276
Burrowing Sod Webworms　277
Granulate Cutworm　278
Striped Grassworm　278
Wireworms　279
Polyphylla spp. Grubs　279
Vegetable Weevil　280
March Flies　281

23　Turfgrass-Associated Arthropods and Near Relatives　283
Turfgrass-Associated Invertebrates　283
Turfgrass-Associated Arthropods　284
Turfgrass-Associated Insects　287
Effect of Insecticides on Nontarget Arthropods　290

24　Vertebrate Pests　291
Birds　291
Mammals　293

25　Principles of Integrated Pest Management　299
General Introduction　299
Site Assessment　300
Scouting or Monitoring　301
Setting Thresholds　301
Predicting Pest Activity　301
Stress Management　302

Biological Control Strategies   305
Chemical Control Strategies   306
Finding Compatible Strategies   307
Evaluation   307
Advantages   307

26   Sampling Techniques and Setting Thresholds   309
Accurate Diagnosis   309
Scouting or Monitoring   310
Collecting and Labeling Specimens for Identification   310
Population Survey Techniques   312
Setting Thresholds   320

27   Biological Control Strategies   326
Use of Natural Enemies in Turfgrass Management   326
Endophytes   337
Pheromones   338
Insect Growth Regulators   338
Botanicals   339

28   Chemical Control Strategies   341
Accurate Diagnosis   341
Threshold Populations   341
Influence of Thatch   342
Timing of Application—Seasonal   343
Timing of Application—Daily   347
Windows of Opportunity   347
Names of Insecticides   347
Chemical Properties of Insecticides   348
Chemical Classes of Insecticides   350
Insecticide Resistance   352
Effect of Insecticides on Nontarget Organisms   353
Insecticide Formulations   353
Toxicity of Insecticides   355
Application Technology   356
Environmental Issues   356
Selecting an Insecticide   358

Appendix 1. English and Metric Units of Measure and Conversions   361
Units of Measure   361

Appendix 2. Abbreviations Used in the Text   365

Glossary   367

References   379

Index   405

Preface to Second Edition

At the time the first edition of *Turfgrass Insects of the United States and Canada* was released (1987), the field of turfgrass entomology was just beginning to expand. During the 1960s and 1970s, there had been only a few people conducting research in this area—notably Haruo Tashiro (Cornell University) and Harry Niemczyk (Ohio State University). However, during the 1980s several newly trained entomologists began to conduct research in turfgrass entomology and the knowledge base began to increase exponentially.

In this revision, we incorporate the critically important research in turfgrass entomology that has been conducted since 1986. Much of this work has enhanced our understanding of turfgrass insect behavior and strategies for managing pest populations. Examples include the soil radiography studies of white grubs conducted by Mike Villani and others; studies of the ecology of cutworms in low-mown bentgrass conducted by Chris Williamson, Dan Potter, Dave Shetlar, and others; the identification of new pests, such as the buffalograss chinch bug, by Fred Baxendale; an understanding of the role of endophytes in providing tolerance to many turf insects, including chinch bugs, webworms, and billbugs; and the advances in insecticide management techniques, including application technology, expanded use of thresholds in decision-making, and the development of new insecticide chemistry.

As in the first edition, the information on each insect is presented in the same manner—comments on taxonomy, importance in fine turfgrass, history and distribution, host plants and damage, description of stages, seasonal history and habits, miscellaneous features, and natural enemies. We include a distribution map and line sketches of several life stages (courtesy of the Entomological Society of America) for each pest we discuss. The distribution maps are based on references in the literature and informal personal observations from numerous colleagues, and are intended to be guides only. There may be several situations where a turf insect is found in an area not delineated by our distribution maps.

We made a few changes in the order of presentation. For example, we moved the chapter titled "Insects and Mites: Turf Association" from the end of the book (formerly Chapter 21) in the first edition to the front (Chapter 3) in this edition so that the discussion of turf habitat and insect habits receives greater emphasis. We replaced the chapters previously titled "Detection and Diagnosis of Infestations and Damage," "Population Survey Techniques," and "Insect Control: Principles and Strategies" with four chapters discussing various aspects of integrated pest management. Much of the original content remains in these new chapters, but we adjusted the order of presentation to emphasize certain aspects and included expanded coverage of several areas. The new layout, with "Principles of Integrated Pest Management" (Chapter 25), "Sampling Techniques and Setting Thresholds" (Chapter 26), "Biological Control Strategies" (Chapter 27), and "Chemical Control Strategies" (Chapter 28), provides a stronger presentation of issues of insect pest management, and makes it easier to find specific information. We stress identification and quantification of pest populations and discuss some of the concepts involved in setting thresholds, because this is the core of any integrated pest management program. All of these chapters provide greatly expanded information on aspects of pest management, particularly the chapter on biological control, and should prove useful to turf managers.

Some of the chapters discussing specific insects are the same or slightly different from those in the first edition. Others, we expanded considerably. For example, we reorganized the discussion of cutworms and armyworms (Chapter 9) so that each species is discussed separately, which should make it easier for turf managers or other students of turfgrass to find the desired information. We reorganized and restructured slightly the chapters on white grubs (Chapters 11–16) to provide information a bit more concisely. At the same time we incorporated results of numerous laboratory and field studies conducted in the past 10 years, greatly expanding the body of knowledge of scarabs. We substantially revised the chapter on billbugs and weevils (Chapter 18) and provide an explanation of the current understanding of the taxonomy of billbugs and weevils. We expanded the discussion of hunting billbugs and added Phoenician and Denver billbugs. We added sections on the buffalograss chinch bug (Chapter 6) and two-lined spittlebug (Chapter 7), as well as on a few vertebrate pests in Chapter 24. We also expanded the discussion of ants (Chapter 20) and of turfgrass-associated arthropods (Chapter 23).

One of the strengths of the original edition was the quality and quantity of full-page color plates. The color plates in the first edition have proved very useful to turf managers for identifying insect pests in the field, and we felt further expansion was appropriate. This edition includes 72 full-page color plates, presented in a visually pleasing manner, including numerous photographs of beneficial organisms and biological control agents, as well as examples of sampling techniques.

We hope you enjoy using this book as much as we enjoyed preparing it.

PATRICIA J. VITTUM, MICHAEL G. VILLANI, AND HARUO TASHIRO

Geneva, New York

Acknowledgments to Second Edition

$\mathbf{S}$uch an undertaking requires the support and cooperation of numerous colleagues. We imposed on several friends and colleagues to review the chapters from the original edition and solicited their input to determine what, and how much, new information should be included in the new edition. Many of these "preliminary reviewers" were extremely helpful and supplied reprints or unpublished data that we incorporated into the text. After we completed our revisions, we sent the chapters back to the same reviewers, and invited their comments on the readability and accuracy of our revisions. Without the patience and dedication of these reviewers, this revision would undoubtedly be less complete and accurate.

All these revisions further strengthen a text that already had come to be recognized as "the bible of turfgrass entomology". We are indebted to Haruo Tashiro, who had the perseverence to complete the massive undertaking of preparing the first edition. His style and organizational plan worked very well, and we have endeavored to preserve it as much as possible.

The chapter reviewers included Steven R. Alm, University of Rhode Island, Kingston; Arthur L. Antonelli, Western Washington Research and Extension Center, Puyallup; Frederick P. Baxendale, University of Nebraska, Lincoln; S. Kris Braman, University of Georgia, Griffin; Rick L. Brandenburg, North Carolina State University, Raleigh; Patricia P. Cobb, Auburn University, Auburn, Alabama; Rich S. Cowles, Connecticut Agricultural Experiment Station, Windsor; Whitney S. Cranshaw, Colorado State University, Fort Collins; Robert L. Crocker, Texas A&M Agricultural and Extension Center, Dallas; Bastiaan M. Drees, Texas A&M Extension, Bryan; Roch Gaussoin, University of Nebraska, Lincoln; Wendy Gelernter, Pace Consulting, San Diego, California; Timothy J. Gibb, Purdue University, West Lafayette, Indiana; Arnold H. Hara, University of Hawaii, Honolulu; Paul R. Heller, Pennsylvania State University, University Park; Will G. Hudson, University of Georgia, Tifton; Richard Hull, University of Rhode Island, Kingston; Albert W. Johnson, Clemson University, Florence, South Carolina; Jennifer M. Johnson-Cicalese, New Brunswick, New Jersey; Michael G. Klein, U.S. Department of Agriculture Japanese Beetle Laboratory, Wooster, Ohio; Daniel A. Potter, University of Kentucky, Lexington; James A. Reinert, Texas A&M Agricultural and Extension Center, Dallas; Paul S. Robbins, New York State Agricultural Experiment Station

(NYSAES), Geneva; Frank Rossi, Cornell University, Ithaca, New York; David J. Shetlar, Ohio State University, Columbus; David R. Smitley, Michigan State University, East Lansing; Gwen Stahnke, Washington State University, Puyallup; Stanley R. Swier, University of New Hampshire, Durham; Mike P. Tolley, Dow Elanco, Indianapolis, Indiana; and R. Chris Williamson, TrueGreen-ChemLawn, Columbus, Ohio.

Several of these individuals were extraordinarily helpful: Fred Baxendale and Pat Cobb reviewed several chapters on short notice, while Dave Shetlar provided tremendous input in the webworm, cutworm, and billbug sections (Chapters 8, 9, and 18). Steve Alm, Fred Baxendale, Robert Crocker, Tim Gibb, Paul Heller, and Dan Potter reviewed all the grub chapters and provided valuable insight on reorganizing them (Chapters 11–16). Jennifer Johnson-Cicalese provided volumes of material and suggested revisions for the billbug section (Chapter 18). Bastiaan ("Bart") Drees provided a very thorough discussion of red imported fire ants (Chapter 20), most of which was incorporated directly into the text. His revisions have greatly enhanced the coverage of RIFAs and their role in turfgrass management.

Just as Haruo Tashiro noted in the first edition, we too are indebted to the many people, including some who were also reviewers, who freely provided us with color transparencies to be included in this edition or insect specimens that were then photographed at the NYSAES in Geneva. Those contributors include W. F. Berliner, G. C. Hickman, L. L. Masters, and T. Yates, all from the American Society of Mammalogists Mammal Slide Library; Arthur L. Antonelli; Frederick P. Baxendale; Rick L. Brandenburg; Leland R. Brown, University of California, Riverside; R. Scott Cameron, Texas Forest Service, Lufkin; Patricia P. Cobb; Sharon J. Collman, Cooperative Extension Service, Seattle, Washington; Cornell University Laboratory of Ornithology, Ithaca, New York; Robert L. Crocker; David N. Ferro, University of Massachusetts, Amherst; Ken Gray Slide Collection, Oregon State University, Corvallis; M. P. Johnson, University of Kentucky, Lexington; James Kalisch, University of Nebraska, Lincoln; Michael G. Klein; Jeffrey Kollenkark; C. Kouskalekas, Auburn University, Auburn, Alabama; W. Mesner; Charles M. Murdoch, University of Hawaii, Honolulu; Harry D. Niemczyk, Ohio Agricultural Research and Development Center, Wooster; Asher K. Ota, Hawaii Sugar Planters' Association, Aiea; F. B. Peairs; Daniel A. Potter; Michael F. Potter, University of Kentucky, Lexington; Roy W. Rings, Ohio Agricultural Research and Development Center, Wooster; Mark K. Sears, University of Guelph, Guelph, Ontario, Canada; David J. Shetlar; Herbert T. Streu, Rutgers University, New Brunswick, New Jersey; Mike P. Tolley; D. M. Tsuda, University of Hawaii, Honolulu; and Louis M. Vasvary, Rutgers University, New Brunswick, New Jersey.

We incorporated several new figures and sketches into this second edition, including many that were drawn by Robert Jarecke (NYSAES) and first appeared in the Handbook of Turfgrass Insect Pests (R. L. Brandenburg and M. G. Villani, editors, 1995), published by the Entomological Society of America (ESA), Lanham, Maryland. We are indebted to the ESA and to Robert Jarecke for granting us permission to reproduce these sketches and distribution maps. In addition, Robert Jarecke provided several new figures, including sketches of the life stages of the fiery skipper and the European crane fly (Figures 10-2 and 19-2) and the lacinia of southern and tawny mole crickets (Figure 5-3). Terrence McSharry (University of Massachusetts, Amherst) provided a new sketch of the turf profile (Figure 3-1). Stephen Thomas (University of Massachusetts, Amherst) provided three new sketches: Berlese funnel (Figure 26-3), pitfall trap (Figure 26-4A), and linear pitfall trap (Figure 26-4B).

Many of the photographic and illustrative services for NYSAES, and for this book in particular, were provided by Joseph Ogrodnick and Robert Way. Rob Way worked indefatigably with Haruo Tashiro to arrange and lay out the color plates. Many of the black-and-white diagrams were drawn by Rose McMillen-Sticht and Haruo Tashiro for the first edition, and were retained in this edition because of their clarity of presentation.

Special thanks go to Daniel Potter and Dave Shetlar, who reviewed the entire text before we began our revisions. Their suggestions, particularly regarding overall organization, proved to be invaluable. In addition, Rick Brandenburg read through the entire revision, double-checking for accuracy of figure and plate numbers, references, readability, consistency, and accuracy of text. Without their efforts, this edition would undoubtedly be much less complete and polished.

We also thank the New York State Turfgrass Association (Latham, New York) for supplying funds to help offset the cost of including the color photographs.

Preface to First Edition

This book attempts to fill a long-standing need for a comprehensive text-reference on turfgrass insects that will bring together under one cover a discussion of practically all insects and other arthropods that are destructive to turfgrass in the United States, including Hawaii but excluding Alaska, and in southern Canada bordering the United States. Turfgrass insects of the latter region are included because climates near the border are similar.

I have provided substantial technical detail to make the book useful to the professional entomologist who might seek background information on biology and behavior as a basis for further study and research. Measurements in most cases are given in SI (metric) units, followed by their American equivalents in parentheses. Conversion to the metric system in the United States is fully expected to take place, however slowly, at least for the more commonly used measurements. I hope that the technical nature of the text does not dissuade turfgrass managers with limited or no entomological training from using this book to help them understand some aspects of the turfgrass insects they encounter. They will find many distribution maps and life history charts, as well as other illustrations that are self-explanatory.

The full-page color plates, which depict some phase of practically all turfgrass insects found in the United States and southern Canada, should be helpful to both professional entomologists and laymen, as they aid in the identification of many turfgrass insects and promote understanding of the insects' habits.

Chapter 1, an introduction, discusses the turfgrasses that are most important agronomically. Chapter 2 provides fundamental information about insects and related arthropods. In Chapters 3 through 21, the orders and families of pest arthropods are covered in the same sequence used in most introductory textbooks in entomology; this sequence affords a logical framework for the treatment of the entire insect fauna. Chapters 3 through 20, and the appropriate sections within those chapters, discuss the taxonomy, importance, history and distribution, host plants, stages, life history and habits, and finally natural enemies of each insect or group of closely related species. Treatment of these topics in the same sequence throughout the book facilitates comparison of the habits of different species or groups. Chapter 22 considers vertebrate pests of turfgrass, since their presence and destruction of turf in most cases is directly related to the presence of insects. Finally, Chapters 23 through

26 provide a general overview of all the interrelationships between insects and the turf-grass ecosystem.

Control recommendations for insects were purposely omitted because these change frequently as new insecticides are approved and as older insecticides become ineffective for various reasons. Chapter 26, a discussion of insect control principles and strategies, should provide sufficient background to permit effective management of turfgrass insects.

Many persons and organizations have contributed to the realization of this book. The inclusion of the full-page color plates was made possible by private funds that helped subsidize their cost. Generous financial support for them, which I wish to acknowledge with gratitude, has been granted by the following organizations: Ciba-Geigy Corporation, Greensboro, North Carolina; Mobay Chemical Corporation, Kansas City, Missouri; New York State Turfgrass Association, Massapequa Park, New York; O. M. Scott & Sons, Marysville, Ohio; The Lawn Institute, Pleasant Hill, Tennessee; and Union Carbide Company, Inc., Research Triangle Park, North Carolina.

The manuscript review process was conducted with help from entomologists and other biologists who have been or are presently involved in turfgrass research or extension activities. I owe much to the following reviewers for their time and effort: entomologists Sami Ahmad and Herbert T. Streu, Rutgers—The State University of New Jersey, New Brunswick; William A. Allen, Virginia Polytechnic Institute and State University, Blacksburg; Arthur L. Antonelli, Western Washington Research and Extension Center, Puyallup; James R. Baker, North Carolina State University, Raleigh; Paul B. Baker, Paul J. Chapman, Timothy J. Dennehy, Charles J. Eckenrode, and Michael G. Villani, New York State Agricultural Experiment Station (NYSAES), Geneva; Robert L. Crocker, Texas A&M Agricultural and Extension Center, Dallas; Paul R. Heller, Pennsylvania State University, University Park; John L. Hellman, University of Maryland, College Park; Milton E. Kageyama, O. M Scott & Sons, Marysville, Ohio; James A. Kamm, U.S. Department of Agriculture at the Oregon State University, Corvallis; M. Keith Kennedy, S. C. Johnson & Sons, Inc., Racine, Wisconsin; S. Dean Kindler, U.S. Department of Agriculture at the University of Nebraska, Lincoln; Thyril L. Ladd and Michael G. Klein, U.S. Department of Agriculture at the Ohio Agricultural Research and Development Center, Wooster; Kenneth O. Lawrence, ChemLawn Services Corporation, Boynton Beach, Florida; Wallace C. Mitchell and Charles L. Murdoch, University of Hawaii, Honolulu; Harry D. Niemczyk, Ohio Agricultural Research and Development Center, Wooster; Daniel A. Potter, University of Kentucky, Lexington; Roger H. Ratcliffe, U.S. Department of Agriculture, Beltsville, Maryland; Donald L. Schuder, Purdue University, West Lafayette, Indiana; Mark K. Sears, University of Guelph, Guelph, Ontario, Canada; David J. Shetlar, ChemLawn Corporation, Columbus, Ohio; Patricia J. Vittum, University of Massachusetts, Suburban Experiment Station, Waltham; Joseph E. Weaver, West Virginia University, Morgantown; and Turf Management Specialist A. Martin Petrovic, Cornell University, Ithaca, New York.

Many entomologists, including some who were also reviewers, lent color transparencies of turfgrass insects for color plates, especially in the case of the southern and western insects; the captions indicate the source of borrowed transparencies. I am most grateful to the following individuals, who have contributed to the plates: William A. Allen, Arthur L. Antonelli, Leland R. Brown, University of California, Riverside; R. Scott Cameron, Texas Forest Service, Lufkin; Patricia P. Cobb and Costas Kouskolekas, Auburn University, Auburn, Alabama; Sharon J. Collman, Cooperative Extension Service, Seattle, Washington; Robert L.

Crocker; M. Keith Kennedy; Michael G. Klein; Charles L. Murdoch; Harry D. Niemczyk; Asher K. Ota, Hawaii Sugar Planters' Association, Aiea; Daniel A. Potter; Roy W. Rings, Ohio Agricultural Research and Development Center, Wooster; Mark K. Sears; David J. Shetlar; Herbert T. Streu and Louis M. Vasvary, Rutgers—The State University of New Jersey; and Michael G. Villani (for slides he prepared before joining the Department of Entomology, NYSAES). Roughly half the color slides attributed to the NYSAES were photographed by Gertrude Catlin (now retired) and Joseph Ogrodnick, and the remaining slides were photographed by myself except for single slides by John Andaloro, Shiu-Ling Chung, and James Larner, all former employees of the NYSAES.

Various individuals supplied live subjects for many photographic transparencies. Some specimens were available from routine rearing colonies, while for others, concerted efforts were made to locate infestations so that live specimens could be collected and forwarded. These contributions have significantly improved the scope of the color illustrations. In this connection I must thank James R. Baker; William C. Buell, veterinarian, Geneva, New York; Frank Consolie, NYSAES; Kenneth O. Lawrence; Wayne C. Mixson, O. M. Scott & Sons, Opapka, Florida; Charles L. Murdoch; Roger H. Ratcliffe; David J. Shetlar; Constance Strang, Cooperative Extension Service, Plainview, New York; Clyde Sorensen, North Carolina State University, Raleigh; and John Zukowski, Eisenhower Park, East Meadow, New York.

Important turfgrass insects for which neither color transparencies nor live specimens were available for photography included mainly pyralid and noctuid moths and the more common beetles of the genus *Phyllophaga*. Illustration of these was possible thanks to Paul J. Chapman and Siegfred E. Lienk, NYSAES, who collected and prepared the noctuid moths as museum specimens. Thanks are also due to James K. Liebherr, curator of the Cornell University Collection at Ithaca, who supplied museum specimens of pyralid moths, *Phyllophaga* adults, and a variety of secondary insect pests. These contributions are indicated in the source notes in the captions.

Mary Van Buren, librarian of the NYSAES, has most helpfully assisted in searching the literature, obtaining publications through interlibrary loan, and tracking down publications and dissertations, often with the vaguest of clues. I am immensely grateful for her time and efforts in adding to the fount of information available to me.

The photographic and illustrative services for the Departments of Entomology and Plant Pathology, NYSAES, are provided by Joseph Ogrodnick, Bernadine Aldwinckle, and Rose McMillen-Sticht. Gertrude Catlin also provided this service until her retirement. In addition to supplying color slides, they photographed many of the black-and-white illustrations. Bernadine Aldwinckle also did much of the intermediate work involved in the preparation of the color plates. Rose McMillen-Sticht's artistry and illustrative skills are evident in the many drawings she prepared in their entirety or adapted from earlier publications. The many significant contributions made by these station staff members have greatly increased the value of the book.

Two members of the secretarial staff of the Department of Entomology, NYSAES, made essential contributions to this book. I owe a debt of gratitude to Janice Allen for typing almost the entire manuscript, entering it on the word processor, and preparing the many revisions. Preparation and completion of the book also entailed much correspondence, particularly to obtain permission to use copyrighted material. I am grateful to Donna Price for doing most of this work.

To the Department of Entomology, NYSAES, I express my most sincere gratitude for allowing me unlimited access to the services of all the staff members whose special skills I sought and for permission to use the fruits of their labor—the many color slides, black-and-white photographs, and drawings, as well as the secretarial and stenographic output so efficiently and willingly rendered. I also thank the department for the use of its facilities and supplies.

Finally, I am grateful to Helene Maddux and Robb Reavill of Cornell University Press, who were most patient and helpful during the preparation and completion of this book.

Haruo Tashiro

Geneva, New York

Turfgrass Insects of the United States and Canada

1

Turfgrass in the Modern Environment

O f the more than 7,500 species of plants in the grass family Poaceae (formerly Gramineae), about 40 species are considered "major" turfgrasses in various ornamental, recreational, and functional uses throughout the world. *Turfgrass* typically refers to an individual plant and *turf* refers to a uniform stand of grass or a mixture of grasses mowed at a relatively low height, usually less than 10 cm and serving any of the above-mentioned purposes. In accordance with common usage, turfgrass maintained around a residential property is called a *lawn*. *Sod* typically refers to plugs, squares, or strips of turf, often grown commercially and used for vegetative planting. A *green* is a smooth, grassy area maintained at the lowest cut for golf, lawn bowling, tennis, croquet, or other sports (Beard 1973, Hanson et al. 1969, Smiley 1983).

Healthy turf is aesthetically pleasing as well as functional. Residential lawns, golf courses, athletic fields, cemeteries, parks, and arboretums all benefit from healthy, lush stands of turfgrass. Healthy root systems reduce soil erosion and allow for enhanced infiltration of water into the soil (thereby reducing surface runoff). Normal plant growth enriches the soil by adding organic matter and other decomposition products. Turfgrass also improves air quality by absorbing carbon dioxide (and, in some cases, carbon monoxide) and releasing oxygen. Through evapotranspiration, turf also reduces air temperatures, particularly in urban and suburban areas.

Many grass species used for turf are also found in pasture, field, and forage production associated with the livestock industry. In this usage, particular grasses often have the same pest problems that are associated with them when used as turf. This text does not, however, cover grasses grown for livestock consumption.

Grass Structure

All turfgrasses have a similar basic structure with some variations. A typical grass plant (Figure 1-1) has a crown (an unelongated stem composed of leaf primordia and buds) located at or near the soil surface where most of the meristematic activity occurs. It is this basal growth that enables the plant to tolerate repeated mowing without sustaining damage. A fibrous adventitious root system originating from the crown permeates throughout the surface soil. The stem, also arising from the crown, is enclosed in a leaf composed of the leaf

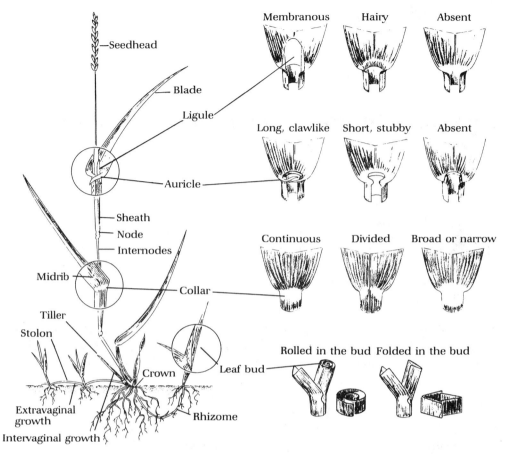

Figure 1-1. Schematic diagram of a grass plant. (Adapted from Scott's information manual for lawns 1979, p. 46, courtesy of the O. M. Scott & Sons Company, Marysville, Ohio.)

sheath and a wider unfolding leaf that is called the *blade*. If it is allowed to grow naturally without mowing, the stem usually terminates in a cluster of flowers (inflorescence), sometimes called a *seedhead*. Following termination, the tiller dies, often resulting in discoloration of the turf. Some mowed turf may also produce flowers below the cutting height.

Lateral growth and maturation of a grass plant consists of *tillers* (primary lateral stems) that emerge directly from the original crown and/or *stolons* (aboveground stems) or *rhizomes* (long underground stems) that produce secondary and tertiary crowns at their internodes. The tissues in the crown are the most vital portion of the turfgrass. A plant can recover from loss of roots or death of leaves and stems but not from the death of its crown.

Grass Identification

Seemingly minute differences found at the junction of the leaf sheath and blade are of utmost importance in the vegetative identification of grass species (Figure 1-1). These include various differences found in structures called the *ligule,* the *auricle,* and the *collar.* The vernation of leaf buds (e.g., rolled or folded) is also an important taxonomic character (Beard 1973, Anonymous 1979, Turgeon 1991).

Turfgrass Climatic Adaptations

Turfgrasses grown in the United States are designated as either cool-season or warm-season grasses, depending on their climatic adaptations. Cool-season grasses grow best at temperatures between 15.5° and 24.0°C and their growth is markedly slower outside this range. Warm-season grasses grow best at temperatures between 26.6° and 35.0°C and their growth is usually inhibited at temperatures below 10°C. Warm-season grasses are best adapted to the climate in roughly the southern third of the United States, designated by a line that proceeds east of the 100th meridian, then westward through the southern half of each state bordering Mexico to the Pacific Ocean (Figure 1-2).

Each of the two zones is further divided according to precipitation into the humid and arid-semiarid zones. The dividing line lies near the 100th meridian. Thus the four distinct turfgrass adaptation zones in the United States (Figure 1-2) are governed by temperature and precipitation. Two distinct coastal influences also occur. Along the Atlantic Coast the warm, humid zone extends north as far as Delaware. Along the Pacific Coast, a narrow cool, humid zone extends from the Canadian border to southern California. The four zones and their dominant features as given by Beard (1973, 1975, 1984) are listed below.

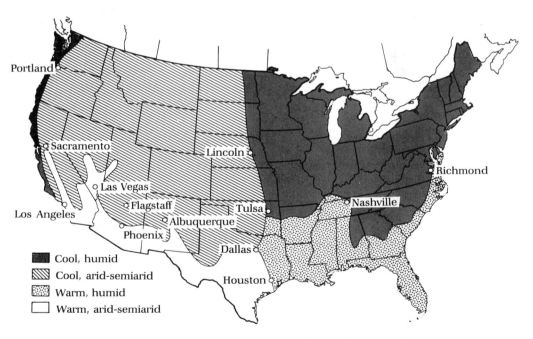

Figure 1-2. Turfgrass adaptation zones in the United States. (Adapted from Beard 1975, p. 4, courtesy of Beard Books, College Station, Tex.)

Cool, Humid Zone

The cool, humid zone is characterized by cold winters and mild to hot summers in the east and by mild temperatures along the Pacific Coast. Rainfall ranges from 50 to 115 cm throughout the zone in the East and from 50 to 250 cm along the Pacific Coast. Kentucky bluegrass is the dominant species, followed by bentgrasses, tall and fine fescues, perennial ryegrass, and annual ryegrass. Although not planted, annual bluegrass is a major species in the region, found on heavily utilized and wet or shady sites. Tall fescue, zoysiagrass, and bermudagrass are sometimes planted in the transition zone.

Cool, Arid-Semiarid Zone

This zone consists of two contrasting regions, the vast central plains area to the east and the mountains to the west. Both are characterized by cold winters and hot summers. Rainfall ranges from less than 25 cm in the intermountain plateaus to about 65 cm along the eastern edge of the zone. In the drier areas, irrigation is required to maintain turf of good quality. With irrigation, Kentucky bluegrass, perennial ryegrass, and creeping and colonial bentgrasses are dominant through most of the region, with bermudagrass and zoysiagrass used sparingly in the transition zone in Kansas, Oklahoma, and Texas. Where there is no irrigation, buffalograss, *Buchloe dactyloides* (Nutt.) Engelm., and wheatgrass, *Agropyron smithii* Rydb., are used in the transition belt because both have good to excellent drought resistance.

Warm, Humid Zone

The climate of this zone is characterized by mild temperatures in the northern portion and subtropical areas along the Gulf of Mexico and in Florida. Rainfall ranges from 100 cm along the Atlantic seaboard to 180 cm along the eastern Gulf Coast, and is as low as 65 cm in parts of Texas and Oklahoma. Bermudagrass is grown throughout the zone. Zoysiagrass is planted mostly in the north, while bahiagrass, centipedegrass, and St. Augustinegrass are used primarily in the southern third of the region. Buffalograss is becoming increasingly popular in the western parts of this region.

Hawaii is in this zone. The turfgrasses grown there are nearly the same as those of southern Florida except for bahiagrass, which is not grown in Hawaii.

Warm, Arid-Semiarid Zone

This zone is a belt occupying the southern half of each state from Texas into southern California, with northward projections into Nevada and central California (Figure 1-2). Precipitation ranges from less than 13 cm to about 50 cm; summers are generally dry. Bermudagrass is used widely with and without irrigation. Zoysiagrass is used to a limited extent. St. Augustinegrass, centipedegrass, and dichondra are grown in southern California. Cool-season grasses such as perennial or annual ryegrass and rough bluegrass are used extensively for temporary cover of dormant warm-season species during the winter months. Irrigation is generally required to maintain lawns of good quality.

Transition Zone

A relatively narrow belt 160–480 km wide extending along the line of cool-season and warm-season adaptation zones constitutes a transitional climatic zone. In this region, either group of turfgrasses (warm season or cool season) may survive and can do well under weather conditions favorable to the agronomic requirements of the given species. However, under adverse conditions, a stressed species may not survive at all. This transition zone is widely regarded as the most challenging place to maintain turf in the United States. Turf managers in these areas often overseed dormant warm-season species with high rates of cool-season species, particularly in late fall and early spring.

Major Turfgrasses in the United States and Canada

Cool-Season Grasses

The most commonly used cool-season grasses include some 12 species in five genera, as listed below (Plate 1). All except the ryegrasses and buffalograss are native to Europe or Eurasia. The ryegrasses are native to North America and Asia Minor (Beard 1973, Hanson et al. 1969).

Bluegrasses	*Poa* spp.
Kentucky	*P. pratensis* L.
Annual	*P. annua* L.
Rough	*P. trivialis* L.
	P. supine
Ryegrasses	*Lolium* spp.
Perennial	*L. perenne* L.
Italian (annual)	*L. multiflorum* Lam.
Fescues	*Festuca* spp.
Red (creeping)	*F. rubra* L.
Chewings	*F. rubra commutata* Gund.
Tall	*F. arundinacea* Schreb.
Bentgrasses	*Agrostis* spp.
Creeping	*A. palustris* (Huds.)
Colonial	*A. tenuis* Sibth
Velvet	*A. canina* L.
Buffalograss	*Buchloe dactyloides* (Nutt.) Engelm.

All of the bentgrasses and bluegrasses listed (except some forms of annual bluegrass), as well as red fescue, have extravaginal growth and produce rhizomes or stolons, which produce creeping, spreading habits of growth. Some ecotypes of annual bluegrass, the ryegrasses, and the remaining fescues have intravaginal growth, resulting in a tufted or bunch habit of growth (Beard 1973). Growth habit has an important bearing on the ability of the turf to recover from a mild attack by insects. Patches of dead grass in recovery are filled in much more rapidly with grasses of extravaginal growth because of their creeping, spreading growth habits.

The **bluegrasses** (most commonly Kentucky bluegrass) are the most widely distributed turfgrasses in the cool, humid region. The most distinctive vegetative character of the genus *Poa* is the boat-shaped leaf tip. Most bluegrasses are adapted to moist, fertile soils of pH 6.0 – 7.0. Except for annual bluegrass, *Poa* spp. show their best quality at a cutting height of 2.5 – 5.0 cm.

Kentucky bluegrass is the most widely used general-purpose turf and lawn grass in cool-season zones. It has an extensive rhizome system that allows rapid recovery from damage. Most general-use sod produced commercially in the cool-season region is composed primarily of Kentucky bluegrass. There have been many new cultivars developed that are very dense, low growing, and dark green and that have good disease resistance. More than 50 improved cultivars are evaluated annually in the National Turfgrass Evaluation Program (NTEP).

Annual bluegrass is generally considered a weed in turfgrass culture, but many golf courses in the Northeast manage *Poa annua reptans* as a perennial grass. It is typically light green to greenish yellow, with shorter, broader, softer leaves than Kentucky bluegrass. It is well adapted to moist, shaded environments and compacted soils, and grows best in fertile soils of pH 6.5 – 7.5. A cutting height of 2.5 cm or less makes it a very aggressive, competitive turfgrass. Its prolific flower production, even at a cutting height of 0.6 cm or less, makes it an undesirable grass on putting greens, which it often invades.

Rough bluegrass produces turf of relatively poor quality in full sun and does best in shaded areas that are wet and poorly drained. It is included in many lawn and turf seed mixes for such areas. Rough bluegrass often invades Kentucky bluegrass much as creeping bentgrass does, producing a light-green patch. It is often used as a temporary grass in winter overseeding of warm-season grasses.

Ryegrasses establish more rapidly than any other commonly used cool-season turfgrasses and are often used to produce a quick stand. *Perennial ryegrass* is used for many general purposes, including home lawns and shoulders alongside roads, as well as golf course tees and fairways in the Middle Atlantic states. It has excellent tolerance to wear, so it is a desirable turfgrass for athletic fields and fairways with little or no irrigation. Many improved cultivars are compatible with Kentucky bluegrass in color, fineness of leaf, and shoot growth, and therefore often make excellent polystands with Kentucky bluegrass blends. Perennial ryegrass is commonly used in the fall and winter for winter overseeding where warm-season turfgrasses are going into dormancy. Some cultivars contain endophytic fungi, which provide a level of resistance to some insects, such as chinch bugs, cutworms, billbugs, and webworms.

Italian ryegrass, often called *annual ryegrass*, has limited usage in general-purpose seed mixtures for obtaining a quick stand and has largely been replaced by perennial ryegrass.

The **fescues** are adapted to the cool, humid regions and tolerate droughty, infertile acid soils with pH 5.5 – 6.5. Their establishment rate lies between that of the slowly germinating Kentucky bluegrass and the rapid ryegrass. *Red fescue* is one of the three most widely used turfgrass species in the cool, humid region. Seed mixtures with Kentucky bluegrass are used to supplement the latter in the shade and in droughty, sandy areas.

Chewings and *hard fescues* are also used in seed mixtures with Kentucky bluegrass for lawns and for general-purpose turf. In appearance they resemble red fescue and have similar performance characteristics. Hard fescue can be grown as a monoculture, whereas red and Chewings fescues are most often mixed with Kentucky bluegrass or perennial ryegrass. *Tall fescue* has a bunch-type growth habit, with leaf blades generally coarser than many other cool-season turfgrasses (Plate 5). New cultivars of tall fescue produce a turf of good quality and can be mixed with other grasses. Tall fescue is well adapted to the transition belt between the cool, humid zone and the warm, humid zone and is used mainly in this area. It is often used in shady sites and on athletic turf.

Bentgrasses are the cool-season grasses that most tolerate the continuous, close mowing heights characteristic of golf greens, lawn bowling greens, and lawn tennis courts, where they are most frequently found. Several improved cultivars are available. *Creeping bentgrass* is one of the most vigorous cool-season grasses, with rapid stolon growth. It forms a high-quality turf at a cutting height of 1.8 cm or less and is among the outstanding grasses available for putting and bowling greens maintained at 0.5–0.8 cm. It also performs well on tees and fairways maintained at 1.2–2.5 cm if irrigated.

Colonial bentgrass, the second most widely used *Agrostis* species, is in very limited usage today. It requires a high maintenance level (including water management and fertility) for the best turf and does well at a cutting height of 0.8–2.0 cm. *Velvet bentgrass* is one of the finest textured grasses, with almost needle-like leaves. It is used primarily for putting and bowling greens and does best in New England. Turf of the highest quality is produced with frequent mowing at a height of 0.5–1.0 cm.

Warm-Season Grasses

The most commonly used warm-season grasses include 10 species in six genera, listed below (Plate 2). They are somewhat more diverse in origin than the cool-season turfgrasses; the bermudagrasses come from Africa, the zoysiagrass and centipedegrass originate in East Asia, St. Augustinegrass is native to the West Indies, bahiagrass is from South America, and buffalograss is a native of the American dry prairies (Beard 1973, Hanson et al. 1969).

Bermudagrasses	*Cynodon* spp.
Common	*C. dactylon* (L.) Pers
African	*C. transvaalensis* Burtt-Davey
Tifway hybrids	*C. dactylon* × *C. transvaalensis*
Magennis	*C. magennisii* (Hurcombe)
Zoysiagrass	*Zoysia* spp.
Japanese lawn grass	*Z. japonica* Steud
Manilagrass lawn grass	*Z. matrella* (L.) Merr.
Korean velvet lawn grass	*Z. tenufolia* Willd. ex Trin.
Centipedegrass, Chinese lawn grass	*Eremochloa ophuiroides* (Munro.) Hack
St. Augustinegrass	*Stenotaphrum secundatum* (Walt.) Kutze
Bahiagrass	*Paspalum notatum* Flugge
Buffalograss	*Buchloe dactyloides* (Nutt.) Engelm.

All of the above-mentioned warm-season grasses have extravaginal growth, with rhizomes and/or stolons producing a creeping, spreading habit.

Bermudagrass is the most important and widely adapted of the warm-season turfgrasses. The improved turf-type bermudagrasses form a vigorous and aggressive turf of high shoot density with fine-textured leaves. Bermudagrass tolerates low temperatures poorly; discoloration occurs at soil temperatures below 10°C. It tolerates a wide range of soil acidity (pH 5.5–7.5) and can withstand flooding and high levels of soil salinity. Bermudagrass is used in the warm, humid and the warm, semiarid regions in lawns and in other general-purpose turf areas. Medium to high maintenance is required, with a cutting height of 1.3–2.5 cm.

Common bermudagrass is the only turf-type bermudagrass that is propagated by seed. Most of the others are hybrids that are sterile and must be propagated vegetatively. Common bermudagrass is relatively coarse in texture compared with the hybrids of *C. dactylon* × *C. transvaalensis* (the Tifway series). *African bermudagrass* has the finest texture and highest shoot density of any bermudagrass. *Magennis bermudagrass* is a natural hybrid between *C. dactylon* and *C. transvaalensis*, resembling the latter, and does not produce viable seed. In Florida, more than 80% of all the bermudagrass turf was devoted to golf courses in 1974 (Cromroy and Short 1981).

Zoysiagrasses are adapted to the warm, humid, and transitional regions. Three species used in turf settings are dense and low growing, with tough and stiff stems and leaves. All go dormant at 10°–13°C. They perform well in full sun on well-drained fertile soils of pH 6–7. They require medium intensity of culture with a cutting height of 1.3–2.5 cm.

Centipedegrass is a medium, coarse-textured, slow-growing species that spreads by thick, leafy stolons. It is propagated vegetatively or by seed and prefers a soil pH of 4.5–5.5. It is used on lawns and in other turfgrass areas where traffic is light and a relatively low intensity of culture is desired.

St. Augustinegrass is a versatile, sod-forming warm-season grass of very coarse leaf texture. Propagation is vegetative, with sprigs, plugs, or sod. It is the least hardy of the warm-season turfgrasses at low temperatures but has outstanding shade tolerance. This grass is used primarily in the warmer portions of the warm, humid region for lawns where a fine-textured turf is not required. A cutting height of 4–6 cm is preferred. St. Augustinegrass produces thatch readily. More than 90% of the entire St. Augustinegrass turf acreage in Florida was devoted to home lawns in 1974 (Cromroy and Short 1981).

Bahiagrass forms a very coarse-textured erect turf and propagates by seed. It is adapted to the warmer areas of the warm, humid region and is used for turf of relatively low quality that requires a low level of maintenance. About 90% of the entire bahiagrass turf acreage in Florida was devoted to home lawns in 1974 (Cromroy and Short 1981).

Buffalograss forms a fine-textured turf of gray-green color (Plate 5) and is especially well adapted to semiarid regions on alkaline soils. It is being used extensively in low-maintenance areas and home lawns in the central Plains states (Nebraska, Kansas, Oklahoma, and Texas). It is very drought tolerant.

Summary of Grass Characters

The several minute but distinct characters found at the junction of the stem and leaf blades—ligule, auricles, and collar—plus the vernation of the leaf bud and growth characteristics are useful diagnostic characters (Figure 1-1). Their differences are summarized for the 13 most common cool- and warm-season grasses (Table 1-1). The appearance of these structures in live green plants should help identify these grasses (Plates 3–5).

Drought Dormancy and Its Relationship to Insect Damage

Turfgrasses can survive extended periods of drought by reducing their growth rate (becoming dormant). Cessation of shoot growth is accompanied by the death of leaves; dor-

Table 1-1. Characteristics of 13 Major Turfgrasses

	Vernation	Ligule and Length	Collar	Leaf Blade and Width	Growth and Spread
COOL-SEASON GRASSES					
Kentucky bluegrass	Folded	Membranous, short, blunt, 0.2–1.0 mm	Medium broad, divided	V-shaped to flat, boat-shaped tip, 2–4 mm	Rhizomes
Annual bluegrass	Folded	Membranous, white, thin, acute tip, 1–3 mm	Conspicuous, divided	V-shaped, light green, boat-shaped tip, 2–3 mm	Bunch and stolons
Perennial ryegrass	Folded	Membranous, blunt, 0.5–1.5 mm	Distinct, divided, smooth	Flat dull above, glossy below, 2–5 mm	Bunch, noncreeping
Red fescue	Folded	Membranous, blunt, 0.2–0.5 mm	Narrow, indistinct, smooth	Narrow, folded to involute, 0.5–1.5 mm	Rhizomes
Creeping bentgrass	Rolled	Membranous, acute to oblong, 1–2 mm	Narrow to medium broad, continuous	Narrow, flat, acuminate tip, 2–3 mm	Stolons
Velvet bentgrass	Rolled	Membranous, acute to oblong, 0.4–0.8 mm	Medium broad	Flat, smooth, acuminate tip, <1 mm	Stolons
WARM-SEASON GRASSES					
Bermudagrass	Folded	Fringe of white hairs, 1–3 mm	Continuous, smooth, ciliate	Flat, stiff, acute tip, 1.5–3.0 mm	Stolons and rhizomes
Buffalograss	Rolled	Fringe of hairs, 0.5–1.0 mm	Broad, continuous, smooth	Flat, wavy, gray-green, margins rough, 1–3 mm	Stolons
Zoysiagrass	Rolled	Fringe of hairs, 0.2 mm	Continuous, broad, ciliate at margins	Flat, stiff, 2–4 mm	Stolons and rhizomes
Centipedegrass	Folded	Membranous, short, ciliate, 0.5 mm	Continuous, broad, ciliate at margins	Flat, blunt, ciliate, 3–5 mm	Stolons
St. Augustinegrass	Folded	Fringe hairs inconspicuous, 0.3 mm	Continuous, broad, smooth	Flat, petioled, smooth, 4–10 mm	Stout stolons
Bahiagrass	Rolled or folded	Membranous, blunt, entire, 1 mm	Continuous, broad to divided	Flat to folded, ciliate margin at base, 4–8 mm	Flattened rhizomes and stolons
Tall fescue	Rolled	Membranous, very blunt, 0.2–0.8 mm	Conspicuous, broad, divided, often hairy on margins	Veins prominent above, flat, stiff, 5–10 mm	

Note: Auricles are absent in all listed species except perennial ryegrass, where auricles are small, soft, and clawlike, and tall fescue, where auricles are small and narrow.
Source: Adapted from Beard 1982.

mant turfgrass often appears dead (Plate 6). Buds in the leaf axils of the crown, stolons, and rhizomes of dormant grasses survive and initiate new growth when soil moisture is replenished. Kentucky bluegrass, for example, can resume growth rapidly, with recovery visible in 3–5 days (Plate 6) (Beard 1973).

Summer drought- or heat-induced dormancy is often associated with turfgrass insect damage that is nearly impossible to detect or differentiate except under very close scrutiny. See Chapters 6, 8, and 25 for further discussion of this phenomenon. Reduced root growth in cool-season grasses during mid to late summer predisposes the turf to serious injury from root-feeding insects.

Dichondra Lawns

In addition to grasses, at least one broadleafed plant is used as a lawn (Plate 5). Dichondra, *Dichondra* spp., a member of the morningglory family, is a native of the southern coastal plains of North America. While tall fescue is being used increasingly in such areas, dichondra still occurs in pure stands in lawns in southern California and finds limited usage in central California and north to the San Francisco Bay and Sacramento areas. Mowing height determines some of its characteristics. When it is mowed at 2.0 cm, a small-leaved wear-resistant stand develops. Cutting at 3.8–5.0 cm produces a mixed stand of large- and small-leaved dichondra. Without mowing, large-leaved dichondra dominates as a prostrate ground cover of about 7.6 cm in height (Gibeault et al. 1977; V. A. Gibeault, Cooperative Extension, University of California, Riverside, personal communication, 1985).

Economic Impact of Turfgrass Culture

Unlike most agricultural crops, whose production costs and sales values can be estimated readily, turfgrass has a value that is difficult to measure, primarily because most turfgrass acreage is not grown for sale. However, turfgrass plays a major role in our daily life. Turfgrass often serves at least one of three roles—beautification, recreation, and utility. Utility turf provides many functional purposes, including dust control, soil erosion control, and glare reduction (Watson et al. 1992). Such turf also produces oxygen, filters some air pollutants, provides some natural cooling, and reduces noise in urban settings (Roberts and Roberts 1988). Recreational turf is used for parks, golf courses, athletic fields, and lawns. It provides a suitable setting for athletic activities, as well as physical exercise and training. Turf in home lawns, industrial park landscaping, or golf courses is expected to be aesthetically pleasing as well as functional.

A major shift in perception of the turf industry has occurred during the last quarter of the 20th century, so that the industry now emphasizes "facilities for people" and stresses use and appearance. This is in contrast to the former mindset that turf production was the primary objective. Now use and appearance of turf are critical considerations and goals. There is an increasing emphasis on the conditioning and grooming of turf (as well as other landscaping considerations)—commercial buildings, resorts, and housing complexes all strive to provide beautifully maintained turf as a sales tool. Thus the turfgrass industry is now envisioned as a critical service industry that provides specialized grasses and other ground covers for the improved health and welfare of a rapidly increasing population (Watson et al. 1992).

Since 1975 many states have conducted surveys of the turf industry, attempting to quantify expenditures related to growing and maintaining the turfgrass of lawns, golf courses, parks, athletic fields, cemeteries, and other turf settings. Each survey has used different techniques, and it is difficult if not impossible to assess values for land, buildings, and the turfgrass itself since the values change annually depending on tax rates, inflation, age of facilities, and other factors. Nevertheless, it is instructive to note that in many cases, surveys have shown dollar values for the turf industry in a given state to approach the values for the largest agricultural crop in that state. For example, in New York State a 1982 survey indicated the estimated expenditure of the state turfgrass industry to be approximately $1 billion a year. New York's largest agricultural industry, dairy products, generated $1.5 billion in sales (56% of farm receipts) in 1981 (Gruttadaurio et al. 1978, Smiley 1983). According to 1983 estimates, turfgrass ranks as the third largest crop in Virginia and Texas (Watson et al. 1992).

The studies cited as well as earlier studies and surveys plainly show that turfgrass culture in its entirety as an industry contributes significantly to the economy of the country. Protection of the existing turfgrass plantings from various pests, including insects and other closely related organisms, is thus an important concern.

2

Insects and Near Relatives

Phylum Arthropoda

Insects and mites belong to a larger category of related animals of the phylum Arthropoda. It includes insects; arachnids (mites, ticks, spiders, and scorpions); chilopods, or centipedes; diplopods, or millipedes; and crustaceans (crabs, lobsters, shrimps, and barnacles). Except for the crustaceans, all arthropods are fairly similar in size, occupy relatively the same niche in the environment, and are fairly similar in general appearance. Arthropod bodies and legs are jointed; the word *arthropod* is derived from two Greek words meaning "jointed feet." The exterior of the body (the integument) is covered by a noncellular horny layer known as *chitin,* which functions as the framework of the body and is the *exoskeleton.* In addition to structural integrity, this exoskeleton provides the base for muscle attachments.

Of all arthropods, mites and spiders are the most closely related to insects in form and function and in their status as pests that damage turfgrass. Compared with insects, they are relatively minor turfgrass pests. The following general discussion is based primarily on four sources: Borror et al. (1989), Jeppson et al. (1975), Krantz (1978), and Romoser and Stoffolano (1998). These excellent sources provide more detailed information on insects and mites.

Form and Function of Insects and Mites

External Characteristics of Adults

Insects are distinguished from other arthropods by a body that has three distinct divisions—the head, thorax, and abdomen (Figure 2-1). The insect head typically possesses mouthparts, simple eyes (ocelli) and/or compound eyes, and a pair of antennae that have a sensory function. The thorax includes three segments, the *prothorax, mesothorax,* and *metathorax* (from anterior to posterior); if an insect has legs, each segment of the thorax has one pair. Some immature insects have no legs.

The most primitive insects and some of the more specialized are wingless, but the vast majority have two pairs of wings as adults, the front pair being attached to the mesothorax and the hind pair to the metathorax. In insects with a single pair, the wings are attached to

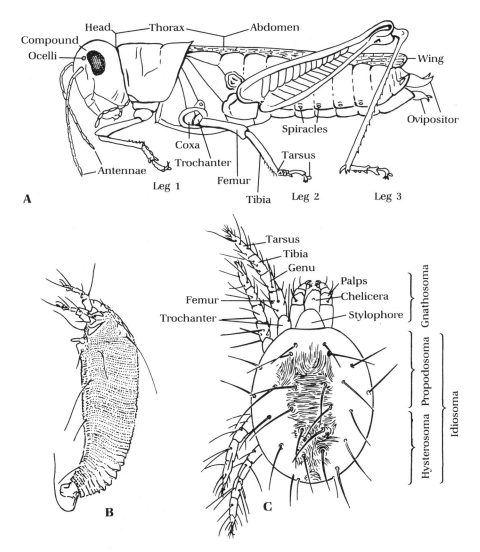

Figure 2-1. External anatomy of arthropods. **A.** Grasshopper (insect). **B.** Body of eriophyid mite. **C.** Body divisions of typical mite. (Part A adapted from Matheson 1951, fig. 31, courtesy of Cornell University Press; parts B and C adapted from Jeppson et al. 1975, figs. 3, 121A, courtesy of the University of California Press.)

the mesothorax. The flies appear to have a single pair attached to the mesothorax, but the metathorax has a pair of minute stubs known as *halteres* that aid in balance and are actually minute wings. The abdomen typically has 11 segments, with the terminal 3 segments modified for reproductive functions.

The body of a typical mite is separated into an anterior gnathosoma and a posterior idiosoma, which is further divided into the propodosoma and the hysterosoma (Figure 2-1). The gnathosoma resembles the head of a typical insect only in that the mouthparts are attached to it. Chelicerae and palps are the main external feeding organs that rasp and pierce the epidermis of host plants, causing damage. The function of the idiosoma parallels that of the abdomen, the thorax, and portions of the head of insects. Division of the propodosoma and hysterosoma may or may not be apparent, but the anterior two pairs of legs are attached to the propodosoma and the posterior two pairs of legs are attached to the hysterosoma. Mites of the superfamily Eriophyoidea (with only one family, Eriophyidae) have only two pairs of legs (Figure 2-1). These mites are basically wormlike and minute, varying from 0.1 to 0.3 mm in length and essentially invisible to the unaided eye. This family contains at least three turf pests—the bermudagrass mite, the buffalograss mite, and the zoysiagrass mite.

Internal Structure and Function

The central nervous system of insects and mites consists of a brain located in the head and a ventrally located double nerve cord that extends the entire length of the body (Figure 2-2). A series of ganglia (nerve centers composed of cell masses) unite the nerve cord at various regions along the nerve cord. Nerve cords radiate out from each ganglion to other parts of the body.

The digestive system may be a relatively simple tube, extending from the mouth to the terminally located anus, or may be a highly convoluted, complex system (Figure 2-2). Whether simple or complex, the digestive tract has three sections: the foregut, midgut, and

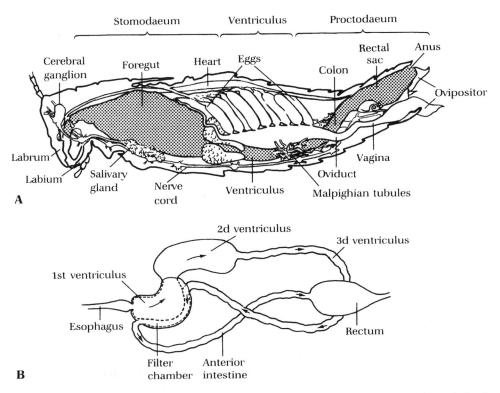

Figure 2-2. Internal anatomy of insects. **A.** Grasshopper, a chewing insect. **B.** Digestive system of piercing, sucking insect (Homoptera). (Part A adapted from Matheson 1951, fig. 51, courtesy of Cornell University Press; part B adapted from Snodgrass 1935, fig. 209A.)

hindgut. Excretory organs called *Malpighian tubules* arise at the anterior end of the hindgut and usually contain many long, stringlike structures that permeate the body cavity.

The circulatory system consists of the dorsal vessel (Figure 2-2), which runs the full length of the body, and all the space between the various organs in the body cavity. The heart is merely a chambered dilation of the dorsal vessel. Hemolymph may be clear or any of various shades of yellow or green and provides the dual function of vertebrate blood and lymph. Hemolymph enters the dorsal vessel at the posterior end of the arthropod and is pumped forward to the head, where it flows back into the body cavity and into all the appendages. As an insect moves, hemolymph moves through the body cavity passively.

Insects have no lungs. Oxygen is delivered to cells throughout the body by diffusion. Oxygen enters through small openings called *spiracles*, which are located laterally on the body (Figure 2-1), and moves through a network of small tubes called *trachea*. These branch and become increasingly small (*tracheoles*), finally diffusing into the tissues and reaching cells. Much of the preceding discussion of the internal structure and function in insects applies to mites as well.

Growth and Development

In insects and mites, growth and development are accomplished through a process known as *metamorphosis*. There are two major types of metamorphosis that turf insect pests undergo. Those with gradual metamorphosis develop through three stages: the egg, the immature form (*nymph*), and the adult (Figure 2-3). The immatures look like small, wingless adults. Relatively simple reorganization occurs between the nymphal stage and the adult. Nymphs and adults occupy a similar niche, often competing for the same resources (e.g., food, water, space). Insects with complete metamorphosis develop through four stages: the egg, the immature form (*larva*), the resting stage (*pupa*), and the adult (Figure 2-4). In complete metamorphosis the immature insects differ radically from the adults in form and function, and adults often occupy a niche very different from that of the larva.

Growth of insects from egg to adult occurs in discrete stages, or steps. Molting for growth occurs in the nymphal or larval stage. Insects molt twice to as many as eight or nine times after hatching from the egg, depending on the species. An insect's stage of growth is des-

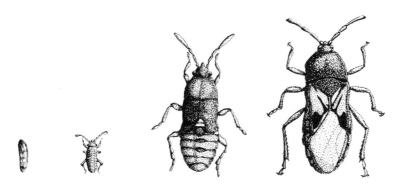

Figure 2-3. Gradual metamorphosis, illustrated by hairy chinch bug: Egg, first-instar nymph, final-instar nymph, and adult. (Courtesy of R. Jarecke, NYSAES.)

Figure 2-4. Complete metamorphosis, illustrated by sod webworm: Egg; first, third, and final larval instars; pupa; and adult. (Courtesy of R. Jarecke, NYSAES.)

ignated as an instar between each molt as an immature: first instar, second instar, and so forth. The pupa ("resting" stage of insects that go through complete metamorphosis) is usually immobile, but many physiological changes occur internally during this stage. For example, the reproductive system often will develop during this time.

Mites develop through a slightly different metamorphosis. The egg hatches into a six-legged larval stage. Upon molting, the larval stage becomes an eight-legged nymph that resembles an adult except that it is smaller. There are generally two nymphal instars, the protonymph and the deutonymph, before the adult stage.

Insects and mites are cold-blooded (ectothermic); that is, their body temperature is near that of the surrounding environment. Their activity is strongly influenced by the ambient temperature.

Reproduction

Reproductive capacities of insects are highly variable and specialized. Egg production of insects may vary between species, with as few as several dozen eggs produced during the life of a female and as many as several thousand eggs in other species. The biology of egg production varies greatly, depending on the insect. Copulation and transfer of sperm are required in most insects. However, in some insects, parthenogenetic reproduction is the rule, and no males are present in the species.

Many species lay eggs (are oviparous) on a substrate such as leaves or in the soil. Some species deposit live young that have hatched from an egg within the female (such insects are viviparous). Most insect eggs develop into a single offspring. A group of parasitic wasps, however, produces multiple offspring from a single egg (a phenomenon known as *polyembryony*) by one of the most specialized and highly efficient reproductive systems known.

Generations

Insects that have one generation each year are said to be *univoltine,* while those with two or more generations each year are said to be *multivoltine.* Certain ambiguities arise in discussions of generations of multivoltine insects. The terms *generation* and *brood* often have been used interchangeably and inaccurately. Schurr and Rings (1964) attempted to clarify this ambiguity and others. In their system, the stage that overwinters is designated as the *overwintering stage,* whether it involves adult, egg, larva, or pupa (e.g., overwintering larvae). With the arrival of spring, the overwintering stage and any succeeding stage in that generation (ending with adult) are called the *spring generation.* A new generation is considered to start when eggs are laid. These offspring, as eggs and in all succeeding stages through the adult, are called *first-generation* eggs, larvae, and so forth. First-generation adults lay eggs, thereby starting the second generation. Generations are thus numbered consecutively through the rest of the season. When winter returns, the hibernating insect again consists of the overwintering stage, whether its members belong to the first generation, the second generation, or any later one. These authors consider the term *brood* applicable only to different generations of insects that have life cycles longer than a year and yet have annual adult emergences. May or June beetles and periodic cicadas are examples of such organisms.

Types of Mouthparts and Turf-Feeding Damage

Since insects and mites damage turf principally in their quest for food, a brief review of the principal types of mouthparts will provide an understanding of the types of damage caused.

Chewing Insects

Chewing mouthparts are most common among insect pests of turf (Figure 2-5). The main mouthparts of chewing insects consist of (1) the upper lip (labrum), (2) a pair of opposed laterally moving jaws (mandibles), (3) a second pair of opposed laterally moving jaws (maxillae), and (4) a lower lip (labium).

The labrum acts as an organ for manipulating and moving food into the jaws. Mandibles lie immediately behind the labrum. They are usually triangular, strongly sclerotized appendages that gradually taper outward to a cutting edge. Maxillae are more complex in structure, lie directly behind the mandibles, and act as a second pair of jaws to manipulate and break up the food. The labium lies at the back of the mouthparts and forms a typical lower lip. Collectively, these mouthparts form the preoral cavity.

Symptoms of turfgrass damaged by insects with biting-chewing mouthparts are characterized by physical removal of plant tissues, such as stripping away of the epidermis of leaves, notching of leaves and stems, severing of plant parts, hollowing out of stems and crowns, and pruning of roots. Most of the chewing insect pests of turfgrass are the immature forms (larvae and nymphs), but the adults of some species (e.g., mole crickets, bluegrass billbugs, annual bluegrass weevils) also feed on turf.

Sucking Insects

The second most common method by which insects feed involves the piercing and sucking of plant tissue (Figure 2-5). Piercing and sucking mouthparts are thought to represent a highly modified form of the more "primitive" chewing mouthparts. The labium forms a beak that surrounds the needle-like mandibles and maxillae. The maxillae unite along their margins to form a tube with two channels, the food channel and the salivary channel. Mandibles that lie outside the maxillae act primarily as the cutting and piercing organ. Once

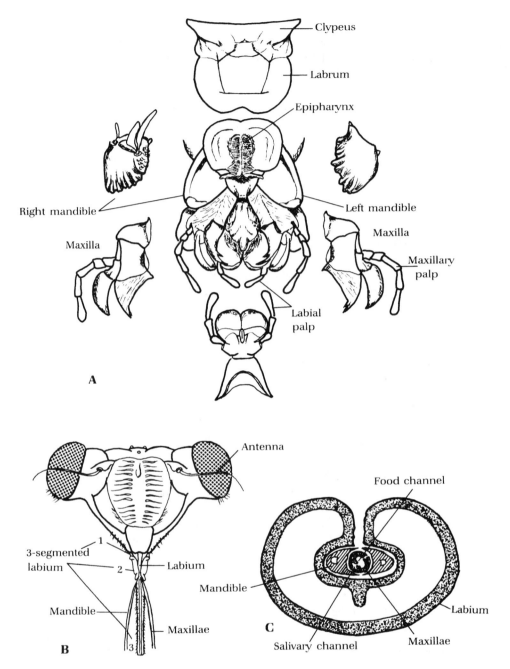

Figure 2-5. Mouthparts of insects. **A.** Chewing insect (grasshopper). **B.** Piercing-sucking insect (cicada). **C.** Cross section of cicada mouthparts. (Adapted from Matheson 1951, part A: fig. 37, parts B and C: fig. 64, Courtesy of Cornell University Press.)

the mandibles and maxillae have been inserted deep into the plant tissues, feeding commences. Salivary secretions pumped into the plant help the insect suck up the plant sap and the cell contents.

Plants injured by this method of feeding generally remain completely intact. The entire plant starts to deteriorate because of the loss of plant sap or in response to the injection of toxic salivary secretions. Chinch bugs and greenbugs are two insects that damage plants by both methods. Early symptoms may be yellowing, wilting, blasting of leaves, and necrosis, followed eventually by browning and death. Both adults and nymphs of insects with piercing and sucking mouthparts are involved in turf injury.

Some insects that feed by piercing and sucking also serve as vectors of plant diseases, primarily those caused by viruses. Fortunately, insect-caused virus transmission in turfgrasses has not been a serious problem to date.

Other Types of Mouthparts and Feeding

Other, less common types of mouthparts are found in flies and mites. Fly maggots have mouth hooks that move vertically rather than horizontally and are used primarily to break and tear plant tissues. The liquified plant tissues and small particles are then sucked in. Mites have still another type of mouthpart, a rasping, sucking type. Their feeding process ruptures the epidermal cells, giving injured tissues a characteristic silvery to gray appearance.

Table 2-1. Orders, Families, and Representative Groups of Some Turfgrass-Infesting Arthropods

Class and Order	Family	Type of Feeding	Examples
Class Arachnida			
Order Acari	Eriophyidae	Rasping-sucking	Bermudagrass mite
			Zoysiagrass mite
	Tetranychidae	Rasping-sucking	Banks grass mite
			Clover mite
Order Araneida			Spiders
Class Insecta			
Order Orthoptera	Gryllotalpidae	Chewing	Mole crickets
Order Homoptera	Aphididae	Piercing-sucking	Greenbug
	Cercopidae	Piercing-sucking	Two-lined spittlebug
	Margarodidae	Piercing-sucking	Ground pearls
	Pseudococcidae	Piercing-sucking	Mealybugs
Order Hemiptera	Lygaeidae	Piercing-sucking	Chinch bugs
Order Lepidoptera	Acrolophidae	Chewing	Burrowing sod webworm
	Hesperiidae	Chewing	Fiery skipper
	Noctuidae	Chewing	Armyworms, cutworms
	Pyralidae	Chewing	Webworms
Order Coleoptera	Chrysomelidae	Chewing	Dichondra flea beetle
	Curculionidae	Chewing	Annual bluegrass weevil
			Billbugs
	Scarabaeidae	Chewing	White grubs
Order Diptera	Bibionidae	Rasping	March flies
	Chloropidae	Rasping	Frit flies
	Tipulidae	Rasping	Crane flies
Order Hymenoptera	Formicidae	Chewing	Fire ants
			Harvester ants
	Sphecidae	Chewing	Cicada killers
	Vespidae	Chewing	Yellow jackets, wasps

Orders of Turfgrass-Damaging Insects and Mites

Relatively few orders of insects and mites have species destructive to turfgrass. In number of destructive species present, the orders Lepidoptera and Coleoptera are by far the most important. The orders Orthoptera, Heteroptera, and Homoptera contain relatively few species of turf pests but those pests can be devastating where they occur. The orders Diptera and Hymenoptera are less important or occur less frequently. Table 2-1 condenses information relating to orders of turf-infesting arthropods.

3

Insects and Mites: Turf Association

Habitats of Turfgrass Insects

Destructive turfgrass insects and mites can be grouped according to the habitat in which the destructive stage of the arthropod spends most of its life in the turfgrass ecosystem. In a great many cases this is the larval or nymphal stage, but in a few instances the adults are also destructive.

This method of grouping has merit, since the control tactics employed for each group, especially in insecticide treatments, have a direct bearing on the ease or difficulty with which satisfactory results may be accomplished. Arthropods living and feeding exclusively on leaves and stems are most readily controlled, in part because it is easier to achieve good coverage and contact with insecticide applications.

Soil-inhabiting insects that feed exclusively on the roots are the most difficult to control, because many insecticides cannot penetrate the thatch and reach the target insect. The thatch adsorbs many insecticides, so the water solubility and persistence of the insecticide become important considerations in selecting control options. In addition, some of the biopesticides that are available commercially do not penetrate thatch readily, so they are less likely to reach the target insect.

Three turf habitats occupied by turfgrass-damaging insects were described by Niemczyk (1981): (1) leaf and stem, (2) thatch, and (3) root zone/soil. A given insect may occupy more than one of these habitats during its development, and may even occupy all habitats in different stages of its development. Shetlar (1995a) observed that many insects spend part of their life in thatch or above it, or in soil and occasionally forage in the thatch, and modified the designation of turf habitats: (1) foliar and stem inhabitants, (2) stem and thatch inhabitants, and (3) thatch and soil inhabitants (Figure 3-1).

Leaves and stems of turfgrasses are above the growing point (meristem), and therefore the plant can tolerate some damage from insects without a significant loss of vigor. Thatch is the layer of dead and living organic matter that occurs between stems and roots. This layer provides insulation and moderates soil temperatures and moisture. Thatch has high organic matter content, which provides refuge for insects and mites and increases the rate of adsorption of pesticides. The root zone is a region of constant renewal—production of roots and stolons, activity of earthworms and various arthropods, production of new organic matter from decomposition.

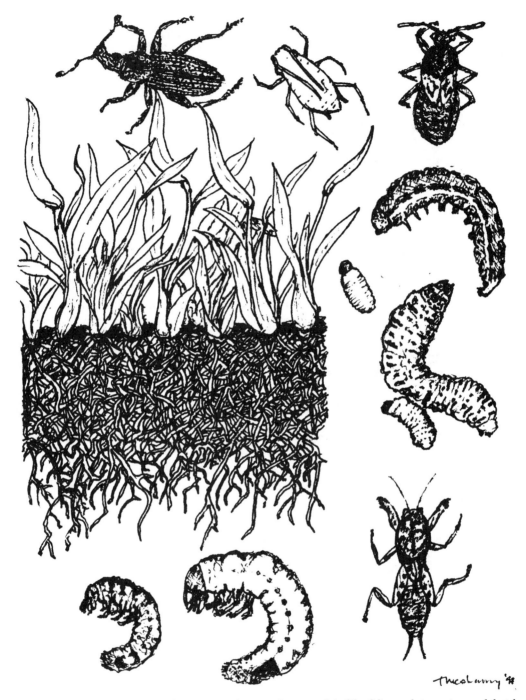

Figure 3-1. Typical turf profile with representative pests (not to scale) of the foliar and stem, stem and thatch, and thatch and soil zones. Clockwise from top left: annual bluegrass weevil adult, aphid, chinch bug, webworm larva, annual bluegrass weevil larva, armyworm larva, billbug larva, mole cricket, and two white grubs. (Drawn by T. McSharry, University of Massachusetts.)

Leaf- and Stem-Infesting Pests

The leaf and stem habitat includes insects and mites that feed on upper leaves and stems. Some of these pests may hide in the thatch part of the day or spend part of their development in the thatch or in nearby protected locations, but most of the life cycle is spent in or on foliage. Others remain exposed on the leaf surface while others hide beneath leaf sheaths.

Major insects and mites inhabiting leaves and stems include the following:

Bermudagrass mite, *Eriophyes cynodoniensis* Sayed
Winter grain mite, *Penthaleus major* (Duges)
Clover mite, *Bryobia praetiosa* Koch
Banks grass mite, *Oligonychus pratensis* (Banks)
Greenbug, *Schizaphis graminum* (Rondani)
Rhodesgrass mealybug, *Antonina graminis* (Maskell)

Bermudagrass mites are destructive as adults and nymphs. They settle in the leaf sheaths of bermudagrass and remain in the internodal areas, producing rosetted internodes (Chapter 4).

Adults and nymphs of the *winter grain mite* feed on the leaves of grasses. They migrate on a daily basis, moving up to the leaves to feed at night and giving the grass the appearance of winter dessication. Mites descend to the base of the grasses during the day. The winter grain mite, most appropriately named, flourishes only at cold temperatures lower than 24°C and aestivates as eggs from May to October (Chapter 4).

Clover mites are most active in spring and fall, and migrate to upright surfaces just before molting. They feed on a variety of grasses, including Kentucky bluegrass and perennial ryegrass, particularly near building foundations (Chapter 4).

Banks grass mite is a pest of bluegrass, bermudagrass, and St. Augustinegrass. It can complete a generation in 2–4 weeks. Adults and nymphs feed, usually on the underside of foliage. It is more adapted to warm weather than other turfgrass mites (Chapter 4).

Greenbug adults, whether winged, wingless, or both, feed on the upper leaf surface of bluegrasses, fescues, and ryegrasses. They produce living young that mature to wingless females to repeat the cycle for many overlapping generations. More than 50 individuals can occur on a single grass blade (Chapter 7).

Rhodesgrass mealybug, called a *scale* until recently, is destructive at the adult and nymphal stages primarily on bermudagrass and St. Augustinegrass. Crawlers (larvae) settle near the base of plants or lower nodes, inserting their mouthparts into the stem to feed. They remain sessile throughout the nymphal stage and adult life. Secretions by the insect produce clusters of cottony masses of about 0.3 cm in diameter on the stems, and this is the main diagnostic feature of the infestation (Chapter 7).

Stem- and Thatch-Inhabiting Pests

The stem and thatch habitat is complex, including stems, stolons, and organic matter, some of which is decaying. Insects with sucking mouthparts (such as chinch bugs) tend to prefer the thatch for feeding and hiding sites. Some insects with chewing mouthparts spend part of their development inside stems, stolons, or crowns (e.g., early instars of annual bluegrass weevils).

Some insects that inhabit the thatch use it as their primary resting place for the destructive stage or stages, but also feed in part on stems or stolons. They include the following:

Hairy chinch bug, *Blissus leucopterus hirtus* Montandon
Southern chinch bug, *B. insularis Barber*
Buffalograss chinch bug, *B. occiduus* (Barber)
Two-lined spittlebug, *Prosapia bicincta* (Say)
Temperate-region sod webworms, major species
   Bluegrass webworm, *Parapediasia teterrella* (Zincken)
   Corn root webworm, *Crambus caliginosellus* Clemens
   Cranberry girdler, *Chrysoteuchia topiaria* (Zeller)
   Larger sod webworm, *Pediasia trisecta* (Walker)
   Pretty crambus, *Microcrambus elegans* (Clemens)
   Silver-barred lawn moth, *Crambus sperryellus* Klots
   Silver-striped webworm, *Crambus praefectellus* (Zincken)
   Striped sod webworm, *Fissicrambus mutabilis* (Clemens)
   Western lawn moth, *Tehama bonifatella* (Hulst)
Tropical-region sod webworms
   Grass webworm, *Herpetogramma licarsisalis* (Walker)
   Tropical sod webworm, *H. phaeopteralis* Guenee
Cutworms
   Black cutworm, *Agrotis ipsilon* (Hufnagel)
   Variegated cutworm, *Peridroma saucia* (Hubner)
   Bronzed cutworm, *Nephelodes minians* Guenee
Armyworms
   Armyworm, *Pseudaletia unipuncta* (Haworth)
   Fall armyworm, *Spodoptera frugiperda* (J. E. Smith)
   Lawn armyworm, *S. mauritia* (Boisduval)
Fiery skipper, *Hylephila phyleus* (Drury)
Annual bluegrass weevils, *Listronotus maculicollis* Dietz
Frit fly, *Oscinella frit* (L.)

Chinch bugs are destructive to their host grasses as adults and as nymphs. They pierce stems, suck plant sap, and inject toxic salivary enzymes that disrupt the water-conducting system of plants. Injury is most severe when heavy infestations occur on plants under drought stress. The hairy chinch bug is most damaging to bluegrasses, fine fescues, and bentgrasses in the cool, humid zone, while the southern chinch bug is most damaging to St. Augustinegrass, bermudagrass, and zoysiagrass (in that order) in the warm, humid zone. Buffalograss chinch bugs feed primarily on buffalograss (Chapter 6).

The *two-lined spittlebug* is destructive to bermudagrass and several other warm-season turfgrasses. Adults and nymphs pierce stems and remove sap. Large populations often are associated with thick or dense thatch, which provides the high humidity that enhances spittlebug survival (Chapter 7).

All of the thatch-inhabiting larvae whose adults are moths have a number of important common features. Adults of all these species fly and deposit their eggs at night. Turfgrass damage is caused only by the larvae. All the larvae are nocturnal, feeding on grass blades and stems at night and remaining hidden in the thatch or soil surface during the day. All deposit copious quantities of green fecal pellets in the thatch (Chapters 8–9).

All of the *webworms*, whether of the temperate or tropical regions, derive their name from their habit of webbing together plant debris, soil particles, and fecal pellets to form a tunnel, in which they hide during the day. Young larvae feed on the surface of tender leaves, but as they grow, they feed on entire leaves and stems. Practically all turfgrasses are susceptible to their feeding. Extensive feeding during dry weather causes the most serious damage. At least one species, the cranberry girdler, feeds on the roots of their host grasses (Chapter 8).

Cutworm larvae are at least twice the size of webworms and produce green fecal pellets at least three times the size of webworm pellets. They do not prepare tunnels to hide in during the day but rest in the thatch or in other secluded areas in a curled position. Their habit of cutting off whole plants near the ground level without consuming any more of the remaining plant gives them their name. All turfgrasses are susceptible (Chapter 9).

Armyworms are very similar to cutworms in size, appearance, and destructive habits. When large numbers of armyworms decimate host plants in one area, they move en masse from the decimated area to a fresh food supply with a definite advancing front. This habit gives them their name. All turfgrasses are susceptible (Chapter 9).

The *fiery skipper* is a butterfly, not a moth, and differs in its habits from the moths. Eggs are deposited on leaves during the day. After larvae hatch, they feed on the blades while they are young. Older larvae feed on leaves and stems and spend most of their life in the thatch. Practically all turfgrasses serve as host plants (Chapter 10).

The *annual bluegrass weevil* is exclusively a golf course problem in the Northeast, damaging only short-cut annual bluegrass and occasionally bentgrass. Callow adults emerge from the soil and remain in the thatch to mature. They move to the surface at night to feed on leaves and mate, but they return to the thatch, where they spend most of their life. Adult feeding appears to be insignificant. Eggs are inserted in the lower stem. Young larvae feed in the stems. Older larvae feed externally on the crowns, causing the most serious damage. The grass first yellows but soon dies in large patches (Chapter 18).

Only the larvae of the *frit fly* cause damage. Adults lay eggs on turfgrass leaves, primarily on bentgrass. Larvae migrate down to the upper portion of the crown to feed on the primordial leaves until pupation (Chapter 19).

Thatch- and Root-Infesting Pests

Many soil-inhabiting insects also spend significant periods of time in the thatch, particularly at the soil-thatch interface. The scarabaeid grubs dominate this group. Beetles deposit eggs in the soil, and the entire larval and pupal life is spent in the soil, although larvae often feed at the soil-thatch interface or in the lower portions of thatch. Adults are predominantly soil inhabitants. They leave the ground only to feed or mate, and return to the soil daily for diurnal or nocturnal oviposition, depending on the species.

Mole crickets and ground pearls are also predominantly soil bound, although large nymphs and adults of mole crickets burrow to the surface and forage at times. Those with multiple habitats but still primarily soil inhabitants include the dichondra flea beetle, billbugs, and the European crane fly. The entire group includes the following:

Southern mole cricket, *Scapteriscus acletus* Rehn & Hebard
Tawny mole cricket, *S. borellii* Giglio-Tos
Ground pearls, *Margarodes meridionalis* Morrison, and other species
Black turfgrass ataenius, *Ataenius spretulus* (Haldeman)

Turfgrass aphodius, *Aphodius granarius* (L.) and *A. pardalis* LeConte
Scarabaeids normally of 1-year life cycles
  Green June beetle, *Cotinus nitida* (L.)
  Northern masked chafer, *Cyclocephala borealis* Arrow
  Southern masked chafer, *C. lurida* Bland
  Asiatic garden beetle, *Maladera castanea* (Arrow)
  European chafer, *Rhizotrogus majalis* (Razoumowsky)
  Japanese beetle, *Popillia japonica* Newman
  Oriental beetle, *Exomola orientalis* Waterhouse
Scarabaeids normally of 2- to 3-year life cycles
  May or June beetles, *Phyllophaga* spp.
Dichondra flea beetle, *Chaetocnema repens* McCrea
Billbugs
  Bluegrass billbug, *Sphenophorus parvulus* Gyllenhal
  Hunting billbug, *S. venatus vestitus* Chittenden
  Phoenician billbug, *S. phoeniciensis* Chittenden
  Denver billbug, *S. cicatristriatus* Fabraeus
European crane fly, *Tipula paludosa* Meigen

Apart from adult mating flights in the spring, *southern* and *tawny mole crickets* live in the soil their entire life. Both adults and nymphs feed on grass roots and burrow in loose soil, disrupting the turf and causing the soil to dry. Bahiagrass and bermudagrass are most susceptible, but all common warm-season turfgrasses are attacked. Mole cricket activity is especially damaging to newly planted turf (Chapter 5).

Ground pearls live in the soil their entire lives and feed on turfgrass roots. Bermudagrass, St. Augustinegrass, zoysiagrass, and centipedegrass are commonly infested, causing the turf to yellow, become unthrifty, and eventually die (Chapter 7).

Damage by the *black turfgrass ataenius* is restricted almost exclusively to golf courses, where fairways and greens are most commonly injured. The grasses most frequently damaged are annual bluegrass, bentgrass, and Kentucky bluegrass. The larvae feed on the roots at the soil-thatch interface or in the soil, causing the grass to wilt even under irrigation because of the loss of roots. The insect has been the most troublesome where there are two generations each year. North of the latitude of New York State and Michigan there is normally one generation a year. The *turfgrass aphodius* causes identical damage and is often associated with the black turfgrass ataenius (Chapter 12).

The group of *scarabaeid beetles (white grubs)*, normally of a 1-year life cycle, have much in common in their life cycles and in their injury to turf. Damage is caused by third-instar grubs during the fall and again during the spring. All species of turfgrasses are susceptible. Adults are present during early summer to midsummer. Eggs are deposited in moist soil in earthen cells. Upon hatching, they molt twice before becoming the large, destructive third instars (Chapters 13–16).

May or June beetles, genus *Phyllophaga*, comprise more than 200 species in North America, with different species being troublesome in different regions of the country. Most of these beetles and their larvae are much larger than many scarabaeid species that complete a generation each year. Species with a 3-year life cycle produce minor turfgrass injury, feeding on roots as second-instar larvae during the first summer. Major injury is produced by third-instar larvae during the second year. There is little spring feeding during the third year, since the insect spends this time primarily completing its life cycle as pupae and adults.

In many species the adults feed on tender, expanding leaves of many deciduous trees (Chapter 15).

In southern and central California, where the broad-leaved dichondra is used as a lawn, the *dichondra flea beetle* can be a devastating pest. Adults produce a characteristic crescent-shaped feeding injury to the upper surface of the leaves, but this injury is secondary to the root-feeding damage done by the larvae. Damage occurs throughout the growing season and has also been reported on bermudagrass (Chapter 17).

The four most common species of *billbugs* have similar life cycles and appear nearly alike but have different primary host plants. Bluegrass billbugs prefer bluegrasses, the hunting billbug prefers zoysiagrass, the Phoenician billbug prefers bermudagrass, and the Denver billbug prefers Kentucky bluegrass and perennial ryegrass. Adults feed by chewing holes in the lower stems, where they deposit their eggs. The young larvae develop within the stems, but the older larvae feed on the crowns and roots, causing the most serious damage (Chapter 18).

The *European cranefly* is found primarily in the Pacific Northwest in North America and can be a very serious turfgrass pest on lawn grasses. Larvae cause all the damage by feeding on leaves and stems during the night and on roots and crowns of turfgrass during the day (Chapter 19).

Seasonal Presence of Injurious Stages

Most turfgrass insects in the tropical regions of the United States do not have a true winter diapause stage but continue to develop through the winter months at a slower pace if it is cool and at a faster pace if it is abnormally warm.

In the temperate regions there is usually a true diapause stage, in which development ceases until spring. These insects show a distinct, predictable life stage during each of the seasons. Having this information, we can determine which insects are most likely to be causing damage at a given time on the basis of seasonal habits alone. Figure 3-2 lists temperate-region insects and the season when they normally cause their major damage in relation to their oviposition period. Mole crickets, while tropical and subtropical in distribution, have been added as an exception because they have only one generation a year (except in south Florida, where some complete two generations each year), and the seasonal occurrence of each stage is as predictable as it is for temperate-region insects.

Winter

The winter grain mite causes turf damage during the winter (Chapter 4). Some cutworms can cause problems in the winter, particularly in southern states (Chapter 9). March flies can cause damage in late winter and very early spring (Chapter 22).

Spring

Only one species, the European crane fly, damages turf only during the spring, as over-wintering third instars become fourth instars in April and feed vigorously until mid-May. While larvae are present in autumn, damage normally is negligible (Chapter 19).

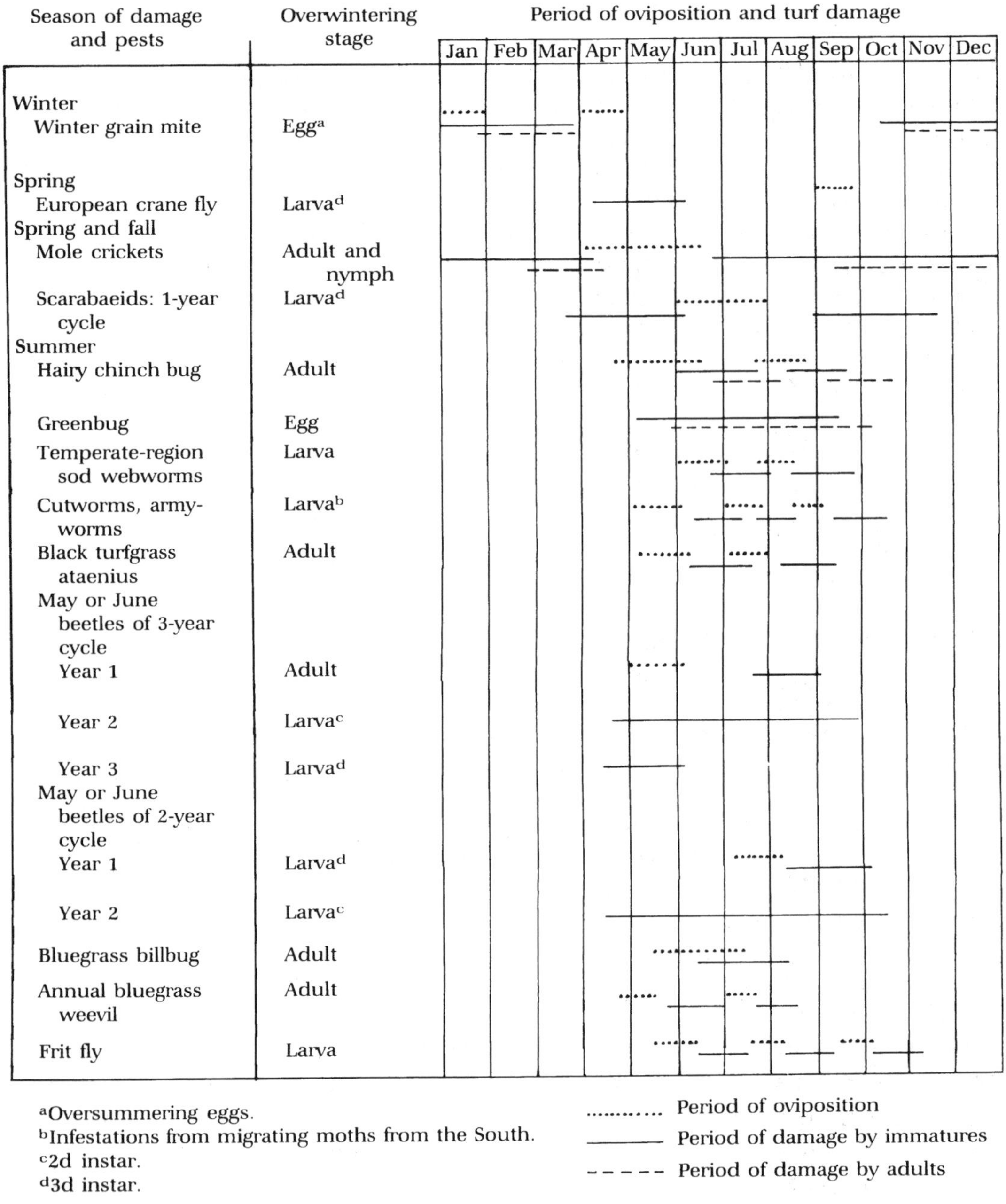

Season of damage and pests	Overwintering stage	Jan	Feb	Mar	Apr	May	Jun	Jul	Aug	Sep	Oct	Nov	Dec
Winter													
Winter grain mite	Egg[a]												
Spring													
European crane fly	Larva[d]												
Spring and fall													
Mole crickets	Adult and nymph												
Scarabaeids: 1-year cycle	Larva[d]												
Summer													
Hairy chinch bug	Adult												
Greenbug	Egg												
Temperate-region sod webworms	Larva												
Cutworms, army-worms	Larva[b]												
Black turfgrass ataenius	Adult												
May or June beetles of 3-year cycle													
Year 1	Adult												
Year 2	Larva[c]												
Year 3	Larva[d]												
May or June beetles of 2-year cycle													
Year 1	Larva[d]												
Year 2	Larva[c]												
Bluegrass billbug	Adult												
Annual bluegrass weevil	Adult												
Frit fly	Larva												

Period of oviposition and turf damage

[a]Oversummering eggs.
[b]Infestations from migrating moths from the South.
[c]2d instar.
[d]3d instar.

.... Period of oviposition
_____ Period of damage by immatures
– – – – – Period of damage by adults

Figure 3-2. Periods of oviposition and major turfgrass damage by temperate-region pests or groups and by mole crickets, tropical-region pests. (Drawn by H. Tashiro, NYSAES.)

Spring and Fall

Two groups of insects are destructive to turf during both the spring and the fall. These are the mole crickets in the warm, humid region and the scarabaeid species that normally have one generation a year in the cool, humid region. Either the adult or the nymph of the southern and tawny mole crickets is present throughout the year. Mole crickets deposit eggs from March through June, depending on local conditions and species. Since adults and nymphs are present throughout the year, turf damage can occur any time, but the period of greatest damage is late summer and early fall, when nymphs of both species are in their later instars and are foraging actively for food. Adult and late nymph foraging in spring also can produce significant damage (Chapter 5).

The scarabaeid grubs with a 1-year life cycle include at least seven species of importance as pests of turfgrass in the cool, humid region. All deposit eggs during a 4- to 6-week period during midsummer. The resulting larvae reach the third instar by late summer and early fall and become highly destructive, feeding on the grass roots near the soil surface. The cold weather of October and November forces them to migrate deeper into the soil for winter hibernation. They resume feeding again during early spring. The period of most vigorous feeding occurs as the grubs reach maturation 2–3 weeks before pupation. During May into early June, large patches of turf may be killed (Chapters 13–16).

Summer

All other major temperate-region turfgrass insects (Figure 3-2) damage turfgrass during the summer. The billbugs do so, developing through a single generation. The hunting billbug and Phoenician billbug may cause damage in autumn in warm-season grasses (Chapter 18).

Many other turf-damaging insects (e.g., chinch bugs) develop through multiple generations and cause damage throughout the summer. May or June beetles, with a 3-year life cycle, cause minor damage during late summer of the first year as second instars. Their major damage occurs from midspring throughout the summer of the second year as third instars. May or June beetles with a 2-year life cycle cause their major damage throughout the second summer (Chapter 15).

Ants, bees, and wasps (not listed on Figure 3-2) are not destructive to turfgrass directly. They do not feed on turfgrass but nest in the drier, well-drained turfgrass soils and make unsightly mounds that interfere with mowing and with other maintenance practices. Their burrowing activity may lead to desiccation of the soil, increasing plant stress. Some of the ants eat seeds, which is a relatively minor problem. Colony formation and mound-making activity can occur throughout the summer and into fall in the areas of the cool-season grasses but happen throughout the year in the Gulf Coast states (Chapter 20).

4

Acarine Pests

Mites are arthropods in the class Arachnida, order Acari, and have several morphological and physiological characteristics that differ from insects. Adult mites have two body regions (the cephalothorax, which includes the head and a region similar to an insect thorax, and the abdomen). Mite metamorphosis in tetranychid mites includes a 6-legged larva, two 8-legged nymphal stages (protonymph and deutonymph), and an adult. Mites never develop wings, but can be blown on wind currents. Many mites of economic importance in agriculture develop very rapidly (often in less than 2 weeks from egg to adult) and complete several generations each year. Mouthparts of mites are described as rasping and sucking—the mite rasps a wound on the surface of plant tissue and then laps or sucks up the resulting plant exudate.

Several mites attack turfgrass in various regions of North America. Some of these are *eriophyid mites* (family Eriophyidae), which are tiny, often no more than 0.3 mm long. These mites tend to be torpedo shaped with distinct folds or "rings" encircling the body and have two pairs of legs situated just behind the head. Other mites that attack turfgrass (family Tetranychidae and others) are larger than eriophyid mites, often about 1 mm, and oval shaped, with two pairs of legs pointed forward and two pairs of legs projecting to the rear.

Bermudagrass Mite

Taxonomy

The bermudagrass mite (BGM), *Eriophyes cynodoniensis* Sayed, family Eriophyidae, was previously named *Aceria neocynodonis* Keifer. It is also called the *bermudagrass stunt mite* (Keifer et al. 1982, Reinert 1982a).

Importance

This mite is a serious pest wherever bermudagrass is grown, particularly on golf courses. The BGM is found throughout Florida and is active all year in areas roughly south of Palm Beach. During the 1970s, Florida golf courses often spent as much as $25,000 each year to control the mite. Pesticide treatments are not always dependable, and repeat applications may be necessary (Cromroy and Short 1981, Johnson 1975, Reinert and Cromroy 1981, Tuttle and Butler 1961).

Figure 4-1. Distribution of bermudagrass mite. (Drawn by E. Gotham, NYSAES, adapted from Handbook of Turfgrass Insect Pests, Brandenburg and Villani 1995, Entomological Society of America.)

History and Distribution

The BGM is widely distributed in New Zealand and North Africa and probably originated in Australia. In the United States it was first discovered in 1959 infesting a bermudagrass lawn in Phoenix, Arizona. It was found later the same year in Tucson and Yuma, where it was causing extensive damage to bermudagrass. It was first reported in Florida in 1962 and now is found in the southern states wherever bermudagrass is grown (Figure 4-1) (Butler 1963, Hudson et al. 1995, Keifer et al. 1982, Reinert 1982a, Reinert et al. 1978).

Host Plants and Damage

Across the southern United States, the BGM feeds only on bermudagrass; most commonly used varieties are highly susceptible. Florida golf course infestations were most abundant where close mowing was not practical, for example, at the edge of bunkers, on the lips of sand traps, and around trees. In Arizona, lawns under flood irrigation were injured less than lawns irrigated with sprinklers (Butler 1963).

Damage is caused by mites that feed under the leaf sheaths. The first injury to appear is a slight yellowing of the leaf tips and a twisting of the leaves, with the margins rolling upward and inward. Shortening of the internodes produces a thick rosette or "witch's broom" (Plate 7). When rosettes are numerous, clumps that resemble cabbage heads develop, and the grass no longer appears to have internodes. Finally all leaves die back to the point of their insertion on the stem. The distorted growth is believed to be caused by a toxin that is injected into the developing grass buds. Death of leaves, stems, and stolons soon follows (Butler 1963, Johnson 1975, Keifer et al. 1982).

In the West, damage is first noticed in the spring when lawns fail to begin normal growth in spite of irrigation and fertilization. Well-fertilized lawns are more attractive to mites than starved grass. Normally mite injury is more pronounced during hot, dry conditions, when grass is under stress. Severe injury causes large areas of turf to become thinned or killed, resembling water or nutrient stress (Cromroy and Short 1981, Tuttle and Butler 1961).

Figure 4-2. Bermudagrass mite adult. Silhouette shows actual size. (Courtesy of R. Jarecke, NYSAES.)

Description of Stages

Adult

Mites are 165–210 μm long, wormlike in shape, creamy white to yellowish, and barely visible with a 10× to 20× magnifying lens. Under higher magnification, two pairs of legs are apparent (Figure 4-2, Plate 7). Males can be distinguished by the presence of five-rayed feather claws, while those of females are six-rayed. The abdomen has minute rings that resemble, but are not true, segments (Baker 1982, Johnson 1975, Keifer et al. 1982, Niemczyk 1981, Shetlar 1995a).

Egg

Eggs are about 70 μm long and transparent to opaque white (Figure 4-2). Eggs are laid on the underside of the leaf sheath (Cranshaw and Ward 1996).

Nymph

The first nymph is tapered and almost transparent, while the second nymph is slightly larger, 0.12 mm, and more white. Both nymphal stages have two pairs of legs (Shetlar 1995a).

Life History and Habits

Life Cycle

BGMs are active primarily during late spring and summer. They require only 5–10 days to complete their development from eggs to adults. After hatching, they pass through two nymphal instars and molt, becoming egg-laying adults in 7–10 days. Therefore populations can build up very rapidly, especially when temperatures are optimal. All life stages are found together under leaf sheaths (Plate 7). As many as 200 individuals may exist in a single leaf sheath, sucking fluids from the plant and injecting toxins. Optimum temperatures for growth and reproduction are 26.5°–44.0°C.

Mites are spread through normal cultural operations, such as mowing that spreads clippings, and may hitchhike on other insects present in the bermudagrass. Wind is one major means of dispersal; movement by water is another. Mites also are spread on grass clippings or by clinging to other turf insects (Baker 1982, Cranshaw and Ward 1996, Hudson et al. 1995, Johnson 1975, Reinert 1982a, 1983c).

Adult Activity

Adults infest bermudagrass by settling under leaf sheaths and depositing eggs. A few mites or more than 100 can congregate under a single leaf sheath. Large numbers of mites are found in the rosettes (Baker 1982, Johnson 1975, Keifer et al. 1982).

Miscellaneous Features

Population Sampling

One technique that has been suggested is to construct a sampler that defines 0.1 m². The sampler is tossed on the ground at random and the number of rosettes (tufts or "witches brooms") within the sampler is recorded. Normally at least 10 counts should be taken every 40–45 m on a fairway, 20 counts from a tee, and 40 counts from a green or apron. If the average count exceeds 8 tufts per 0.1 m², chemical control strategies may be necessary to suppress the population, while if counts are less than 4 tufts per 0.1 m², cultural controls may be sufficient (Shetlar 1995a).

Plant Resistance

Evaluations of many cultivars and accessions of bermudagrass for resistance to the BGM have been made in Arizona and Florida. 'FloraTeX' (FB-119) showed no infestation in laboratory tests, and during 6 years of field observations. This common type of bermudagrass was released by the Florida Agricultural Experiment Station (Dudeck et al. 1994). 'Midiron' and 'Tifdwarf' also showed a high degree of resistance. 'Tifway' and 'FB-141' showed moderate resistance to the mite, with most of the plants tested exhibiting susceptibility (Reinert 1982a, Reinert et al. 1978).

More than 200 golf course greens were examined in Florida over a 2-year period. None of the greens planted with 'Tifgreen' (328) showed any evidence of mites or damage. When bermudagrass was cut at 2.8–5.0 cm, 'Common' and common types were severely infested, while 'Tifway' (419) was observed to be free of infestations (Johnson 1975).

Irrigation

The BGM thrives in warm, dry conditions. Irrigation at regular intervals may reduce mite populations or minimize the damage caused by mite activity (Shetlar 1995a).

Natural Enemies

A predacious mite, *Neocunoxoides andrei* (Baker and Hoffman), attacks the BGM and is widely distributed in Florida. It belongs to the suborder Prostigmata, family Cunaxidae. In Arizona the mite *Stenotarsonemus spirifex* (Marchal) of the suborder Trombidiformes, family Tarsonemidae, has been most frequently associated with reduced BGM populations (Butler 1963, Johnson 1975).

Zoysiagrass Mite

Taxonomy and Importance

The zoysiagrass mite (ZGM), *Eriophyes zoysiae* Baker, Kane, and O'Neill, is native to Japan and Korea but was introduced accidentally to the United States in 1982. It is now established in Maryland, Florida, and Texas (Hudson et al. 1995) and probably occurs in other areas where zoysiagrass is grown.

Host Plants and Damage

The ZGM apparently restricts feeding to zoysiagrass. Symptoms are virtually identical to those of the BGM on bermudagrass, and new leaf tips are twisted or snagged in partially unrolled older leaves, resulting in terminal arches. Damage to the host normally is not debilitating, but seed production can be adversely affected (Hudson et al. 1995).

Description

Adult females are creamy white and slightly larger than BGMs, 0.25–0.28 mm.

Life History and Habits

The life cycle of the ZGM has not been fully characterized but is presumed to be very similar to that of the BGM. All stages can be found throughout the growing season.

Buffalograss Mite

Taxonomy and Importance

The buffalograss mite (BFM), *Eriophyes slyhuisi* (Hall), is native to North America and is found throughout the Central Plains and midwestern United States, presumably throughout the range of buffalograss (Hudson et al. 1995).

Host Plants and Damage

The BFM restricts its feeding to buffalograss. Symptoms are very similar to those of the BGM on bermudagrass—thinning of the stand and the appearance of "witch's brooms" (tufts, Plate 7), especially in late summer when the grass is under drought stress. Damage can be more severe on certain cultivars.

Description

The adult female is white, slightly arched in a lateral view, and 0.19–0.24 mm long (Hudson et al. 1995).

Life History

Little is known about the life cycle but it is presumed to be similar to that of the BGM (Hudson et al. 1995).

Winter Grain Mite

Taxonomy

The winter grain mite (WGM), *Penthaleus major* (Duges), is a member of the family Eupodidae (Penthaleidae), which consists of seven genera. *Penthaleus* is easily separated from the other genera by the presence of a dorsal anus. It was first described as *Notophallus dorsalis* n. sp. by Banks (1904). Other common names ascribed are *red-legged earth mite, pea mite,* and *blue oat mite.* This mite occurs during the winter and causes more damage to small grain than to any other crop. It is not blue. Its habits make *winter grain mite* an appropriate name (Chada 1956, Streu 1981).

Importance

The WGM is a common and important pest of small grains and certain cool-season vegetables throughout the world. It affects grass seed production in Oregon and was discovered relatively recently causing winter injury to cool-season grasses in the midwestern and eastern United States. Its widespread distribution in the United States (Figure 4-3) makes it more than an incidental turfgrass pest (Baxendale 1997, Krantz 1957, Streu 1981). There are indications that the use of the insecticide carbaryl against other turfgrass insects increases the severity of WGM buildup. Reduction of predatory mites is the probable cause (Streu and Gingrich 1972).

History and Distribution

The WGM is widely distributed throughout both northern and southern temperate zones of the world. In the United States it was originally reported primarily west of the Mississippi River. Although the WGM was first described from specimens collected in Washington, D.C., it was rarely recorded east of the Mississippi River. It was not known to infest turf-

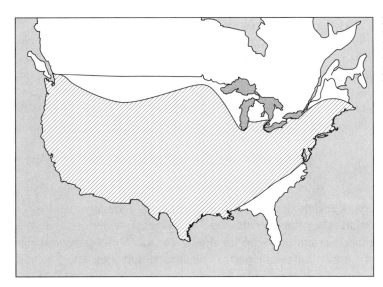

Figure 4-3. Distribution of winter grain mite. (Drawn by E. Gotham, NYSAES, adapted from Handbook of Turfgrass Insect Pests, Brandenburg and Villani 1995, Entomological Society of America.)

grasses in the Northeast prior to 1968. Pitfall trap catches on a red fescue–Kentucky blue-grass utility-type turf confirmed the presence of the WGM in New Jersey for the first time during October and November and again during February 1968–70. Of all the arthropods collected, the WGM accounted for more than 95% of the total specimens trapped during peak population periods when feeding damage was apparent (Streu and Gingrich 1972).

During March and April 1978–80, several incidents involving significant WGM injury to turfgrasses in the Midwest and Northeast were noted. Reports included damage to bent-grass golf course fairways in Pennsylvania; bentgrass fairways and greens in Cincinnati, Ohio; Kentucky bluegrass and fine fescue home lawns south of Cleveland, Ohio; and Kentucky bluegrass parkways on Long Island, New York. WGM occasionally reaches pest status in Nebraska and Colorado, as well (Baxendale 1997, Cranshaw and Ward 1996). From its widely scattered locations, reported from 19 states (Figure 4-3), it is logical to assume that the WGM now is present in practically every mainland state (Streu and Niemczyk 1982).

Host Plants and Damage

Kentucky bluegrass, bentgrass, Chewings red fescue, and perennial ryegrass are damaged. In Oregon seed production, perennial ryegrass is more susceptible to damage than Kentucky bluegrass and bentgrass (Kamm and Capizzi 1977, Streu and Niemczyk 1982).

The host plants recorded, exclusive of turfgrasses, include the small grains barley (*Hordeum vulgare*), oats (*Avena sativa*), rye (*Secale cereale*), and wheat (*Triticum sativum*); the legumes alfalfa, several clovers, lupine, and peas; the vegetables lettuce and potatoes; and various weeds (Chada 1956).

The rasping of the leaf surface with mouthparts called *chelicerae* and the sucking up of the plant sap produce a silvered, scorched appearance caused by loss of chlorophyll (Plate 8). The most severe grass damage appears from mid-December through mid-March, the period of highest mite population. Heavy damage to turf resembles typical winter freezing damage due to desiccation, a resemblance that no doubt contributed to the failure to identify this problem until relatively recently. Feeding by the WGM does not cause the yellowing that is characteristic of feeding by most tetranychid mites. The highest WGM populations are found in turfgrass that was treated with carbaryl during the previous summer (Figure 4-4) (Niemczyk 1978, Streu 1981, Streu and Gingrich 1972).

A laboratory host preference study that used petri dishes as containers for isolating the WGM and plants and holding them at 10°C and an 8 hr–16 hr light-dark period showed that mites survived the longest on Kentucky bluegrass, living for about 7 weeks and depositing eggs at the base of the plants and in the root system. Perennial ryegrass and Chewings red fescue supported mites for about 4 weeks. On bentgrass, the least suitable host, the mites died in 1–2 days (Streu and Gingrich 1972).

Description of Stages

Adult

Males have not been observed, and their occurrence has not been definitely established. Females average 1 mm in length. They have a dark-brown or reddish-brown body that is tinged with green and often has two light spots on the dorsal surface. Mouthparts and legs are orange to reddish-brown. Adults have four pairs of six-segmented legs. The first and

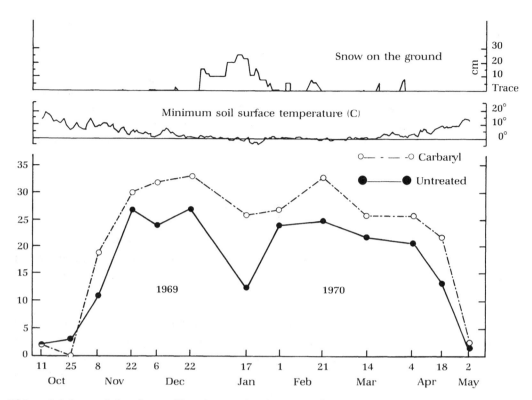

Figure 4-4. Seasonal abundance of the winter grain mite recorded in turf with and without summer carbaryl treatment in New Jersey, 1969–70. (Adapted from Streu and Gingrich 1972, fig. 1, courtesy of the Entomological Society of America.)

fourth pairs are longer than the middle two pairs. Two silvery eyes are located just behind the second pair of legs. The entire body is sparsely covered with very small, white setae, and each leg is covered with more pronounced setae. The most unique feature is a dorsal anus, present in no other mite associated with cool-season grasses. It is surrounded by a conspicuous reddish-orange spot (Plate 8). A droplet of fluid that is clear to light yellow is frequently seen exuding from the anus. The genital opening is ventral (Baxendale 1997, Chada 1956, Streu 1981).

Egg

Freshly laid eggs are plump and kidney shaped. They average 0.25 mm in length and 0.14 mm in diameter. They are orange to reddish brown and are glued singly to the base of grass plants, to the root system, or to the thatch (Plate 8). The brightly colored smooth surface becomes opaque and wrinkled within 1–2 days (Chada 1956, Streu 1981).

Larva

The sole larval instar is readily recognizable by its three pairs of legs. Larvae are 0.18 mm in length, 0.11 mm in width, and sparsely covered with white setae. They are reddish orange when first hatched but after a day become light brown, with legs and mouthparts yellowish orange. Just before molting the body becomes darker brown and tinged with green (Chada 1956).

Nymph

First-instar nymphs resemble full-grown larvae except that they have four pairs of legs. The body is dark brown, with a light-brown spot surrounding the dorsal anus. Legs and mouthparts are yellowish orange. The three nymphal instars have bodies that change from dark brown to black, with legs and mouthparts that change to reddish orange as they complete their nymphal stage (Chada 1956, Streu 1981).

Life History and Habits

Seasonal Cycle

The WGM has an unusual biology in that it feeds on grasses only during the cold winter months and oversummers as eggs. Two generations a year are reported in north central Texas and in New Jersey; the second-generation mites deposit eggs that aestivate (Chada 1956, Streu 1981).

In Texas, mites are normally present from early November to mid-April. Maximum populations occur in early February for the first generation and in early April for the second. In Nebraska the first-generation eggs hatch in October and develop completely to adults by late May. Mites of the second generation are most active in late February through early April (Baxendale 1997). In New Jersey, eggs hatch in early October, and females are present by November. Populations increase rapidly during November and early December, decline slightly in January between generations, and then peak at several thousand mites per 0.1 m² by late February or early March. Mites decline in April and disappear by May after having deposited aestivating eggs (Figure 4-4) (Chada 1956, Streu 1981).

Daily Activity

Most feeding by adults (and presumably also by larvae and nymphs) takes place during the night. As soon as the sun rises, mites move to shaded areas or descend into the base of the plants. During the heat of the day they are found on the moist soil surface under foliage. Mites migrate upward as the sun declines. After sunset the entire plant becomes covered with feeders. Mites may also feed on cloudy days or under snow cover (Chada 1956).

Oviposition

Females lay an average of 1–2 eggs a day for a total of 30–65 eggs during a nearly 40-day oviposition period. A secretion cements eggs to sheath leaves, to the stems, or on or in the soil. For the second generation, egg laying begins in February or March. By May all the mites are dead except for the aestivating eggs. The WGM is not seen again until the following October (Chada 1956, Niemczyk 1978).

Environmental Effects

Cold conditions rather than warmth favor the development of the WGM. Individuals are not harmed by short periods of sleet, ice, or frozen ground, although they are less active during periods of extremely cold weather (Baxendale 1997). Oviposition is heaviest between 10.0° and 15.5°C, while optimum temperatures for hatching are 7°–13°C. Adult activities are greatest at 4°–24°C. When temperatures go beyond these optima, mites stop feeding and descend to the ground or burrow into the soil. Hot, dry conditions force mites to penetrate 10–13 cm into the soil to seek moisture and relief from heat (Chada 1956). Mites often become inactive and shrivel up if moved to a dry location (Shetlar 1995a).

Dispersion

The mite is probably spread by transportation of aestivating eggs on organic debris or soil. Eggs may also be wind-borne (Chada 1956).

Natural Enemies

Many predatory mites have been implicated. Larvae *of Chrysopa californica* (Cog.) and the predatory mite *Balaustium* sp. feed on several stages of mites (Chada 1956, Wildermuth 1916).

Clover Mite

Importance, Host, and Damage

The clover mite (CM), *Bryobia praetiosa* Koch (Acari: Tetranychidae), is often concentrated in the turf next to the foundation of a building (Plate 8). The common name indicates only one of its preferred host plants; it also feeds on a variety of turfgrasses, particularly perennial ryegrass and Kentucky bluegrass. The grass may be killed about 0.3–1.0 m out from a building foundation. CM feeding resembles that of many tetranychid mites—characteristic silvery streaks or a speckled appearance on aboveground parts of the plant because of the extraction of plant sap and the drying of cells. During the spring and fall when they are plentiful, mites will often invade homes and are first seen on windowsills. They do not bite, transmit any disease, or feed in the house, but when crushed, they leave a reddish stain. They are normally more of a nuisance than a pest but occasionally can cause significant damage in turf (Cranshaw and Ward 1996).

The CM is a cosmopolitan species that can be found throughout North and South America, Europe, Asia, Africa, and Australasia (Shetlar 1995a).

Description

Adult CMs are less than 1 mm in length, with a reddish-brown to greenish body and four pairs of legs (Figure 4-5). The unusually long front pair of legs, extending in front of the body, is their most prominent feature (Plate 8). Eggs are bright red and spherical and are laid on the walls of buildings, on the bark of trees, and on other plants (Plate 8). A 10× hand lens is sufficient for viewing these mites and their eggs.

Life Cycle and Habits

CMs prefer to be active during cool spring and fall periods. Like the WGM, the CM usually survives the summer in the egg stage. When day temperatures regularly rise above 21°C, this mite becomes less active, and the adult population is reduced. The summer eggs hatch

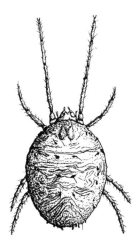

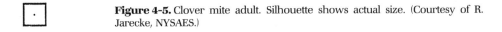

Figure 4-5. Clover mite adult. Silhouette shows actual size. (Courtesy of R. Jarecke, NYSAES.)

when cool fall temperatures return (often induced by freezing temperatures). The mites will remain active until the turf becomes dormant. Several generations can occur during the fall, winter, and spring seasons (Jeppson et al. 1975).

In many locations, the CM oversummers in the egg stage, but adults and nymphs are common throughout the winter months. Some eggs that are produced in the spring may not hatch until autumn (Shetlar 1995a). Arid conditions favor survival—supplemental irrigation in winter can reduce populations (Cranshaw and Ward 1996).

As a CM prepares to molt, it often returns to upright surfaces such as a tree trunk, fence post, or side of a building. The mites also congregate in these settings during unfavorable conditions (e.g., after mowing or during extremely hot or cold weather). Injury and activity normally are concentrated on the south and west (more exposure to sun) sides of buildings, trees, and shrubs, and numbers are greatest within 3 m of a building (Cranshaw and Ward 1996).

Banks Grass Mite

Importance, Host, and Distribution

The Banks grass mite (BKM), *Oligonychus pratensis* (Banks) (Acari: Tetranychidae), is reported as a pest of bluegrass and bermudagrass and has become a pest of St. Augustinegrass in Florida. Feeding produces stippling and resembles mildew or St. Augustinegrass decline, a viral disease (Cromroy and Short 1981). It is a common pest of wheat, corn, sugar cane, sorghum, and turfgrass in the Central Plains and Intermountain regions (Cranshaw and Ward 1996). The most serious damage occurs in water-stressed turf. It is not considered to damage irrigated turf. Bluegrass is attacked in the Pacific Northwest, and bermudagrass is damaged in Arizona, New Mexico, and Texas (Jeppson et al. 1975).

Damage

Early mite infestations produce plants with small yellow specks along the grass blades. These leaves become straw colored, and wither and die. The BKM is most damaging when drought stress is also present (Cranshaw and Ward 1996).

Description

Adult

Adult females are greenish yellow, with a very light salmon color over the palpi and front pair of legs (Plate 8). They resemble the two-spotted spider mite (*Tetranychus urticae*), and often have two dark spots on the dorsal surface. They are oval shaped and 0.4–0.8 mm long. Males have a more tapered abdomen and are shorter (0.33 mm). Mite coloration often makes detection difficult on turfgrass plants. The BKM is smaller than most spider mites. It sometimes forms webbing near the base of host grasses, often the first indication of the presence of the mite (Baxendale 1997, Shetlar 1995a).

Egg

Eggs are spherical, about 0.125 mm in diameter (Plate 8). They are pearly white when first produced but later darken to a straw-yellow color (Shetlar 1995a).

Larva

The larva is oval shaped and has three pairs of legs. Initially the larva is salmon colored but later the body becomes light green. The front pair of legs often is light orange (Shetlar 1995a).

Nymph

Both nymphal stages resemble the adult (four pairs of legs, oval shape) and are bright green (Shetlar 1995a).

Life Cycle and Habits

The life cycle lasts 8–25 days in field conditions, depending on the temperatures. Adults live an average of 23 days (Cromroy and Short 1981, Jeppson et al. 1975, Malcolm 1955). The BKM overwinters as adult females and late nymphs, which resume feeding as temperatures rise in the spring. All stages (eggs, larvae, nymphs, and adults) can be found throughout the growing season, often feeding in colonies within a network of fine webbing on the underside of foliage. There may be 7–10 generations in a year (Baxendale 1997).

The BKM is more adapted to warm weather than most other turfgrass mites and reproduces very rapidly in hot, dry conditions. Populations often peak in late summer, although they can be active any time during the year when temperatures are high enough. The BKM normally is damaging to turf in the spring and yet is a serious pest of corn in summer in the High Plains (Baxendale 1997, Cranshaw and Ward 1996, Shetlar 1995a).

5

Orthopteran Pests: Family Gryllotalpidae

Southern and Tawny Mole Crickets

Taxonomy

The southern mole cricket (SMK), *Scapteriscus borellii* Giglio-Tos (formerly *S. acletus* Rehn and Hebard), and the tawny mole cricket (TMK), *S. vicinus* Scudder, are members of the order Orthoptera, family Gryllotalpidae, subfamily Gryllotalpinae (Plate 9). Two genera are listed in this subfamily. The genus *Scapteriscus* can be distinguished by the front tibia with two dactyls; the genus *Gryllotalpa* has front tibia with four dactyls. Before 1984, the TMK was called the *changa mole cricket*, but this name is now reserved for *S. didactylus* (Latreille), a species that was introduced into Puerto Rico. The common name *changa* originated in Puerto Rico, where the face of the TMK is thought to resemble that of a pet monkey called *chango*. The TMK has also been called the *Puerto Rican* and *West Indian mole cricket*. Both the SMK and the TMK are sufficiently similar in appearance, habits, and destructive nature to be discussed together (Blatchley 1920, Nickle and Castner 1984, Reinert 1983b, Van Zwaluwenburg 1918).

Another species introduced into Brunswick, Georgia, in about 1904 and into at least four Florida coastal areas (Tampa, Key West, Miami, and Fort Myers) is the short-winged mole cricket, *Scapteriscus abbreviatus* Scudder (Plate 10). It is a serious pest of turf along the coast of Southeast Florida and in pockets around Fort Myers to Tampa, as well as a few inland areas, where it apparently spread in sod (W. Hudson, University of Georgia, Tifton, Georgia, personal communication, 1998). This species is easily recognized by its short tegmina, which cover only about one-third of the abdomen. The hind wings are vestigial. The pronotum is more elongated than in other species of *Scapteriscus*, and it has a distinctly mottled color pattern (Nickle and Castner 1984, Walker and Nickle 1981). Little is known about the short-winged mole cricket.

Importance

The SMK and the TMK are the most destructive insect pests of bahiagrass and bermudagrass turfgrass and pastures in the southeastern United States, and attack and damage most other warm-season grasses as well. Florida has millions of acres of these two favorite host grasses, with sandy soils that promote the development and spread of mole crickets. Golf courses are among the areas hardest hit, but home lawns, athletic fields, and other areas also sustain serious damage. Conservative estimates suggest mole cricket control costs exceeded $350 million in Florida during 1990–96. In some locations virtually 100% of the bahiagrass, including that in lawns, is damaged severely (Koehler and Short 1976, Reinert and Short 1981, Short and Reinert 1982).

Economically significant damage appears when nymphs are too old to be controlled easily. In North and South Carolina, Georgia, and Alabama, mole crickets have become the major insect problem on bermudagrass since about 1976, with the TMK considered the more serious (Cobb 1982, Reinert and Short 1981, Short and Koehler 1979, Short and Reinert 1982). Serious damage from both species now occurs consistently from North Carolina along the Atlantic and Gulf coasts to eastern Texas. Although SMK is primarily an insect predator, with little vegetation in its diet, it causes significant damage to turf through its burrowing activity, resulting in mechanical damage to roots and desiccation.

History and Distribution

Neither SMK nor TMK is native to the United States. The common belief that the TMK was introduced into the United States from Puerto Rico has been refuted; the species in the United States have calling songs that differ greatly in pulse rate from those of species in Puerto Rico. The TMK's place of origin is not known for certain but is thought to be somewhere in South America. Uruguay, Argentina, and Brazil seem particularly likely original sources of the U.S. population. This assumption is strengthened by the fact that the bahiagrass cultivars attacked by the TMK in Florida originated in these areas of South America (Walker and Nickle 1981).

The TMK was introduced into Brunswick, Georgia, about 1899, presumably in ship ballast, from a source other than Puerto Rico. By 1960 the original colony had spread into southernmost South Carolina, across southern Georgia and all of Florida, and into southeastern Alabama. It is now found in North Carolina, Mississippi, Louisiana, and eastern Texas as well (Figure 5-1) (Hudson 1995). The distribution of the TMK usually is slightly

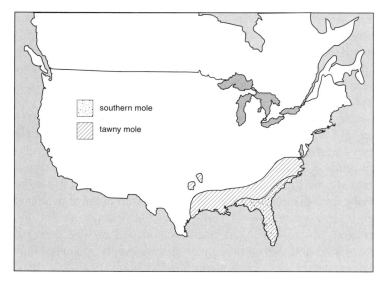

Figure 5-1. Distribution of southern and tawny mole crickets. (Drawn by E. Gotham, NYSAES, adapted from Handbook of Turfgrass Insect Pests, Brandenburg and Villani 1995, Entomological Society of America.)

southern mole

tawny mole

more coastal than that of the SMK, but in South Carolina, the TMK has spread farther inland, perhaps as a result of the movement of sod from infested areas (A. Johnson, Clemson University, Florence, South Carolina, personal communication, 1998).

The SMK was introduced at Brunswick, Georgia, in about 1904, also presumably in ship ballast. The subsequent spread has been slow. Individuals were present in Jacksonville, Florida, in 1920, in southern Florida in the 1940s, and in western Florida in about 1955. During the 1960s the SMK spread northward into North Carolina and westward into Louisiana. Currently the SMK is distributed throughout the Coastal Plains of the Southeast, from southeastern North Carolina through to eastern Texas (Figure 5-1). The range of SMK normally extends farther inland than that of TMK and includes Puerto Rico (Hudson 1995). In addition, a disjunct population has been confirmed in southwestern Arizona (Nickle and Frank 1988). Morphological variations in the SMK from various areas support the hypothesis that a number of introductions occurred (Crocker and Beard 1982, Walker and Nickle 1981).

Host Plants and Damage

Many of the warm-season turfgrasses are attacked by the TMK, and SMK burrowing activity also results in severe damage. Although bahiagrass and bermudagrass are the two damaged the most severely, zoysiagrass, centipedegrass, and bentgrass also experience damage. The texture of the grass may be related to the degree of damage, particularly when mole crickets have a choice in feeding. In a study of turfgrass varieties with potential resistance, the coarser selections of St. Augustinegrass, bermudagrass, and bahiagrass sustained the least damage, while the greatest reduction in growth was exhibited by 'Tifway' and 'Tifgreen', two fine-textured bermudagrasses (Short and Reinert 1982). More recently, 'Tif 94' bermudagrass, a fine-textured cultivar, showed almost no mole cricket activity. This cultivar is appropriate for golf course fairways, athletic fields, parks, and lawns. 'TW72', a new dwarf bermudagrass for use on golf greens, also showed significantly less damage than 'Tifdwarf' (Hanna, 1997).

The TMK also tunnels and feeds in seedling vegetables, ornamentals, and tobacco; in seed beds; and in recent transplants. In addition, mole cricket feeding sometimes accompanies increased activity of *Rhizoctonia* root rots in infested seed beds (Hudson 1995).

Mole crickets are seldom found in heavy soil, although the range of SMK suggests a wider tolerance for soil conditions. They usually occur in soils that range from light sand to loam; the soil must be compressible enough to allow tunneling without the need to remove loosened material (Van Zwaluwenburg 1918).

The most severe damage occurs during late summer and early fall, when the nymphs are approaching maturity and are actively foraging for food. Overwintering adults often cause severe damage in March and April. Burrowing in the upper soil levels mechanically dislodges roots and leaves mounds of soil on the surface, causing the soil to dry out excessively (Plate 10). Tawny and short-winged mole crickets also feed directly on the roots and other plant parts and seriously weaken the turf. Growth habits and cultural practices influence damage. Bahiagrass with its open growth allows greater dessication of the disturbed root system. Because golf course bermudagrass is maintained at low mowing heights, it has a reduced root system and therefore is more susceptible to uprooting and dessication. In Alabama, the presence of mole crickets on turf has been noted for many years, but turf-damaging populations started in about 1975. The severely damaged bermudagrass turf appears to have been plowed. On thick sod the tunneling activities produce a fluffy turf. On Jekyll Island off Georgia, a population of more than 1 cricket per 0.1 m² had the capacity of destroying a tee overnight (Cobb 1982, Duff 1982, Koehler and Short 1976, Reinert 1983b, Reinert and Short 1981).

Description of Stages

Adult

Mole cricket bodies, well adapted for burrowing, have strong shovel-like forelegs and a greatly enlarged, heavily chitinized prothorax for shaping and packing the soil in tunnels (Figure 5-2). The forewings overlap and are shorter than the abdomen (Plate 9). Wings of the short-winged mole crickets cover only about a third of the abdomen; individuals cannot fly. Males have a dark spot on the forewings that is made by the coalescence of wing veins, which also form the stridulating organ (Plate 9). The sex of a short-winged mole cricket cannot be determined by wing vein patterns. The sexes are about equal in ratio in the TMK and are presumed to be similar in the SMK (Nickle and Castner 1984, Short and Reinert 1982, Van Zwaluwenburg 1918).

Adults of both the SMK and the TMK are similar in appearance. Both average about 3.2 cm in length and about 1 cm in width (Reinert and Short 1981, Short and Reinert 1982).

Separation of Species. The TMK is slightly larger and more robust than the SMK and has a broader thorax. The TMK is golden brown with a mottled brown pronotum, while the SMK is grayish brown with a blue tinge and four pale spots on the pronotum (Plate 9).

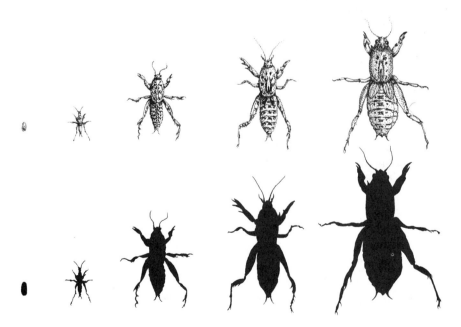

Figure 5-2. Stages of development of southern mole cricket: egg, early instar, middle instar, late instar, and adult. Silhouettes show actual size of each stage. (Courtesy of R. Jarecke, NYSAES.)

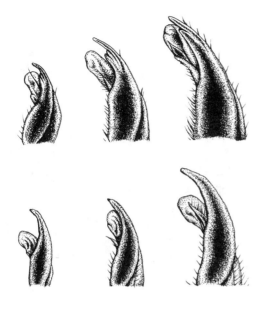

Figure 5-3. Tawny mole cricket lacinia (top) with tooth-like process and southern mole cricket lacinia (bottom) lacking such a process. (Adapted from Matheny and Kepner 1980, fig. 1, drawn by R. Jarecke, NYSAES.)

The short-winged mole cricket resembles the TMK in coloration but the wings cover only one-third of the abdomen (Hudson 1995).

The most reliable diagnostic character for separation of the two species relates to the tibial dactyls on the foreleg (Plate 9). In the TMK the two dactyls are separated by a V-shaped space narrower than the width of one dactyl. In the SMK the space is U-shaped and about as wide as one dactyl (Hudson 1995).

The two species can also be distinguished by differences in the maxillary laciniae (Figure 5-3). The lacinia bears a secondary toothlike process in the TMK, which is absent in the SMK. This difference is noted in the first instar, in the middle instar, and in adults (Matheny and Kepner 1980).

Egg

Freshly deposited eggs of the TMK are gray and change to a yellowish-white or brownish color as they mature. They are oblong-oval, with a shiny, unsculptured surface. As eggs develop, they change to milky white or light brown. At maturity the reddish-brown appendages become visible through the chorion (eggshell). Young eggs are about 3.0 mm by 1.7 mm and increase about 25% in size to about 3.9 mm by 2.8 mm in width before hatching (Hayslip 1943, Van Zwaluwenburg 1918).

Nymph

First instars of the TMK are about 6 mm long. Budlike wing pads are first noticeable in second instars, increase in size with each succeeding molt, and become plainly visible in the fifth instar. The number of nymphal instars is not certain, but six to eight are recognizable (Short and Reinert 1982, Van Zwaluwenburg 1918). A laboratory study revealed 10 instars for SMK, as determined by variation in pronotal length, ranging from 1.81 mm (first instar) to 3.94 mm (fifth instar) to 8.50 mm (tenth instar) (Hudson 1987a).

Seasonal History and Habits

Seasonal Cycle

As determined in Florida, the SMK and the TMK have generally similar life cycles and require about a year to complete a generation (Figure 5-4), although the SMK completes two generations per year in South Florida and other variations occur (Hudson 1987a). Both overwinter as adults and nymphs in northern and central Florida. However, approximately 85% of TMKs overwinter as adults compared with 25% of SMKs. The rest of the populations overwinter as late-instar juveniles and mature the following spring (Forrest 1986).

In Florida, oviposition usually begins in the latter part of March, and 75% of the eggs are deposited during May to mid-June. Eggs deposited in May and June require about 20 days to hatch. Peak hatching occurs during the first half of June in northern Florida and continues through August in southern Florida. Some oviposition occurs throughout the year in southern Florida, as evidenced by the presence of a few first instars every month of the

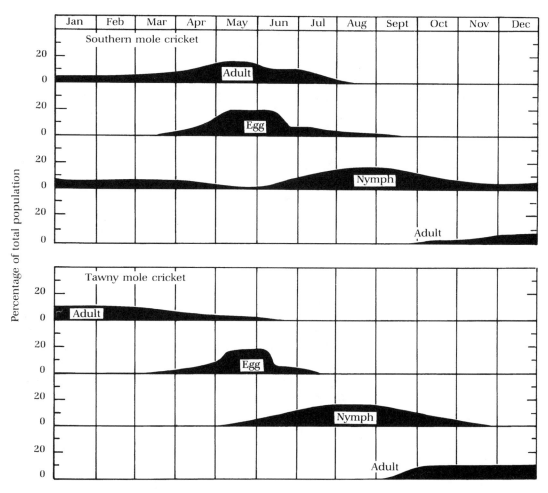

Figure 5-4. Life cycles of the southern and tawny mole crickets. (Adapted from Hayslip 1943, fig. 1.)

year (Short and Reinert 1982). In other locations (the Carolinas, Gulf States) most oviposition occurs in June or July (Hudson 1995).

The seasonal cycle in other southern states is very similar (univoltine, overwintering as adults or late instars) with some variation in the exact timing of specific developmental stages depending on local climate and species. These variations in timing of development are critical when designing effective management strategies (Brandenburg 1997).

Adult Activity

Flight. Both SMK and TMK adults have a major dispersal flight in spring and a minor autumn flight. While large numbers of both species often fly on the same night, flights of the two species are seasonally separated during part of the spring. The TMK flies from March to May, and the SMK flies from April to July. However, large flights of both species on the same night are not uncommon. Fall flights normally occur on warm evenings from October into December (Ulagaraj 1975, Walker et al. 1983). Dispersal flights consist mainly of females and peak in March in northern Florida with peak oviposition in May, while flights occur 1–2 weeks later in Georgia (Braman and Hudson 1993, Forrest 1986).

While oviposition periods for the SMK and TMK are similar in northern Florida, flight seasons are different. The TMK completes most flight activity before oviposition begins, but SMK flights occur before and during the oviposition period. The willingness of SMK females to fly between clutches may be one factor in that species's more rapid spread after its introduction into southeastern United States. Flying females usually have small oocytes, suggesting they may be starting a new clutch cycle (typically 7–12 days) (Forrest 1986).

Heavy flights normally occur after heavy rains during warm weather, starting soon after sunset and continuing for about 1 hr. The insects are strongly attracted to fluorescent, incandescent, and mercury-vapor lights (Ulagaraj 1975, Short and Reinert 1982). This attraction to lights can lead to severe problems in lighted turf areas, such as athletic fields and driving ranges, as adult females settle and oviposit.

Adult Sound Production and Trapping. Males of the SMK and the TMK produce calling songs after sunset for 1.0–1.5 hr in specially constructed subterranean chambers. Tegminal stridulation produces the sound. Adults also produce sound by tapping the soil with their forelegs. The specially constructed bulbous earthen chambers are 2.5 cm by 1.0 cm by 2.0 cm, about 3–5 cm below the soil surface. The function of male calling songs (as with all crickets and katydids) is to attract sexually responsive females (Ulagaraj 1976).

The natural calling songs produced by males of each species were tape-recorded in the field (natural environment) and in the laboratory (synthetic environment). Taped songs were broadcast from 0.5 hr after sunset to the end of the flight period, about 1 hr. It was found that crickets flying toward lights 100 m away would alter direction to fly toward the source of the broadcast sound. Playbacks of taped natural songs and synthetic songs showed that both species were attracted to their own songs. Many more females than males were trapped, and more than 60% of the trapped females bore sperm in their spermatheca (Ulagaraj and Walker 1973).

Marking and release of captured adults showed that at least 2% of the adults fly more than once. Some were recaptured twice, indicating at least three flights as much as 6.5 weeks after the first flight. Some were recaptured at least 0.7 km away from their first capture (Ulagaraj 1975).

Traps broadcasting synthetic calls of SMK and TMK were used to determine flight activity in three locations in Georgia. Peak flight activity at the southernmost location preceded that of the northern sites by 1–8 weeks, depending on species and year. Both SMK and TMK were collected at all three sites, but TMK comprised 45.3%, 12.5%, and 1.3% of the total catch at the south, south central, and north central locations, respectively. Minor fall flights and larger spring flights were recorded for each species and at all locations (Braman and Hudson 1993).

Mating and Oviposition. Mating flights occur during spring. By mid-June, the majority of the crickets have mated. After mating, the females enter the soil for oviposition. Eggs are deposited in oval chambers about 3.8 cm long by 2.5 cm high by 2.5 cm wide off the main galleries. The entrances to the chambers are concealed by a packing of loose earth after eggs are deposited. Most chambers are in the upper 15 cm of warm, moist soil, but cool temperatures and dry soil force females to construct chambers to a depth of about 30 cm.

Each female excavates three to five chambers and deposits about 35 eggs in each. The range for both species is 10–59 eggs (Hayslip 1943, Short 1973, Short and Reinert 1982, Van Zwaluwenburg 1918, Walker 1984). For each ovipositing female, the dry weight of eggs, number of eggs per clutch, and percentage that hatch decrease with successive clutches. Average investment in eggs per clutch ranges from 1% to 24% of female body weight for SMK and from 5% to 16% for TMK. There is no relationship between female size and number of clutches produced or the size of eggs produced for either species (Forrest 1986).

Hudson and Saw (1987) used a radioisotope labeling technique to determine spatial distribution of adult female TMKs that were maintained in field cages. They found that TMK females do not aggregate, even for feeding, and may be trying to avoid each other in the spring. They also found that adult activity in the fall (as determined by captures in pitfall traps) is much less than that of nymphs or spring adults. Adult crickets recovered from pitfall traps in the fall and winter were predominately male.

Nymphal Activity

Immediately upon hatching, the young nymphs search and fight for food in the egg cell, eating eggs, empty chorions, and weaker living nymphs. Many young nymphs perish through cannibalism. Soon the young nymphs escape from their egg cells and burrow to the soil surface to begin feeding on roots, organic matter, other insects, and other small organisms. Nymphs develop rapidly throughout the summer, and the first adults appear in September (Hayslip 1943, Reinert and Short 1981, Short 1973).

Most crickets, remaining as nymphs in November when cold weather arrives, overwinter as large nymphs or adults. Those that overwinter as nymphs become adults the following spring. About 75% of the SMK and 15% of the TMK overwintered as nymphs in central Florida during a 2-year observation (1970–72). Eggs deposited late during the oviposition periods tend to produce overwintering nymphs, while those deposited early become adults in the fall (Hayslip 1943, Reinert and Short 1981).

Feeding Activity

TMK nymphs as well as adults come to the surface at night to feed. Tunneling of more than 3–6 m a night has been observed. Most of the feeding occurs during warm nights following rain or irrigation. Crickets return to permanent burrows during the day and may remain there for long periods during dry periods or cold weather. At low densities, each adult has its own burrow, which may extend to a depth of 36 cm (Hayslip 1943, Reinert and Short 1981, Short and Reinert 1982).

The TMK does more plant feeding; its gut contains mostly plant food material, while that of the SMK primarily has remains of insects (Plate 10) (Reinert and Short 1981, Ulagaraj 1975).

Miscellaneous Features

Sampling Techniques

Soapy Flush (Irritating Drench). A soapy flush (irritating drench) can be used to force nymphs and adults of both species to burrow to the surface, where they can be counted. The disclosing solution normally consists of 30 ml of a lemon-scented dish detergent in 4–8 l of water. This mixture is poured over an area approximately 0.6 m on each side. (A synergized pyrethrin can be used in place of detergent.) Normally mole crickets will surface within a few minutes. This technique has proved to be as accurate as using a tractor-mounted soil corer (Hudson 1988). The technique is more effective at bringing nymphs to the surface than adults (Hudson 1994).

When pyrethrins are used as the disclosing solution, as many as 65% of the crickets die before they reach the surface, resulting in an underestimation of the population (Hudson 1988, Ulagaraj 1975). Soap flushing is approximately 50% efficient when soil moisture is 13.3%, and 90% efficient at 19.5% soil moisture (Hudson 1989).

The soapy flush does not provide an absolute population estimate, but it does provide critical information about the life stages present at the time of sampling, which in turn helps turf managers plan management strategies. In addition, the technique can be used to confirm the presence or absence of mole crickets after controls have been applied.

Grid-Square Rating System. A rating system can be used to determine mole cricket abundance and activity. Using PVC pipe or wood, a frame that is 75 cm on each side and divided into nine equal, square-shaped sections is constructed (Plate 70). The frame is placed on the ground and, using visual and manual inspection, the number of sections that contain mounds or tunnels is counted. In each square, any activity (whether one or several tunnels) is scored as "1", while squares with no activity are scored as "0". The total score for the nine squares is recorded. Thus damage rating ranges from 0 (no damage in any of the sections) to 9 (damage in all of the sections). The damage rating system is linearly related to soap flush counts (Cobb and Mack 1989).

Linear Pitfall Trap. The linear pitfall trap (Lawrence 1982) is used mainly for collecting mole crickets for research needs. The trap (Plate 10) works equally well in sod or bare ground to capture mole crickets of all stages as well as other insects that crawl over the soil surface. This trap is described in detail in Chapter 26.

Sound Trap. Routine monitoring of mole cricket flights is accomplished using a timer-operated sound synthesizer for each species. A standard trapping station consists of one SMK trap and one TMK trap, each with a wading pool 1.5 m in diameter that has been partially filled with water (Plate 10). Captured crickets remain floating on the surface of the water and are routinely removed the next morning (Walker 1982). An alternative arrangement, used to infect crickets with entomopathogenic nematodes, consists of a large funnel and 20-l bucket (Parkman et al. 1994).

Sound traps are used regularly in the Southeast to monitor flight activity (e.g., date of first flight, peak flights) and population trends, and to collect large numbers of adults for research purposes. In addition, they are used to disseminate mole crickets that are infested with entomopathogenic nematodes.

Mapping. Using a combination of monitoring techniques, turf managers can determine where mole cricket populations are likely to be highest and concentrate control efforts in those areas. When adults are present and active (as determined by flight activity and presence of tunneling damage), a map of the area should be sketched, noting where damage (or activity) is greatest. In an Alabama study, such a map accounted for 90% of the total area damaged or treated the following growing season (Cobb and Lewis 1990).

When egg hatch is anticipated, soapy flushes should be conducted at least weekly until small nymphs are found. Most traditional management strategies are most effective when applied at this time. For best results, night temperatures should be above 15.6°C, and the soil should remain moist when insecticides are applied (Reinert and Short 1981, Short 1973, Short and Koehler 1979, Short and Reinert 1982). Late June through July is considered the optimum period for insecticidal control in central Florida or southern Alabama, because most eggs have hatched, nymphs are small and close to the surface, and extensive damage has not yet occurred. Some biological control alternatives, however, are more effective against adults and should be applied when that stage is present.

Threshold Populations

Most researchers believe that there are no reliable damage thresholds, although turf managers often indicate that a damage rating of 3–4 reflects their tolerance level for mole cricket activity. As with any pest insect, thresholds (or tolerance levels) vary depending on many factors, such as size (age) of crickets, turf vigor and susceptibility, mowing height, soil type, availability of water, and expectations, so thresholds are site specific.

Natural Enemies

Microorganisms

Several fungi have been observed infecting mole crickets. *Metarrhizium anisopliae* (Metchnikoff) infections produce a carcass covered with white hyphae that are later covered with light-green spores (Plate 57).

The fungus *Sorosporella uvella* (Kass.), which produces a distinct brick-red carcass, causes some mortality in mole crickets (Hayslip 1943, Short and Reinert 1982). Infected mole crickets are found throughout the year. The overall rate of infection from August through December is 4% in field conditions, although rates may be higher in crowded conditions, as are encountered in laboratory studies (Pendland and Boucias 1987).

Strains of *Beauveria bassiana* (Balsamo) and occasionally *Entomophthora* sp. have been collected from mole crickets in southern Brazil and Uruguay. While *Entomophthora* is a weak pathogen of *Scapteriscus*, various Brazilian isolates of *B. bassiana* and *M. anisopliae* are highly virulent in the laboratory (Boucias 1984).

Steinernema scapterisci Nguyen and Smart was identified as an entomopathogenic nematode that was fairly specific to mole crickets in South America (Nguyen and Smart 1991). The nematode was released in three bahiagrass pastures in Alachua County, Florida, in 1985 by burying nematode-infected mole cricket cadavers 2–4 cm beneath the soil surface or by sprinkling an aqueous solution of nematodes on the soil surface. The nematodes were established within 1 week after release, regardless of the method of release, as confirmed by the collection of nematode-infected mole crickets in pitfall traps (Parkman et al. 1993).

Hudson and Nguyen (1989) reported that TMK nymphs are less susceptible to infection and mortality than adults. SMKs and TMKs were equally susceptible to infection by the nematode in laboratory conditions (Hudson et al. 1988), but a significantly higher percentage of SMKs than TMKs captured during a study were infected (Parkman et al. 1993). Parkman and Frank (1992) reported similar results for adult SMKs and TMKs recovered from sound traps and suggested that the difference in flight behavior of the two species might account for the greater levels of infection for the SMK. In addition, the SMK is primarily predaceous and, therefore, more active above and below the soil surface. As a result, individual SMKs may be more likely to come in contact with infective-stage nematodes (Parkman et al. 1993).

Parkman et al. (1994) reported that single inoculative applications of *S. scapterisci*, applied to golf courses in north central Florida (Alachua County) and southeastern Florida (Broward County), resulted in significantly lower mole cricket damage a year after application, as reported by turf managers. The rate of infection of adult mole crickets was significantly greater than that for nymphs, and the infection rate for SMKs (25.0%) was significantly greater than that for TMKs (11.0%). The authors suggested that *S. scapterisci* can be an effective biopesticide that suppresses mole cricket populations relatively rapidly.

Invertebrate Parasitoids and Predators

Ormia depleta (Wiedemann), the red-eyed fly, is a tachinid fly that originated in South America. Females are attracted to natural and recorded songs of the TMK and other mole crickets. Adult females produce larvae directly, which feed on all three introduced *Scapteriscus* species (Hudson et al. 1988). The fly was released in Alachua and Manatee counties, Florida, in 1988 and 1989, respectively. These and subsequent releases produced a continuous population of the fly in 38 contiguous counties in Florida, demonstrating great potential for quick dispersal (Parkman et al. 1996).

Another promising parasitoid is the sphecid wasp, *Larra bicolor (americana)*, found in abundance in Brazil, Venezuela, Dutch Guiana, and Puerto Rico, where it is an effective parasitoid (Plate 63). There have been recent attempts to colonize it in Florida (Short and Reinert 1982, Wolcott 1941). The adult wasp enters a mole cricket tunnel and chases the cricket to the surface, where she captures it and stings it at the base of the prothoracic legs, temporarily paralyzing it. She also stings the cricket at the base of the mandibles, preventing the cricket from biting in defense. The wasp then deposits an egg ventrally behind the prothoracic legs. After about 3–5 min, the mole cricket recovers briefly and burrows back into the soil. The parasitoid larva emerges in 5–7 days and develops externally on the host. It grows slowly for 10–11 days, then doubles in size within a day. Pupation of the parasitoid occurs in the mole cricket gallery and lasts about a week (Anonymous 1983, Hudson et al. 1988). The wasp was established in southeast Florida in 1981 and in north central Florida in 1988 and 1989 (Parkman et al. 1996).

Two genera of bombardier beetles (*Stenaptinus* and *Pheropsophus*) contain species that prey on mole crickets. Adults of *S. jessoensis* (Morawitz) are generalist predators, while larvae are specialist predators on mole cricket eggs. In laboratory trials, larvae tunneled through sand to find mole cricket eggs in a simulated egg chamber (Hudson et al. 1988).

Many ground-dwelling predatory insects and spiders that are found in Florida pastures take mole crickets as prey. These include fire ants, earwigs, ground beetles, tiger beetles, and spiders, especially in the families Lycosidae (wolf spiders) and Salticidae (jumping spiders). Several predators are mentioned by Hudson et al. (1988), including *Lycosa annexa* Gertsch (a wolf spider), *Megacephala virginica* (a tiger beetle, Plate 60), and *Pasimachus* sp. (a ground beetle). *Sirthenea carinata*, a subterranean assassin bug, is distributed widely in the eastern United States, but adults and large nymphs show a decided preference for mole crickets as prey. Small nymphs of the reduviid show no such preference (Hudson 1987b).

Vertebrate Predators

Raccoons, skunks, red foxes, armadillos, and toads feed on mole crickets. Of the birds, grackles, cow egrets, and white ibis will search for crickets on turf (Short and Reinert 1982). Cattle egrets often take advantage of local flooding, plowing, or other disturbances that expose mole crickets on or near the soil surface. Loggerhead shrikes (*Lanius ludovicianus* L.) sometimes deposit mole cricket carcasses on fence posts (Hudson et al. 1988).

6

Hemipteran Pests: Suborder Heteroptera

Chinch Bug Taxonomy

Chinch bugs infesting turfgrasses are members of the family Lygaeidae, subfamily Blissinae. The family Lygaeidae was originally known as the chinch bug family, with about 1,400 known species. Since only four chinch bug species are of any economic importance to humans, the common name was not considered appropriate for the family. The common name for this family that is currently accepted is *lygaeid bugs*. The subfamily Blissinae, previously considered to have only two genera, now has three recognized genera. Only the genus *Blissus* contains serious turfgrass pests.

The *leucopterus* Complex

Five species or subspecies make up the *leucopterus* complex (Blatchley 1926; Leonard 1966, 1968; Werner 1982), including two incidental species of *Blissus* that inhabit the Atlantic coastal dunes. The two turfgrass-infesting chinch bugs, the hairy chinch bug (HCB), *Blissus leucopterus hirtus* Montandon, and the southern chinch bug (SCB), *B. insularis* Barber, plus the common chinch bug (CCB), *B. leucopterus leucopterus* (Say), complete the complex. The CCB frequently is a serious pest of small grains, sorghum, and corn in the heartland of the United States and throughout the central Atlantic region. All three turfgrass-infesting species occur in the eastern half of the United States, with some overlapping of geographic ranges between *B. l. hirtus* and *B. l. leucopterus* and between *B. insularis* and *B. l. leucopterus* (Figure 6-1). Much has been published on the subspecies *leucopterus,* and considerable literature has been developed on the two primary turfgrass-infesting species (Leonard 1966). Another species, *Blissus occiduus* (Barber) has been identified damaging buffalograss in Nebraska, but its range is believed to be somewhat restricted (Baxendale et al. 1997).

It is difficult to distinguish between HCB, SCB, and CCB but they can be separated when they are treated as populations. Morphological similarities and large variations among individuals make it very difficult to differentiate between individual specimens or within a short series (Leonard 1966).

Differences

A comparison of mean width and length of CCB, HCB, and SCB adults from various geographic regions shows that the HCB is generally more robust than the SCB but is similar in

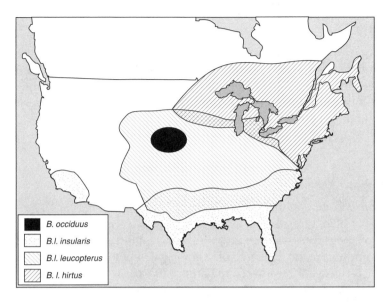

Figure 6-1. Distribution of chinch bugs: hairy chinch bug (*B. leucopterus hirtus*), southern chinch bug (*B. insularis*), common chinch bug (*B. l. leucopterus*), buffalograss chinchbug (*B. occiduus*). (Drawn by E. Gotham, NYSAES, adapted from Handbook of Turfgrass Insect Pests, Brandenburg and Villani 1995, Entomological Society of America; new species on buffalograss adapted from Baxendale 1997.)

B. occiduus
B.l. insularis
B.l. leucopterus
B. l. hirtus

length or somewhat shorter (Leonard 1966). The setae of the HCB are golden yellow, while the setae in the CCB are silver or light straw yellow. The abdomen of the HCB is darker than that of the CCB (Leonard 1966).

The much higher incidence of brachypters (short-winged adults) in the HCB and SCB than in the CCB is the most characteristic difference. Macroptery (a long-winged form) in the CCB is apparently associated with agricultural ecosystems where migration is often necessary. Such migration is rarely necessary in the turf habitat of the HCB and SCB (Leonard 1966).

Speciation

In isolation, both HCB males and females mate successfully with CCB adults, supporting the concept of two subspecies. Copulation occurs between the SCB and CCB, resulting in production of fertile eggs, but no nymphal development occurs. The genetic inviability of the SCB when crossed with other members of the *leucopterus* complex provides strong evidence that the SCB is a distinct species (Leonard 1966).

Hairy Chinch Bug

Importance

The HCB is a major pest of turfgrasses in Connecticut, New Jersey, New York, Pennsylvania, and Ohio. July through August is the single most critical time for turf damage throughout most of the insect's range. The semidormancy of unirrigated turf due to drought often masks HCB damage. Damage may not become apparent until turfgrass previously infested with the HCB fails to respond to late summer rains (Maxwell and McLeod 1936, Niemczyk 1982, Polivka 1963, Schread 1970a, Streu and Vasvary 1966). In New York it was considered a serious pest only in the southeastern portion of the state, but since about 1970 it has become a serious home lawn pest in much of upstate and western New York (H. Tashiro, personal observation).

History and Distribution

The HCB is found in all the northern states in and east of Minnesota and throughout all the New England and mid-Atlantic states south into Virginia (Figure 6-1). In Canada, the HCB occurs in all of the provinces that border the United States from Ontario to the Atlantic Coast. In Ontario the first HCB damage to lawns was reported in 1971. Damage has since occurred in many locations (Leonard 1966, Mailloux and Streu 1981, Maxwell and McLeod 1936, Sears 1978).

Host Plants and Damage

The HCB feeds on most of the cool-season turfgrasses, including red fescue, perennial rye-grass, bentgrass, and Kentucky bluegrass, as well as on zoysiagrass, a warm-season grass that is occasionally grown in cooler locations. Creeping bentgrass maintained at lawn height is most susceptible to injury because of the thick mat of stolons, but at putting-green height, with the stolons eliminated, it is relatively immune from damage (Baker et al. 1981, Maxwell and McLeod 1936). HCB infestations often occur in turfgrass with thick thatch. The HCB is a major home lawn pest, but also occasionally causes damage on golf courses, athletic fields, and sod farms.

In a Michigan study, HCB population density was higher in plots where a thick thatch lay-er was induced with applications of a fungicide, mancozeb. In addition, in a survey of home lawns, thatch thickness was greater in HCB-infested lawns than in uninfested lawns. The rate of development was similar in treated (thatchy) and untreated plots (Davis and Smitley 1990a). However, other turf environment parameters (such as dry weight of grass clippings and chlorophyll content) were not significantly correlated with chinch bug populations over a 3-year period. Infested lawns had more fine fescue and less Kentucky bluegrass than un-infested lawns, suggesting that HCBs either respond to or influence species composition in home lawns (Davis and Smitley 1990b).

Chinch bugs have piercing-sucking mouthparts and suck the sap from the crown and stems of grasses. They tend to aggregate. Their feeding results in localized turfgrass injury— as yellow grass that soon turns brown. Coalescence of localized injury causes large patch-es of dead or dying grass (Plate 12). Most susceptible to damage are lawns on sandy loca-tions in direct sunlight, where moisture deficiency leaves the grass less tolerant to injury (Maxwell and McLeod 1936, Niemczyk 1980a). Symptoms of HCB feeding closely resem-ble drought injury or sunscald (Reinert et al. 1995).

Description of Stages

Much of the description of chinch bugs comes from accounts of *B. l. leucopterus* and is con-sidered to describe the HCB reliably. When chinch bugs are crushed, they emit an odor re-sembling that of stink bugs.

Adult

HCB adults are black with shiny white wings. They are slightly more than 1.0 mm in width and 3.0 to 3.6 mm in length (Figure 6-2, Plate 11). The four-segmented antennae are mostly black but have reddish proximal segments. The legs vary from red to yellowish red. The most conspicuous feature of adults is the pattern of the forewings, which are folded

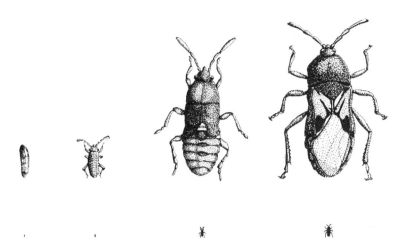

Figure 6-2. Stages of development of hairy chinch bug: egg, first-instar nymph, final-instar nymph, and adult. Silhouettes show actual size of each stage. (Courtesy of R. Jarecke, NYSAES.)

over the body. Near the middle of the costal margin of each hemelytron (front wing), there is a black spot with a somewhat Y-shaped black line extending diagonally toward the head. The black spot and line separate the front wings into a large median posterior white area and two smaller laterally located white areas (French 1964, Leonard 1968). HCBs can occur as macropterous (long-winged) or brachypterous (short-winged) adults (Plate 11). In large HCB populations, brachyptery dominated by about 64% (Leonard 1966).

Male HCBs are slightly smaller and less robust than females. The most conspicuous differences are evident when individuals are viewed ventrally (Plate 11). The abdomen of the male is rounded in cross section, while that of the female forms an inverted V produced by a distinct median ridge that is part of the ovipositor (Luginbill 1922).

In general there were more female than male macropterous adults, but there was an excess of male brachypterous adults in a 1974 New Jersey field population. The largest difference in macropters occurred in spring adults, which showed a male-female ratio of 1:1.5. In brachypters the largest difference occurred in summer adults, which had a male-female ratio of 1.4:1. The sex ratio of the overall population was 1.1:1 male to female (Mailloux and Streu 1981).

Egg

Eggs are elongate, ovate, slightly reniform, rounded at the anterior end, and truncated at the other (Plate 11). The truncated end has three to six, usually four, micropyles 0.1 mm in length. Eggs average 0.31 mm in width and 0.86 mm in length. Freshly deposited eggs are whitish and turn yellow in a few days, becoming deep red several days before hatching. Just before hatching, the embryo is visible through the smooth, shiny, somewhat iridescent chorion (Chambliss 1895, Choban and Gupta 1972, Luginbill 1922).

Nymph

Chinch bugs have five nymphal instars (Figure 6-2, Plate 11). They grow from a width of 0.23 mm and length of 0.90 mm in first instars to 0.96 mm and 2.97 mm in fifth instars (Luginbill 1922).

Morphological differences distinguish the nymphal instars. In the first instar, the head width is greater than or subequal to the thoracic width. In the second instar, the head is narrower than the thoracic width. In the third instar, mesothoracic wing pads appear. In the fourth instar, the wing pads extend over the abdomen no farther than the posterior area of the first abdominal segment, which is white. In the fifth and final nymphal instar, the wing pads become prominent and conspicuous and extend at least onto the second abdominal segment, which is white, and sometimes onto the third abdominal segment, which is brown. There is also a distinct color variation as nymphs grow. The first and second instars are bright red, with a distinct white band on the anterior two abdominal segments. The red changes to orange in the third instar, then orange brown in the fourth instar, then blackish in the fifth instar (Luginbill 1922, Mailloux and Streu 1981, Niemczyk 1981).

Seasonal History and Habits

Seasonal Cycle

The HCB has two generations a year in southern New England, in the mid-Atlantic states, including New Jersey and Long Island, and westward through Ohio (Figure 6-3). It normally has only a single generation in Upstate New York around Rochester and in southern Ontario, Canada. Eggs from overwintering females were present during May and June and

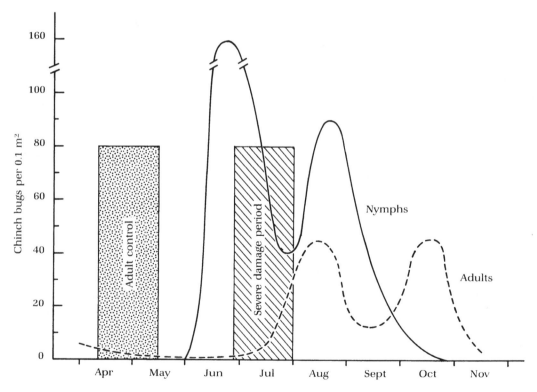

Figure 6-3. Generalized life history and seasonal abundance of the hairy chinch bug at Wooster, Ohio, showing the time of severe damage and the period when adult control programs may be most effective. (Adapted from Niemczyk 1982, fig. 3.)

Table 6-1. Hairy Chinch Bug, Mean Numbers per 0.1 m² in Turf in Wooster, Ohio, in 1978

Month	Chinch Bug Stages					Total No. of Nymphs
	Adult	1st	2nd	3rd and 4th	5th	
April	3	0	0	0	0	0
May	2	0	0	0	0	0
June	4	63	30	2	<1	95
July	9	23	31	27	30	111
August	40	14	23	11	19	81
September	19	7	11	10	17	45
October	26	<1	<1	<1	2	3
November	4	0	0	<1	<1	1

Note: Based on 12 samples of turf, 10.7-cm diameter by 7.6-cm deep (using a standard golf cup cutter), collected weekly.
Source: Compiled and adapted from Niemczyk 1982.

from summer females from mid-July through August and into September on Long Island during 1934–35. In New Jersey, overwintering females lay eggs from the third week in April to the end of May, and summer females do so from the third week in July to the end of August, precisely the same period as that recorded on Long Island (Liu and McEwen 1979, Mailloux and Streu 1981, Maxwell and McLeod 1936, Niemczyk 1981).

The predominance of nymphs during the second half of June and the second half of August in Ohio (Table 6-1) corresponds to the pattern seen on Long Island. Second-generation nymphs complete their development by early October, becoming overwintering adults that seek hibernation quarters (Mailloux and Streu 1981, Maxwell and McLeod 1936).

Adult Activity
Adults seek hibernation sites in late summer and fall. Common sites include infested turf with sufficient undamaged grass to supply shelter and food before hibernation. Thatch or tall grass near the edge of lawns and putting greens is sought (Plate 12). Other overwintering sites include plant debris and space around the foundations of houses, under shingles and clapboards (Plate 12) (Leonard 1966, Maxwell and McLeod 1936).

Overwintering adults become active and leave their hibernation sites when a threshold temperature of 7°C is reached. This threshold temperature was evident in New Jersey, where there are two generations a year, and in southern Ontario, Canada, where there is one. The HCB migrates primarily by crawling, but the presence of specimens on clothes on a clothesline suggests a limited flight capability. Individuals immediately start to feed and copulate. The HCB mates repeatedly at 5- to 8-day intervals. Copulation lasts 12 hr or more, with the female alone being active, dragging the male behind her. Females start to lay eggs after a preoviposition period of about 2 weeks (Leonard 1966, Liu and McEwen 1979, Luginbill 1922, Mailloux and Streu 1981).

Eggs are deposited in leaf sheaths and in the ground on roots of host plants. Estimates of egg production vary considerably, ranging from an average of 1.5 eggs per day, for a total of 54 eggs, to 20 eggs per day for 2–3 weeks. During observation in New Jersey in Spring 1974, overwintering females averaged 15.6 eggs and summer adults, 6.9 eggs per female (Kennedy 1981, Mailloux and Streu 1981).

During 1974, a peak population of 196 eggs per 0.1 m² was found during the period from May 6 to June 1, when two common legumes associated with turf, white clover, *Trifolium repens* L., and bird's-foot trefoil, *Lotus corniculatus* L., were in early bloom. Field-collected females are reported to lay as many as 170 eggs, with a female longevity of about 100 days (Baker et al. 1981, Mailloux and Streu 1981).

Nymphal Activity

During early spring, eggs may not hatch for a month or more, but during midsummer, eggs hatch in as short a time as 7–10 days. Immediately upon hatching, young nymphs begin feeding on the sap of plants by piercing the turfgrass stems and leaves. Development through the five nymphal instars requires about 4–6 weeks during the summer. Nymphs may seek hibernation quarters but will die unless they transform to adults (Kennedy 1981, Leonard 1966).

In areas where two generations occur each year, summer adult emergence coincides with well-developed inflorescence of sumac (*Rhus copallina*). Those adults subsequently lay eggs (usually mid-July through late August) that hatch into second-generation nymphs and continue feeding. They complete development by late summer or early autumn (Reinert et al. 1995).

Miscellaneous Features

Degree-Day Developmental Relationship

Degree-day (DD) accumulations and relationship to HCB development have yielded useful information for the management of HCB infestations. In Ontario, Canada, there is one generation per season. When 7°C was used as the base temperature for development, third-instar nymphs peaked at 750–900 DD in the thatch. In 1977, this occurred in early July. Sampling for treatment or nontreatment decisions was started at 897 DD and terminated at 950 DD. The peak third-instar period was chosen for the decision because at that time most eggs have hatched but no adults have developed to disperse out of the area. Populations of third-instar nymphs per 225 cm² of turf were determined. A count of fewer than 20 nymphs per sample (approximately 80 nymphs per 0.1 m²) required no treatment, 80–120 nymphs per 0.1 m² required a decision to treat or not, and more than 120 nymphs per 0.1 m² required treatment. When no damage is yet evident, a single insecticidal treatment during the third-instar period provides maximum control (Liu and McEwen 1979).

By comparison, in New Jersey, where two generations occur per season, an air temperature of 14.6°C was used as the base temperature for egg development. Degree-day accumulations (in degrees Celsius) for development of the two generations of the HCB in relation to calendar periods were determined. The egg hatch of the first generation was completed at 115 DD during early June (14.6°C base), and the peak adult presence occurred at about 630 DD in late July. The second-generation egg hatch was completed near 850 DD after mid-August, and peak adult presence occurred at above 1,159 DD during mid-October (Mailloux and Streu 1981). These determinations are very useful in estimating the development of various stages of the HCB in turf without the need for laborious and continuous field sampling.

In northeastern Ohio, where two complete generations usually occur annually, a degree-day model accurately predicts the occurrence of second instars of the first generation. The model, which uses a 7°C base temperature, estimates the number of degree-days for half the population to complete the second instar to be 715 DD for the first generation. Similar attempts to predict second-generation development are less accurate. First instars are most abundant during late June (first generation) and late August (second generation); second instars, during early July and early September; third instars, during late July and mid-September; fourth instars, during early August and mid-September; and fifth instars, during mid-August and late October (Niemczyk et al. 1992).

Mortality Factors

Studies conducted in New Jersey during 1974 and 1975 revealed that a high incidence of mortality occurs in HCB eggs—59% and 48% in the spring and summer generations, respectively. Six mortality factors with regard to the overall population were identified: (1) infection by *Beauveria bassiana* (Bals.) Vuillemin, (2) parasitism by *Eumicrosoma benefica* Gahan, (3) predation by *Amara* sp., (4) desiccation, (5) failure to hatch, and (6) wet conditions at eclosion (Mailloux and Streu 1981). The three biotic factors are discussed more thoroughly in connection with natural enemies.

The winter mortality of adults can be significant; it was 68% and 28% in 1974 and 1975, respectively. Winter mortality is considered to relate to the presence of moisture in the overwintering sites. Lack of snow cover increases mortality. Generally speaking, the survival of chinch bugs is promoted by high humidity at low temperatures and by low humidity at high temperatures (Guthrie and Decker 1954, Mailloux and Streu 1981).

Laboratory Rearing

HCB rearing has been accomplished on a year-round basis on sections of young corn (*Zea mays*) treated with 2% sodium hypochlorite. Egg survival was greatest following treatment with the same sterilant. The rearing environment had a constant temperature of 26°C, with 40–75% RH and a 16-hr photoperiod. The mean preoviposition period was about 11 days, with about 80% of the females ovipositing between days 5 and 14. Under these conditions, the nymphal periods averaged 12.3, 5.4, 5.2, 4.9, and 7.1 days for the first through the fifth instars, respectively. Diapause in the HCB can be broken by 10–14 days of continuous exposure to a temperature of 29.5°C in the absence of light (Baker et al. 1981, Leonard 1966).

Population Management

The commonly accepted period for insecticidal control of the HCB is during the summer, when chinch bugs are feeding actively. Tests conducted during 1978 and 1979 in Ohio support the theory that the HCB can be controlled by an early-spring application of an insecticide to eliminate overwintering adults before they oviposit. The life history chart (Figure 6-3) illustrates the periods when the two generations of adults and nymphs are in their greatest numbers and shows why elimination of a small population of adults in early spring may destroy the potential for severe damage (Niemczyk 1982).

Sampling Techniques

Several sampling techniques have been employed to determine populations in turfgrass. The most common involves a metal flotation cylinder that is open at both ends. This cylinder is driven into the soil to a depth of 5 cm or more, and is filled with water. Nymphal and adult chinch bugs that are present will float to the surface within 10 min and can be counted (Streu and Vasvary 1966). In locations where thatch density makes it difficult to drive such a cylinder into the soil, a 10.8-cm-diameter core can be collected and placed in a bucket, which is then filled with water (P. J. Vittum, personal observation).

A modification of this technique, using water pressure to remove thatch, made it possible to collect eggs as well. Centrifugation separated the chinch bug stages from thatch. A 90–100% recovery of artificially introduced eggs demonstrated the efficiency of this field-sampling technique (Mailloux and Streu 1979). A turfgrass area can also be flooded and then covered with a piece of white cloth. If chinch bugs are present, they will crawl to the underside of the cloth (Schread 1970a).

In a third sampling technique, soil plugs 10.8 cm in diameter and 7.6 cm deep, obtained with a standard golf cup cutter, are placed grass side down in a Berlese funnel fitted with a 25-watt lamp. During a 16-hr period, chinch bugs will be forced downward by the heat and may be collected in a jar of 70% ethyl alcohol or similar preservative (Niemcyzk 1982).

Plant Resistance

Laboratory and field evaluations for selection of HCB-resistant strains indicated preferential response of several cool-season grasses. In Kentucky bluegrass, 'Baron' and 'Newport' were more tolerant to feeding injury, as compared with 'Adelphi', which was the most susceptible. In perennial ryegrasses, 'Score', 'Pennfine', and 'Manhattan' had significantly lower infestations than did 12 other cultivars in field plots during 2 years of observation (1979–80). In fine-leaf fescues, 'Jamestown' had the heaviest HCB populations of 11 cultivars in field plots (Ratcliffe 1982).

More recent studies demonstrated that 'Repell', a perennial ryegrass cultivar infected with a fungal endophyte, *Neotyphodium* (formerly *Acremonium*) *lolii* Latch, Christensen, and Samuels, was highly resistant to HCB. Adults and first- and third-instar nymphs reared on endophyte-infected plants had significantly lower survival rates than did insects reared on endophyte-free 'Repell'. In addition, nymphs and adults either avoided infected plants or resided only in the leaf blade, where the endophyte does not occur. Apparently use of endophytic ryegrass can lead to significant reductions in HCB infestations (Mathias et al. 1990).

Natural Enemies

Microorganisms

Warm, moist weather is necessary before the entomophagous fungus *B. bassiana* becomes active, killing adults and nymphs (Plate 57). Infection can occur at any time from May to October. During 1973 and 1974, the maximum incidence of infection ranged from 80 to 90%, with a minimum incidence of about 20%. Early stages of infection are not externally detectable, but when the atmosphere is very moist, the insects become covered with white mycelium that later sporulates on the surface of the dead insect (Mailloux and Streu 1981).

Insect Parasitoid

Eumicrosoma benefica Gahan (Hymenoptera: Scelionidae) is a recently discovered egg parasitoid of the HCB in New Jersey. Adults are present throughout the year. Parasitized

HCB eggs can be distinguished easily by the parasitoid pupa, which is visible through the host chorion (Mailloux and Streu 1981).

Insect Predators

Some eight species of arthropods in the turfgrass fauna feed on early stages of the HCB when taken to the laboratory. A predatory mite and the lygaeid, *Geocoris bullatus* (Say), a big-eyed bug (Plate 60), are important predators. In Connecticut, *G. uliginosus* (Say) was found in association with *G. bullatus*. Adults of the latter fed on bentgrass, bluegrass, and ryegrass but did little damage. Big-eyed bugs resemble HCB but big-eyed bugs are more robust and have large eyes that protrude at the sides of the head. They move more rapidly and more readily than the HCB. *Amara* sp., the most common adult carabid of the HCB habitat, was the only egg predator detected (Dunbar 1971, Mailloux and Streu 1981).

Southern Chinch Bug

Importance

The SCB was formerly known as the *lawn chinch bug*. Its present common name was adopted when it was designated a distinct species (Stringfellow 1969). The SCB is the most injurious pest of St. Augustinegrass turf in Florida. More than 368,700 ha of St. Augustinegrass grows in Florida, accounting for about 37% of the total turfgrass area. More than $25 million is spent annually to control this pest in Florida lawns, with as many as six insecticide applications to a given lawn (Kerr 1966, McGregor 1976, Reinert 1978, Stringfellow 1969, Strobel 1971).

In Louisiana, the SCB is also considered the most injurious pest of St. Augustinegrass, which has been adopted as a lawn grass throughout the state. In Texas, the SCB is the most injurious turfgrass pest wherever St. Augustinegrass is grown (McGregor 1976, Oliver and Komblas 1981).

Insecticide Resistance

The SCB's continuous development of resistance to insecticides greatly increases the cost of control and makes necessary the constant evaluation of new compounds. With the need for as many as six applications of insecticide to manage 7–10 generations of SCB infestations each year, the species's development of resistance to the chemicals is not surprising. In Florida, resistance developed to chlordane and DDT after 8–12 years, to parathion after 7 years, to diazinon after about 20 years, and to chlorpyrifos (Dursban) after 11 years. Propoxur, a carbamate, was used for more than 15 years without an indication of resistance (Reinert 1982b), but is no longer available. Field trials in Texas indicate the SCB in that region has not developed resistance to organophosphates (Crocker 1993).

In spite of resistance, chlorpyrifos is still the standard chemical widely used by the lawn spray industry in Florida. SCB populations vary, showing no resistance in some localities and as much as 3.19×10^8-fold resistance in others, as measured by LC_{50} values. The greatest problem areas are on the east coast of Florida, from Fort Lauderdale and Palm Beach south through Miami (Reinert and Portier 1983, Reinert et al. 1995). In Texas, the efficacy of chlorpyrifos varied from 34% to 94% control in studies conducted in 1985–90 (Crocker 1993).

History and Distribution

The SCB is distributed wherever suitable habitats exist, from southern North Carolina southward throughout most of South Carolina, Georgia, and all of Florida. Westward it is present in most of Alabama and Mississippi, in all of Louisiana, and over the southeastern third of Texas (Figure 6-1). In Florida, severe damage occurs throughout the state. In Texas, damaging populations have been present since the early 1950s, and populations increased considerably during the early 1970s. The St. Augustinegrass–growing areas of Texas extend south from San Antonio to the lower Rio Grande Valley and eastward to Louisiana (Hamman 1969, Stringfellow 1969). The SCB is also established in Hawaii, on the island of Oahu (R. Tsuda, University of Hawaii, personal communication, 1998).

Host Plants and Damage

St. Augustinegrass is the preferred host plant of the SCB. It also feeds to some extent on centipedegrass, zoysiagrass, bahiagrass, and bermudagrass but only where St. Augustinegrass may be growing in close association. Because little damage occurs on these other grasses, the SCB is considered a serious threat only to St. Augustinegrass (French 1964, Kerr 1966).

SCB damages grasses by sucking the sap from the nodes and basal parts of the plant, causing it to become dwarfed, to yellow, and eventually to die. Both nymphs and adults are destructive to the grass. Damage occurs any time from May to November but is most evident during dry conditions, when populations of as low as $25-30$ insects per 0.1 m^2 can cause severe damage. Many of the nymphs remain hidden for as long as 10 days, feeding where the grass blades come together at the nodes (French 1964, Kerr 1966, Oliver and Komblas 1981, Stringfellow 1969). By midsummer, populations of $500-1,000$ SCBs per 0.1 m^2 are common in parts of Florida (Reinert et al. 1995).

Since the SCB aggregates in scattered patches rather than being evenly distributed, small spots of damaged grass initially become noticeable. The aggregated colonies do not move to new areas until the infested patch has been killed completely, so that there is ample time to apply controls before large patches of grass are dead (Kerr 1966).

Influence of the Turfgrass Environment

The condition of the host plant has a marked effect on injury. A heavily fertilized turfgrass with lush growth suffers the greatest injury from SCBs. Populations develop more rapidly and cause injury more quickly on heavily fertilized grass than on grass that receives moderate amounts of nitrogen (Kerr 1966).

Response to Fertilization

SCB populations often increase rapidly after heavy fertilization of St. Augustinegrass. In a Florida study, insect densities increased 65%, 28%, and 0%, with ammonium nitrate isobutylidene diurea (IBDU), and Milorganite applications, respectively. Application rates of nitrogen at 5.0 and 10.0 g/m^2/month increased SCB densities 45% and 49%, respectively. Early population regulation effects (e.g., eggs per female per week, rate of development of early instars) may explain the observed response of SCBs to high fertilization with readily available forms of nitrogen (Busey and Snyder 1993).

Weather and the thickness of the thatch appear important in affecting chinch bug development. In areas of Florida where frost rarely occurs, the grass grows continuously, creating a thick, spongy thatch. It is commonly 10–15 cm deep and may be as much as 30 cm deep. Such thatch provides an ideal habitat for chinch bugs (Plate 72) Reinert and Kerr 1973).

Moisture has a paradoxical effect on chinch bug damage potential. While an abundance of moisture makes the grass lush and susceptible to greater damage from chinch bugs, the moisture also has an effect in suppressing chinch bug populations (Kerr 1966).

Description of Stages

The description of adults, eggs, and nymphs of the SCB is very similar to that of the HCB and is derived, for the most part, from that of the chinch bug *B. l. leucopterus.*

Brachypterous adults dominate, but macropters are also found. Macropters can fly, but there is no spring dispersal flight like that of *B. l. leucopterus* (Reinert and Kerr 1973).

Seasonal History and Habits

Seasonal Cycle

In southern Florida, six to seven generations occur each year. During the winter, adults comprise 80–90% of the population, followed by nymphs, with eggs also present in smaller numbers. In northern Florida and Louisiana, there may be three or four generations, with only adults present during the winter (Reinert and Kerr 1973).

The first large surge of first-instar nymphs occurs during February in southern Florida and in late March to early April in northern Florida. In sunny open areas of lawns, aggregations of 500–1,000 per 0.1 m² are common, and as many as 2,300 per 0.1 m² have been found (Reinert and Kerr 1973).

During warm weather, a generation passes from the egg stage to the egg-laying stage in 5–8 weeks in southern Florida and in 7.5–8.0 weeks in northern Florida. In the laboratory at constant temperatures, a generation is completed in 6 weeks at 28°C and in 17 weeks at 21°C (Reinert and Kerr 1973).

In Louisiana, where four generations per year are considered normal, first-generation eggs are deposited in early April, second-generation eggs in early June, third-generation eggs in August, and fourth-generation eggs during August into September (Oliver and Komblas 1981).

Adult Activity

Migratory flights do occur but appear to be of minor importance. The SCB moves short distances, mainly by walking. Literally streams of bugs can be seen moving from heavily infested areas, where they may walk several hundred feet in a half hour. They are numerous in grass on sandy soil but not on muck soils. Adults tend to be aggregated more than evenly dispersed. They occur throughout the turf thatch and into the upper, largely organic layer of soil and in cracks in the soil. When populations are large, and on hot days, adults may be seen running over St. Augustinegrass blades, but they are not feeding or resting on the blades (Crocker and Simpson 1981, Kerr 1966).

Members of the genus *Blissus* have a definite courtship behavior. Males and females approach each other and make first contact with their antennae. Once they have paired, the bugs characteristically face in opposite directions, with the female the more active, walking about and sometimes feeding. Copulation, which occurs first in spring after warm weather has prompted activity, may last as long as 2 hr (Leonard 1966).

Females begin depositing eggs 7–10 days after mating. Eggs are inserted into the crevices of grass nodes and at the junction of blades and stems, either singly or in groups of two or three (Kerr 1966). Adults live for about 70 days, the females depositing a few eggs a day over several weeks for a total of 100–300 eggs before they die. Since chinch bugs mate repeatedly in a caged environment, they are assumed to mate repeatedly in the field as well. The male-female ratio is considered to be about 1:1 (Kerr 1966).

Nymphal Activity

Eggs hatch in about 2 weeks under normal conditions, but development and hatching vary with temperature. In Louisiana, where there are four generations from April into September, incubation periods average about 29 days for the first generation and 14–15 days for the remaining three because of warmer weather (Oliver and Komblas 1981).

After hatching, first-stage nymphs feed largely on tender basal growth and nodes of runners. Thick, spongy thatch that is 10–15 cm thick is an ideal habitat for SCB development. Nymphs require about 30 days to transform through five instars and reach maturity with a total life cycle of 7–8 weeks in Louisiana. First instars found during the winter indicate oviposition during warm winter days (Oliver and Komblas 1981, Reinert and Kerr 1973).

Miscellaneous Features

Threshold Populations

Insecticide efficacy has been increased by monitoring populations in order to identify an economic threshold. Treatment only when populations exceeded 22–28 bugs per 0.1 m^2 reduced the need for pesticide applications by 90% (Reinert 1982c).

Plant Resistance

Work by Reinert et al. (1980) indicates that the most practical method for managing the SCB results from the development of insect-resistant cultivars of St. Augustinegrass. In laboratory feeding tests, 'Floratam', 'Floralawn', and several other accessions produced 66–80% adult mortality, compared with only 11% mortality when feeding on 'Florida Common'. Also, significantly fewer eggs were produced when adults fed only on the resistant genotypes compared to 'Florida Common'.

More recently a Florida study found that populations of SCBs could survive and cause damage to plantings of 'Floratam'. In laboratory studies of this population, "damaging" and "standard" SCB populations were provided with either 'Floratam' or a susceptible cultivar of St. Augustinegrass. Standard SCB confined on 'Floratam' or 'Floralawn' survived only 12 and 22 days, respectively, but survived 77 days on 'Florida Common'. In contrast, damaging SCB populations survived 62 days on 'Floratam'. Standard females laid 2 eggs per female on 'Floratam' and 'Floralawn', while damaging SCBs produced 30 eggs per female on 'Floratam' and 85 eggs on 'Floralawn'. Both damaging and standard SCBs developed to adulthood on 'Florida Common', whereas only the damaging SCBs did so on 'Floratam'. Thus it appears that some populations of SCBs have overcome the resistance of 'Floratam' and 'Floralawn', which will have major implications for sod production and lawn maintenance in Florida (Busey and Center 1987).

Natural Enemies

Reinert (1978) made a comprehensive study of the natural enemies of the SCB during 1971 to 1977. Parasites and predators associated with or observed feeding on various stages of the SCB in the field were collected, were taken to the laboratory for observation, and were confined with life stages of the SCB on stolons of St. Augustinegrass. Those found to be natural enemies of the SCB are listed in Table 6-2.

Microorganisms

The fungus *B. bassiana,* pathogenic on all life stages of the SCB, produced epizootics only when high populations and high moisture levels were present (Reinert 1978).

Insect Parasitoids

The only parasitoid of the SCB observed in Florida is an egg parasitoid, *E. benefica* (Reinert 1972). The wasp, on finding an egg, examines the entire egg, first hurriedly, then carefully, by tapping it with her antennae. If it is found suitable, she thrusts her ovipositor into the host egg and deposits her own egg inside. The developing parasitoid usually consumes

Table 6-2. Natural Enemy Complex of the Southern Chinch Bug and the Life Stages Atacked by Each

| Organism | Eggs | Nymphal Stage | | | | | Adults |
		1	2	3	4	5	
Fungus							
Moniliaceae							
Beauveria bassiana Vuillemin	×	×	×	×	×	×	×
Parasitoid							
Scelionidae (Hymenoptera)							
Eumicrosoma benefica Gahan	×						
Predator							
Lygaeidae (Hemiptera)							
Geocoris uliginosus (Say)	×	×	×	×	×	×	×
Geocoris bullatus (Say)	×	×	×	×	×	×	×
Nabidae (Hemiptera)							
Pagasa pallipes Stal.		×	×	×	×	×	×
Anthocoridae (Hemiptera)							
Xylocoris vicarius (Reuter)	×	×	×	×	×		
Lasiochilus pallidulus Reuter	×	×	×	×	×		
Reduviidae (Hemiptera)							
Sinea sp.		×	×	×	×	×	×
Labiduridae (Dermaptera)							
Labidura riparia Pallas		×	×	×	×	×	×
Formicidae (Hymenoptera)							
Solenopsis geminata (F.)		×	×	×	×	×	×
Lycosidae (Araneida)							
Lycosa sp.		×	×	×	×	×	×

Source: From Reinert 1978, table 1.

the entire content of the host egg before pupation. Unlike a normal SCB egg, which turns reddish when it is about 3 days old, a parasitized egg remains tan until the parasitoid pupates, then becomes blackish, with the parasitoid pupa clearly visible. This parasitoid has been observed throughout the year in southern Florida. An average population of about 35 wasps was associated with about 90 SCBs per 0.1 m² of St. Augustinegrass (Reinert 1972).

Insect Predators

The most numerous and most frequently encountered predator in turf in Florida is *G. uliginosus*, a species of big-eyed bug. In size, shape, and color, it superficially resembles its host, for which it is often mistaken. Big-eyed bugs are more robust, with large eyes that protrude at the sides of the head, which is the widest part of the body. They move more rapidly among the grass blades and stolons than chinch bugs. Densities as high as 17 per 0.1 m² may be present. In the laboratory each predator fed on an average of 9.6 nymphs per 5-day period. A less frequently observed predator of SCB, *G. bullatus*, did not exceed 2 per 0.1 m² and was found in thinner grass with frequent bare areas. Both species of *Geocoris* fed on all stages of the SCB, including adults, but only when they had molted recently (Reinert 1978). In addition, occasionally both species were observed feeding on bermudagrass (Crocker and Whitcomb 1980), but this activity is presumed to be inconsequential.

A dermapteran, *Labidura riparia*, was one of the most active predators of the SCB (Plate 59). Although only one or two were present per 0.1 m², they range over a wide area in search of prey. In the laboratory an adult consumed 50 adult chinch bugs in 24 hr (Reinert 1978).

In a 1971 observation in Fort Lauderdale, Florida, where an area damaged by the SCB was not spreading at the typically rapid rate during the summer, the four most numerous biotic agents present were the egg parasitoid, *E. benefica*, and predators *G. uliginosus*, *Xylocoris vicarius*, and *Lasiochilus pallidulus*. The latter two are predators on eggs and on early-instar nymphs. The combined parasitoid-predator complex is considered to have been responsible for the relatively low density of SCBs during the summer and to have caused the population to collapse during August (Reinert 1982b).

In light of many observations of this type, parasitoids and predators are now considered to prevent rapid population increases where insecticides are not applied. Conversely, lawns receiving repeated insecticide applications continue to have chinch bug problems, presumably because the broad-spectrum insecticides typically used have a significant detrimental effect on parasitoids and predators. On lawns where SCB populations were monitored and were treated only when economic thresholds of slightly more than 20 bugs per 0.1 m² had been exceeded, the need for pesticide applications was reduced by 90% (Reinert 1978, 1982c).

Buffalograss Chinch Bug

History and Distribution

The buffalograss chinch bug (BCB), *Blissus occiduus* (Barber), has become one of the most serious insect pests of buffalograss in the Central Plains (Plate 12). The BCB was first observed damaging a buffalograss lawn in Lincoln, Nebraska, in 1989. Subsequent surveys confirmed its presence throughout Nebraska (Baxendale et al. 1997). It is likely the range extends through much of the range of buffalograss. As buffalograss is used increasingly as a low-input turfgrass in the Central Plains, the range and importance of the BCB are likely to increase.

Host Plants and Damage

While relatively little information is available on the BCB, it is presumed to feed primarily on buffalograss. Damage resembles that of other chinch bugs—nymphs and adults insert their strawlike stylets into plant tissue (often near the crown) and withdraw plant juices. They also inject a salivary toxin that damages plant tissue and inhibits the translocation of water and nutrients.

Initial damage appears as patchy yellow areas. Close inspection reveals a reddish discoloration of plant tissues. As feeding progresses, the turf takes on a dried, straw-brown appearance. High population density can result in severe thinning or outright death of the buffalograss stand (Baxendale et al. 1997).

Description of Stages

Adult
Adults closely resemble HCB adults in size and shape, but are slightly smaller. A significant portion of the population of BCBs are brachypterous. These individuals appear to be wingless (segments of the abdomen are highly visible), but in fact "vestigial" wings are present and appear as small nubs near the base of the thorax.

Nymph
Like the HCB, the immature BCB resembles adults in body shape and feeding habits. Early instars are bright red with a white band across the abdomen. Later instars are orange brown to nearly black.

Seasonal History and Habits

The BCB overwinters primarily as brachypterous adults in semiprotected areas in or near buffalograss stands. In early spring, adults emerge from their overwintering sites and mate. Females deposit eggs in the crowns of plants or in surrounding soil. Eggs hatch in May and nymphs feed until early July. Adults, many of which have fully developed wings and are capable of dispersing to new feeding sites, begin to appear in early July. A second generation is initiated, with eggs laid in mid to late July and nymphs feeding well into September. Short-winged adults emerge in late September or early October and prepare to overwinter. At least some of these brachypterous adults mate in the fall (Baxendale et al. 1997).

Miscellaneous Features

Sampling Techniques
The flooding technique (see Chapter 26) is the traditional method used to survey BCB activity. Alternatively, a small section of turf can be removed with a trowel or a knife and shaken vigorously over a white sheet of paper. Baxendale et al. (1997) indicated that populations exceeding 20 BCBs per 0.1 m^2 may cause visible damage if left untreated.

Thatch

As with other chinch bug species, the BCB prefers turf areas with considerable thatch and organic debris, so cultural practices that minimize thatch should reduce or delay initial infestations (Baxendale et al. 1997).

Common Chinch Bug

The CCB is primarily a pest of corn, sorghum, millet, rye, and various bunchgrasses. However, it also feeds on several turf species, including both cool-season and warm-season species. Damage to turf can occur in a short period of time when large numbers of CCBs migrate from maturing grain fields to bermudagrass or Kentucky bluegrass turf (Reinert et al. 1995).

The range of CCB overlaps that of the HCB and SCB. Populations have been confirmed in 25 states, from North Carolina through Georgia westward through Texas and Oklahoma, as well as the Midwest and North Central Plains (Reinert et al. 1995).

7

Hemipteran Pests:
Suborder Homoptera

Greenbug

Taxonomy

The greenbug (GB), *Schizaphis graminum* (Rondani), family Aphididae, was originally placed in the genus *Toxoptera*, but since the mid-1960s it has belonged to the genus *Schizaphis*. It was commonly called the *spring grain aphis,* or greenbug (Webster and Phillips 1912), but it is presently known as the *greenbug.*

Much of the description and bionomics of the GB necessarily comes from literature that treats this insect as a small-grain pest. Nevertheless, this information is largely pertinent to the GB as a turfgrass pest.

Importance

Kentucky bluegrass has long been known as a host of GB in the United States and Canada, and occasionally this aphid has caused minor damage to lawns and pastures of this grass species (Garman 1926, Webster and Phillips 1912). Not until 1970 was GB reported in epidemic levels on Kentucky bluegrass lawns in many midwestern states. Since then, certain lawns, particularly in Ohio, have been devastated for at least 4 successive years (Niemczyk 1980b).

The importance of the GB as a serious pest of turfgrass has been accentuated by its development of resistance to some of the common insecticides used on turfgrasses (chlorpyrifos, diazinon, and malathion). A level of resistance to chlorpyrifos as great as ninefold has been reported (Niemczyk and Moser 1982, Potter 1982a).

Few insects have had such a long history of devastation to American agriculture. Since the turn of the century, the GB has caused an estimated annual loss of at least 50 million bushels of oats and wheat in the three states of Kansas, Oklahoma, and Texas. Because of its general distribution and its fecundity, the GB caused a loss of as much as 3% of the annual wheat crop in the early 1960s (Metcalf et al. 1962). While the GB overall can be considered a relatively minor pest of turfgrass in North America, it can and does cause severe damage in localized areas (Gibb 1995).

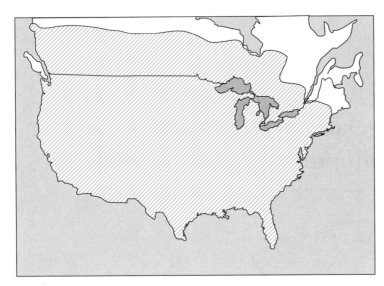

Figure 7-1. Distribution of the greenbug. (Drawn by E. Gotham, NYSAES, adapted from Handbook of Turfgrass Insect Pests, Brandenburg and Villani 1995, Entomological Society of America.)

History and Distribution

The GB is widely distributed in North and South America and in Europe, Africa, and Asia. It was first recorded not as a pest of small grain but, because of its abundance, as a nuisance to humans in Italy in 1847 (Street et al. 1978, Webster and Phillips 1912).

The first serious outbreak of the GB in the United States occurred in Virginia in 1882 on oats. As early as 1907, GB damage to bluegrass lawns was observed in Washington, D.C. By 1912 the insect was known to occur throughout the wheat-growing areas of the United States and southern Canada (Figure 7-1). The New England states are about the only area of the country where it is nearly absent (Gibb 1995).

During the 1970s, GB damage to Kentucky bluegrass was reported in Illinois, Indiana, Kansas, Missouri, Ohio, and Wisconsin. In 1981, Kentucky bluegrass and fine fescue lawns were damaged in the additional states of Iowa, Kentucky, and Minnesota (Figure 7-1) (Niemczyk 1980c, Niemczyk and Power 1982, Street et al. 1978).

Host Plants and Damage

The GB has a wide range of host plants in the family Gramineae. At least 60 species of grasses are listed as hosts. It feeds only rarely on plants outside this family. Turfgrasses that serve as hosts include Kentucky bluegrass, Canada bluegrass, annual bluegrass, fescues, and perennial ryegrass. Reproduction has been observed on Kentucky bluegrass, Chewings fescue, and tall fescue. Small grains serving as the dominant host plants include wheat; oats; rye; rice, *Oryza sativa;* and sorghum, *Sorghum vulgare* (Potter 1982a, Webster and Phillips 1912).

Damage is caused when the aphid pierces grass blades with its needle-like mouthparts and feeds on the phloem tissue (Plate 13). Feeding alone weakens the plants, but the GB also injects its toxic salivary secretions, causing yellow spots with necrotic centers, followed by death of the tissues surrounding the feeding site, which turns burnt orange. The translocation of the salivary toxin weakens the entire plant, including the root system, as the enzymes break down the plant cells (Niemczyk and Moser 1982).

On home lawns, damage usually begins in shaded areas under trees or along the north or east side of a building, in the form of circular to irregular brown patches as much as 4.6 m in diameter (Plate 13). A narrow band of yellow to burnt-orange grass exists just beyond the areas most densely settled by the aphid in live grass at the edge of the damaged turf. From a distance, damaged turf typically has a burnt-orange cast. As feeding continues, dead patches may expand to as much as 3.3 m², and may be confused with early dormancy. Injury can also begin in open areas. In Kentucky, injury begins to appear in May and early June and continues until autumn. In 1981 the most severe damage occurred in mid-November. Some 10–50 aphids may line the midrib on the upper surface of a single grass blade. Populations of 4,000 aphids per 0.1 m² are not uncommon (Gibb 1995, Niemczyk and Moser 1982, Potter 1982a). The GB does not infest trees.

Biotypes

At least four biotypes of GB are recognized on small grain, and good evidence indicates that other biotypes occur on turfgrasses. On small grain, biotypes A and B are recognized by their feeding habits. Biotype A penetrates intercellularly (between cells only) and invariably feeds in phloem tissue. Biotype B penetrates intracellularly and intercellularly (within and between cells) and preferentially feeds in the mesophyll parenchyma of leaves. It causes greater damage than biotype A. Biotype C attacks sorghum. A strain of the GB occurring in Bushland in southeastern Texas was designated as biotype E (Porter et al. 1982, Saxena and Chada 1971). Plant resistance studies conducted in recent years in widely scattered locations demonstrated the probability of additional biotypes (Niemczyk and Moser 1982).

The rather sudden appearance of damaging populations of the GB on Kentucky bluegrass lawns suggests the development of a new biotype adapted to this grass. Not only is there evidence that such an adaptation is occurring, but within the population of GBs that infest Kentucky bluegrass, at least two biotypes, separable by visual differences, are currently recognized. The normal green dorsal-median stripe is present (except in the youngest nymphs) in the Kentucky bluegrass–infesting Maryland biotype (Plate 13). The stripe is absent in the Kentucky bluegrass–infesting Ohio biotype (Plate 13) (R. H. Ratcliffe, U.S. Department of Agriculture, Beltsville, Maryland, personal communication, 1985).

Description of Stages

Adult

GB adults are soft bodied, pear shaped, 1.5–2.5 mm long, and pale yellow to light green, usually with a dark-green dorsal-median stripe. The tips of the legs, the tips of the cornicles, and the antennae are black (Figure 7-2, Plate 13).

Adult Forms. At least three distinct forms of females occur: (1) a winged viviparous female, (2) a wingless viviparous female, and (3) a wingless oviparous female (Figure 7-3). Males occur only as winged insects and can be distinguished from the winged female by the differences in the terminal abdominal segments (Figure 7-3). The male is also smaller. The wingless oviparous female is the largest of the adults. It can be distinguished from the

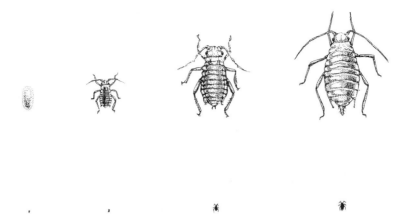

Figure 7-2. Stages of development of greenbug: egg, first-instar nymph, final-instar nymph, and adult. Silhouettes show actual size of each stage. (Courtesy of R. Jarecke, NYSAES.)

wingless viviparous female by the evidence of eggs in the abdomen and by the slightly wider hind tibia (Webster and Phillips 1912).

Overall body length for the winged male is 1.3 mm (wingspan 4.5 mm), for the winged female is 1.5–2.0 mm (wingspan 5–7 mm), for the wingless viviparous female is 1.0–1.8 mm, and for the wingless oviparous female is 2.0–2.5 mm (Hunter 1909, Potter 1982a, Webster and Phillips 1912).

Egg
Freshly deposited eggs, cemented to host plants, are pale yellow and change within a few hours to faint green, with a circular area of darker green at one pole (end) due to the ovarian yolk. After a day this region turns dark green, and during the second day the entire egg turns darker green. By the end of the third day, the egg has become jet black. Eggs are broadly elliptical, slightly reniform, 0.70–0.78 mm long, and 0.33–0.45 mm wide (Webster and Phillips 1912).

Nymph
There are four nymphal instars, and all resemble the wingless adults in general shape and color (Figure 7-2). Upon hatching they are green with legs that have black tips.

Seasonal History and Habits

Seasonal Cycle
In seasonal cycle the GB in the southern states differs considerably from that in the North. These two regions can be separated by the 35° latitude, a line that extends along the southern edge of North Carolina and Tennessee through the midsection of Arkansas and Oklahoma and the Texas panhandle. However, conditions become more severe (i.e., winter temperatures are colder) as one moves west, so the complex life cycle typical of "northern" populations is often seen in GB in the south central plains as well.

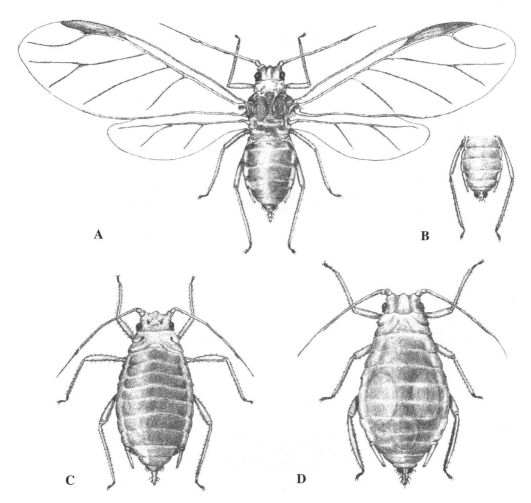

Figure 7-3. Greenbug forms. **A.** Winged viviparous female. **B.** Abdomen of winged male. **C.** Wingless viviparous female. **D.** Wingless oviparous female showing egg development. (From Webster and Phillips 1912, figs. 1, 3, 4, 5; redrawn by R. McMillen-Sticht, NYSAES.)

South of Latitude 35°. The life cycle of the GB in southern states is very simple. Active nymphs and wingless viviparous females reproduce during winter warm spells (Figure 7-4). Reproduction continues during the warmer months. All nymphs become wingless viviparous females in 7–18 days and begin producing 30–40 living young during an average life span of about 35 days. The hot, dry summer periods are the least favorable, with a food shortage as their grain host ripens (Walton 1921, Webster and Phillips 1912).

North of Latitude 35°. The life cycle of the GB is much more complex in the northern states (Figure 7-4). It overwinters primarily as eggs deposited on leaves of host plants. Sexual males and females may also overwinter under the protection of vegetation. Eggs hatch during late winter or early spring, and in 7–18 days nymphs become wingless females, producing living young that may become winged or wingless females. Both types of females

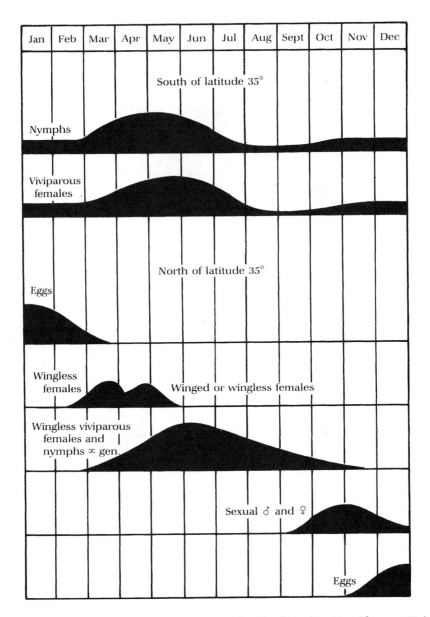

Figure 7-4. Generalized life cycle of the greenbug south and north of latitude 35° in midwestern United States. (Data from Webster and Phillips 1912; drawn by R. McMillen-Sticht, NYSAES.)

give birth to living young that become reproductive in as few as 6 days or as many as 18 days. A single female may produce 1–8 young per day for 14–21 days, giving birth to 50–60 young per female. Parthenogenetic reproduction (reproduction without fertilization) of wingless females continues throughout the growing season, with as many as 20 generations a year in the latitude of central Indiana. The most rapid reproduction occurs at relatively low temperatures in the spring (Potter 1982a, Walton 1921, Webster and Phillips 1912).

During late September and October, winged males and females appear and mate, and females produce eggs that overwinter. From 5 to 14 generations appear each season. All except the last generation are composed entirely of females (Walton 1921, Webster and Phillips 1912).

Infestations of northern lawns were first considered to result from annual migrations from the South. This hypothesis, however, did not explain reinfestations of the same lawns year after year. Examination of previously infested Kentucky bluegrass lawns in Ohio in November 1981 revealed many GB eggs glued to grass blades, debris, and tree leaves. Overwintering eggs in turfgrass plugs collected in March 1982 and held at 24°C hatched and thereby confirmed that the GB can overwinter as eggs in home lawns in the northern states (Niemcyzk and Power 1982).

Adult Activity

Overwintering in Northern States. A small percentage of the GB population can successfully overwinter in northern states when hibernation occurs in tall, rank Kentucky bluegrass in waste areas. The insect prefers to hibernate in dead or dying leaf blades that are buried under several inches of matted leaf cover. In the spring adults crawl out near the tips of the leaves, where they begin to deposit eggs. The temperature under such vegetative cover aided by a snow cover may be 5° to 8°C higher than the ambient air temperature (Webster and Phillips 1912).

Influence of Temperature. GB development takes place at temperatures between 7° and 33°C. The most rapid development occurs at 30°C. Maximum reproduction occurs at about 22°C (Wadley 1931).

Development of Sexual Forms. Development of sexual forms, winged males, and oviparous females depends on and varies with a combination of temperature and day length, the latter being the dominant factor. When days are less than 12 hr long and temperatures average less than 22°C, the sexual forms appear and continue to develop as long as days remain short (Wadley 1931).

Oviparous females deposit an average of only 5.4 eggs per female, far less than viviparous females. Their longevity depends on weather conditions and on the presence or absence of males. When the weather is favorable and males are present, females live 60–70 days, but in the absence of males, females may live nearly 90 days (Webster and Phillips 1912).

Development of Winged Forms. Differences between winged and wingless forms seem less marked than differences between parthenogenetic and sexual forms. Winged forms appear most frequently when the parent aphids received poor nutrition and when the temperature averages about 15°C, typically during spring or fall. In the absence of this temperature, limited nutrition of parents results in production of winged forms and can occur any time during the season. Winged females seem inherently migratory. They appear more restless than the wingless females and leave plants even when nutrition is adequate. Flights seem to require strong effort during calm periods but become easy when there is wind; the wind directly assists flight. The shaking of plants helps the aphids become airborne. Migration from wintering areas to the north appears to occur in stages involving successive generations; no one generation traverses the entire distance (Wadley 1931).

Miscellaneous Features

Plant Resistance

In Kentucky greenhouse trials, tall fescue, Chewings fescue, and three genetically diverse Kentucky bluegrass cultivars—'Kenblue', 'Vantage', and 'Adelphi'—all supported heavy GB populations and suffered severe feeding damage. Perennial ryegrass, creeping bentgrass, zoysiagrass, and bermudagrass suffered no visible feeding damage; virtually no GB survived on these grasses (Jackson et al. 1981).

In Maryland, seedlings of 48 Kentucky bluegrass cultivars screened for resistance to the Maryland biotype of the GB showed that all cultivars were highly susceptible. Individual plants in 36 of these cultivars, however, showed varying degrees of antibiosis or tolerance ranging from 0.1% to 7.1% of individual seedlings. Cultivars with 1% or greater frequency of resistant plants included, in descending order of frequency, 'Piedmont', 'A-34', 'Troy', 'Rugby', 'Adelphi', 'Kenblue', and 'Sydsport'. There was also some indication of GB biotypes on turfgrass. Progeny from the Ohio biotype generally caused more severe feeding injury on susceptible plants than did that of the Maryland biotype. In addition, the latter could be reared on Kentucky bluegrass and barley, *Hordeum vulgare* L., but the Ohio biotype did not establish well on barley (Ratcliffe and Murray 1983).

In Nebraska, GB from the sorghum-feeding biotypes C and E, the two most prevalent biotypes in the United States, were selected for screening Canada bluegrass and 23 cultivars of Kentucky bluegrass to feeding injury. Response to both biotypes was similar. The highest levels of resistance were shown by Canada bluegrass, followed by Kentucky bluegrass cultivars 'South Dakota Common', 'Nebraska Common', 'Vantage', and 'Sydsport' (Kindler et al. 1983)

Recent studies suggested that the endophyte, *Neotyphodium* (formerly *Acremonium*) *lolii* Latch, Christensen, and Samuels, can reduce GB fecundity and survival on perennial ryegrass significantly. The concentration of peramine (which in turn affects insect feeding deterrence) and of endophyte hyphae was greatest in spring and fall (when daily temperatures averaged 15°C the previous 4 weeks), and was significantly lower at higher temperatures (when daily temperatures averaged 23°C the previous 4 weeks) (Breen 1992).

A study in Kentucky demonstrated that GB survival was significantly lower on tall fescue infected by *Neotyphodium* (formerly *Acremonium*) *coenophialum* (Morgan-Jones & Gams) than noninfected tall fescue. Increased fertilization did not reverse the observed dilatory effect (Davidson and Potter 1995).

Natural Enemies

The literature contains no indication that the GB is affected by any entomogenous or entomopathogenic microorganisms. However, GBs are subject to attack by many parasitoids and predators.

Insect Parasitoids

At least five species of parasitic wasps in the genera *Aphidius* and *Lysiphlebus* are recorded as parasitoids of the GB. Probably the most important species is *Lysiphlebus testaceipes* (Cresson), a small black parasitoid that prefers the wingless GB (Figure 7-5). A single female can oviposit in more than 300 GBs but normally lays eggs in about 100 GBs (Webster and Phillips 1912).

The worst outbreaks of the GB occur when mild winters are followed by cool, late

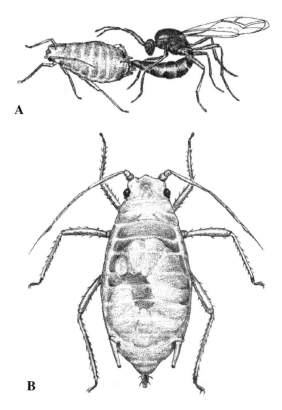

A

B

Figure 7-5. *Lysephlebus testacipes.* **A.** Ovipositing in the body of the greenbug. **B.** Parasite larva developing within the greenbug. (From Webster and Phillips 1912, figs. 8, 9; redrawn by R. McMillen-Sticht, NYSAES.)

springs. Reproduction begins slowly at slightly above 4°C and increases rapidly at 13°–18°C. During such conditions temperatures are too low for *L. testaceipes* to develop (Webster and Phillips 1912).

Insect Predators

Chief among the predatory insects is the coccinellid, *Hippodamia convergens* Guerin-Meneville (Plate 60), with both adults and larvae feeding on the aphid. Other coccinellids include *Coccinella 9-notata* Hbst. and *Megilla maculata* (DeG.). Larvae of a syrphid fly of the genus *Syrphus,* a lacewing, *Chrysopa plorabunda* Fitch, and a cecidomyiid fly, *Aphidoletes* spp., are also natural enemies (Webster and Phillips 1912).

Vertebrate Predators

Birds listed as effective predators include the goldfinch, *Carduelis tristis,* and at least four species of sparrows (several genera) (Webster and Phillips 1912).

Two-lined Spittlebug

Taxonomy

The two-lined spittlebug (TLS), *Prosapia bicincta* (Say), is in the family Cercopidae. Previous scientific names include *Cercopis bicincta, Monecphora bifusca,* and *Tomaspis bicincta,* among others.

Figure 7-6. Distribution of two-lined spittlebug. (Drawn by E. Gotham, NYSAES, adapted from Handbook of Turfgrass Insect Pests, Brandenburg and Villani 1995, Entomological Society of America.)

Importance and Host Plants

The TLS has caused severe damage on coastal bermudagrass and other bermudagrass pastures in the Southeast since the 1950s. More recently it has inflicted damage to most warm-season turfgrasses from the Carolinas through the Gulf States. Adults also rest or feed on a wide variety of woody plants.

History and Distribution

The TLS is native to North America and is found from Maine to Florida, and west to Iowa, Kansas, and Oklahoma (Figure 7-6). It feeds on several grasses, including bermudagrass, St. Augustinegrass, centipedegrass, bahiagrass, and seashore paspalum.

Damage

TLS adults and nymphs resemble each other in shape, and use their needle-like mouthparts to pierce plant tissue and feed on the sap, causing the plants to wither or to stop growing. Adults feed on stems and leaves and inject a salivary toxin, which results in a chlorotic stippling of the leaves. The toxin is translocated up and down the stems from the initial feeding site. Affected areas subsequently coalesce and turn brown. Nymphs feed deeper in the turf, at the crown of the plant, and produce masses of frothy material ("spittle"). When these masses are numerous, the turf feels "squishy" (Braman 1995).

Description

Adult

The adult TLS is wedge shaped, dark brown or black with red eyes and legs, about 6–10 mm long. It has two prominent red or orange lines across the wings and a narrower

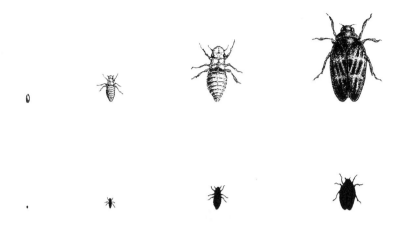

Figure 7-7. Stages of development of two-lined spittlebug: egg, first-instar nymph, final-instar nymph, and adult. Silhouettes show actual size of each stage. (Courtesy of R. Jarecke, NYSAES.)

band across the thorax (Figure 7-7, Plate 14). Some individuals may lack bands. The dorsal side of the abdomen under the wings is bright red (Braman 1995).

Egg
Eggs are bright yellow or orange and oblong, pointed at the end, and about 1 mm long (Plate 14).

Nymph
Nymphs resemble adults in shape but are smaller and lack wings (Figure 7-7). They are cream, yellow, or white, with brown heads and red eyes. The name *spittlebug* is derived from the white frothy mass that nymphs produce (Plate 14).

Life History and Habits

Eggs overwinter in hollow stems, behind leaf sheaths, or among plant debris. Most nymphs emerge in March and April and seek sheltered, humid hiding places among plants, and start feeding. The nymph exudes a white frothy mass resembling spittle that protects it from natural enemies and desiccation (Plate 14). Nymphs develop through four or five instars to become adults in about 1 month. The final molt (to the adult) takes place within the spittlemass of the fifth instar. Adults are active at night or during early morning, spending the warmer part of the day hiding in the grass. They live about 23 days, with females depositing eggs for about 2 weeks. Adults often rest or feed on a variety of woody ornamental plants, including hollies. As hollies are used increasingly in landscape plantings, the likelihood of infestations of TLS nymphs in neighboring turf increases (Braman 1995).

Two generations of TLSs normally occur annually from northern Florida to North Carolina. Adults of the first generation are active in June, while a second peak in adult activity usually occurs in August or September. This second generation deposits overwintering eggs (Baker 1982, Braman 1995, Byers 1965, Fagan and Kuitert 1969).

Miscellaneous Features

Thatch

High spittlebug populations often are associated with thick or dense thatch, which results in the high humidity that enhances spittlebug survival. Thatch management (dethatching and topdressing) are cultural techniques that can reduce spittlebug severity in turfgrass settings (Braman 1995).

Natural Enemies

No parasitoids or predators of nymphs or eggs have been reported. A fungus, *Entomophthora grylii* Fresenius, occasionally attacks adults. Remains of adults have been found in stomachs of some song birds, including southern meadowlarks and red-winged blackbirds. Adults have been trapped in webs of the yellow garden spider, *Argiope aurantia* Lucas, and the golden silk spider, *Nephila clavipes* (L.) (Braman 1995).

Rhodesgrass Mealybug

Taxonomy

The Rhodesgrass mealybug (RMB), *Antonina graminis* (Maskell), family Pseudococcidae, was known until very recently as the *Rhodesgrass scale,* even though it is classified in the mealybug family. Most of the information presented comes from Chada and Wood (1960). Additional work is cited only when it supplements that of these authors.

Importance

Actual damage by the RMB is difficult to assess because drought and close mowing as well as the insect contribute to damage. This is the most widely distributed and damaging of the grass mealybugs or scales in Florida. The RMB is established in a few locations in southern California, but the severity of its damage has not been fully determined (Bowen et al. 1980, Kelsheimer and Kerr 1957).

History and Distribution

The RMB occurs in all tropical and subtropical regions of the world. It was first identified in the United States in 1942 infesting rhodesgrass, *Chloris sayana* Kunth, in southern Texas. The RMB is currently known to occur in all the states bordering the Gulf of Mexico, Mexico, and Hawaii, as well as Georgia (Figure 7-8). No winged forms occur, and dispersion takes place when crawlers are carried by air currents, when sod is moved, and when crawlers hitchhike on animals.

Host Plants and Damage

Rhodesgrass, a coarse-textured pasturegrass widely used in the Gulf States, is the most preferred host. Preferred turfgrasses include bermudagrass and St. Augustinegrass, followed by tall fescue and centipedegrass. Several weed grasses associated with turf are also infest-

Figure 7-8. Distribution of Rhodesgrass mealybug. (Drawn by E. Gotham, NYSAES, adapted from Handbook of Turfgrass Insect Pests, Brandenburg and Villani 1995, Entomological Society of America.)

ed. Lawns and golf course turf mowed at a height of 4 cm are less prone to injury than grass that has been cut shorter. Irrigation and fertilization help prevent damage.

Destruction of grass is not dramatic, but infested plants slowly lose vitality. The loss becomes apparent during periods of drought as the insect feeds on the plant sap and causes cells to collapse. The base of infested plants, including the crown, nodes, and leaf axils, appears covered with tufts of cotton. These tufts result from a waxlike secretion produced by the mealybug that encloses the insect. Eventually, after the plant cells have collapsed, the plant appears to be suffering from drought and fades to a dull, lifeless brown. Injury is most severe during extended hot, dry periods (Converse 1982).

Description of Stages

Adult
The most obvious sign of infested grass is the white, cottony masses that the insect secretes. Adults found inside these tufts are dark, purplish brown, with saclike, broadly oval to subcircular bodies about 3 mm by 1.5 mm. The caudal extremity is strongly chitinized. Appendages include minute, two-segmented antennae, long, stylet-like mouthparts, and a white, waxy, tubular filament protruding from the anal end of the insect and outside the white cottony mass (Plate 15). This is an excretory organ, and some filaments may exude droplets of excrement. The body is enclosed in a waxy sac that turns yellow with age. Openings in the anterior and posterior ends expose the actual insect.

Egg
Eggs are observable only when the female is dissected. They are cream colored and oblong and measure about 500 μm by 180 μm.

Nymph
The first-instar nymphs, or crawlers, are born alive and are the only mobile form of the insect (Plate 15). They are oblong-oval and cream colored, the median area being tinged

with purple. Appendages include the long, stylet-like mouthparts, six-segmented antennae, three pairs of legs, and two long caudal appendages. The second and third instars have saclike bodies bearing little resemblance to the first instars (Plate 15). They become sessile and lose legs and appendages in the first molt. The body becomes enclosed in a waxy sac, so that they resemble adults except for their smaller size.

Life History and Habits

Life Cycle

No males have been reported. The parthenogenetic, ovoviviparous females reproduce over a period of about 50 days. Emerging first-instar nymphs move upward onto leaf sheaths, settle beneath a leaf sheath at a node, insert their mouthparts, and become sessile. A life cycle takes 60–70 days, and there are four or five generations annually. The winter generation in southern Texas requires about 3.5–4.0 months for completion. Spring, summer, and fall generations require about 2 months each. There is no winter diapause.

The RMB thrives under moisture conditions that promote good plant growth. Temperature is most influential in affecting development. Optimum temperatures are 29°–32°C, but the insect does well between 0° and 38°C. Freezing temperatures cause high mortality; −2.2°C for 24 hr or more is fatal to all stages.

Natural Enemies

The only recorded parasitoid is *Anagyrus antoninae* Timb. (Hymenoptera: Encyrtidae), established in Hawaii. Its importation into Texas apparently was not successful (Dean and Schuster 1958).

Ground Pearl

Taxonomy

The ground pearl, suborder Homoptera, family Margarodidae, is a subterranean scale insect infesting the roots of turfgrasses in the southeastern United States. Two species that occur regularly on turfgrass are *Margarodes meriodionalis* Morrill and *Eumargarodes laingi* Jakubski. The name is derived from the pearllike appearance of the cysts that enclose the immature stages. Most of the information comes from Brandenburg (1995a) or Kouskolekas and Self (1974). Other works are cited only when they supplement the work of these authors.

Importance, History, and Distribution

The ground pearl infests the roots of turfgrasses and causes widespread, extensive damage to home lawns and golf courses from the Carolinas to California (Figure 7-9). The insect has the potential for causing serious turf damage in the southwestern states (Baker 1982). The origin of ground pearls is unclear—*E. laingi* was first reported from Australia, while *M. meridionalis* apparently originated in North America (Brandenburg 1995a).

Figure 7-9. Distribution of ground pearls. (Drawn by E. Gotham, NYSAES, adapted from Handbook of Turfgrass Insect Pests, Brandenburg and Villani 1995, Entomological Society of America.)

Host Plants and Damage

The ground pearl attacks the roots of bermudagrass, St. Augustinegrass, zoysiagrass, and centipedegrass, with damage most common on bermudagrass and centipedegrass. Nymphs extract plant sap from the roots. The damage appears as irregular patches. During summer dry spells, the grass yellows, browns, and usually dies by fall (Plate 16). Grass rarely recovers in infested areas, and weeds often encroach. Cysts are present in larger numbers at the interface between damaged and healthy grass (Baker 1982). Ground pearls tend to be sporadic but can cause widespread damage throughout the Southeast and Southwest.

Description of Stages

Adult and Egg

Females are wingless, pinkish scale insects with well-developed forelegs and claws and are about 1.6 mm long (Plate 16). Males, considered rare, are gnatlike and vary from 1 to 8 mm in length (Plate 16). Clusters of pinkish-white eggs are enclosed in a white waxy sac (Plate 16) (Baker 1982).

Nymph

The first instar, about 0.2 mm long, is called a *crawler* and moves until it finds a suitable place to feed. It attaches itself to a root and produces a protective shell, or cyst. This "ground pearl" is hard, globular, and yellowish purple (Figure 7-10) and increases from 0.5 to 2.0 mm in diameter as the nymph grows. The sucking mouthparts extend through the wall of the cyst and are inserted into grass roots (Plate 16) (Baker 1982).

Life History and Habits

Life Cycle

The life cycle of the ground pearl is not completely understood. They overwinter in the cysts. Females mature in late May through early July, emerge from the cysts, and after a

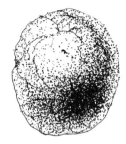

Figure 7-10. Stages of development of ground pearls: egg, nymphal cyst, and adult female. Silhouettes show actual size of each stage. (Courtesy of R. Jarecke, NYSAES.)

short period of mobility (usually about 9 days), secrete a waxy filament that covers the body completely. They remain about 5.0–7.5 cm deep in the soil and deposit eggs within the waxy coat. Oviposition begins in June and continues into July, with the egg hatch extending into August. Young crawlers start feeding on grass roots and develop the globular appearance. There is usually one generation each year, but under unfavorable conditions 2 or 3 years may be required to complete a generation (Baker 1982).

Adult and Nymphal Activities

The ground pearl has parthenogenetic reproduction. Each female deposits about 100 eggs over a period of 7–12 days. Egg hatch begins 9–15 days after the first eggs were laid. After hatching, young crawlers start feeding on the grass roots and form their pearllike cysts. Ground pearls are found as deep as 25 cm in the soil, which often makes control with traditional insecticides difficult (Niemczyk 1981).

8

Lepidopteran Pests:
Family Pyralidae

Temperate-Region Sod Webworms

Taxonomy

The name *sod webworm* refers to a large number of grass-infesting moths and larvae of the family Pyralidae, subfamily Crambinae. Adults are often called *lawn moths* because of their habitat or *snout moths* because of the prominent labial palpi that extend in front of the head. Most of the turfgrass-infesting species originally were placed in the genus *Crambus*, which is distributed practically worldwide. About 100 species are recognized in North America (Bohart 1947).

The genus *Crambus* was revised in 1983. Original members of the genus are now divided into more than a dozen genera. Species reported to infest turfgrass in the United States are listed in Table 8-1. The binomials accepted and used here are those occurring in the *Check List of Lepidoptera of America North of Mexico* (Hodges et al. 1983). Since many of the turfgrass-infesting species of sod webworms have not been assigned common names, the scientific names will be used for sake of uniformity. The common names, if known, appear only in Table 8-1 and in the following paragraph.

The six most important sod webworm species in the eastern temperate regions of the United States include the bluegrass webworm, *Parapediasia teterrella* (Zincken); the striped sod webworm, *Fissicrambus mutabilis* (Clemens); the silver-striped webworm, *Crambus praefectellus* (Zincken); the larger sod webworm, *Pediasia trisecta* (Walker); the corn root webworm, *Crambus caliginosellus* Clemens; and the subterranean webworm, also known as the cranberry girdler, *Chrysoteuchia topiaria* (Zeller). The western lawn moth, *Tehama bonifatella* (Hulst), damages turf from the Rocky Mountain plateau to California, and is most common on the Pacific Coast (Plate 17) (Ainslie 1923a, 1923b, 1927, 1930; Kennedy 1980; Shetlar 1995a).

Burrowing sod webworms, *Acrolophus* spp. (Acrolophidae), occasionally damage turfgrass. These are considered secondary pests and are discussed briefly in Chapter 22.

Importance

Sod webworms restrict their feeding, with rare exceptions, to plants of the family Gramineae, and turfgrasses serve as ideal host plants. The most serious problems with sod

Table 8-1. Sod Webworms of the Subfamily Crambinae Infesting Turfgrass in the United States

Latin Binomial	Common Name	Major References
Agriphila ruricolella (Zeller)	—	Robinson and Tolley 1982
A. vulgivagella (Clemens)	Vagabond crambus	Felt 1894, Werner 1982
Chrysoteuchia topiaria (Zeller)	Cranberry girdler, subterranean webworm	Kamm 1973, Kennedy 1980
Crambus agitatillus Clemens	—	Robinson and Tolley 1982
C. caliginosellus Clemens	Corn root webworm	Dominick 1964, Kennedy 1980
C. laqueatellus Clemens	Paneled crambus	Matheny 1971, Felt 1894
C. leachellus (Zincken)	Leach's crambus	Felt 1894, Robinson and Tolley 1982
C. luteolellus (Clemens)	Yellow crambus	Felt 1894, Robinson and Tolley 1982
C. perlellus (Scop.)	—	Robinson and Tolley 1982
C. praefectellus (Zincken)	Silver-striped webworm	Ainslie 1923a
C. sperryellus Klots	Silver-barred webworm	Bohart 1947
C. tutillus McD.	—	Kamm 1971
Fissicrambus haytiellus (Zincken)	—	Wylie 1944
F. mutabilis (Clemens)	Striped sod webworm	Ainslie 1923b
Microcrambus elegans (Clemens)	Pretty crambus	Robinson and Tolley 1982, Tolley 1983
Parapediasia decorella (Zincken)	—	Robinson and Tolley 1982
P. teterrella (Zincken)	Bluegrass webworm	Ainslie 1930, Robinson and Tolley 1982
Pediasia trisecta (Walker)	Larger sod webworm	Ainslie 1927, Robinson and Tolley 1982
Surattha identella Kearfott	Buffalograss webworm	Sorensen and Thompson 1971
Tehama bonifatella (Hulst)	Western lawn moth	Bohart 1947
Urola nivalis (Drury)	—	Robinson and Tolley 1982

Note: Dashes indicate that no common name has been assigned.

webworms have occurred in the midwestern, eastern, and southeastern portions of the United States, with Illinois, Indiana, Ohio, Kentucky, and Tennessee routinely having serious webworm problems (Ainslie 1930, Heinrichs 1973, Matheny 1971).

In the Pacific Northwest, webworm problems occurring in commercial turfgrass seed production receive greater attention than problems occurring on home lawns. In addition to being turfgrass pests, some species are recognized as being of even greater importance to agriculture, as they feed on field and forage crops. The sod webworm with the most diverse feeding habits is *Chrysoteuchia topiaria,* which destroys the roots of turfgrasses, cranberries, and coniferous seedling plants (Kamm 1971, 1973; Kamm et al. 1983; Kennedy 1980).

Increased problems with sod webworms on turfgrasses occurred several years after the chlorinated hydrocarbon insecticides, such as chlordane and dieldrin, came into general use. The destruction of natural enemies was considered to be the primary cause of severe webworm outbreaks. In addition, development of almost complete resistance to the chlorinated hydrocarbons complicated the webworm problem in the 1960s (Heinrichs and Southards 1970, Schuder 1964).

History and Distribution

A wide-scale drought occurred during 1928–34 affecting much of the United States. After 1929, sod webworms, coupled with the drought, became recognized as serious lawn and golf course turf pests. Outbreaks of unprecedented magnitude occurred in many states during 1931 (Bohart 1947).

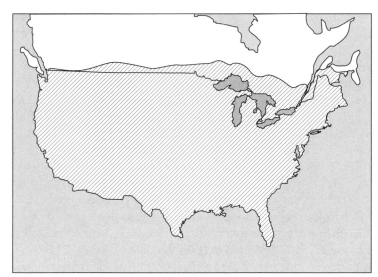

Figure 8-1. Distribution of sod webworms in the United States. (Drawn by E. Gotham, NYSAES, adapted from Handbook of Turfgrass Insect Pests, Brandenburg and Villani 1995, Entomological Society of America.)

Unlike many of our most destructive introduced pests, the webworms are all native to North America. The most common webworm species infesting turfgrass is *Parapediasia teterrella.* Its geographic range covers an area south and west of a line from Massachusetts and Connecticut through southeastern New York, westward into eastern Colorado, and southward to the southern tip of Texas (Figure 8-1). It is more abundant in the bluegrass region of Kentucky and Tennessee than in any other part of its range (Ainslie 1922, 1930).

One of the most voracious feeders is *Pediasia trisecta,* injurious to lawns, pastures, meadows, and cornfields. This is a major species widely distributed from coast to coast in the northern United States and across southern Canada (Figure 8-1). It is most abundant in the region encompassing Ohio and Iowa. Southern extensions include North Carolina, Texas, and New Mexico (Ainslie 1927).

Fissicrambus mutabilis is one of the most common species and is widespread over much of the United States east of the Rocky Mountains (Figure 8-1). It is most abundant in a triangular area encompassing Illinois, Tennessee, and Pennsylvania. Another prominent species, *Crambus praefectellus,* is distributed widely over the eastern half of the United States in practically every state east of the Mississippi River, in at least seven states west of the river, and in two provinces in Canada (Figure 8-1) (Ainslie 1923a, 1923b).

Another webworm of turfgrass that is widespread across the northern portion of the United States is *Crambus caliginosellus.* It can be troublesome in the north central and northeastern states and southern Ontario in Canada. To the south it extends into Virginia and North Carolina (Dominick 1964, Kennedy 1980). It often occurs in equal numbers with *Crambus luteolellus* (D. Shetlar, Ohio State University, personal communication, 1998).

A pest of turfgrasses in Michigan, Ohio, and western Pennsylvania, *Chrysoteuchia topiaria* is an important species that occurs over much of northern United States. It is very important in the Pacific Northwest (including Utah, Idaho, Oregon, and Washington), where greatest attention is paid to it because it infests turfgrasses grown for seed production as well as high-quality lawns. It burrows into the crown of grasses and also eats roots. It is also a serious pest of cranberries because it girdles roots (McDonough and Kamm 1979).

Large numbers of *Agriphila ruricolella* and *A. vulgivagella* adults emerge in late August and September from Pennsylvania to Iowa. The presence of the conspicuous moths often causes alarm in homeowners, but there has been no evidence of damage to lawns by either of these species (D. Shetlar, personal communication, 1998).

Serious turfgrass pests of California are *T. bonifatella* (Hulst) and *Crambus sperryellus* Klots (Plate 17). The former occurs in the Rocky Mountain plateau, in the Great Basin (occupying the western half of Utah and most of Nevada, as well as California, Oregon, and Idaho). Along the Pacific Coast, it is confined to a coastal area that is 80 km wide from Washington to southern California. *Crambus sperryellus* occupies the coastal, inland, and desert valleys of California. In coastal southern California these two species are the most common lawn insect pests and favor bluegrass and bentgrass lawns where larval populations may exceed 22 per 0.1 m² (Bohart 1940, 1947; Jefferson and Eades 1952; Jefferson et al. 1964).

At least 14 species of sod webworms were associated with turfgrass in Virginia as indicated by blacklight trap catches during May to October 1981. Traps were located in Blacksburg in the western part of the state and Virginia Beach on the Atlantic coast. The three most abundant species were, in descending order, *Parapediasia teterrella*, *Microcrambus elegans* (Clem.), and *Pediasia trisecta* (Robinson and Tolley 1982).

Host Plants and Damage

Host Plants
With rare exceptions, sod webworms feed primarily on plants of the family Gramineae. Turfgrasses most commonly recorded as primary host plants include Kentucky bluegrass, perennial ryegrass, fine fescue, and bentgrass. There are relatively few records of damage to warm-season grasses, perhaps because of the aggressive growth habits of these species. *Fissicrambus haytiellus* (Zinck.) damages bermudagrass and zoysiagrass in Florida. In addition to lawns and other turfgrasses, most webworms feed on corn; wheat; rye; oats; timothy, *Phleum* L.; pastures; and meadows, with greatest destruction to areas of permanent sod. Webworm damage becomes most pronounced at times of drought (Ainslie 1930, Bohart 1940, Wylie 1944).

Some webworms feed on entirely different types of host plants. *Crambus caliginosellus* feeds on turfgrasses but also on corn and tobacco seedlings. Its larvae concentrate most in the root zone of the narrow-leaf plantain, *Plantago lanceolata* L., where more than 30 larvae have been found around a single plant (Dominick 1964, Kennedy 1980). *Chrysoteuchia topiaria* feeds in the crown and roots of grasses in fields of commercial seed production and girdles the roots of cranberry plants. Seedlings of Douglas fir, *Pseudotsuga menziesii* (Mirb.) Franco, and several true firs, *Abies* species, that are grown in nurseries are girdled by these larvae (Kamm et al. 1983, Kennedy 1980, McDonough and Kamm 1979).

Damage
First-instar sod webworm larvae feed only on the surface layers of leaves and stems (Plate 18). The first evidence of feeding damage to a normally growing lawn may be small patches of leaves that are yellow to brown during the summer. The patches grow daily. Holes pecked by birds may become evident (Plate 18). Since larvae are nocturnal, they are found during the day only in their burrows in the thatch or in surface soil along with accumulations of green fecal pellets (Plate 18).

Larger caterpillars construct tunnels in the thatch and upper root zone, and forage from these burrows, cutting down individual blades of grass. Closely mown turf shows symptoms more quickly than does poorly maintained turf. Individual caterpillars may produce small depressed marks of brown grass that coalesce into larger brown patches. These large irregular areas of dead or dying grass occur most commonly in sunny locations, with shaded areas and weedy patches other than grasses remaining green (Plate 18) (Shetlar 1995a).

Because sod webworms feed primarily on the surface, well above the growing point of the grass plant, the injury produced is not as damaging to the plant as injury caused by insects feeding within the crown or roots. Bohart (1947) suggested that a count of 12 larvae per 0.1 m², flushed out in 10 min with a pyrethrum or detergent drench, could be considered a serious infestation. Others have flushed 18−24 larvae per 0.1 m² with little or no visible damage in bentgrass (D. Shetlar, personal communication, 1998).

The presence of a large number of moths flying over the lawn at dusk does not necessarily mean that a heavy larval infestation will occur, nor does the presence of flocks of birds feeding on the lawn during the day signify a heavy infestation (Bohart 1947). These phenomena, however, should cause the presence of webworm larvae to be suspected and should not be ignored.

The presence of webworms under a drought condition constitutes the most serious situation for potentially serious turfgrass damage. Not only can the dormancy of the grass restrict the manifestation of early feeding symptoms, but all too often the dead turf does not become evident until fall rains revitalize the normal turf. Normally sod webworm feeding in itself does not kill the crown. A normal lawn can often be revived if the infestation is controlled and if fertilizer and water are added.

Description of Stages

Adults

Moths of sod webworms, often called *snout moths,* are distinguished from all other moths by their appearance. When they are at rest, their very long labial palpi extend, snoutlike, in front of their heads (Figure 8-2, Plate 19). Moths also fold their wings partially around their abdomen, so that they appear very slender, almost tubular, and are often very difficult to see after they have come to rest on a grass stem or leaf (Ainslee 1922).

Moths appear to be whitish or light gray to tan, but on closer observation their forewings often show designs of silver, gold, yellow, brown, and black. These colors occur as longitudinal stripes on a whitish or dull-gray background. The hindwings are usually white or grayish. Both pairs of wings have delicate fringes on the outer margin. Moths of most species are about 12 mm in length, with wing expanses of 20−25 mm. Table 8-2 gives the dimensions of three webworms of common size together with that of *Pediasia trisecta,* one of the largest of webworms, with a wing expanse of 21−35 mm. The larger size is also evident for larvae as well (Ainslie 1923a, 1923b, 1927; Felt 1894).

Eggs

Sod webworm eggs are completely dry and nonadhesive. The many species have eggs that are very similar in shape, size, sculpturing, and coloration (Plate 19). Felt (1894) and Matheny and Heinrichs (1972) measured eggs of a total of 20 species and examined them to eclosion. Depending on species, most eggs are oval to elliptical oval in form, with vary-

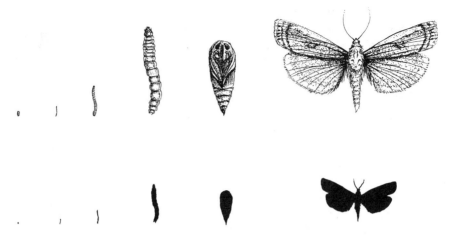

Figure 8-2. Stages of development of sod webworm: egg; first, third, and final larval instars; pupa; and adult. Silhouettes show actual size of each stage. (Courtesy of R. Jarecke, NYSAES.)

ing numbers (from 12 to 30) of longitudinal ridges on the surface from pole to pole. Length and width varied from 0.598 mm and 0.503 mm, respectively, for species with the largest eggs to 0.390 mm and 0.300 mm for species with the smallest eggs. The eggs of most species are white to creamy white when they are first laid and turn to various shades of bright orange to bright red, becoming ocherous at eclosion (Plate 19).

Scanning electron micrographs were taken of the chorion of eggs from 15 species of sod webworms in Tennessee. Sufficient differences in the longitudinal carinae, polar areas, and various other features were present to permit the development of a key to Crambiinae eggs (Matheny and Heinrichs 1972).

Larvae

In general appearance, sod webworm larvae vary in color, from greenish to beige, brown, or gray. Most have characteristic dark circular spots scattered over the entire body (Plate 19). At least two common species lack the darker circular spots, *Crambus caliginosellus* (Plate 19) and *Chrysoteuchia topiaria* (Kennedy 1980).

Head capsules in some species are black during the first two or three instars. In most

Table 8-2. Size of Some of the More Common Sod Webworms in the United States (mm)

Species	Adult Wingspan	Eggs, Width by Length	Larval Head Capsule Width		Larval Total Length	
			First Instar	Final Instar	First Instar	Final Instar
Parapediasia teterrella	18–21	0.31 × 0.51	0.21	1.23	1–2	9–13
Fissicrambus mutabilis	18–24	0.31 × 0.48	0.19	1.73	1.8	18
Crambus praefectellus	18–25	0.31 × 0.51	0.23	1.40	?	?
Pediasia trisecta	21–35	0.34 × 0.52	0.21	2.21	1–2	24–28

Sources: Ainslie 1923a, 1923b, 1927, 1930.

species, mature larvae have light-brown head capsules, with varying shapes of black sculpturing (Plate 19). Table 8-2 shows that head capsule widths of four common species vary from 0.19–0.23 mm in first instars to 1.23–2.21 mm in ultimate instars. Total larval lengths for the final instar vary from 9 to 13 mm in the smallest and from 24 to 28 mm in the largest webworm (Ainslie 1923a, 1923b, 1927, 1930).

Larval instars vary from 6 to as many as 10, with 8 being most common; *Parapediasia teterrella* has 7 instars over a period of about 50 days, and *Pediasia trisecta* has 8 instars (Robinson and Tolley 1982).

Pupae and Cocoons

The cocoons are made from silk attached to soil particles, plant debris, and fecal pellets and may be a part of the larval tunnel or may be constructed separately. In size and shape the cocoon resembles a peanut meat (Figure 8-2). It is firm, smoothly lined with soft gray silk, and outwardly covered with soil and grass. Its depth depends on the looseness of soil, and it may be mistaken for a lump of earth. The pupae developing within the cocoon are pale yellow when first formed and darken to mahogany brown. A day or two before their emergence, the color patterns of the forewing become plainly visible through the integument. The pupae of *Parapediasia teterrella* are about 8–10 mm long and about 2.5 mm wide (Ainslie 1923a, 1923b, 1927, 1930).

Seasonal History and Habits

Seasonal Cycle

Sod webworms overwinter as larvae in their hibernacula in the thatch or soil. Most species do so in their penultimate or ultimate instar, but others may overwinter in earlier instars. Larvae of *Pediasia trisecta* are known to overwinter in the second to the fifth instar. With the advent of warmer weather, larvae resume feeding or, if in the ultimate instar, will pupate. Moths of most species emerge during late spring or early summer, but *Agriphila* sp. normally emerge in late summer and *Chrysoteuchia topiaria* usually emerges in midsummer. Moths of *Crambus praefectellus* are the first to appear in the spring (late April through May) in the eastern United States (Ainslie 1923a, 1927; Shetlar 1995a).

Some species, such as *Chrysoteuchia topiaria* and *Crambus caliginosellus*, are univoltine wherever they occur. Why *Pediasia trisecta* is univoltine in the Pacific Northwest and bivoltine or trivoltine in the eastern states is not known (Figure 8-3). It is reported to have two generations a year in the Midwest, two or three in Virginia, and three in New Jersey (Ainslie 1927, Kamm 1970, Mailloux and Streu 1982, Robinson and Tolley 1982).

Parapediasia teterrella has two generations in Virginia and three in Tennessee, a slightly more southern latitude (Figure 8-3). Figure 8-4 presents a detailed life history of *Parapediasia teterrella* as it often occurs in Tennessee. The overwintering larvae pupate in April and May, and moths make their first appearance in early May, steadily increase throughout the month, and then decrease through June. First-generation eggs are laid during June. Larvae are present during June and July, with pupation in July. Adults are present in July to August, depositing second-generation eggs during August. Larval and pupal development occurs throughout August and September. With favorable weather conditions, moths of the second generation lay eggs into early October for overwintering larvae (Ainslie 1930).

In the Los Angeles area, *Crambus sperryellus* has three generations a year and *T. boni-*

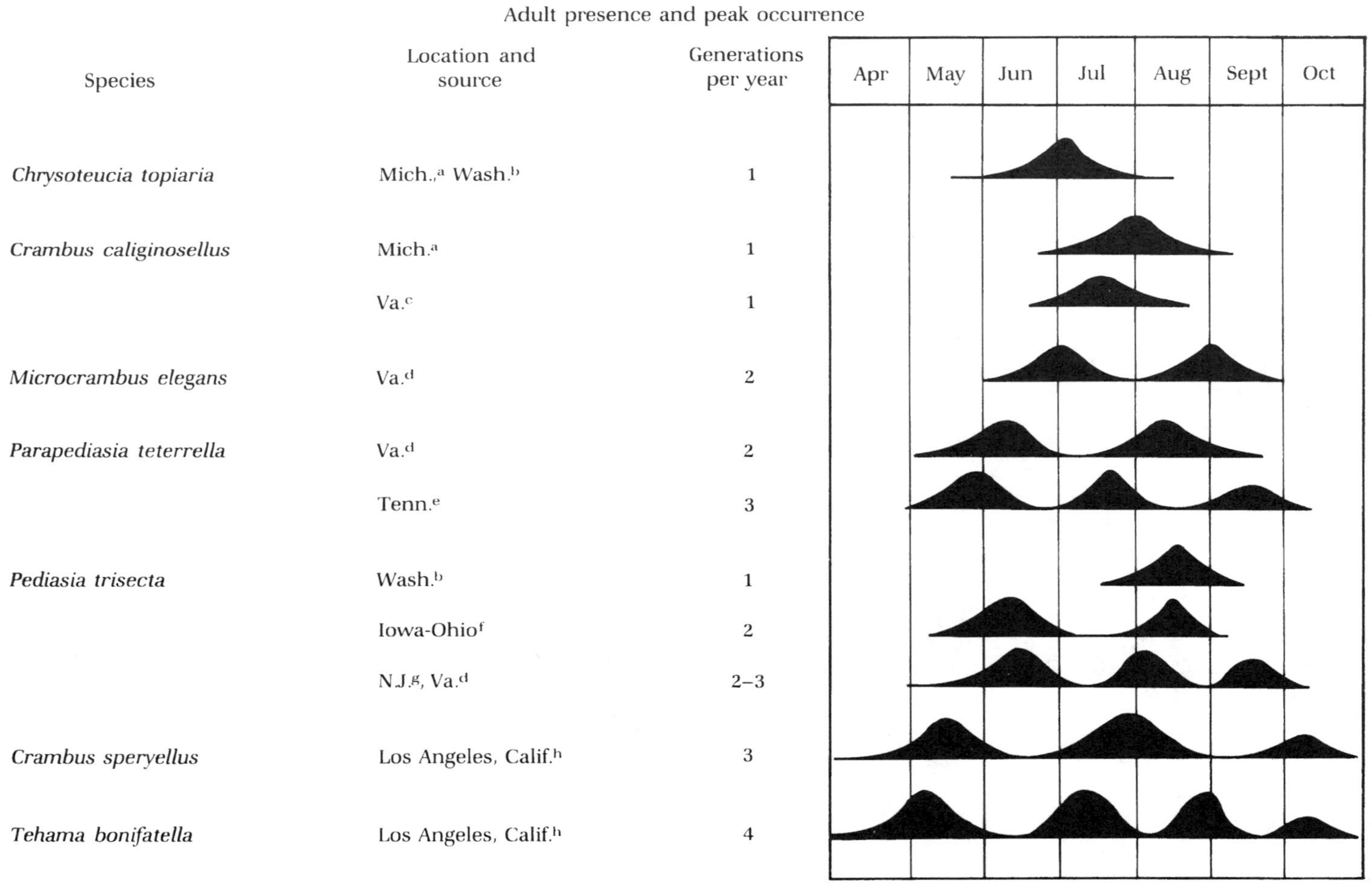

Adult presence and peak occurrence

Figure 8-3. Seasonal presence of common sod webworm adults at various locations in the United States. (Data from [a]Kennedy 1980, [b]Crawford and Harwood 1964, [c]Dominick 1964, [d]Robinson and Tolley 1982, [e]Ainslie 1930, [f]Ainslie 1927, [g]Mailloux and Streu 1982, and [h]Bohart 1947; redrawn by R. McMillen-Sticht, NYSAES.)

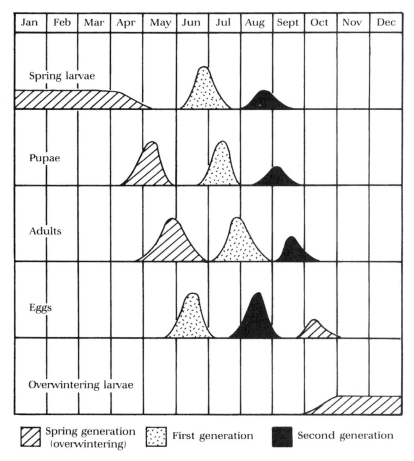

	Jan	Feb	Mar	Apr	May	Jun	Jul	Aug	Sept	Oct	Nov	Dec

Spring larvae

Pupae

Adults

Eggs

Overwintering larvae

Spring generation (overwintering)   First generation   Second generation

Figure 8-4. Life history (approximate) of *Parapediasia teterrella* in Tennessee. (Data from Ainslie 1930 and Robinson and Tolley 1982, fig. 3; drawn by R. McMillen-Sticht, NYSAES.)

fatella has four, completing a life cycle within 6 weeks in warm weather (Bohart 1940). Both species were reared from eggs to adults at 24°C. *Crambus sperryellus* required 6, 38, and 11 days to complete the egg, larval, and pupal stages, respectively, for a total of 55 days. *Tehama bonifatella* required 4.5, 24.0, and 8.0 days to complete the three stages, for a total of 36 days, showing that it is a much more rapidly developing species (Bohart 1947).

Adult Activity

Sod webworm moths are inactive during the day. They rest on turfgrass, on associated weeds, or on the leaves and stems of nearby trees or shrubs (Plate 19). If disturbed they will take flight, but flight is weak. Fluttering moths soon come to rest. In Virginia, daily sampling for moths during August to October 1982 revealed that *M. elegans, Parapediasia teterrella*, and *A. ruricolella* (Zell.) preferred shrubs over turfgrass for daylight resting sites, with no preference for shrub species (Tolley 1983).

Moths of *Parapediasia teterrella* prefer broad-leaved plants to grass for resting places. They often alight on the upper surface of leaves and immediately move underneath and rest, with the head pressed close to the leaf surface and the wings and abdomen elevated about 25 degrees. This habit together with the white scales on top of the head helps to identify the species (Ainslie 1930).

During the day, if one walks across a turfgrass area, webworm moths will fly for a short distance (a few yards or a few meters) in a zigzag pattern before again alighting on the turf. Toward evening the moths are more easily flushed from their resting place and fly freely, just above the turf.

Nightly Abundance. Webworm moths are most active during the early hours of the night, as revealed in light trap catches. Under favorable weather conditions, females of *Parapediasia teterrella, Pediasia trisecta,* and *F. mutabalis* are most abundant from about 30 min after dark to 1.5–2.0 hr after sunset, while males reach maximum flights from 11:00 PM to 1:00 AM (Ainslie 1930, Banerjee 1967a, Banerjee and Decker 1966).

Mating. Females apparently mate only once. Females of some species mate during their first night while others, such as *Parapediasia teterrella, Pediasia trisecta,* and *M. elegans,* mate 2 or 3 nights following emergence. Males may mate more than once but usually not on the same night (Banerjee 1969b, Robinson and Tolley 1982).

Mating habits of different species of webworm moths appear to be similar. Males do not arrive in numbers until the evening is well advanced. In the case of *Parapediasia teterrella,* males appear at about 9:00 PM, and by midnight the two sexes may be present in roughly equal numbers. Mated pairs are tail to tail and generally show little or no movement, but occasionally the female drags the smaller male about. Observations have indicated that the union may last 12 min to more than 2 hr, with an average of nearly an hour. Many of the males die within 12–24 hr after mating (Ainslie 1930).

Sex Pheromones. The presence of species-specific sex pheromones was demonstrated in a turfgrass area in Illinois when virgin females of *Parapediasia teterrella* and *Pediasia trisecta* of known age were placed in traps and were exposed to natural flights of moths. Virgin females were most attractive during the first 3 or 4 days after emergence (Banerjee 1969a).

The sex pheromone of *Chrysoteuchia topiaria* has been identified as (Z)-11-hexadecenal (Z)-9-hexadecenal. When it was used to monitor the presence of males by placing a trap just above the canopy of the grass, it was found to be six times as attractive as a single live virgin female. One trap per 1–2 ha gave a good indication of the population density. Since the larva enters the crown of grasses and its presence is undetectable until damage has occurred, the pheromone is a valuable monitoring tool to determine population densities (Kamm and McDonough 1979, 1980).

A combination of (Z)-11-hexadecenal and (Z)-13-octadecenal is a sex attractant for *T. bonifatella.* Traps baited with the two-component lure caught eight times more males than traps baited with females (McDonough et al. 1982).

Oviposition. Nightfall is the primary stimulus for oviposition. Females of most species drop their dry, nonadhesive eggs indiscriminately as they fly over the turfgrass. Females of *Parapediasia teterrella* start ovipositing during their flights at dusk and continue for 2–3 hr as they fly low across the lawn, scattering their eggs over the grass. They may also drop eggs while at rest. The area below fluttering moths is usually thickly strewn with eggs. Unlike other webworms, *Chrysoteuchia topiaria* females deposit their eggs while resting in the grass (Ainslie 1930, Crawford 1968, Kennedy 1980).

Oviposition by *Pediasia trisecta* begins on the second night after emergence, at the approach of darkness when the light intensity approaches 0 ft-c. About 60% of the eggs are laid the first hour and about 90% during the second hour (Banerjee and Decker 1966).

Fecundity. Fecundity varies greatly among species and is related somewhat to the size of the moths. *Crambus laqueatellus* Clemens, one of the larger, lays more than 480 eggs per female. *Microcrambus elegens* (Clemens), a small moth, lays eggs of the same size as those of *C. laqueatellus* (0.28 mm by 0.42 mm) but averages only 32 per female. Females of *Parapediasia teterrella* normally deposit 200–250 eggs with oviposition of as many as 60 eggs per day until death. Females of *Crambus caliginosellus* deposited an average of slightly over 200 eggs per female in Michigan. In New Jersey, a mated female of *Pediasia trisecta,* common in Kentucky bluegrass and red fescue lawns, lived about 8 days and laid an average of 470 eggs per female (Ainslie 1930, Kennedy 1980, Mailloux and Streu 1982, Matheny 1971).

Adult Longevity. Normally in nature the bluegrass webworm is considered to live 7–10 days. The females live slightly longer than the males. Caged moths had a maximum longevity of 28 days when they were given only water or honey in water. Captive moths without moisture live only a third as long (Ainslie 1930).

Egg Activity

Eggs of various species generally hatch in 6–10 days in the field. Eggs of *Parapediasia teterrella* hatched in 5–7 days during summer but required 41 days at 10°C. Eggs of *Pediasia trisecta* did not hatch below 10°C and required 29 days at 14°C and 5.4 days at 32°C. Eggs of *Crambus caliginosellus* hatched in 5–9 days during the summer (Ainslie 1930, Banerjee 1969a, Dominick 1964, Heinrichs 1973).

Widely distributed species may show adaptations to local climatic conditions. Eggs of *Parapediasia teterrella* from moths collected at Lexington, Kentucky, and Albuquerque, New Mexico, both had a hatch percentage of more than 90% at more than 60% RH. At less than 60% RH, the egg hatch percentage from Lexington moths was decidedly less than that from Albuquerque moths at any and all relative humidity percentages, with no hatch of eggs from either location at 0% RH (Morrison et al. 1972).

Larval Activity

As soon as they have hatched, the first instars of many species find suitable hiding places and begin to conceal themselves by webbing together particles of debris. First instars feed only on the surface tissues of tender leaves (Plate 18).

Habitat and Feeding. As they grow, third and fourth instars prepare burrows by webbing together plant debris, soil particles, and their own excrement and lining the interior with their webbing (Plate 18). A larva of *Crambus caliginosellus* seldom constructs a complete burrow of webbing as the others do. It often constructs only one side, using the stem of the plant on which it is feeding as the other side. Burrows open not to the surface but downward, where much of the feeding takes place. During the day, larvae remain concealed in their burrows but during the night wander out to feed. They notch leaves and during the later instars may cut off entire leaves and draw them into their silk-lined tubes. The tube of *Parapediasia teterrella* measures 10 mm by 30 mm. Larvae of most species feed above the crown of the plant and above ground, but at least two well-known species feed below ground. These are *Crambus caliginosellus* and *Chrysoteuchia topiaria,* the former infesting the crown and roots of grasses and the latter infesting cranberry and fir seedling roots as well as those of grasses (Ainslie 1922, Kennedy 1980).

Young larvae of *T. bonifatella* and *Crambus sperryellus* curl into a compact ball when they are disturbed but after they are half grown try to escape when disturbed. They crawl backward as rapidly as they do forward and thrash wildly when touched. With the aid of a flashlight, larval feeding can be observed between 8 PM and 10 PM; leaves are being notched, and blades are being cut off and drawn into burrows. Larvae feed on the greener portions of the crown but do not molest roots. Fifth instars consume twice their weight in bluegrass daily (Bohart 1947).

Larval Instars. Larval instars vary among species, but 7 or 8 are common. Larvae of *Parapediasia teterrella* normally have 8 instars but may have 6–10. Those of *F. mutabalis* normally have 7 instars, those of *Crambus praefectellus* have 6 instars, and those of *Pediasia trisecta* have 7 or 8 instars (Ainslie 1923a, 1923b, 1927, 1930).

Diapause. Sod webworms overwinter as larvae in their final instar in some species and as larvae in earlier instars in others. The photoperiod was the primary factor in the induction and termination of diapause in larvae of *Pediasia trisecta,* but both the photoperiod and temperature were involved for *Crambus leachellus* (Zincken). Both species experience diapause in the second to fifth instar. Exposure lasting 25–35 days to short days, 10 hr light with 14 hr dark, at 15° and 24°C induced 100% diapause in *Pediasia trisecta* (Kamm 1970).

In a univoltine species, *Crambus tutillus* McD., diapause was induced by exposing seventh instars to short days, 10 hr light and 14 hr dark, which initiated molt to the eighth instar, which went into diapause. Short days and warm temperatures (18° and 24°C) killed larvae. Those exposed to long days and warm temperatures terminated diapause, completed the ninth instar, pupated, and became adults. In early instars, growth rate accelerated as day length decreased (Kamm 1971).

Pupal Activity

Upon completion of the larval stage, *Parapediasia teterrella* abandons its feeding burrows. It then constructs cells in the soil lined loosely with gray silk and forms a cocoon. Pupation lasts from 5 to 15 days (Ainslie 1930).

Temperature influenced pupation of *Pediasia trisecta.* Larvae failed to pupate at 18°C but pupated after 59 days at 21°C and after 29 days at 32°C (Banerjee 1969b).

Miscellaneous Features

Plant Resistance

In New Jersey, resistance of perennial ryegrasses to sod webworm was associated with the presence of a *Lolium* endophyte fungus. Cultivars 'GT-11', 'Pennant', and 'All-Star' have high levels of endophyte and were found to be highly resistant. The fungus is transmitted primarily by seed and vegetative propagation (Funk and Hurley 1984).

Cultivars of tall fescue infected with *Neotyphodium* (formerly *Acremonium*) *coenophialum*

Morgan-Jones & Gams had significantly fewer sod webworm larvae (not identified to species) than did the same cultivars when endophyte free (Murphy et al. 1993). This suggests that endophytic cultivars of perennial ryegrass and tall fescue are resistant to sod webworms generically, and are viable alternatives for suppressing webworm populations.

'Scaldis' hard fescue and 'Dawson' red fescue have some tolerance to sod webworms, as have the Kentucky bluegrasses 'Windsor' and 'Park'. Crossing resistant × resistant, resistant × susceptible, and susceptible × susceptible cultivars yielded highly resistant, moderately resistant, and susceptible strains of grass, respectively, and demonstrated the heritable nature of resistance (Sargent 1982).

Diagnosis
Sod webworm damage sometimes is confused with disease activity, particularly some of the warm-weather patch diseases common in cool-season turfgrass. Webworm-affected turf will show a general thinning in midsummer (often resembling drought damage initially) or will be thin in the spring and fail to come out of spring dormancy. Flocks of birds (especially starlings and blackbirds) may congregate and feed on caterpillars. Close inspection reveals clipped leaves on the surface, as well as clumps of green pellets (frass) or silk lining burrows in the thatch.

Threshold Populations
Since the presence of adults is not a positive indication of damaging larval populations to come, threshold populations of larvae should be determined before a decision is made to apply an insecticidal treatment. There are wide differences of opinion as to what constitutes a threshold population. Reduced to the common denominator of numbers per 0.1 m^2, suggested thresholds include 0.22 larva in Wisconsin, 1.7 larvae in California, 18–24 larvae in Ohio, and 12–16 larvae as a general countrywide recommendation (Bowen et al. 1980, Mahr and Kachadoorian 1984, Shetlar 1995a, Vance and App 1971). Obviously these threshold estimates are much too divergent to be of much practical value. When a decision is made, all other contributing factors must be considered, including turf vigor, temperature, moisture regime, mowing height, and expectations.

Degree-Day Predictions
Prediction of peak moth flights is helpful in managing sod webworm populations. Visual observation or monitoring moth captures in black light traps can be useful, but degree-day predictions enable one to concentrate monitoring efforts when moths are most likely to be flying. A 2-year study in Virginia generated degree-day models for *A. ruricolella*, *Crambus laqueatellus*, second-generation *Peripediasia teterrella*, and first-generation *Pediasia trisecta*. The same study found that calendar dates were as good or better than degree-days in predicting peak flights for first-generation *Peripediasia teterrella* and second- and third-generation *Pediasia trisecta* (Tolley and Robinson 1986).

Natural Enemies

Microorganisms
Larvae and pupae of *Pediasia trisecta* in Illinois were found heavily infested with microsporidia *Nosema* spp. and *Thelohania* spp. These organisms also infected *Parapediasia teterrella* (Banerjee 1968).

The fungus *Beauveria bassiana* was present in all field collections but infected only 3.1% of *Chrysoteuchia topiaria* larvae in Oregon (Kamm 1973).

Insect Parasitoids

Larvae and pupae of *Pediasia trisecta* in Illinois were parasitized by braconid wasps, *Macrocentrus crambivorus* Vier and *Orgilus detectiformis* Vier, and by the tachinid fly, *Stomatomyia floridensis* (Tris). Another tachinid, *Lydina polidoides* (Townsend), is reported to parasitize 8.3% of the mature larvae of *Chrysoteuchia topiaria* in Oregon. In California, webworms in lawns were parasitized by two braconid wasps, *Orgilus* spp. and *Apanteles* spp., and by a tachinid, *Aplomyia confusionis* Sllrs. (Banerjee 1967b, Bohart 1947, Kamm 1973).

Insect Predators

The robber fly, *Erax aestuans* L., has been observed capturing webworm moths. Vespid wasps, native earwigs, carabid beetles, and rove beetles are considered important predators of webworms in California. In Kentucky predators consumed or carried off as many as 75% of sod webworm eggs (*Crambus* and *Pediasia* spp.) within 48 hr of exposure. Four species of ants, including *Pheidole tysoni* Forel as the major predator, and a mite, *Macrocheles* spp., foraged on the eggs in turfgrass (Banerjee 1968, Bohart 1947, Cockfield and Potter 1984).

Vertebrate Predators

Brewer's blackbird is considered to be an important predator of sod webworms in California. In Oregon, overwintering larvae of *Chrysoteuchia topiaria* as prepupae in their hibernacula were prey to starlings, killdeer, sandpipers, and blackbirds, which reduced the population more than 80% and, combined with parasites and diseases, reduced the overwintering population about 91% (Bohart 1947, Kamm 1973).

On lawns, probing holes made by birds may be a clue to webworm presence. These holes are not always a positive indication, but such clues should certainly be pursued.

Tropical-Region Sod Webworms

Grass Webworm

Taxonomy and Importance

The grass webworm (GWW), *Herpetogramma licarsisalis* (Walker), is a member of the order Lepidoptera, family Pyralidae, subfamily Pyrustinae. The insect was placed previously under the genera *Psara* and *Pachyzancla*. Since its discovery in Hawaii, it has been considered the most serious turfgrass pest in the state. At least 90% of all turfgrass insects in bermudagrass have been GWW larvae (Swezey 1946, Tashiro 1976a).

History and Distribution

The GWW is widely distributed in Southeast Asia, in the adjoining islands, and in Australia. It was first found in Hawaii in 1967 on Oahu, where it was damaging pasturegrass. In 1968 it was collected from four other islands, Hawaii, Kauai, Maui, and Molokai (Davis 1969). Hawaii is its only known location in the United States.

Host Plants and Damage

All of the important turfgrasses in Hawaii, including bermudagrass, centipedegrass, and St. Augustinegrass, are attacked. Kikuyugrass, *Pennisetum cladestinum* Hochst ex Chiov, a relatively minor turfgrass but a major pasturegrass, is the most heavily infested, with larval counts as high as 55 per 0.1 m² (Davis 1969).

The GWW feeds on leaves, stems, and crowns of turfgrass. During early feeding, the turf appears ragged but is still green. With continued feeding and the passage of time, large brown patches develop (Plate 20). Silk webbing and frass often are quite apparent in the thatch.

In a species and varietal feeding test, there was no evidence of oviposition preference. Common bermudagrass and its cultivar 'Tifway' were the least injured. Feeding injury developed faster on fine-textured grasses than on those with coarse texture (Murdoch and Tashiro 1976, Tashiro 1977).

Description of Stages

Adult. Moths are uniformly fawn to light brown, with a mean wingspan of 23.9 mm. The only apparent pattern on the front wing is a faint zigzag line of slightly darker scales paralleling the apical margin and near it. Males have a more slender abdomen than females, with seven visible segments, as compared with six visible segments, ending in a recessed opening for females. The body is about 10 mm long. Unlike sod webworm moths of the subfamily Crambinae, which wrap their wings around the body when at rest, the GWW holds its wings horizontally, so that the moth has a triangular outline (Plate 20) (Tashiro 1976a).

Egg. The flat, elliptical eggs, measuring 0.64 mm by 0.91 mm, are laid singly or in masses generally glued to the upper surface of leaves along the midrib or on nonliving objects where there is free moisture (Plate 20). In masses, the eggs are laid so that they overlap each other like shingles. The chorion has fine reticulations. When it is fresh, it is creamy white. It becomes light yellow on the second day. A light-orange embryo shows on the third day. It becomes dark orange, with the black head capsule visible, on the fourth day, the day before hatching (Tashiro 1976a).

Larva. GWW larvae have five instars, with head capsule widths of 0.25 mm in the first to 1.8 mm in the fifth instar. Total body length grows from 2.32 mm in the first to 20 mm in the fifth instar. First-instar larvae have black head capsules; all others have brown (Plates 20, 21). The prothoracic shield is lighter brown than the head. Mature larvae are brown to greenish, depending on the food consumed, and many have a rose tint over part or most of the body. Each body segment has a conspicuous ring of dark brown spots (Plate 21) (Tashiro 1976a).

Pupa. Pupation occurs in a hibernaculum-like case (Plate 21). It cannot be considered a true hibernaculum, since the GWW does not experience diapause. The pupa is creamy white when first formed and turns light brown, then dark brown (Plate 21) (Tashiro 1976a). The sex of pupae can be determined in the way described by Butt and Cantu (1962).

Life History and Habits

Life Cycle. A generation is completed in about 32 days at a mean temperature of 24.5°C and 60–80% RH. Females have a 3–6-day preoviposition period, eggs hatch in about 5 days, larval development takes about 14 days, and the pupal period lasts about 7 days. During an average life span of 13 days, each female lays about 250 eggs. The maximum observed was 556 eggs. There is no diapause, and the rate of development depends mainly on temperature and food supply. The GWW is strictly nocturnal; adult eclosion, mating, oviposition, hatching, feeding, molting, and pupation all occur at night (Tashiro 1976a, 1977).

Adult Activity. During daylight, moths remain at rest, mainly in tall grass, unless they are disturbed. They will then fly for a short distance before returning to the grass. Literally hundreds of GWW moths can be flushed out of tall grass simply by walking through it (H. Tashiro, personal observations).

Moths require some nutrients for maximum longevity and fecundity. In screen cages with no water, females lived less than 4 days and laid no eggs, while females supplied with a 10% honey or sucrose solution as food lived at least 12 days and laid an average of more than 340 eggs (Tashiro 1977).

Larval Activity. First and second instars feed on the upper surface of the leaf, leaving the lower epidermis intact (Plate 20). From the third instar onward, larvae notch leaves, start to eat entire leaves, and spin copious amounts of webbing. Mature larvae in preparation for pupation become slightly shortened and web together bits of grass, feces, and other debris to form cocoons in which to pupate (Tashiro 1976a).

During the day, larvae are seldom, if ever, seen in the field but can be found hidden in the thatch. However, they can be flushed to the surface within minutes with a pyrethrin or detergent solution in copious amounts of water (Tashiro et al. 1983).

A 10% injury to hybrid bermudagrass turf in Hawaii is considered the level at and beyond which insecticidal treatments would be desirable (Mitchell and Murdoch 1974).

Miscellaneous Features

Rearing. The GWW is an easy insect to rear when 60–80% RH is maintained. Field-collected moths placed in a screen cage and supplied with honey or sugar solution oviposit on the grass blades and on any moist surface. Colonies can be maintained on potted 'Sunturf' bermudagrass and on kikuyugrass, but the succulence of the latter makes it the better host plant. Any warm-season grass with soft blades apparently is suitable for rearing. The rapidity of development is governed by temperature, with time between hatching and adulthood being about 22 days at a constant 24.5°C and 16 days at 31°C.

Natural Enemies

Microorganisms. No reports note the presence of any microorganism infecting the GWW.

Insect Parasitoids and Predators. The GWW is susceptible to several parasitoids, some of which were introduced to Hawaii for other lepidopterous pests. *Trichogramma semifumatum* (Perkins), accidentally introduced, has parasitized as many as 96% of GWW eggs. Larval parasitoids include the tachinid fly, *Eucelatoria armigera* (Coq.), and a braconid wasp, *Meteorus laphygmae* Vier. The pupae are parasitized by an ichneumonid wasp. An ant, *Pheidole* sp., destroys GWW eggs (Davis 1969).

Vertebrate Predators. The cattle egret, *Bubulcus ibis,* which forages in infested grass, consumes larvae (Plate 66) (Davis 1969). Other avian predators include the common mynah bird, *Acridothermes tristis tristis* (L.) (Plate 21), the red-crested or Brazilian cardinal, *Paroaria coronata,* and the Pacific golden plover, *Pluvialis dominica fulva.* Head capsules of the GWW have been found in scats of the giant toad, *Bufo marinus* (L.) (Plate 21) (H. Tashiro, personal observation).

Tropical Sod Webworm

Taxonomy

The tropical sod webworm (TSW), *Herpetogramma phaeopteralis* Guenee, is also a member of the order Lepidoptera, family Pyralidae, subfamily Pyrustinae. As in the case of the GWW, the TSW was previously listed under the genera *Pachyzancla* and *Psara* (Hodges et al. 1983, Kerr 1955).

Importance

The TSW is considered to be one of the most destructive turfgrass pests in Florida, along with mole crickets and southern chinch bugs. Caterpillar feeding severely damages all major southern turfgrass species. Late April to December is the period of greatest potential damage. Bermudagrass grown under high-maintenance programs has the highest populations and suffers the most. In mixed larval populations, the TSW is generally the dominant species. Populations of 11–22 larvae per 0.1 m^2 occur frequently, and counts of 85–100 per 0.1 m^2 are sometimes present (Kerr 1955; Reinert 1974, 1983a).

History and Distribution

The TSW has a wide tropical distribution throughout the southeastern United States and the Caribbean archipelago. In the United States it has caused severe turf damage in Louisiana since 1933 and in Georgia since 1953, as well as the Houston, Texas, area since the 1990s. The severe turf damage caused by a mixed population of lepidopteran pests in Florida in 1953 was due primarily to the TSW (Kerr 1955).

Host Plants and Damage

The turfgrasses damaged by the TSW include bermudagrass, centipedegrass, St. Augustinegrass, zoysiagrass, and bahiagrass, with the first three most widely grown as fine turf. Damage is caused when the larvae eat leaves, giving the turf a notched, ragged appearance. The notches along the edge of grass blades are a sure sign that these larvae are present. Continued feeding gives the turf an extremely close-cropped appearance. Damage appears as patches that become yellowish, then brown. Turf adjacent to flower beds and shrubs usually shows the first signs of damage, since adults rest in such foliage and moths lay more eggs in nearby turf (Kerr 1955).

Damage may be seen in southern Florida in the spring. By late summer, webworms are active throughout the state and may continue to cause injury into November. Much of the damage attributed to armyworms is actually caused by the TSW (Kelsheimer and Kerr 1957). Damage is most common during the "rainy season" in south Florida.

Description of Stages

Adult. The dingy brown moths have a wingspread of about 20 mm. Like a near relative, the GWW, the TSW moth can be distinguished from members of the Crambinae by the fact that they do not roll their wings about their bodies when they are at rest (Plate 21). As in the GWW, the males have a slimmer abdomen and six visible segments, while the females have five visible segments, with the terminal segment ending in a large, fusiform opening (Kerr 1955).

Egg. TSW eggs share several characteristics with the GWW. Eggs appear in clusters of 6–15, but some are deposited singly on grass blades or on moist surfaces. Eggs in a group partially overlap each other, like shingles. Each egg is flat, rounded, about 0.70–0.73 mm in widest diameter, and about 0.1 mm high. Fresh eggs are whitish and become brownish red just before hatching. The head of the larva is plainly visible (Plate 21) (Kerr 1955).

Larva. The body is dingy cream but appears mostly green when larvae are feeding. The head is dark yellowish brown. There are seven or eight larval instars that grow from just over 1 mm long at the time of hatch to about 19 mm long at maturity (Plate 21) (Kelsheimer and Kerr 1957, Kerr 1955).

Pupa. Pupae lie freely or partially buried in duff or are enclosed in a shapeless bag that the mature larva spins, using bits of grass and soil particles. At maturity pupae are reddish brown, about 8.5–9.5 mm long and about 2.1–2.9 mm wide (Kerr 1955).

Life History and Habits

Life Cycle. Populations of the TSW are present throughout the year in southern Florida, with the highest numbers present in late summer and fall. In Gainesville (in northern central Florida), peak adult emergence occurs in October and November. It appears that pupae and adults cannot survive the winters north of Gainesville. South of Gainesville there appears to be no single overwintering stage; all stages are present throughout the year. At 25.5°C, the TSW develops from egg to adult in about 6 weeks, spending 6–10 days as an egg, about 25 days in the seven larval instars, and about 7 days as a pupa during the summer (Kerr 1955, Reinert 1973).

Adult Activity. During the day most moths rest in shrubbery around lawns and when disturbed will make weak short flights before settling down again. At dusk, moth flights are stronger and longer. Moths lay their eggs in the grass largely at dusk (Kelsheimer and Kerr 1957, Kerr 1955).

Moths require liquid food for survival and for oviposition. No eggs were laid until moths were fed a sucrose and yeast solution; females lived about 14 days. Females that received only water lived only 7 days (Kerr 1955).

Larval Activity. Larvae feed at night or on overcast days. Newly hatched larvae feed along the midrib lengthwise, and consume only surface cells. After larvae are 10–12 days old, they devour all portions of leaves. Typically, larvae with seven instars require about 25 days for completion of larval life at 25.5°C. Those with eight instars require 45–50 days for completion at 23°C. Larvae typically rest in a tightly curled position. Mature larvae spin shape-less silken bags with bits of grass and soil (Kerr 1955).

Miscellaneous Features

Plant Resistance. There is some evidence that certain bermudagrass strains have resis-tance to TSW oviposition and feeding. When seven clonal selections were exposed for oviposition, fewer adults emerged from two plant introduction selections of South African origin. The least foliage damage occurred to one of these selections plus common bermuda-grass and to the hybrid FB-119 (Reinert and Busey 1983).

Natural Enemies

The only parasitoid mentioned is an ichneumonid wasp, *Horogenes* sp. (Kerr 1955).

9

Lepidopteran Pests: Family Noctuidae

Cutworms and Armyworms

Taxonomy

The popular names *cutworm* and *armyworm* describe larval habits. In many situations *cutworms* sever young plants at or near ground level, do no additional feeding on the plant, and proceed to the next plant to repeat the process. This feeding pattern contributes greatly to the highly destructive nature of these insects.

Larvae of some species under high population pressures move army-like across a field, devouring all tender plants in their path. Species with this habit are called *armyworms.*

Of the many armyworms and cutworms that feed on grasses, relatively few are reported as pests of turfgrass in the United States. Those listed as turfgrass pests belong to the order Lepidoptera, in three subfamilies of Noctuidae (Crumb 1956, Hodges et al. 1983):

> Subfamily Noctuinae
>   *Agrotis ipsilon* (Hufnagel), black cutworm (BCW)
>   *Peridroma saucia* (Hubner), variegated cutworm (VCW)
> Subfamily Hederinae
>   *Pseudaletia unipuncta* (Haworth), armyworm (AW)
>   *Nephelodes minians* Guenee, bronzed cutworm (BZCW)
> Subfamily Amphipyrinae
>   *Spodoptera frugiperda* (J. E. Smith), fall armyworm (FAW)
>   *Spodoptera ornithogalli* (Guenee), yellow-striped armyworm (YAW)
>
>   *Spodoptera mauritia* (Boisduval), lawn armyworm (LAW)

Many of these cutworm and armyworm species are significant pests in agricultural crops and have undergone numerous name changes that may cause confusion when reviewing older literature. The BCW was called the *greasy cutworm* in North America, and in Great Britain its approved common name is the *dark sword grass moth.* The VCW has also been called the *common cutworm* and the *alfalfa cutworm,* and the moth has been called the *unarmed rustic* in the United States. In England the adult is named the *pearly underwing moth.* *Peridroma margaritosa* (Haworth) is a binomial synonym of VCW often used in past and in relatively recent literature. The AW was listed under the genus *Cirphis,* and the FAW has been listed under the genus *Laphygma* (Okamura 1959; Rings et al. 1974a, 1976).

Importance

All seven noctuid species feed on grasses as a primary host. Though BCW larvae can damage lawns and grounds, it is the most significant caterpillar pest found on golf course putting greens and tees (Shetlar 1995b). In turfgrass culture, the VCW, BZCW, and AW are considered as occasional pests that have the potential for serious outbreaks on turfgrass. The FAW and YAW are principal turfgrass pests on the mainland only in the Southeast. These species, as well as the BCW, have a wide host range as major pests of agricultural crops over vast areas of the United States.

The LAW is a serious turfgrass pest in Hawaii. It does not occur on the mainland. Because it is restricted to bermudagrass and to Hawaii, the LAW is discussed separately in the second section of this chapter (Oliver 1982b, Tanada and Beardsley 1958).

History and Distribution

The six species under discussion here have varied but wide distribution over large areas of the continental United States and Canada (Figures 9-1, 9-2, and 9-3).

Host Plants and Damage

Cutworms and armyworms feed on grasses as one of their primary host plants, but most species are also serious pests of other agricultural crops, feeding principally on foliage at the soil surface or on any portion above ground. In some species larvae become subterranean, feeding in the crown and underground fleshy structures.

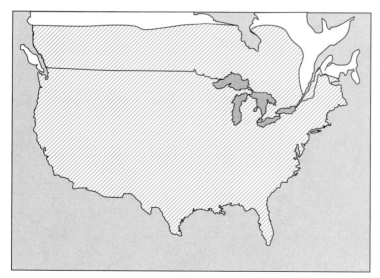

Figure 9-1. Distribution of cutworms. (Drawn by E. Gotham, NYSAES, adapted from Handbook of Turfgrass Insect Pests, Brandenburg and Villani 1995, Entomological Society of America.)

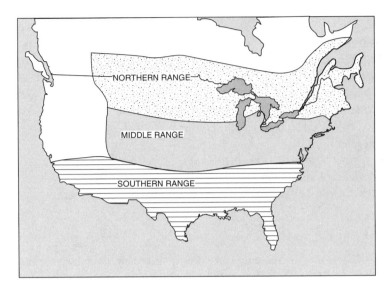

Figure 9-2. Distribution of armyworms, showing northern, middle, and southern ranges. (Drawn by E. Gotham, NYSAES, adapted from Handbook of Turfgrass Insect Pests, Brandenburg and Villani 1995, Entomological Society of America.)

Description of Stages

Adults

Cutworm and armyworm moths have forewings with a background color of dull brown, gray, and black, with few outstanding prominent identifying marks. Figure 9-4, adapted from Forbes (1954) and Rings (1977), showing a schematic wing pattern of noctuid moths, should assist in identifying the various markings on forewings of the species being discussed. The orbicular and reniform spots are rather indistinct but distinguishable in these species and form the single most reliable identifying mark. The descriptions of Chapman and Lienk (1981) appear below in abbreviated form (also see Plate 22).

The hindwings of all six species have a similar appearance, with a whitish to brownish back-

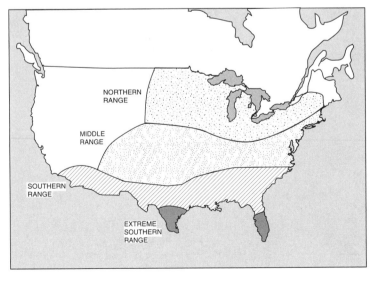

Figure 9-3. Distribution of fall armyworms, showing northern, middle, southern, and extreme southern ranges. (Drawn by E. Gotham, NYSAES, adapted from Handbook of Turfgrass Insect Pests, Brandenburg and Villani 1995, Entomological Society of America.)

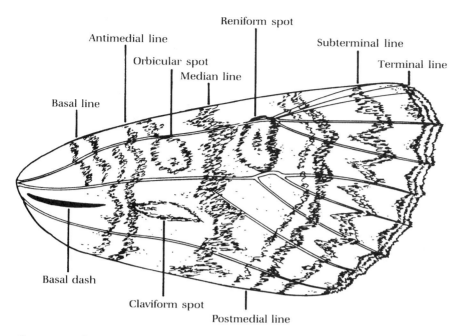

Basal line
Antimedial line
Orbicular spot
Median line
Reniform spot
Subterminal line
Terminal line
Basal dash
Claviform spot
Postmedial line

Figure 9-4. Schematic wing pattern of noctuid moths to show principal lines, spots, and dashes. (Adapted from Rings 1977, fig. 1, courtesy of the Ohio Agricultural Research and Development Center.)

ground color and brownish visible veins. The wingspan of all six species falls within a rather close range and is of little value in separating these species. The FAW, the smallest, has a wingspan of 25–40 mm, while BZCW, the largest, has a wingspan of 34–50 mm (Rings 1977).

Eggs

Cutworm eggs are circular, oblate, slightly wider than tall, and fairly similar in size (Plate 23). They are 0.47–0.58 mm broad by 0.39–0.50 mm high for the three species; the eggs of the FAW are smallest. When freshly laid, the eggs of the BCW, VCW, and YAW are white, while they are pale green to light gray in the BZCW, AW, and FAW. As they age, eggs darken to tan, gray, and dark brown to blackish before hatching (D. Shetlar, Ohio State University, personal communication, 1998).

Eggs usually are deposited generally in masses by the BZCW, VCW, AW, FAW, and YAW. The egg masses of the BZCW and the VCW remain naked, but those of the LAW and YAW are covered with abdominal hairs of the female. Females of VCW, FAW, YAW, and LAW deposit eggs indiscriminately on their host plants, on twigs, on fence posts, on buildings, and on other objects (Baker 1982, Crumb 1929, Luginbill 1928, Oliver 1982b). On creeping bentgrass, the BCW lays most of its eggs singly on the tips of grass blades (Williamson and Potter 1997b).

Larvae

Larvae of the six species are relatively thick bodied, nonhairy caterpillars often marked with stripes but never spotted like webworms. In addition to three pairs of true legs on the thorax, these caterpillars have five pairs of fleshy prolegs on the abdomen (Shetlar 1995a).

The most cryptic of the larvae are those of the BCW, which are dark gray to nearly black on the dorsum and lighter gray on the ventral surface. Other species tend to be brighter, with shades of light pink, yellow, and green. All cutworms and armyworms have a dorsal-median stripe that may be continuous or broken. All except the BCW have conspicuous subspiracular stripes of various widths and shades. All have black spiracles except the FAW, which has pale spiracles surrounded by a whitish ring. When full grown, the larvae are fairly large, ranging in length from about 30 to 45 mm and in maximum width from 4.5 to 9.0 mm (Rings and Musick 1976).

The general color of each species and the color and location of the markings, especially of the stripes, are diagnostic, despite variations within species (Plates 22–24). Specific information on coloration, markings, and measurements as presented by Rings and Musick (1976) appears below.

Life History and Habits

Seasonal Cycles

Four of the six species discussed in this section typically overwinter in the southernmost regions of North America, and the annual infestations in the temperate regions result from annual northward migrations of moths. Winter conditions dictate the date of arrival of moths in the Northeast. The annual migrant species include the BCW, VCW, YAW, and FAW. Only the BZCW and AW are indigenous to areas where they occur in the United States. The BZCW overwinters as eggs and young larvae (Chapman and Lienk 1981, Shetlar 1995a, Walkden 1950), while the AW overwinters as pupae (Shetlar 1995a).

Blacklight trap catches determined the annual appearance of moths in Ohio and New York. Figure 9-5 illustrates the seasonal presence of adults. In Ohio, three species whose adults are present from April into November, the AW, BCW, and VCW, have three peaks of adult abundance, indicating three generations each season. Moths of the same three species are present in New York for a shorter duration because of their later arrival. These three species have only two generations a season in New York (Chapman and Lienk 1981, Rings 1977). Based on pheromone trap catches, BCW moths arrive in New England in late April or early May most years (S. W. Swier, University of New Hampshire, personal communication, 1998).

The number of generations of each species depends primarily on latitude. The period required to complete a full generation varies considerably among the multivoltine species. The AW requires approximately 60 days, the BCW about 40–65 days, and the FAW only 23–28 days to complete a generation. The VCW requires 58–69 days in spring, 57–78 days in summer, and 101–152 days in the fall (Oliver and Chapin 1981, Vickery 1929, Walkden 1950). Stages in the life cycle of the BCW are illustrated in Plate 23.

Adult Activity

Adults of the six species have only a few habits in common. All are attracted to light, especially black light.

Larval Activity

Generally speaking, cutworms are nocturnal in habit, although some late instars of some species are active in daylight. Many armyworm species are active any time of the day. On

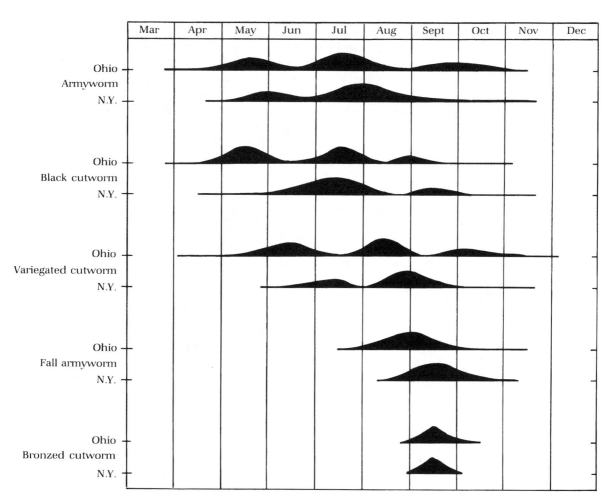

Figure 9-5. Seasonal distribution of cutworm and armyworm moths as determined by blacklight trap catches in Ohio and New York. (Data from Chapman and Lienk 1981 and Rings 1977; drawn by R. McMillen-Sticht, NYSAES.)

maintained turf that is mowed regularly, FAW larvae normally are most active in twilight and after dark (Cobb 1992).

Miscellaneous Features

Rearing Methods

The BCW has been reared by collecting moths from light traps to start cultures. The adult diet consisted primarily of water, beer, and honey. The larval diet was mainly water and pinto beans plus a few "minor" ingredients, which were blended, cooked, and placed in containers. At 21° ± 1°C, 16-hr–8-hr light-dark cycle, moths began ovipositing 4 days after emergence (Harris et al. 1962, Reese et al. 1972). Larvae can be purchased from commercial companies for use in research.

Population Survey Techniques

Blacklight and pheromone traps can be used to monitor adult flight activity. While trap catches are not well correlated with subsequent larval populations, traps serve as important monitoring tools, determining when peak moth flights occur and enabling a turf manager to anticipate when oviposition will begin or larvae will begin to emerge.

Irritating drenches, or soapy flushes (see Chapter 26), are effective at forcing caterpillars to the surface, where they can be counted. The technique is particularly useful for determining the size distribution of a population and enables a turf manager to decide whether control strategies are likely to be effective.

Thresholds

Action thresholds, or tolerance levels, for cutworms and armyworms vary widely and are seldom published. In some instances a few (<10) damage marks on one putting green will be considered unacceptable, while in other less highly maintained turfgrass, one or more caterpillars per square foot might be tolerated.

Effect of Endophytic Grasses on Cutworms and Fall Armyworms

A laboratory study indicated that endophyte-enhanced resistance to the FAW varies among and within species of grasses and endophytes. Third-instar FAWs showed no difference in rate of consumption, body weight, survival, or developmental time when fed tall fescue infected with *Neotyphodium* (=*Acremonium*) *coenophialum* Morgan-Jones & Gams, compared with endophyte-free tall fescue. However, FAW development was delayed when third instars fed on perennial ryegrass infected with *Neotyphodium lolii* Latch Christensen & Samuels, or on hard fescue or Chewings fescue infected with *Neotyphodium* spp. compared with identical endophyte-free genotypes. In contrast, southern armyworm (*Spodoptera eridania* [Cramer]) preferred and had improved development on endophyte-infected tall fescue (Breen 1993).

A laboratory bioassay in Kentucky demonstrated there was no difference in body weight, developmental time, or survival of the BCW when fed tall fescue infected with *N. coenophialum* or perennial ryegrass infected with *N. lolii* compared with identical endophyte-free genotypes. This suggests that endophyte-infected turfgrasses are not likely to suppress BCW populations significantly (Williamson and Potter 1997d).

Turfgrass Species Preferences

A laboratory study suggests that BCW larval survival is significantly poorer on Kentucky bluegrass than on other cool-season grasses. In nonendophyte trials, neonate larvae preferred creeping bentgrass over Kentucky bluegrass or tall fescue. Food preferences of fifth instars were less consistent, but generally they preferred bentgrass to Kentucky bluegrass (Williamson and Potter 1997d).

Natural Enemies

Commercial formulations of the bacterium *Bacillus thuringiensis* v. *kurstaki* and the entomopathogenic nematodes *Steinernema carpocapsae* and *Steinernema riobravis* can be used to manage cutworm and armyworm populations and can be effective, particularly when most of the population is in early instars at the time of application.

All six species have hymenopteran parasitoids, among which *Apanteles* spp. were common. The five species were also parasitized by tachinid flies of many genera (Walkden 1950).

Black Cutworm, *Agrotis ipsilon*

History and Distribution

One of the most cosmopolitan of species, the BCW is distributed throughout much of the Americas, Europe, Asia, and Africa. Nevertheless, until recently little was known of the ecology and life history of this important pest. Williamson and Potter (1997a,b,c) have identified oviposition habits and larval behavior of the insect on turfgrass. Both adults and larvae are commonly encountered in the field, but there is a poor correlation between blacklight trap or pheromone trap catches and the subsequent abundance of larvae at any given location. The BCW is most troublesome in temperate regions where bentgrass is grown. It appears to be a subtropical species incapable of overwintering where soils freeze, but spring migratory flights of adults can lead to damaging infestations as far north as Canada (Chapman and Lienk 1981, Okamura 1959, Shetlar 1995a, Troester et al. 1982).

Host Plants and Damage

The BCW, an economically important pest, feeds primarily on grasses. It is a serious pest of golf course greens, where the larvae make linear or irregular sunken spots in the turf surface. There is a very low threshold for the BCW on golf course greens because of the damage it causes. On California lawns it feeds on all common grasses, on dichondra, and on white clover. It is a major pest of corn in the United States (Harris et al. 1962, Rings 1977, Troester et al. 1982).

Description

BCW moths are dark gray mottled with black, some having brown also. The male has pectinate antennae and the female has filiform antennae. On the forewings a dagger-shaped black marking appears at the outer edge of the black-lined reniform spot.

The general color of the BCW caterpillar above the spiracles is nearly uniform, ranging from light gray-green to nearly black. The lower half of the body is lighter gray but not strikingly so. A pale and indistinct middorsal line is evident. The spiracles are black. Under 15× magnification, the BCW integument has a diagnostic rough surface resembling a cobblestone road (D. Shetlar, personal communication, 1998). Body length of late instars is 30–45 mm, and the width is 7 mm (Figure 9-6, plate 23).

Seasonal Cycles

The BCW in Louisiana has five to six overlapping generations each year, with adults found every month. Larvae are present from March into December and overwinter primarily as pupae. Four broods are reported in northern Tennessee and three in the central Great Plains and in Missouri. New England often experiences two to three generations per year, depending on overwintering conditions and spring and summer temperatures (Crumb 1956, Oliver and Chapin 1981, Satterthwait 1933, Walkden 1950).

Adult Activity

In field crops, adults of the BCW have an oviposition preference for curled dock, *Rumex crispis* L., and yellow rocket mustard, *Barbarea vulgaris* R. and Br. The basis for the attrac-

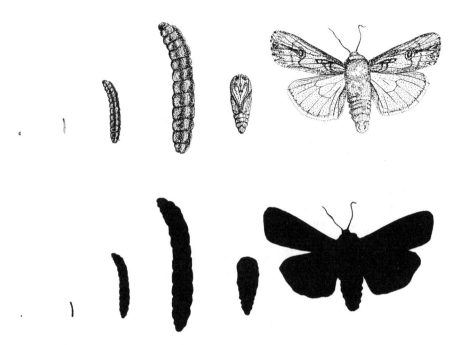

Figure 9-6. Stages of development of black cutworm: egg; first, third, and final larval instars; pupa; and adult. Silhouettes show actual size of each stage. (Courtesy of R. Jarecke, NYSAES.)

tion appears to be the low, dense growth forms of these plants. The petioles and the lower surface of leaves receive the most eggs. In turf BCW females attach one to several eggs to the distal ends of grass blades, especially bentgrass, and show no preference between areas mowed at 5 mm and 13 mm (Shetlar 1995c, Williamson and Potter 1997b, Williamson and Shetlar 1995). During August and September in western New York, dirty brown to black moths that are flushed out when the lawn is being mowed and fly only 0.5 to 1 m before alighting are most commonly BCW moths (Busching and Turpin 1976; Chapman and Lienk 1981; H. Tashiro, personal observations).

Larval Activity

Most BCW eggs are laid singly on the distal 25% of creeping bentgrass blades. Because most golf course superintendents collect clippings when mowing putting greens, many BCW eggs are removed with mowing and discarded with the clippings. Therefore, most BCW infestations observed on greens probably originate from BCW oviposition in peripheral areas, such as collars, aprons, approaches, fairways, and roughs (Williamson and Potter 1997b, Williamson and Shetlar 1995).

Plots where BCW females were caged over turf maintained at 3.2, 4.8, or 13.0 mm were mowed 48 hr after oviposition. Mowing removed 75–91% and 81–84% of eggs at the 3.2-mm and 4.8-mm cutting heights, respectively, while 5–10% of the eggs were dislodged from grass blades by the mower roller. Hatching rates of dislodged eggs approached 90%, demonstrating that BCW eggs survive the mechanical mowing process. Manipulation of mowing height does not deter oviposition. Clippings with eggs that are discarded near greens or tees may serve as a source of subsequent larval reinfestation of those sites (Williamson and Potter 1997b).

Small larvae (first through third instars) feed primarily on the surface, crawling on putting greens and remaining exposed while feeding. Large larvae (fourth through sixth instars) feed primarily from within a turf burrow. This feeding pattern often results in characteristic pockmarks resembling ball marks on putting greens. Turf managers often report increased damage from BCW shortly after aerification. However, studies show that the BCW does not prefer (seek out) aerified over nonaerified creeping bentgrass, although a large number will exploit aerification holes as burrows (Williamson and Potter 1997a,c).

Fifth instars are strongly nocturnal, with feeding occurring mainly from 1 hr after sunset to 1 hr after sunrise. Williamson and Shetlar (1995) suggested that sampling with an irritating solution might be most effective shortly after sunrise, just before larvae terminate feeding. Williamson and Potter (1997a) suggested that during times when late-instar BCWs are feeding actively, mowing putting greens well before sunrise may result in significant mechanical control of BCWs.

BCW larvae that are crowded are highly cannibalistic. Pupal cells are formed in the soil by rolling movements (Archer and Musick 1976, Crumb 1929, Harris et al. 1962, Satterthwait 1933).

Variegated Cutworm, *Peridroma saucia*

History and Distribution
Another widely distributed species, the VCW is found throughout North and South America and over much of Eurasia. Adult trap catches in New York correlated well with its pest status as larvae (Chapman and Lienk 1981, Okamura 1959, Rings et al. 1976).

Host Plants and Damage
On lawns, the VCW feeds on bentgrass and white clover in California, but since it has a wide host range, it probably feeds on most grasses, whether grown in lawns or elsewhere. It attacks many field and forage crops, vegetables, and ornamental plants. At times it becomes a climbing cutworm, feeding on buds and leaves of fruit trees (Okamura 1959, Rings 1977, Walton 1929).

Description
The ground color of the VCW moth is gray to reddish brown except for the blackish costal margin. Although they appear obscure to the unaided eye, under some magnifications the reniform, orbicular, and the claviform spots appear well formed and are outlined incompletely with black and some white. These sets of "quotation marks" are present along the front margin of the wing.

VCW caterpillars range in ground color from dark brown to gray. The middorsal line is broken, leaving four to seven whitish to yellowish dashes. Both a narrow orange-brown spiracular stripe and a black W-shaped mark on the dorsum of the eight abdominal segments are usually present. Spiracles are black (Plate 22). The body length of late instars is about 40 mm; the width is 6 mm.

Seasonal Cycles
Adults of the VCW are active throughout the year in Louisiana and have four or five generations annually. The VCW overwinters primarily as pupae. Four generations occur in Tennessee, and three with a partial fourth occur in Kansas. Pupal survival is low even dur-

ing mild southern winters, but the VCW's enormous reproductive capacity ensures high annual populations. The VCW is capable of overwintering in cool-season turf areas (Crumb 1929, Oliver and Chapin 1981, Shetlar 1995a, Walkden 1950).

Adult Activity

In Kansas, adult VCW moths can be collected every month except during December to February. It is difficult to identify generations, because all stages are found the rest of the year. Of all the cutworms and armyworms observed, the VCW has the greatest fecundity of all. Each female lays 1,185–2,696 eggs, with a mean of 2,111 eggs.

Larval Activity

When large numbers are present, the VCW assumes armyworm habits. Larvae are not strictly nocturnal and frequently feed during cloudy days. A population of late-instar VCWs can completely destroy all the grass blades over several hundred square feet in one night (D. Shetlar, personal communication, 1998).

Armyworm, *Pseudaletia unipuncta*

History and Distribution

The AW extends south of the United States into Mexico and northwestern South America. It is found throughout the United States east of the Rocky Mountains and rather sparingly in the Pacific Coast states and into British Columbia (Crumb 1956, Okamura 1959, Walkden 1950).

Host Plants and Damage

Larvae of the AW feed almost exclusively on grasses, including all lawn grasses and also white clover in California, and are especially abundant in damp situations. The AW is a more serious pest of corn and small grains and is known to cut off the seedheads of wheat. When grass and small grains are not available, it will eat vegetables and ornamental plants. Periodically large numbers of larvae develop and move army-like from decimated fields to new fields, in search of food (Baker 1982, Chapman and Lienk 1981, Okamura 1959, Oliver and Chapin 1981, Rings 1977).

Description

Forewings of AW moths are a fairly uniform light tannish to reddish brown. The most distinctive diagnostic character is the small conspicuous white diamond-shaped dot in the center of the forewings in the lower angle of the reniform spot.

The yellowish or gray ground color of the AW caterpillar is more or less tinged with pink. The dorsum of the larva is greenish brown to black, with a narrow, broken, light median stripe. A dark stripe includes the black spiracles in its lower edge. The subspiracular stripe is pale orange, mottled, and edged with white. The body tapers posteriorly and is about 35 mm in length and 5 mm wide in the middle in late instars (Figure 9-7; Plate 22).

Seasonal Cycles

The AW has up to four generations in Louisiana and overwinters primarily as pupae, with the largest populations of late-instar larvae appearing from late March into May (Oliv-

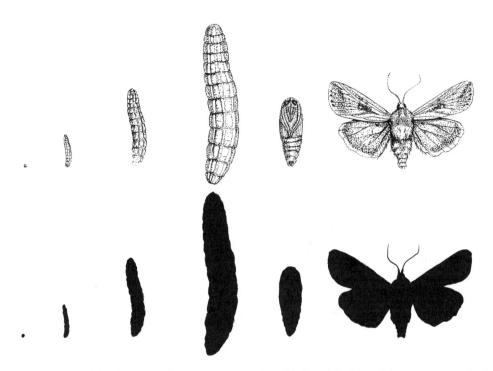

Figure 9-7. Stages of development of armyworm: egg; first, third, and final larval instars; pupa; and adult. Silhouettes show actual size of each stage. (Courtesy of R. Jarecke, NYSAES.)

er and Chapin 1981). It has at least five generations each year in North Carolina, where outbreaks appear and disappear suddenly, while it overwinters as larvae in northern locations (Baker 1982, Shetlar 1995a).

Adult Activity
AW moths oviposit at night in clusters with as many as 130 eggs between the sheath and the blade of grass. Each female can lay as many as 2,000 eggs (Baker 1982, Walkden 1950).

Larval Activity
The AW invades grass in wet areas and may become most abundant after flooding has occurred. Plants that have lodged to make a dense canopy are often infested with the highest numbers. This species is a true armyworm; the larvae migrate en masse from a decimated area to enter an area of abundant food supply (Oliver and Chapin 1981, Walkden 1950).

Bronzed Cutworm, *Nephelodes minians*

History and Distribution
The BZCW is found generally in North America east of the Rocky Mountains between latitudes 35° and 50°. Its greatest abundance occurs from Missouri to Maine and from Min-

nesota to West Virginia, with Colorado, Kansas, Missouri, Tennessee, and Virginia its approximate southern range (Chapman and Lienk 1981, Rings et al. 1974b, Walkden 1950).

Host Plants and Damage

The BZCW prefers bluegrass and ryegrass lawns and frequently appears on turf and pastures throughout its region. On bluegrass it may be found deep within the crown. It also feeds on clover, small grains, and corn. Often the BZCW becomes a climbing cutworm to eat the buds and leaves of fruit trees (Chapman and Lienk 1981, Rings 1977, Walkden 1950). Turf damage can occur under cover of snow and in early spring (D. Shetlar, personal communication, 1998).

Description

The ground color of the BZCW adult is somewhat variable but is usually rose, purplish gray, or brown. A wide, rich, darker-brown band crosses the center of the wing. Large diffuse patches make up the orbicular and reniform spots. This moth is found only from early September to early October.

The first three instars of the BZCW larvae are green (D. Shetlar, personal communication, 1998); later instars are light to dark brown to blackish above and paler below, usually with a distinct bronzy sheen. Its middorsal stripe is yellow, broad, and sharply defined. The subspiracular stripe is broad and pale. All spiracles are black. The pronotum has a distinctive set of dark and light stripes (Plate 22). The body length of late instars is 35–45 mm and the width is 9 mm at the middle.

Seasonal Cycles

Moths of the BZCW are present from mid-September to early October in Kansas to New York, oviposit shortly thereafter, and overwinter as eggs (Walkden 1950). Occasionally eggs will hatch under the cover of snow and small larvae may remove all the green leaves by the time the snow melts. Larvae mature from mid-May to mid-June, depending on spring temperatures. Mature larvae dig into the soil and form summer aestivating pupae (D. Shetlar, personal communication, 1998). The BZCW has moths appearing mainly in September and peaking at the same time in Ohio and New York (Chapman and Lienk 1981, Rings 1977).

Fall Armyworm, *Spodoptera frugiperda*

History and Distribution

The FAW is very susceptible to extreme cold. During mild winters it is capable of surviving in the United States only in the southernmost regions of Florida and Texas. The FAW is a permanent resident in South and Central America and in the West Indies. Like the BCW, every season the FAW spreads from these areas throughout the United States east of the Rocky Mountains and westward into southern California, Arizona, and New Mexico (Baker 1982, Luginbill 1928, Vickery 1929).

Host Plants and Damage

FAW and YAW larvae prefer grasses and feed on bermudagrass, fescue, ryegrass, and bluegrass, consuming all plant parts that are above the ground. The tips of the grass blades may appear transparent where the plant cells have been eaten, leaving only a thin membrane. Damage on bermudagrass resembles drought damage and recovery is possible, while feeding on newly established ryegrass or fescue may result in severe stunting or death (Cobb 1992).

Description

Moths of FAW show distinct sexual differences on the forewings similar to those of the LAW discussed in the next section, with the wings of the male being more vividly marked. Wings of both sexes are generally gray with white markings, including a teardrop-shaped light mark running from the middle of the forewing to the hind tip margin. Males have a prominent, pale diagonal marking over the orbicular spot and extending to the costa. Forewings of the female are dull gray brown with small oblique, oblong inconspicuous orbicular spots.

The general color of the FAW caterpillar ranges from pinkish to yellowish, greenish, and dull gray to almost black. There is a faint, narrow, pale middorsal stripe. A broad, sharply defined, yellowish or whitish subspiracular stripe is mottled with reddish brown. The body length of late instars is about 30 mm and the width about 4.5 mm, with all abdominal segments about equal in width (Figure 9-8, Plate 24). The top of the cranium has a light-colored inverted Y. The dorsum of the eighth abdominal segment has four distinct spots.

Seasonal Cycles

The FAW is the species that is most susceptible to the cold. The ability to survive mild winters in Louisiana is so uncertain that more southern tropical areas are the main source of annual migrants. As few as 4 generations occur each year in Louisiana, while 9–11 generations occur in Brownsville, Texas, where all stages are present during the entire year. There is only 1 generation in Kansas, and larvae are most numerous during late August

Figure 9-8. Stages of development of fall armyworm: egg; first, third, and final larval instars; pupa; and adult. Silhouettes show actual size of each stage. (Courtesy of R. Jaecke, NYSAES.)

and September (Oliver and Chapin 1981, Vickery 1929, Walkden 1950). The FAW, dependent on annual northward migration of moths, has a single generation rather late in the season in both Ohio and New York (Chapman and Lienk 1981, Rings 1977).

Development from egg to full-grown larva takes 2–3 weeks in Alabama. Larvae then burrow into the soil to pupate and emerge as adults 10–14 days later. The FAW overwinters in Alabama as pupae (Cobb 1992).

Adult Activity

Moths of the FAW are noticed mostly at night, as they are attracted to lights. Each female lays about 1,000 eggs in masses of as many as 400 eggs two to four layers thick and covered with abdominal hairs. Females attach egg masses to goal posts, flag poles, and flags on golf courses, or glue them to the undersides of tree leaves that overhang grassy fields. Oviposition begins shortly after dark and lasts until about midnight. Adults live 1–3 weeks (Cobb 1992, Luginbill 1928, Oliver 1982b, Vickery 1929, Walkden 1950).

Larval Activity

First-instar FAW larvae spin silken threads as soon as they have hatched. The threads assist in wind dispersal or allow the insects to lower themselves to the turf below. First instars feed only on the undersurface of leaf blades, leaving behind the upper, clear epidermal layer. Larvae feed any time of the day but are most active early in the morning or late in the evening. Later-instar larvae eat the entire leaf area, giving the turf a ragged appearance. The larvae prefer lush, dense grass. They leave cut leaf particles strewn about, and when abundant, the FAW assumes the true armyworm's habit, moving en masse from a heavily damaged area to a fresh source of food (Baker 1982, Oliver 1982b).

Yellow-Striped Armyworm, *Spodoptera ornithogalli*

History and Distribution

The YAW is very susceptible to extreme cold, and overwinters only in the southernmost regions of Florida and Texas. Like the FAW, it is a permanent resident in South and Central America and in the West Indies.

Host Plants and Damage

The FAW and YAW are major turfgrass pests in the southern United States. Home lawns and golf course fairways of bermudagrass have been damaged in Louisiana, southern Texas, and California, but the FAW also feeds on bentgrass, bluegrass, and white clover. Both species migrate northward over the summer and can cause damage similar to the BCW on northern golf courses in late July through September (Shetlar 1995a). In addition, they are serious pests of small grains and corn. Moths attracted to light may become so numerous around buildings that they soil windows and white walls, so that the surfaces require scrubbing or even repainting (Baker 1982, Okamura 1959, Oliver 1982b, Oliver and Chapin 1981).

Description

Like the FAW, YAW moths have different wing markings depending on sex. Males are similar to the FAW in having a teardrop-shaped light mark running from the middle of the forewing and coalescing with a large pale area along the hind margin. There is also a distinct light V-shaped mark at the wing base. Females resemble FAW females.

YAW larvae vary from green to brown. Most have a yellow or cream-colored stripe on each side of the dorsum. Between this stripe and the middorsum, several dark triangles point inward on each abdominal segment. A distinctive dark area surrounds the first abdominal spiracle in light-colored specimens.

Seasonal Cycles
The YAW, dependent on annual northward migration of moths, has a single generation rather late in the season in Ohio and New York (Chapman and Lienk 1981, Rings 1977).

Lawn Armyworm

Taxonomy

The LAW, *Spodoptera mauritia* (Boisduval), belongs to the order Lepidoptera, family Noctuidae, subfamily Amphipyrinae. Most of the information about it comes from Tanada and Beardsley (1958). Other work is cited only when it supplements that by these authors.

Importance

Within 1 year of its discovery in Hawaii, the LAW became the most serious pest of bermudagrass lawns in that state. It continued to be the most severe lawn and turfgrass pest during the 1960s, but in more recent years its populations have stabilized, apparently because of numerous parasitoids and predators. The LAW is now considered one of four major turfgrass pests in Hawaii (LaPlante 1966a; H. Tashiro personal observations).

History and Distribution

The LAW is a native of the oriental, Indo-Australian, and Pacific regions. It apparently arrived on the island of Oahu in Hawaii well before 1953, when it was first recorded correctly. It is now present on all the Hawaiian islands but is not known to occur anywhere else in the United States (Fletcher 1956, Pemberton 1955, Tanada 1955).

Host Plants and Damage

In Hawaii, the LAW has inflicted most of its damage on bermudagrass lawns and golf courses. It will also feed on sedges, sugarcane seedlings, zoysiagrass, and several grassy weeds. Severe damage to lawns is characterized by a sharply defined front of undamaged turf and a more or less completely denuded area (Plate 25). With heavy populations of actively feeding larvae, the front may move about 30 cm each night. The insects leave no leaves or stems behind them in their path (LaPlante 1966a).

Description of Stages

Adult
Males of the LAW are more vividly marked than females and have a conspicuous white diagonal mark in the anterior median area of the forewing between the whitish to buff orbicular spot and the roughly reniform dark spot (Plate 25). In the female the dark reniform

spot on the forewing is well defined. Hindwings are pale except for a darker costal margin and outer margins in both sexes. The dorsum of the thorax is covered with elongate gray-ish- to reddish-brown scales. The front legs, clothed with brushes of hairlike scales, are much more strongly developed in the males than in the females. Wingspan is 30–37 mm in males and 34–40 mm in females.

Egg

Egg masses (Plate 25), cemented to leaves of trees, on buildings, or on other objects, are elongate-oval in outline, with five or more irregular layers of eggs. Young females cover their egg masses with long, light-brown hairs from their abdomen, so that individual eggs are not visible. As the female ages and her abdominal hairs are exhausted, the last egg mass-es have somewhat naked eggs (Plate 25). There may be 600–700 eggs in a mass. Some egg masses acquire a greenish or pinkish cast (LaPlante 1966a).

Individual eggs are light tan with a pearly luster and darken to gray or dark tan before hatching. Eggs are circular and somewhat flattened and sculptured, with fine longitudinal striations. Each egg is about 0.5 mm in diameter and about 0.4 mm through the polar axis.

Larva

The larvae of the LAW have seven or eight instars in either sex. First instars are about 1.24 mm long, with a head capsule width of about 0.30 mm. Mature larvae are 35–40 mm long, with head capsule widths of about 2.8 mm in the seventh instars and 3.5 mm in the eighth instars.

First instars become greenish soon after feeding and remain predominantly green as sec-ond and third instars (Plate 25). Patterns and stripes characteristic of mature larvae devel-op in the fifth instar. Mature larvae are typically smooth skinned and vary considerably in color, from brown to purplish brown and even blackish (Plate 25). The head capsule and pronotal shield are dark brown. The dorsal-median stripe varies in color. A pair of promi-nent jet-black marks occur on each body segment except the prothorax and terminal-seg-ment. The spiracles are black.

Pupa

Pupae of the LAW have the same general appearance as those of other armyworms and cutworms and are reddish brown when fully hardened (Plate 25). They average 16 mm in length and 4.5 mm in width.

Life History and Habits

Life Cycle

Development of the LAW is continuous; there is no overwintering stage. The entire life cycle from egg to adult requires about 42 days. Moths have a preoviposition period of near-ly 4 days, eggs hatch in about 3 days, the larval period lasts nearly 28 days, and the pupal period averages nearly 11 days.

Adult Activity

Moths mate within a day after eclosion and start laying eggs about 4 days later. Oviposi-tion begins shortly after dusk and is generally completed before midnight. Eggs are deposited on the foliage of shrubs or small trees, on the lower leaves of tall trees, or on buildings. The moths rarely lay eggs on grass. Since adults are attracted to light, egg masses are often on buildings and foliage near outdoor lights. When the moths are fed sugar water, they live 9–14 days.

Larval Activity

Only first to fifth instars are seen feeding during the day. The older larvae are nocturnal and hide during the day. No cannibalism has been reported.

Natural Enemies

Microorganism

A polyhedrosis virus was found infecting larvae a year after discovery of this insect in Hawaii, and the virus was probably introduced by the insect. Attempts to obtain reciprocal infections with a virus of the armyworm *P. unipunata* that was present in Hawaii were unsuccessful. A microsporidian, *Nosema sp.*, was found in eggs (Bianchi 1957, Tanada and Beardsley 1957).

Insect Parasitoids

Two species of hymenopteran egg parasitoids attacking LAW eggs in Hawaii are *Telenomus nawai* Ashmead and *Trichogramma minutum* Riley. *Apanteles marginiventris* (Cress.), a larval parasitoid, appears to be one of the most important natural enemies of the LAW in Hawaii. Three species of tachinid flies were found parasitizing larvae (Laigo and Tamashiro 1966).

Insect Predators

Two species of ants have been observed attacking eggs of the LAW, and coccinellids are found feeding on eggs.

Vertebrate Predators

The giant toad and the common mynah bird have been observed feeding on LAW larvae.

10

Lepidopteran Pests:
Family Hesperiidae

Fiery Skipper

Taxonomy

The fiery skipper (FS), *Hylephila phyleus* (Drury), belongs to the order Lepidoptera, family Hesperiidae, subfamily Hesperiinae (Hodges et al. 1983). Members of this family, called *skippers*, were named for their fast, erratic flights. Members of the subfamily are known as *tawny skippers*, and their larvae are chiefly grass feeders (Borror et al. 1989).

Importance

In Hawaii, the FS is considered the third or fourth most serious lepidopterous pest of turfgrass but has the potential of being the most serious pest during the warmest period of the year. It is considered one of the five most injurious lepidopterous pests attacking lawns in California. Turfgrass damage by the FS has not been reported from any other state (Okamura 1959, Tashiro and Mitchell 1985).

History and Distribution

The FS is a wide-ranging butterfly found from South America northward to Connecticut, Michigan, and Nebraska and is abundant in much of the southeastern United States (Figure 10-1). It is probably a permanent resident south from the coastal Carolinas and lower Mississippi Valley and regularly migrates to more northern areas each summer. In California it is abundant from the lowlands of southern California to the San Francisco Bay area, and it is generally distributed throughout the rest of the state in residential and agricultural areas (Comstock 1927, Klots 1951, Okamura 1959, Opler and Krizek 1984).

In Hawaii, the FS was first found on the island of Oahu in 1970, and since 1973 it has been reported on all the islands except Lanai (Tashiro and Mitchell 1985).

Host Plants and Damage

The larvae of the FS feed on all common lawn grasses but appear to prefer bermudagrass for oviposition as well as for food. Other foods include St. Augustinegrass, bentgrass, and

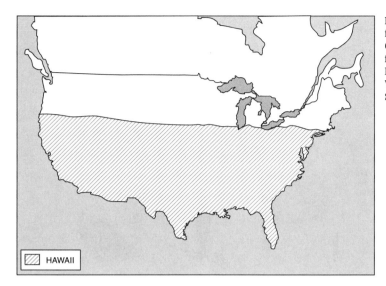

Figure 10-1. Distribution of fiery skipper. (Drawn by E. Gotham, NYSAES, adapted from Handbook of Turfgrass Insect Pests, Brandenburg and Villani 1995, Entomological Society of America.)

HAWAII

weedy grasses, especially crabgrass. Early stages of infestation are marked by isolated round spots measuring 2.5–5.0 cm in diameter where grass blades have been removed by a single larva. Coalescence of individual spots causes large areas of lawn to die (Bohart 1947, Okamura 1959, Opler and Krizek 1984).

Adults at rest were observed most frequently on the bermudagrass fairways of golf courses on the island of Oahu and on bermudagrass lawn bowling greens in Honolulu as females landed for oviposition. Flying adults were most frequently observed as they visited the flowers of lantana, *Lantana camara* L., and the flowers of other plants to feed on nectar (Tashiro and Mitchell 1985).

Description of Stages

Adult

The FS adults are predominantly orange, yellow, and brown butterflies with a wingspread of about 25 mm (Plate 26). The males are slightly smaller than the females but can be distinguished more readily by coloration. The males are predominantly bright orange-yellow above and pale yellow, with submarginal dark spots on the underside of both forewings and hindwings. The females are predominantly dark brown on the upper surface, with coloration similar to that of males on the underside but much overlaid with olivaceous dusting (Bohart 1947, Klots 1951).

Egg

The hemispherical eggs (Plate 26), glued singly to grass blades, turn from white to powder blue to greenish blue in 1–2 days. Before hatching, they become nearly white again, and the black head becomes plainly visible. Eggs range from 0.70 to 0.75 mm in breadth and from 0.50 to 0.55 mm in height (Bohart 1947, Tashiro and Mitchell 1985).

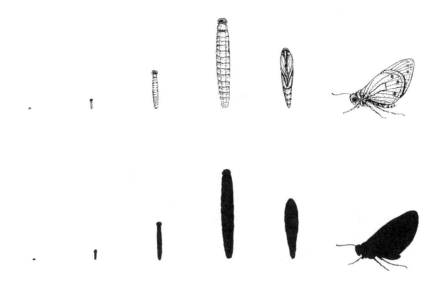

Figure 10-2. Stages of development of fiery skipper: egg; first, third, and fifth larval instars; pupa; and adult. Silhouettes show actual size of each stage. (Drawn by R. Jarecke, NYSAES.)

Larva

The body of the first-instar larva (Plate 26) is pale greenish yellow, with a granular surface appearance that becomes more pronounced in the second-instar larva. The most distinctive feature of a skipper larva is the strongly constricted neck and the presence of a narrow but prominent pronotal shield in all five instars (Plate 26). Both head and shield are coal black in all five instars. The black head of a mature larva is finely pitted, with minute setae, and the front bears several reddish-brown markings, two elongate spots near the base and two parallel lines, one on each side of the epicranial suture (Okamura 1959, Tashiro and Mitchell 1985).

In later instars, the body becomes yellow-brown to gray-brown, with an indistinct to distinct median longitudinal stripe. Even fainter lateral stripes may be present. Many short secondary setae cover the entire body, and the granulated appearance is maintained (Figure 10-2). The prepupa differs little from the mature larva except that it becomes rigidly straight (Bohart 1947, Okamura 1959, Tashiro and Mitchell 1985).

In size, FS larvae grow in head capsule widths from a mean of 0.44 mm in first instars to 1.75 mm in fifth instars. Mean body lengths increase from 2.5 mm in first instars to 24–25 mm in fifth instars (Tashiro and Mitchell 1985).

Pupa

Pupae of the FS are 15–18 mm long and have abdominal and head areas covered with sparse, bristly hairs (Plate 26). Young pupae have a head and thoracic area colored light green and an abdomen of light tan. As they mature, they turn an overall brownish shade, with an olivaceous abdomen and considerable black in the head. The color of the forewings becomes apparent before eclosion. The pupae are found either free in the grass-root zone or partially enclosed in loosely webbed debris in the grass near the soil surface. Males and females can be separated according to the methods of Butt and Cantu (1962) (also see Plate 26, Bohart 1947, Tashiro and Mitchell 1985).

Life History and Habits

Seasonal Cycle

In areas where the FS is a permanent resident, it appears to have three to five generations per year. It may be seen throughout the year in Florida but is not seen during January and February in Mississippi. In the northern states, the FS is not seen until May and is not common until late August. In California, adults were most numerous in late August and September, especially around low flowers such as red clover, *Trifolium pratense* (Bohart 1947, Opler and Krizek 1984). In Hawaii during 2 years, 1975 and 1982, they were observed more frequently from May to July than earlier in the year (H. Tashiro, personal observation).

When the FS was reared on bermudagrass at 24°C, it required an average of 48 days to pass from egg to adult. When reared at 27.5°–29.0°C, it required an average of only 23 days to complete the same development (Bohart 1947, Tashiro and Mitchell 1985).

Adult Activity

Adults of the FS are most often seen as rapidly flying butterflies frequenting the flowers of lantana, honeysuckle, alfalfa, clover, and other plants to feed on nectar. Their most rapid flights occur when males are pursuing females but they are much too rapid in flight for easy netting, even when they are flying more slowly to seek flowers (Tashiro and Mitchell 1985).

Males perch close to the ground all day, waiting for females. The flight of a passing female elicits a male response. Males mate primarily with virgin females, and mating pairs have been seen in late afternoon. The union lasts about 40 min, and if the female is disturbed, she takes flight, carrying the male. During the heat of the day, females alight on the turfgrass to deposit an egg before flying a short distance to repeat the process. Eggs are generally cemented singly to the underside of grass blades, but smaller numbers, at least in captivity, were placed on stems and on the upper surface of blades. When ovipositing, females can be caught if they are approached slowly and a net is dropped over them (Shapiro 1975, Tashiro and Mitchell 1985).

Larval Activity

Newly hatched larvae notch the edge of leaves (Plate 26) and in later instars eat entire leaves. Starting in the third instar, the larvae actively spin profuse amounts of strong webbing. Larvae are seldom seen even when they are abundant because they remain concealed in the thatch area in lightly woven silken shelters. During two separate 6-month periods in Hawaii, January to July, no larvae were observed in undisturbed turf except those forced to the surface by survey techniques (Tashiro and Mitchell 1985).

Pupal Activity

Pupation occurs in loosely woven shelters (Plate 26), or pupae are free near the soil surface if debris is not readily available (Bohart 1947, Tashiro and Mitchell 1985).

Table 10-1. Oviposition and the Viability of Eggs of the Fiery Skipper from Field-Collected and Laboratory-Reared Butterflies

Days from Oviposition to Hatching	Field (586 eggs)	Laboratory	
		Large Cage[a] (149 eggs)	Small Cage[a] (212 eggs)
2	0	1	0
3	271	50	2
4	181	71	7
5	114	0	5
6	15	1	0
Total hatched	581	123	14
Viable eggs (%)	99	83	7

[a]Large cage, 61 cm by 99 cm by 66 cm; small cage, 24 cm by 24 cm by 24 cm.
Source: Adapted from Tashiro and Mitchell 1985.

Miscellaneous Features

Greenhouse Rearing

Field-collected females, all presumed to have mated, laid no eggs under artificial light in the laboratory. They readily laid viable eggs when held in screen cages, measuring 24 cm by 24 cm by 24 cm, in the greenhouse under natural light at diurnal temperatures of 26.5°–35.0°C. Oviposition began about 10:00 AM, continued throughout the afternoon, and conformed with limited observations made in the field. Potted 'FB-137' bermudagrass was placed in each cage as an oviposition medium, and blooming potted lantana was used as a source of nectar.

Adults from laboratory-reared larvae (under artificial illumination) were paired and were placed in screen cages of two sizes and held in the greenhouse. Potted 'FB-137' bermudagrass was placed in each cage as an oviposition medium and potted lantana in bloom as a source of nectar. The pairs held in a cage of 0.4 m³ oviposited readily, producing 83% viable eggs, and compared favorably with field-collected females, which deposited 99% viable eggs. Pairs held in a cage of 0.014 m³ produced only a few eggs of only 7% viability (Table 10-1). This difference seemingly indicates that greater flight activities are needed for normal reproductive development (Tashiro and Mitchell 1985).

Laboratory-reared females began ovipositing on the third day following eclosion and peaked between days 5 and 9. Their egg production varied from 9 eggs to 71 eggs per female, while field-collected females laid 14–168 eggs per female. The maximum life of laboratory-reared females was 11 days (Tashiro and Mitchell 1985).

Development of eggs was relatively rapid. When they were held at 26.6°C and a saturated atmosphere, they began to hatch in 2 days, with more than 80% hatching on the third and fourth days; by the sixth day, hatching was completed (H. Tashiro, personal observations).

Natural Enemies

Very little is known about natural enemies of the FS. Only two hymenopteran insect parasitoids have been reported—a braconid, *Apanteles* sp., and an ichneumonid, *Amblyteles* sp. (Bohart 1947).

11

Coleopteran Pests:
Family Scarabaeidae

Overview

The larvae of turfgrass-infesting species of the family Scarabaeidae constitute a large complex whose members (white grubs) are similar in general appearance, in habits, and in the turfgrass damage they cause. Because of their similarities, a general account of these pests will serve as a basis for understanding the entire group.

Taxonomy and Nomenclature

At least 10 species of scarabs, belonging to five subfamilies, are pests of turfgrass in the United States. The larvae of this family are known also as *grubs,* a term applied to the larvae of Coleoptera and Hymenoptera in general. *White grub* is the common name applied in many countries to larvae of the family Scarabaeidae and in particular to those of agricultural importance. However, the larvae belonging to the subfamily Melolonthinae are known universally as white grubs (Ritcher 1966). In current usage, many individuals engaged in turfgrass research and management refer to larvae of the masked chafers (genus *Cyclocephala,* subfamily Dynastinae) as *annual white grubs.* This confusion regarding *white grubs* poses no real conflict if we recognize that the general term is used loosely.

Some authors have divided the family into two groups, calling them *lamellicorn scavengers* and *lamellicorn leaf chafers.* This designation reflects the most notable morphological character, the lamellate antennae (described later) of the family, and the varied food habits of both adults and larvae.

Importance

Grubs of the Scarabaeidae are the most serious turfgrass pests in the northeastern United States, and are considered a major pest in the Midwest, Southeast, and parts of the southwestern United States. They are also a problem in eastern Canada, particularly in the provinces of Ontario and Quebec. Their subterranean habits make them among the most difficult of turfgrass insects to manage. Chemical insecticides to control grubs normally must be applied to the surface of established turf, after which gravity and precipitation (including irrigation) are needed to move the chemicals into the root zone. One exception to this rule is third-instar green June beetle grubs, which spend considerable time on the turf surface and are easily controlled at this time.

Although the adults of some species are serious defoliators of diverse woody plants and herbs, the larvae are the more general problem owing to the destruction they cause to the roots and other underground parts of turfgrasses and other plant species. Both the adults and the grubs of some turfgrass pests (such as the Japanese beetle and certain June beetles) are also serious pests of agricultural crops (Fleming 1972, Ritcher 1940).

History and Distribution

Some of our most serious scarabaeid pests of eastern U.S. and Canadian turfgrass are introduced species, with the most notable of these being the Japanese beetle, followed (in order of importance) by the oriental beetle, the European chafer, and the Asiatic garden beetle. Each was introduced accidentally into North America from the Orient or Europe, as their various names imply, and was discovered along the eastern seaboard between 1916 and 1940. With the exception of the Japanese beetle, which is a serious pest east of the Mississippi River and throughout eastern Canada, these introduced species currently are problems predominantly in the northeastern United States and southern Ontario.

Some native Scarabaeidae are also serious turfgrass pests. In general, the native species have much wider geographic distributions than do introduced species. The most notable native species are the northern and southern masked chafers, present as destructive pests from the Northeast through the Midwest, extending south into Florida, and west into Texas, and California. The black turfgrass ataenius currently is an important turfgrass pest in the northern United States from Colorado to the Atlantic Coast and parts of southern Canada, as well as parts of southern California. The most widespread species are the group known commonly as May or June beetles in the genus *Phyllophaga*. There are 152 species that occur in the United States and Canada. They are distributed throughout the continent but are most serious in the northeastern quarter of the United States, in Texas, and in Ontario and Quebec. Not all species, however, are serious pests (Luginbill and Painter 1953; Ritcher 1940, 1966; Ratcliffe, 1991).

Host Plants and Damage

Adult Feeding

While most turfgrass-infesting scarabaeid grubs cause similar damage, the adults have food habits that vary widely according to species. Some (e.g., black turfgrass ataenius) feed on dead and decaying organic matter, including carrion and dung. Others (e.g., Japanese beetle) feed on foliage, flowers, and fruit. Some adults (e.g., masked and European chafers) feed little or have not been observed feeding at all.

Of the turfgrass-infesting species that damage the parts of plants that are above the ground, the Japanese beetle is by far the most notorious. Adult Japanese beetles feed on nearly 300 species of plants, damaging the foliage, flowers, and fruit. Many of the May or June beetle adults feed on foliage of various deciduous shade and forest trees as well as on diverse forbs and to some extent on grass leaves (R. Crocker, personal communication, 1998). Some other scarab adults (e.g., the oriental beetle) also feed on petals of flowers. Asiatic garden beetle adults will feed on the foliage and flowers of more than 100 diverse host plants, with heavy infestations causing destruction of all leaf tissue except the midribs (Heller 1995). Scarabaeids that injure turfgrass as grubs, but do not feed as adults, or feed so little that they do not damage plants, include the black turfgrass ataenius, the European chafer, and the northern and southern masked chafers (Fleming 1972, Hammond 1940, Johnson 1941, Ritcher 1940, Tashiro et al. 1969).

Larval Feeding

Turfgrass-infesting scarabaeids are considered among the most important group of turfgrass pests because they are able to feed on the roots of all species and cultivars of the commonly used turfgrasses. Extensive pruning of the roots at or just below the soil-thatch interface or exclusively in the thatch (Plate 72) is the primary cause of turf damage. When severe root pruning is accompanied by drought, death of the turfgrass can be rapid.

Damage: Seasonal Aspects. The occurrence of turfgrass damage is directly related to life cycle. For species having 1-year life cycles, the most severe turfgrass damage occurs during late summer and early fall when large third-instar grubs, stimulated by autumn rains, move upward to the soil-thatch interface and feed actively before overwintering deeper in the soil profile to escape killing frost. Damage continues the following spring as overwintering grubs return to the root zone. Feeding during the spring, prior to the final maturation of third instars and pupation, is of relatively short duration but can be vigorous in some scarab species. Turfgrass damaged by scarabaeid grubs during midsummer cannot be attributed to actively feeding members of this group except for the black turfgrass ataenius and *Phyllophaga* spp. (Fleming 1972, Tashiro et al. 1969).

Feeding by the black turfgrass ataenius grubs causes damage to golf course fairways during June, July, and August in regions where there are two generations a year, and during late July and August where there is only one generation a year (Vittum 1995b, Wegner and Niemczyk 1981).

Members of the genus *Phyllophaga* have 1-, 2-, or 3-year life cycles, with the last being the most common in most of the United States and Canada. Major turf damage is caused throughout the summer of the second year by the mature second instars and young third instars of species that have a 3-year life cycle. Minor turf damage occurs during late summer of the first year and also in late spring of the third year, as mature third instars complete their feeding preparatory to pupation during midsummer (Hammond 1940, Ritcher 1940).

Damage: Progression. Turf-damaging populations of grubs produce symptoms varying from weakness to death of large patches of turf (Plate 27). A lack of growth followed by gradual thinning and weakening of the stand are among the early symptoms of grub activity. When the grub populations are large, there may be a sudden wilting of the grass caused by the severe root pruning, even with adequate soil moisture. As damage continues, the surface of the soil tends to become very spongy, with the sponginess being evident as one walks over the infested area. Birds and mammals often prey on grubs, causing further damage as they tear the turf in search of grubs.

With high grub densities, root pruning in the soil-thatch interface may be so complete that the sod can be lifted in large sections to reveal the actively feeding grubs. These exposed grubs may be only a part of the population, since many more will often be present just below the soil surface or in the thatch (Plate 27).

Description of Stages

Adults

Beetles of this large family have a number of distinguishing characters (Figure 11-1). Most are stout bodied and are usually convex with somewhat shortened elytra, the leathery or chitinous forewings that serve to cover to the hindwings (Figure 11-1). The membranous hindwings are folded under the elytra and are visible only when the beetle is in flight. In flight, the elytra simply are spread, elevated, and fixed above the body. The tibia of each foreleg is fossorial and fitted with broad teeth on the outer edge. Most beetles (e.g., orien-

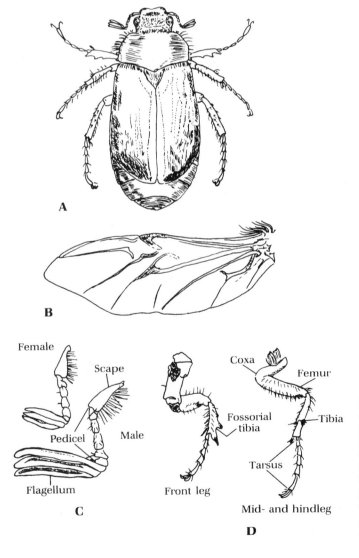

Figure 11-1. Adult scarabaeid characteristics, illustrated by the European chafer. **A.** Adult. **B.** Membranous hindwing folded under elytra at rest. **C.** Lamellicorn antennae. **D.** Legs of beetles. (Adapted from Butt 1944, plates 6b, 7c, 10c, 10d, 10f, courtesy of the Cornell University Agricultural Experiment Station, Ithaca.)

tal beetle) are predominantly dull colored, but some (e.g., Japanese beetle) are brightly colored, often with a metallic sheen.

The last segment of the antenna consists of three or more (depending upon species) flattened elongate parts (lamellae) that normally are held close together, but that can be spread like a Chinese fan (Plate 44), thus exposing olfactory receptors on their inner surfaces. This lamellate club is the most distinguishing characteristic and is the origin of the name, *lamellicorn beetles* (Figure 11-1). In many species the antennal club of males is about twice as long as that of the female, making separation of sexes fairly simple (Figure 11-1). In other species, sexual differentiation of adults (exclusive of sex organs) occurs on other parts of the body. For example, the tibial spur on the front leg is used to differentiate male and female Japanese beetles.

Eggs

All turfgrass-infesting scarabaeids deposit eggs that are nearly identical in shape and appearance except for size. Freshly deposited eggs are shiny, milky white, ellipsoidal, and about 1.3–1.5 times as long as they are wide. Eggs absorb water and become more spherical as they mature (Potter 1983, Regniere et al. 1981). The surface of the egg (chorion), which may be smooth or textured, is elastic to accommodate the growing embryo. In fully mature eggs, the tan mandibles of the larva are visible through the translucent chorion (Plates 41, 44).

Larvae

Range in Size. Turfgrass-infesting scarabaeid grubs in the United States vary in size, from relatively small (black turfgrass ataenius, less than 10 mm in length) to large (green June beetle nearly 50 mm long) (Figure 11-2). Species such as the Japanese beetle, with medi-

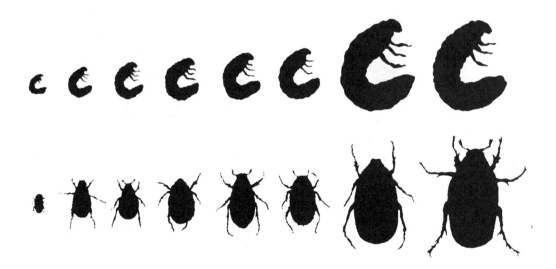

Figure 11-2. Relative sizes of turfgrass-infesting scarabaeid grubs. Left to right: black turfgrass ataenius (smallest), Asiatic garden beetle, Japanese beetle, oriental beetle, European chafer, northern masked chafer, May or June beetle, and green June beetle (largest). Silhouettes show actual size of each stage. (Courtesy of R. Jarecke, NYSAES.)

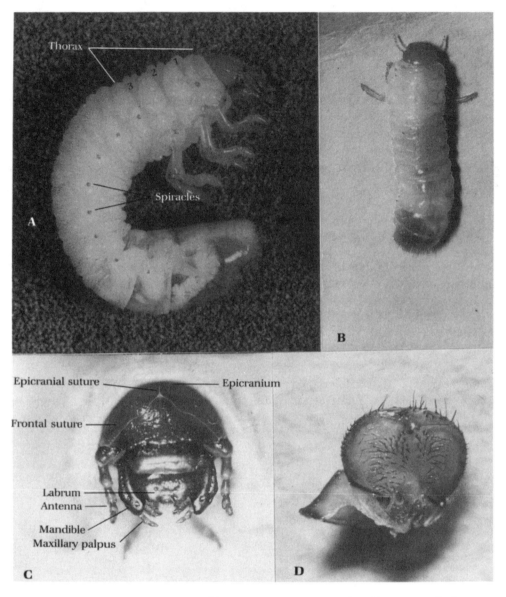

Figure 11-3. The scarabaeid grub, illustrated by various species. **A.** Natural position. **B.** Stretched-out unnatural position. **C.** Front view of head. **D.** The epipharynx, or inner surface of the labrum. (Photos by B. Aldwinckle, NYSAES.)

um-sized mature grubs, are about 20–25 mm in length and weigh 90–270 mg (Fleming 1972).

General External Features. Newly hatched larvae are translucent white, but after feeding, the posterior region turns various shades of gray or brown, depending on the color of the ingested soil and the plant material in the rectal sac.

A distinguishing feature of scarabaeid larvae is the C-shaped contour of their body (Figure 11-3). The head is scleritized and pigmented without eyes but with a pair of simple antennae and sharp, strongly developed mandibles. The three thoracic segments, each with a pair of legs, are contiguous with a 10-segmented abdomen. The grub's integument is transversely wrinkled and may have pigmented hairs scattered over its surface. In some species, there is a slight constriction of the midabdominal segments (e.g., masked chafers; Plate 35), while other species (e.g., oriental beetle), show no evidence of constriction. Common to all members of this family is a pair of spiracles on the prothoracic segment and on each of the first eight abdominal segments. Several features of the larval head and epipharynx (Figure 11-3) are useful in distinguishing species. To observe characters on the epipharynx, the labrum can be lifted back, thus exposing the epipharynx without injuring a living grub.

All scarab species have three larval instars (or stadia). Freshly transformed larvae of each stadium have head capsules distinctly wider than the thorax and abdomen. With time, the width of the growing thorax and abdomen greatly expands and soon exceeds that of the hard, chitinized head capsule, which maintains a fixed size during each stadium.

Rastral Pattern. In scarabaeid larvae, the ventral area of the last (10th) abdominal segment (just anterior to the anus) has a definitely arranged pattern of spines, hairs, and bare spaces and is called the *raster* (Figure 11-4). Boving (1942) designated the median longitudinal bare area the *septula*. A longitudinal line of *pali* (spines) lying on either side of the septula is called the *palidium,* with the area lateral to each palidium and covered with scattered spines and hairs known as the *tegillum.*

The anal slit is surrounded by the upper and lower anal lobes. The anal slit may be transverse (as in the Japanese beetle); Y-shaped, with the stem generally shorter than the arms (as in May or June beetles); or essentially longitudinal (as in the Asiatic garden beetle).

The characters of the rastral patterns and anal slits are useful for identification of species (Plate 28), especially in the field. A hand lens 10× or stronger with good illumination is adequate for identification of all three stadia of the medium-sized to larger larvae common in turfgrass.

Internal Anatomy. The alimentary tract of scarabaeid grubs is quite similar for all species. It consists of three anatomically differentiated areas: the foregut (stomodaeum), the midgut (ventriculus), and the hindgut (proctodaeum). The midgut occupies the largest area, extending approximately from the mesothoracic segment to the eighth abdominal segment. Cecal diverticula are present at the anterior and posterior ends of the midgut. Malpighian tubules originating at the junction of the anterior intestine (of the hindgut) and midgut extend anteriorly the length of the midgut, then posteriorly to the ventral surface of the rectal sac (Splittstoesser et al. 1973).

Prepupae

The prepupa (the terminal period of the final larval instar) is an active but nonfeeding stage of some insects with complete metamorphosis. The prepupa of scarabaeids occupies an earthen cell (those living in soil) prepared by the grub (Plate 30). Some prepupae retain the C-shaped contour of the larvae, while others straighten out except for a slight crook at the posterior end of the abdomen (e.g., European chafer; Plate 37).

Pupae

Scarabaeid pupae are similar in appearance except in size, having legs and wings that are free from the body and clearly visible (Plate 30). As pupation begins, members of some sub-

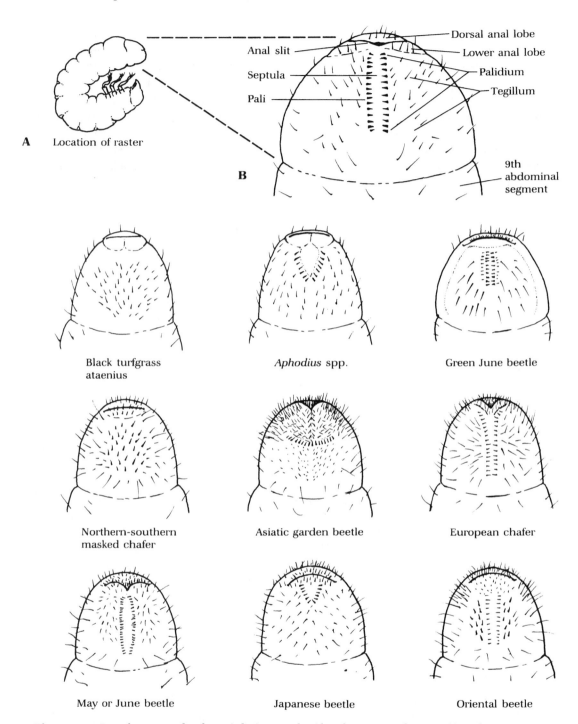

Figure 11-4. Rastral patterns of turfgrass-infesting scarabaeid grubs, not to scale. **A.** Position of raster. **B.** Details of raster and anal area. (Adapted in part from U.S. Department of Agriculture 1951–1980; drawn by R. McMillen-Sticht, NYSAES.)

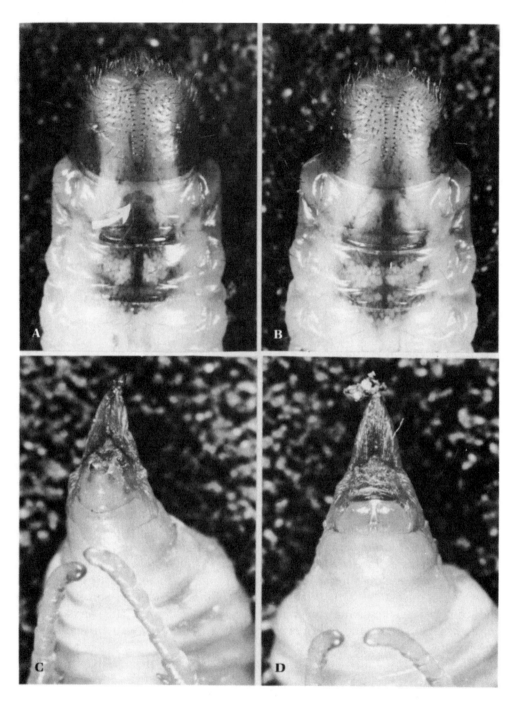

Figure 11-5. Typical sexual differentiation in the larvae and pupae of scarabaeids, illustrated by the European chafer. **A.** Male larva with terminal ampullae visible. **B.** Female larva with ampullae not visible. **C.** Terminal ventral view of male pupa. **D.** Terminal ventral view of female pupa. (From Tashiro et al. 1969, fig. 11, courtesy of the NYSAES.)

families (e.g., Melolonthinae: European chafer; Plate 30) slough off the larval exuviae; these remain attached to the last abdominal segment. In other subfamilies, pupation occurs within the thin, meshlike exuviae, which eventually split to release the pupa (e.g., the Rutelinae: Japanese beetle; Plate 43). Young pupae are creamy white and extremely fragile. They gradually assume adult coloration just before transforming to adults.

Sexual Differentiation: Larvae and Pupae

Both males and females of scarabaeid grubs have internal sex organs called *terminal ampullae* that lie along the ventral median line of the ninth abdominal segment in males and the eighth abdominal segment in females. The ampullae are not visible externally in the females but generally are visible in males, except when a layer of fat obscures these organs. When the terminal ampullae are visible, they appear as two ovoidal structures (Figure 11-5) in the ninth abdominal segment, and the grub is certain to be male (Hurpin 1953). The terminal ampullae are more clearly evident in the European chafer than in several other common turfgrass species (H. Tashiro, personal observation).

The sex of pupae can be determined easily by wholly external differences in the ventral portion of the terminal abdominal segment. Male genital organs are more pronounced than those of the female (Figure 11-5) (Tashiro et al. 1969).

Seasonal History and Habits

Seasonal Cycles

Turfgrass-infesting scarabaeids, depending on species, have one or two generations a year or require 2–4 years to complete a generation. The black turfgrass ataenius has two generations a year from the latitude of southern New England southward and one generation a year north of this latitude. In either case, grubs do their major damage during the summer, and adults overwinter.

Turfgrass-infesting scarabaeids that normally have a 1-year life cycle in much of their geographic range may have a few individuals that require 2 years to complete their life cycle in the most northern latitudes of their range. Such is the case with the Japanese beetle and the European chafer. Those that complete a generation in 1 year are most damaging in the fall and again in the spring, and overwinter as third instars (Fleming 1972, Vittum 1986).

During the 3-year cycle, a May or June beetle spends the first winter as a second instar, the second winter as a third instar, and the third winter as an adult in the soil. Major turf damage occurs during the summer and early fall of the second year. Some May or June beetles complete a generation a year in their most southern range. They are suspected of requiring 2 years in their most northern range. Such is the case with *Phyllophaga crinita* Burmeister in Texas (Frankie et al. 1973).

Adult Activities

General Habits. Adults may be nocturnal or diurnal (depending on species) and are present above ground only for short periods during late spring, summer, or early fall. Many of the nocturnal species are strongly attracted to light and are captured easily in light traps. A female sex pheromone (to attract potential mates) has been demonstrated to be present in several important turfgrass species, including the Japanese beetle (Ladd 1970), the northern and southern masked chafers (Potter 1980), the oriental beetle (Zhang et al. 1994), and the June beetle, *Phyllophaga anxia* (Zhang et al. 1997).

Adults may feed on foliage, blossoms, and fruit or on organic matter or may not feed at all, depending on the species. Feeding habits are discussed more thoroughly in chapters describing the individual species.

Oviposition. All turfgrass pests of this family deposit their eggs singly in earthen cells formed in moist soil, or in groups of about 10 eggs in the soil or in the soil-thatch interface. The process by which they form an earthen cell and deposit an egg has not been reported in the literature. In their actions, European chafer females are considered to be essentially the same as other scarabaeids that form cells for individual eggs (Plate 29). Evagination of the vaginal tract creates a balloon-like organ that compresses the moist soil and forms a cell. An egg is deposited simultaneously with the invagination of the vagina. The smooth wall of the earthen cell protects the egg against physical injury. Moisture from the surrounding soil is essential for growth and maturation of the egg.

Larval Activities

Upon hatching, first instars of phytophagous species begin feeding on root hairs and the fine roots of grasses and other plants, their primary source of food. They can also feed on the organic matter present in the thatch and soil, but when they feed in this manner, they do not grow as rapidly.

Vertical movement of the grubs is governed by soil moisture and temperature (Villani and Wright 1988a). When the surface soil becomes dry, grubs migrate downward to seek moisture. After a drought, when grubs may be well below 10 cm, precipitation or irrigation that wets the soil to the depths of the grubs will induce them to move toward the soil surface within 24 hr. During spring when soil moisture is generally constant and high, most grubs are in the soil-thatch interface.

Cold temperature in the fall forces grubs to migrate downward before the ground freezes. They will migrate as deep as necessary to avoid the frozen soil and will remain below the frost line all winter. Upward migration begins in the spring when all soil frost has disappeared and the soil warms. There appears to be a species-specific temperature threshold for movement of grubs from the soil depths to the thatch-soil interface.

Villani and Wright (1988a) radiographed soil blocks in the laboratory to study the response of three scarab grub species (Japanese beetle, European chafer, and oriental beetle) to temperature. Fluctuations in temperature had very little impact on the position of European chafer grubs. Population distributions of European chafer grubs within the soil profile were nearly identical in temperature regimes that were either constant (20°C throughout the profile over the duration of the experiment) or fluctuating (reducing temperature from 20° to 2°C in 6° increments, then returned to 20°C). This lack of response to temperature conforms with field observations indicating that European chafer grubs are often found in the upper turf root zone well into early winter and again in early spring; at times they feed in the upper root zone under snow if this zone is not frozen.

In contrast, the other two scarab species responded rapidly to shifting temperatures. Japanese beetle grubs fed in the upper root zone in the stable temperature regime whereas in the shifting temperature regime, grubs moved from the upper root zone downward with the onset of cooling soil (14°C) and returned to the surface as temperatures increased. Oriental beetle movement appeared more variable, but again grubs tended to remain in the upper root zone in the stable treatment and to respond to lower soil temperatures (8°C) by moving down in the soil profile. A portion of the oriental beetle population moved back to the upper root zone when soil temperature increased. Similar movement behavior of oriental beetle populations in response to warming has been observed in the field.

Radiographs of soil blocks containing third-instar Japanese beetle, oriental beetle, European chafer, and northern masked chafer grubs indicated that these species respond to simulated irrigation and drought. Individual grubs of all species studied moved upward after the addition of moisture in dry soils. European chafers showed the least sensitivity to decreased soil moisture; this fact may be related to their ability to escape rapidly from extreme conditions, such as sudden sharp freezes.

Larva-to-Adult Sequence

When the third instar is mature and feeding has been completed, the larva moves downward in the soil and forms an earthen cell, oscillating to compress and smooth the cell wall. Much occurs in this cell during the metamorphosis of the individual. This sequence is illustrated by the transformation of the European chafer in Plate 30.

The larva ejects its accumulated excrement and becomes a pale, flaccid prepupa with no power of locomotion. It can move only by flexing and reflexing its abdomen, which further aids in smoothing the earthen cell. After a few days it becomes a pupa, and in the case of the European chafer, sloughs off the larval exuvia down to its abdominal tip. In other species (e.g., Japanese beetle), pupation occurs within the exuvia, which later splits and frees the pupa. Over several days the pupa gradually turns from being creamy white to having a brown head, wings, and legs. In time it becomes a teneral adult with thin, lightly colored elytra and fully expanded hindwings. Within a few more days the elytra harden, the membranous hindwings are folded, and the insect becomes a mature adult.

In most species with a 1-year life cycle, the beetle digs its way to the soil surface and makes its first flight a few days after adulthood. In most *Phyllophaga* spp. with a 2- or 3-year life cycle, the adult remains in its earthen cell throughout the rest of the season (late summer and autumn) and the entire winter before emerging in the spring (Ritcher 1940).

More detailed accounts of the most important scarabaeid pests of turfgrass in the United States appear in Chapters 12–16.

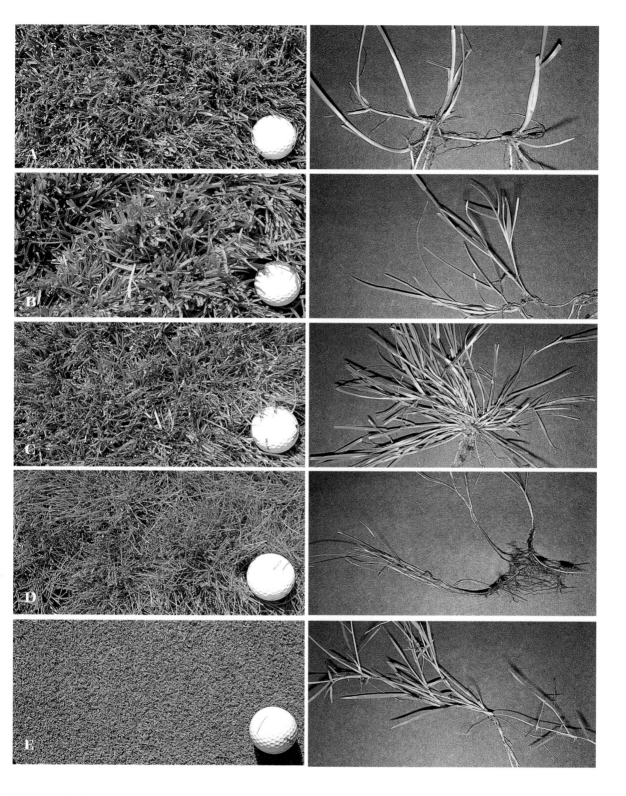

Plate 1. Cool-season grasses cut at 4 cm except bentgrass. Turf (left) and vegetative growth characteristics (right). **A.** Kentucky bluegrass. **B.** Annual bluegrass. **C.** Perennial ryegrass. **D.** Fine fescue. **E.** Creeping bentgrass on putting green. (Photos: New York State Agricultural Experiment Station [NYSAES], Geneva, NY.)

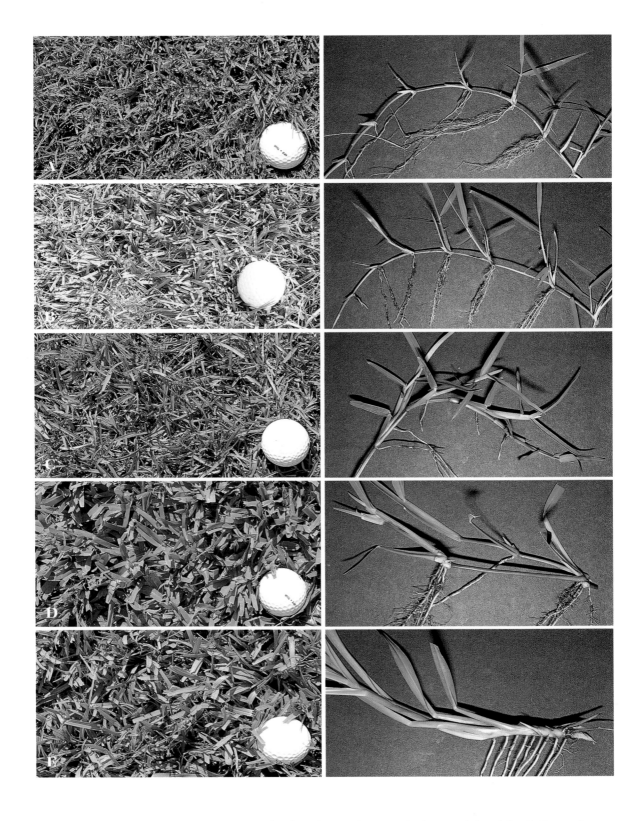

Plate 2. Warm-season grasses cut at 4 cm. Turf (left) and vegetative growth characteristics (right). **A.** Bermudagrass. **B.** Zoysiagrass. **C.** Centipedegrass. **D.** St. Augustinegrass. **E.** Bahiagrass. (Photos: NYSAES.)

Plate 3. Cool-season grasses. Leaf bud vernations (left), ligules (center), and stems and leaves (right). **A.** Kentucky bluegrass. **B.** Annual bluegrass. **C.** Perennial ryegrass. **D.** Fine fescue. **E.** Creeping bentgrass. (See Table 1-1 for descriptions of characteristics.) (Photos: NYSAES.)

Plate 4. Warm-season grasses. Leaf bud vernations (left), ligules (center), and stems and leaves (right). **A.** Bermudagrass. **B.** Zoysiagrass. **C.** Centipedegrass. **D.** St. Augustinegrass. **E.** Bahiagrass. (See Table 1-1 for descriptions of characteristics.) (Photos: NYSAES.)

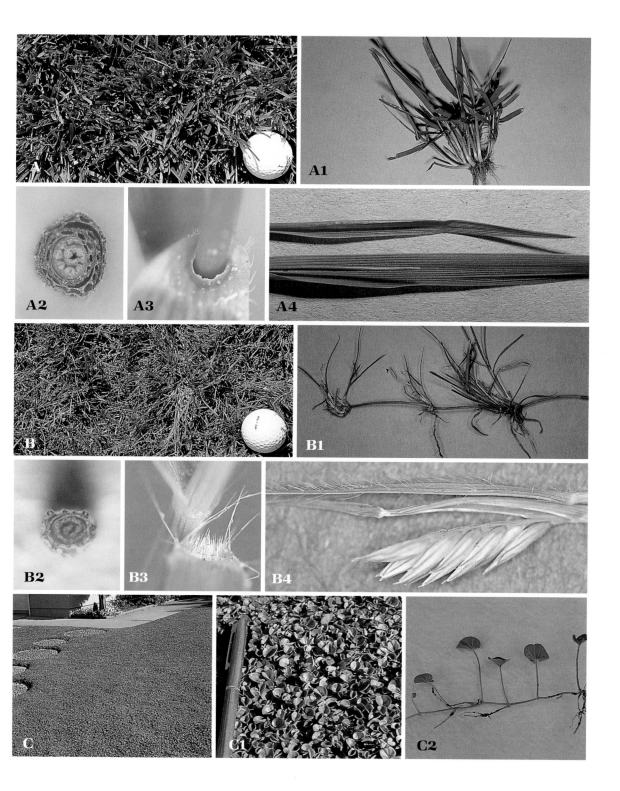

Plate 5. Minor turf plants. **A.** Tall fescue cut at 4 cm. **A1.** Vegetative growth characteristics. **A2.** Leaf bud vernation. **A3.** Ligule. **A4.** Stem and leaves. **B.** Buffalograss cut at 4 cm. **B1.** Vegetative growth characteristics. **B2.** Leaf bud verna- tion. **B3.** Ligule. **B4.** Stem and leaves. **C.** Dichondra home lawn. **C1.** Close-up of dichondra lawn. **C2.** Dichondra vegeta- tive growth characteristics. (Photos: NYSAES.)

Plate 6. Drought dormancy of Kentucky bluegrass–perennial ryegrass lawn with rejuvenation following a 2.5-cm precipitation. **A.** Drought dormancy. **B.** Rudimentary shoots (7–8 visible) from one crown 1 week after precipitation. **C.** Rudimentary roots (2 visible) from one crown. **D.** Rejuvenation of turf 1 week following precipitation. **E.** Fully rejuvenated turf 2 weeks after precipitation. (Photos: NYSAES.)

Plate 7. Mites injurious to turfgrass: Eriophyidae: Bermudagrass mite. **A.** Mites under leaf sheath. **B.** Mites magnified about 100x. **C.** Mite magnified about 170x. **D.** Normal (left) and mite-injured stem (right), referred to as "witch's broom." (Photos: NYSAES.)

Buffalograss mite. **E.** Mites on buffalograss stem. **F.** Mite-injured buffalograss stem, also referred to as "witch's broom." (Photos: courtesy of J. Kalisch.)

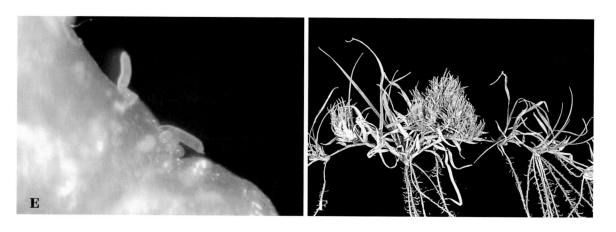

Plate 8. Mites injurious to turfgrass: Eupodidae: Winter grain mite. **A.** Adult. **B.** Eggs. **C.** Mite injury to grass stem and blade. **D.** Mite-injured golf course fairway. (Photo D courtesy of H. T. Streu; photos A–C: NYSAES.)

Tetranychidae: **E.** Clover mite adult. **F.** Eggs. **G.** Injury to turf at base of building. **H.** Banks grass mite adult and egg. **I.** Brown wheat mite. (Photo G courtesy of H. D. Niemczyk; photos H, I: courtesy of F. B. Peairs; photos E, F: NYSAES.)

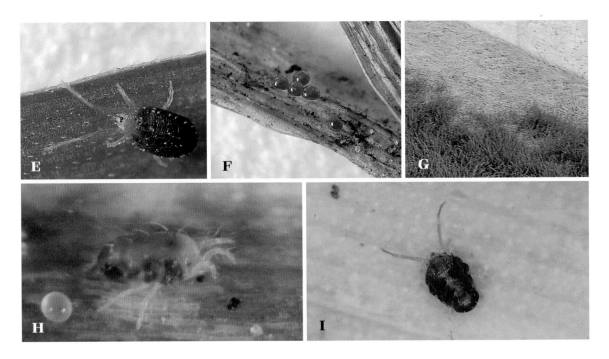

Plate 9. Mole crickets, *Scapteriscus* spp. **A.** Tawny mole cricket, male. **B.** Female. **C.** Southern mole cricket, male. **D.** Female. **E.** Tibial dactyls of forelegs of southern mole cricket (left) with U-shaped separation and tawny mole cricket (left) with V-shaped separation. **F.** Eggs. **G.** Nymphs. (Photos: NYSAES.)

Plate 10. Mole crickets. **A.** Tawny mole cricket turf damage. **B.** Burrowing mounds and ridges. **C.** Short-winged mole cricket. **D.** Pitfall trap for capturing mole crickets, designed by K. Lawrence. **E.** Sound traps for capturing southern and tawny mole crickets. **F.** Cannibalism by the southern mole cricket. (Photo A courtesy of P. P. Cobb; photo B courtesy of R. L. Crocker; photos C–F: NYSAES.)

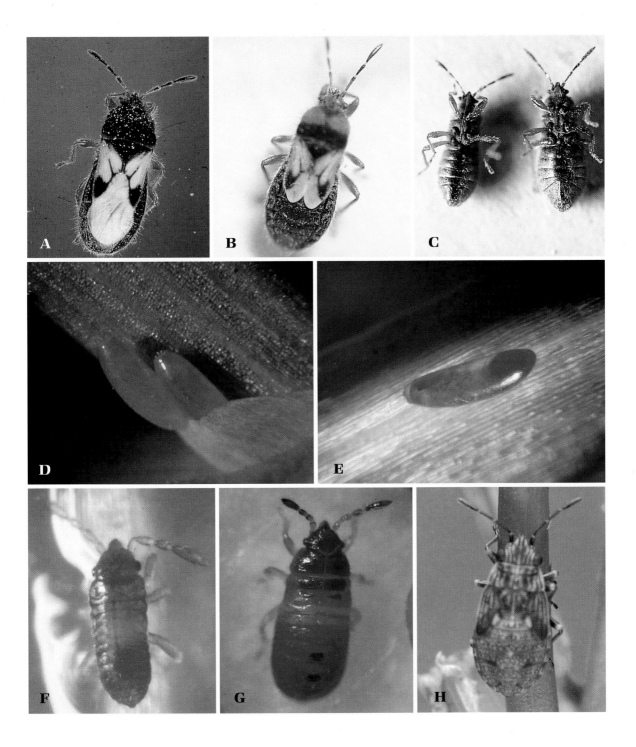

Plate 11. Hairy chinch bug, *Blissus leucopterus hirtus*. **A.** Macropterous adult. **B.** Brachypterous (short-winged) adult. **C.** Sex differences shown ventrally with male (left) and female with keeled sternum (right). **D.** Young eggs. **E.** Mature egg. **F.** Young nymph. **G, H.** Older nymphs. (Photos A, D, E–G: courtesy of H. T. Streu; photos B, C, H: NYSAES.)

Plate 12. Hairy chinch bug. **A.** Adult feeding on grass stem. **B.** Fescue turf harboring overwintering hairy chinch bugs. **C.** Hairy chinch bug–damaged lawn. (Photo A courtesy of H. T. Streu; photos B, C: NYSAES.)

Common chinch bug, *Blissus l. leucopterus.* **D.** Adult and cluster of nymphs. (Photo courtesy of J. Kalisch.)

Southern chinch bug, *B. insularis.* **E.** Nymph. **F.** Adult. (Photos courtesy of D. J. Shetlar.)

Buffalograss chinch bug, *Blissus* spp. **G.** Winged and brachypterous females. **H.** Chinch bug-damaged buffalograss. (Photos courtesy of J. Kalisch.)

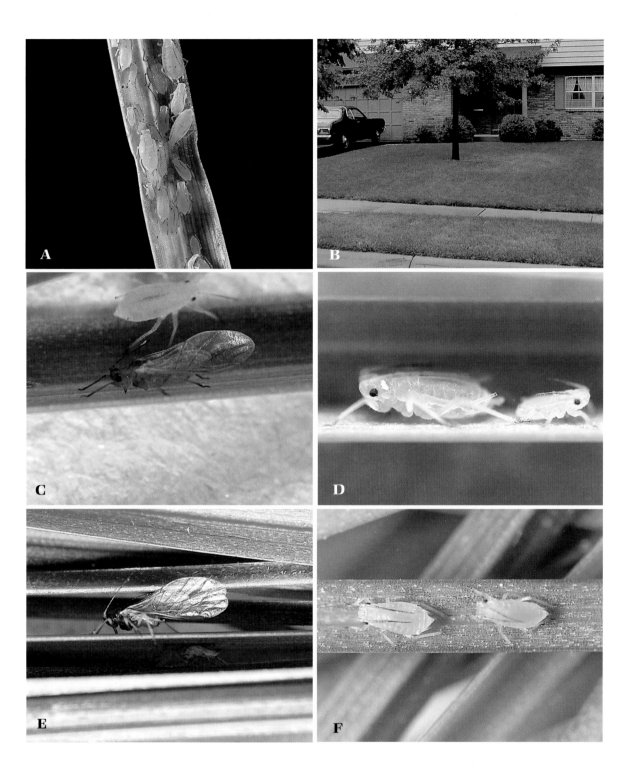

Plate 13. Greenbugs, *Schizaphis graminum*. **A.** Feeding on grass blade. **B.** Lawn damage. **C.** Maryland biotype winged female. **D.** Maryland biotype wingless greenbug with green median dorsal stripe. **E.** Ohio biotype winged female. **F.** Ohio biotype wingless greenbugs without green median dorsal stripe. (Photo A courtesy of D. J. Shetlar; photo B courtesy of H. D. Niemczyk; photos C–F: NYSAES.)

Plate 14. Two-lined spittlebug, *Prosapia bicincta*.
A. Adult. **B.** Turf damage, yellowing grass. **C.** Eggs.
D. Nymph without spittle. **E.** Nymph encased in
spittle. **F.** Relative size of nymph. **G.** Location of spittle-
encased nymphs just above soil line. (Photos A, B, F:
courtesy of P. P. Cobb; photos C–E, G: NYSAES.)

Plate 15. Rhodesgrass mealybug, *Antonina graminis*. **A.** Young adult female. **B.** Anal tubes of adults. **C.** Young larva. **D.** Nymph. (Photo B courtesy of D. J. Shetlar; photos A, C, D: NYSAES.)

Buffalograss mealybug. **E.** Female nymph. **F.** Winged adult male. (Photo E courtesy of F. P. Baxendale; photo F courtesy of J. Kalisch.)

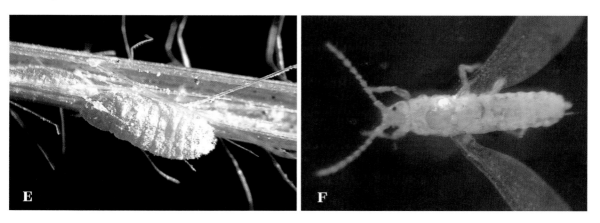

Plate 18. Sod webworm damage. **A.** First-instar larvae feeding on grass blade. **B.** Early stages of turf damage. **C.** More extensive damage with bird feeding peck holes. **D.** Larval burrow in thatch and fresh fecal pellets. **E.** Serious but local-ized damage. **F.** Extensive, irreversible damage with much of green areas consisting of clover and weeds. (Photos: NYSAES.)

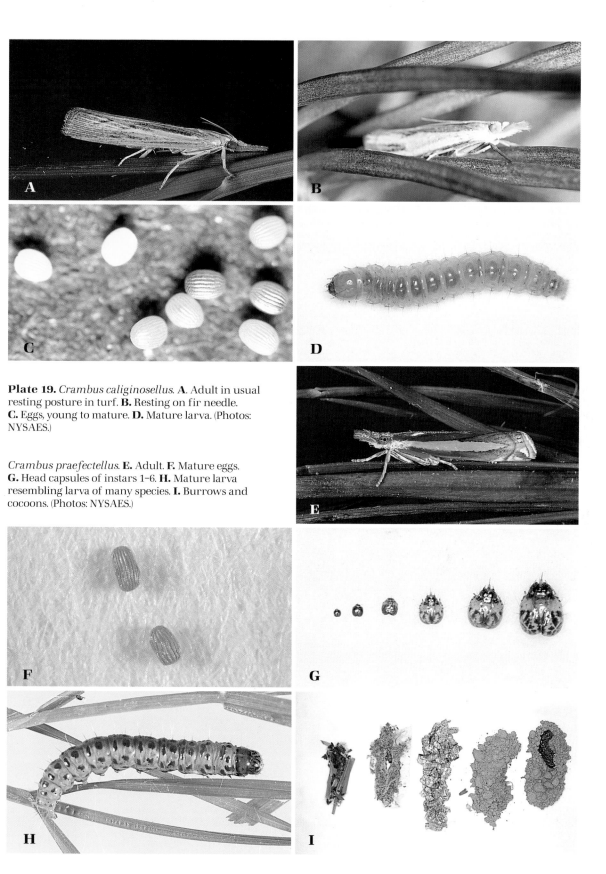

Plate 19. *Crambus caliginosellus.* **A**. Adult in usual resting posture in turf. **B.** Resting on fir needle. **C.** Eggs, young to mature. **D.** Mature larva. (Photos: NYSAES.)

Crambus praefectellus. **E.** Adult. **F.** Mature eggs. **G.** Head capsules of instars 1–6. **H.** Mature larva resembling larva of many species. **I.** Burrows and cocoons. (Photos: NYSAES.)

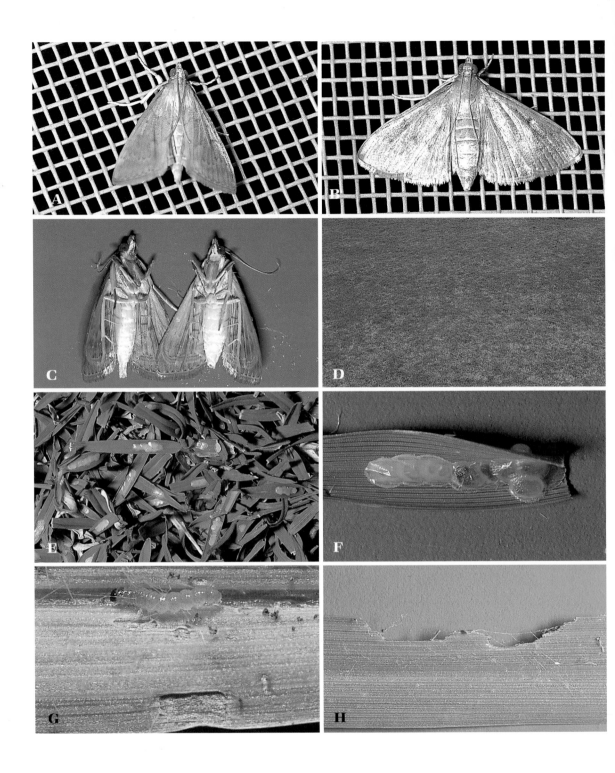

Plate 20. Grass webworm, *Herpetogramma licarsisalis.* **A.** Male. **B.** Female. **C.** Male (left) and female (right) ventral views. **D.** Damage to bermudagrass. **E.** Mass of eggs on bermudagrass turf. **F.** Eggs, young (left) and mature (right). **G.** First instar feeding on epidermis. **H.** Feeding habit of older larvae. (Photos A–C, E–H: courtesy of D. M. Tsuda; photo D courtesy of C. L. Murdoch.)

Plate 21. Grass webworm. **A.** Mature fifth instar. **B.** Cocoon of debris and fecal pellets. **C.** Pupa. **D.** Pupal terminalia, male (left) with undivided eigth segment and female (right) with divided eigth segment. **E.** Myna birds. **F.** *Bufo marinus* toad. (Both E and F major predators of grass webworms in Hawaii.) (Photos A, B, D: courtesy of D. M. Tsuda; photos C, E, F: NYSAES.)

Tropical sod webworm, *Herpetogramma phaeopteralis*. **G.** Adult. **H.** Mature eggs. **I.** Larva in typical daytime resting pose. (Photos G–I: courtesy of D. J. Shetlar.)

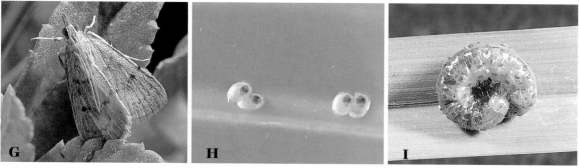

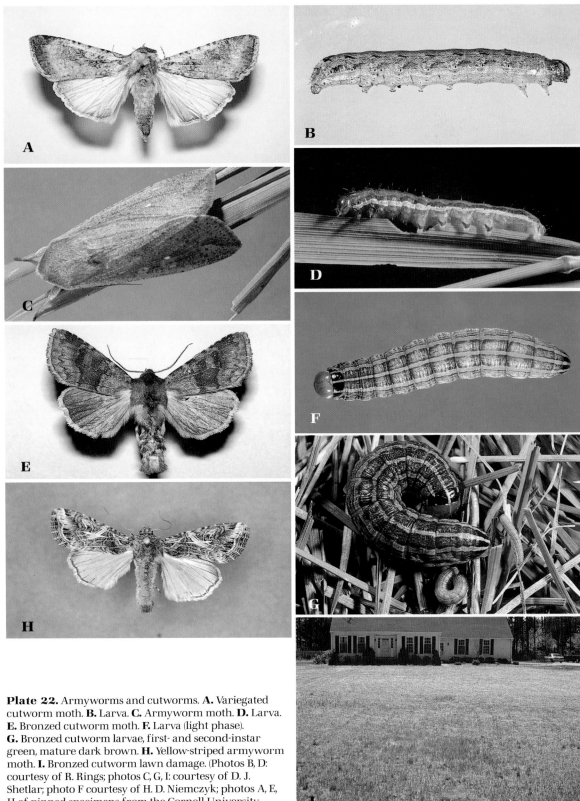

Plate 22. Armyworms and cutworms. **A.** Variegated cutworm moth. **B.** Larva. **C.** Armyworm moth. **D.** Larva. **E.** Bronzed cutworm moth. **F.** Larva (light phase). **G.** Bronzed cutworm larvae, first- and second-instar green, mature dark brown. **H.** Yellow-striped armyworm moth. **I.** Bronzed cutworm lawn damage. (Photos B, D: courtesy of R. Rings; photos C, G, I: courtesy of D. J. Shetlar; photo F courtesy of H. D. Niemczyk; photos A, E, H of pinned specimens from the Cornell University collection: NYSAES.)

Plate 23. Black cutworm, *Agrotis ipsilon*. **A.** Male moth (pectinate antennae). **B.** Female moth (filiform antennae). **C.** Eggs. **D.** Egg just hatching. **E.** First instar feeding on epidermis. **F.** Mature larva. **G.** Frass-filled tunnel in bentgrass turf. **H.** Characteristic larval feeding in aeration hole on golf green and on grass surrounding hole. **I.** Pupal terminalia male (left) and female (right). (Photo H courtesy of D. J. Shetlar; photos A–G, I: NYSAES.)

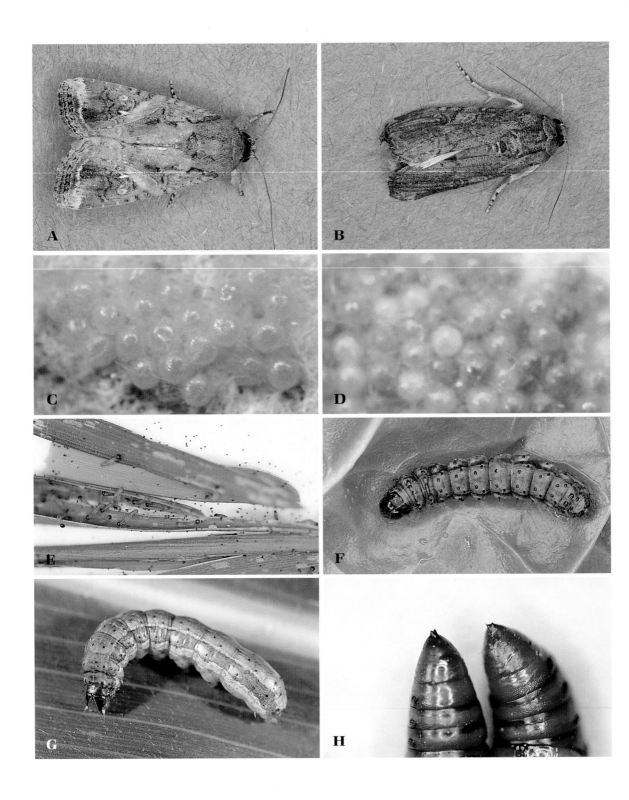

Plate 24. Fall armyworm, *Spodoptera frugiperda*. **A.** Male moth. **B.** Female moth. **C, D.** Eggs, color variation. **E.** First instar feeding on epidermis. **F, G.** Mature larvae, color variations. **H.** Pupal terminalia, male (left) and female (right). (Photos: NYSAES.)

Plate 25. Lawn armyworm, *Spodoptera mauritia*. **A.** Mating pair (male left). **B.** Characteristic damage to bermudagrass. **C.** Egg mass covered with abdominal hairs by young female. **D.** Aging female with abdominal hairs nearly spent. **E.** Naked egg mass deposited by aging female. **F.** Young larvae. **G.** Mature larva. **H.** Pupa. (Photo B courtesy of A. Ota; photos A, C–H: NYSAES.)

Plate 26. Fiery skipper, *Hylephila phyleus*. **A.** Male. **B.**. Female feeding on lantana, a favorite nectar plant. **C.** Eggs. **D.** First instar. **E.** Second instar. **F.** Mature larva. **G.** Young pupa. **H.** Male (left) and female (right) pupal terminalia. **I.** Pupa in loosely woven cocoon. (Photos: NYSAES.)

Plate 27. Progression of turf damage by scarabaeid grubs, similar for various species. **A.** Slight thinning of turf (lower left) in autumn. **B.** Noticeable thinning in early spring. **C.** Extensive thinning but still recoverable with fertilization and irrigation. **D.** Irreversible damage in need of reseeding. **E.** Grubs at soil-thatch interface with about half the grubs invisible in the soil. **F.** Typical renovation of small areas by removal of thatch and dead grass and reseeding. (Photo E courtesy of D. A. Potter; Photos A–D, F: NYSAES.)

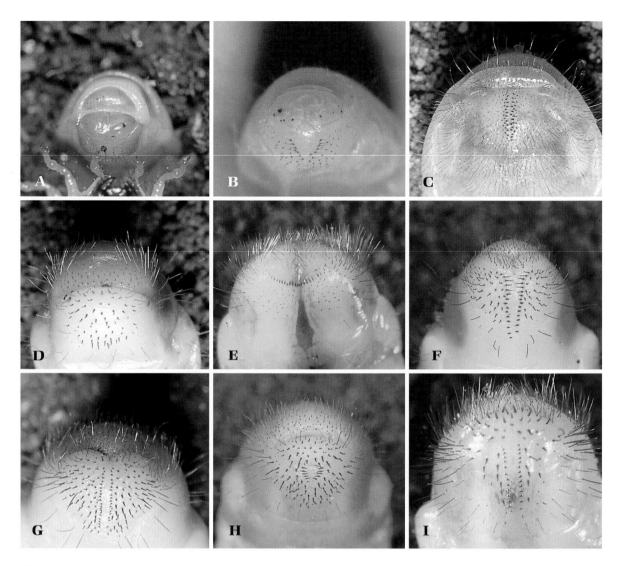

Plate 28. Rastral patterns in turf-damaging scarabaeid grubs. **A.** Black turfgrass ataenius. **B.** *Aphodius granarius*. **C.** Green June beetle. **D.** Masked chafers. **E.** Asiatic garden beetle. **F.** European chafer. **G.** May-June beetle, *Phyllophaga anxia*. **H.** Japanese beetle. **I.** Oriental beetle. Not to scale. (Photos: NYSAES.)

J, K, L: Rastral variations in May-June beetle (*Phyllophaga* spp.) grubs. (Photos: NYSAES.)

Plate 29. Scarabaeid oviposition sequence illustrated by the European chafer, dorsal views (left) and ventral views (right). **A.** Evagination of vagina as bulbous organ to compress soil to make an earthen cell. **B.** Egg deposition, with simultaneous invagination of vagina. **C.** Egg in partial cell in ventral view only. (Photos: NYSAES.)

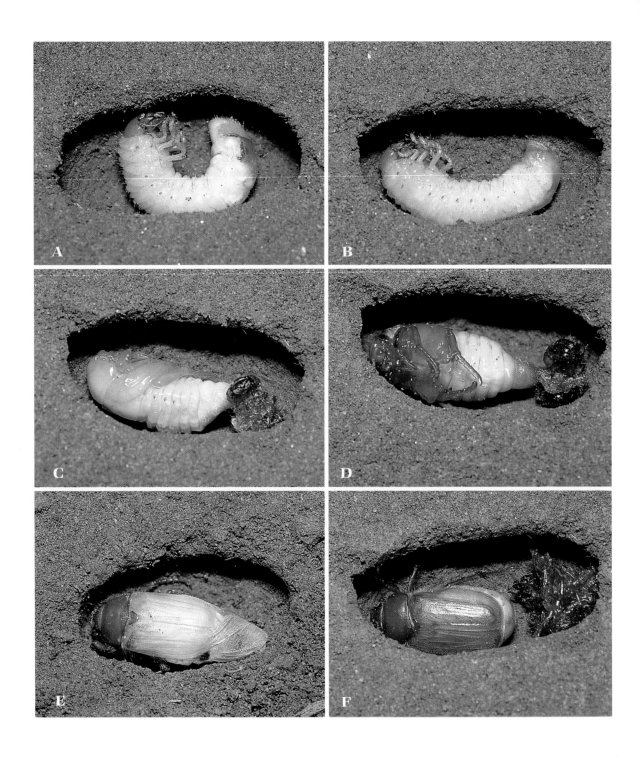

Plate 30. Scarabaeid larva-to-adult sequence in earthen cell illustrated by the European chafer. **A.** Larva in earthen cell. **B.** Prepupa. **C.** Young pupa. **D.** Mature pupa. **E.** Teneral adult with transparent elytra and unfolded wings. **F.** Mature adult ready to leave earthen cell. (Photos: NYSAES.)

Plate 31. Black turfgrass ataenius, *Ataenius spretulus*. **A.** Teneral to mature adults. **B.** Larva, pupa, and adult stages. **C.** Egg cluster. **D.** Fairway turf damage by larval feeding. **E.** Grubs in 0.1 m² at soil-thatch interface with about half the grubs visible and the remainder in soil and invisible. **F.** Actively feeding grubs. **G.** Pupae. (Photo C courtesy of H. D. Niemczyk; photos A, B, D–G: NYSAES.)

Plate 32. Black turfgrass ataenius phenological relationships. Start of egg deposition by overwintering beetles when: **A.** Vanhouette spirea in full bloom. **B.** Flowers. **C.** Horse chestnut in full bloom. **D.** Flower cluster. **E.** Black locust in early bloom. **F.** Flower cluster. Start of egg deposition by first generation beetles when: **G.** Rose of Sharon in full bloom. **H.** Flower. (Photos: NYSAES.)

Plate 33. Green June beetle, *Cotinis nitida*. **A.** Adult. **B.** Feeding on apricot. **C.** Eggs, young to mature. **D.** First instar crawling on its back. **E.** Mature larva. **F.** Turf damage caused by grubs entering and surfacing. **G.** Grub at bottom of vertical burrow with Japanese beetle grub on either side. (Photos A, E by M. P. Johnson and courtesy of D. A. Potter; photo F courtesy of R. L. Crocker; photo G courtesy of D. J. Shetlar; photos B–D: NYSAES.)

Plate 34. Green June beetle, *Cotinis nitida*. **A.** Earthen cell prepared by mature grub. **B.** Prepupa. **C.** Pupa. **D.** Teneral adult. **E.** Soil ball containing eggs. **F.** Same soil ball showing eggs in cells. **G.** Eggs (20) from soil ball in **E.** (Photos: NYSAES.)

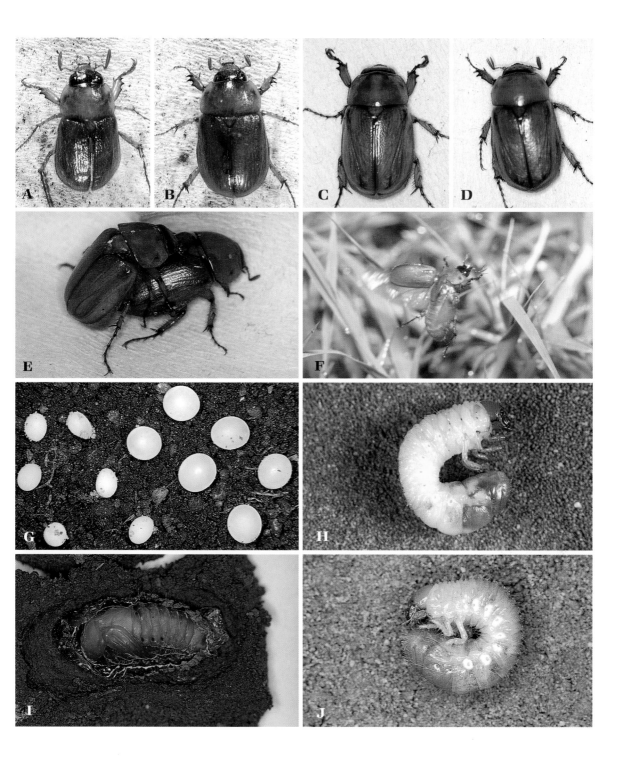

Plate 35. Masked chafers, *Cyclocephala* spp. **A.** Northern masked chafer, male. **B.** Female. **C.** Southern masked chafer, male. **D.** Female. **E.** Mating pair of southern masked chafer. **F.** Start of flight. **G.** Young to mature eggs. **H.** Mature grub of northern masked chafer showing typical midsection constriction. **I.** Pupa with split exuvia and in earthen cell cemented by secretion. **J.** Mature grub of *C. pasadenae* from California with typical white "halo" encircling each spiracle. (Photos: NYSAES.)

Plate 36. Asiatic garden beetle, *Maladera castanea*.
A. Adult. **B.** Ventral abdominal segments 5, 6, and 7, each
with a row of spines. **C.** Egg clusters and cemented cell.
D. Third instar. **E.** Bulbous stipe of left maxilla.
F. Prepupa. **G.** Pupa in cell cemented by secretion.
(Photos: NYSAES.)

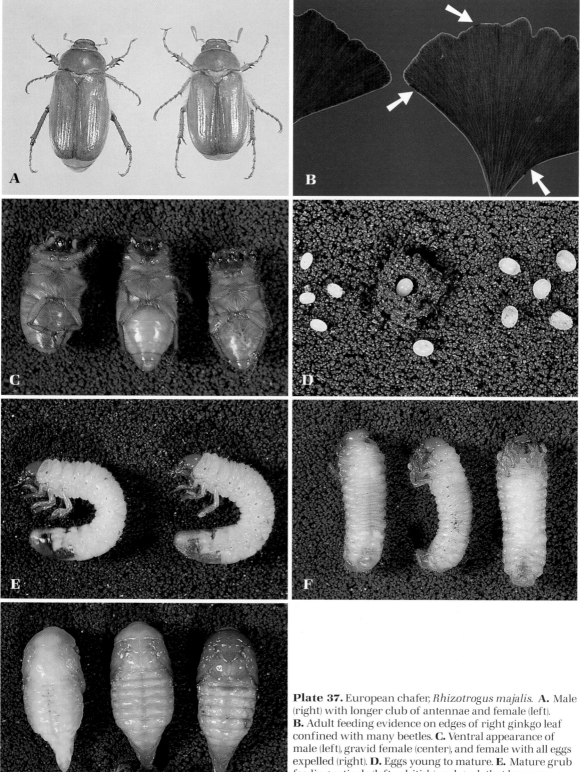

Plate 37. European chafer, *Rhizotrogus majalis*. **A.** Male (right) with longer club of antennae and female (left). **B.** Adult feeding evidence on edges of right ginkgo leaf confined with many beetles. **C.** Ventral appearance of male (left), gravid female (center), and female with all eggs expelled (right). **D.** Eggs young to mature. **E.** Mature grub feeding actively (left, whitish) and grub that has completed feeding (right, yellowish). **F.** Prepupae. **G.** Pupae young (left) to mature (right). (Photos: NYSAES.)

Plate 38. European chafer peak flights. **A.** Common catalpa in full bloom, period of peak beetle flights. **B.** Catalpa flower cluster. **C.** Individual flower. **D.** Beetle climbing grass stem to take flight. **E.** Mating flight to locust tree shortly after sunset. **F.** Mating pair. **G.** "Chains" of males formed under single females. **H.** Mass of beetles shaken from tree about 90 minutes after sunset. (Photos: NYSAES.)

Plate 39. Adults of common species of May-June beetles, *Phyllophaga* spp. **A.** *P. anxia.* **B.** *P. crinita.* **C.** *P. crenulata.* **D.** *P. ephilida.* **E.** *P. fraterna.* **F.** *P. fusca.* **G.** *P. gracilis.* **H.** *P. hirticula.* **I.** *P. implicata.* **J.** *P. inversa.* **K.** *P. rugosa.* **L.** *P. tristis.* To scale. (Photos of pinned specimens from the Cornell University collection: NYSAES.)

Plate 40. May-June beetle, *Phyllophaga latifrons*. **A.** Beetle. **B.** Male antennae (left) and female (right). **C.** First instars on penny. **D.** Third instar. **E.** Raster with appearance of straight line anal slit. **F.** Larval injury to spruce nursery stock. (Photos: NYSAES.)

Plate 41. Japanese beetle, *Popillia japonica*. **A.** Adult.
B. Forelegs of male beetle (left) with pointed tibial spur
and first tarsal segment wider than long, and female
(right) with spatulate tibial spur and first tarsal segment
longer than wide. **C.** Males attracted to single female as
she emerges to surface of turf. **D.** Mating pair. **E.** Eggs
young to mature. **F.** Mature grub. **G.** Pupa after splitting
of meshlike exuvia. (Photos: NYSAES.)

Plate 42. Japanese beetle adults feeding on favorite host plants. **A.** Yew. **B.** Larch. **C.** Linden. **D.** Linden tree after extensive feeding. **E.** Ornamental cherry. **F.** Peach. **G.** Rose. (Photos C, D: courtesy of D. A. Potter; photo G by W. Mesner and courtesy of D. A. Potter; photos A, B, E, F: NYSAES.)

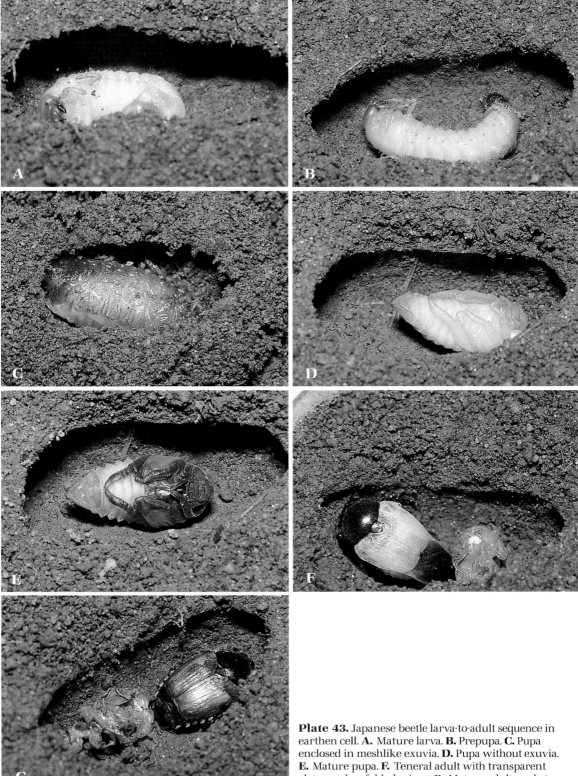

Plate 43. Japanese beetle larva-to-adult sequence in earthen cell. **A.** Mature larva. **B.** Prepupa. **C.** Pupa enclosed in meshlike exuvia. **D.** Pupa without exuvia. **E.** Mature pupa. **F.** Teneral adult with transparent elytra and unfolded wings. **G.** Mature adult ready to emerge. (Photos: NYSAES.)

Plate 44. Oriental beetle, *Exomala orientalis*. **A.** Canada hemlock nursery planting with heavy grub infestation to soil depths of 30 cm. **B.** Adult of the most common color. **C.** Antennal differences, male (left) and female (right). **D.** Adults with color extremes. **E.** Mating (occurs readily in captivity). **F.** Male beetle in wind tunnel showing response (antennae) when pheromone released. **G.** Eggs, young to mature. **H.** Prepupa. **I.** Pupa with split exuvia. (Photos: NYSAES.)

Plate 45. Dichondra flea beetle, *Chaetocnema repens*.
A. Adult. **B.** Normal dichondra leaves. **C.** Adult feeding damage. **D.** Lawn damage primarily by larval feeding. **E.** Unusually large femur of beetle. **F.** Eggs. **G.** Larva. (Photo D courtesy of L. Brown; photos A–C, E-G: NYSAES.)

Dichondra flea beetle on bermudagrass. **H.** Feeding damage on leaves. **I.** Turf killed by flea beetle. (Photos H, I: courtesy of J. Kollenkark.)

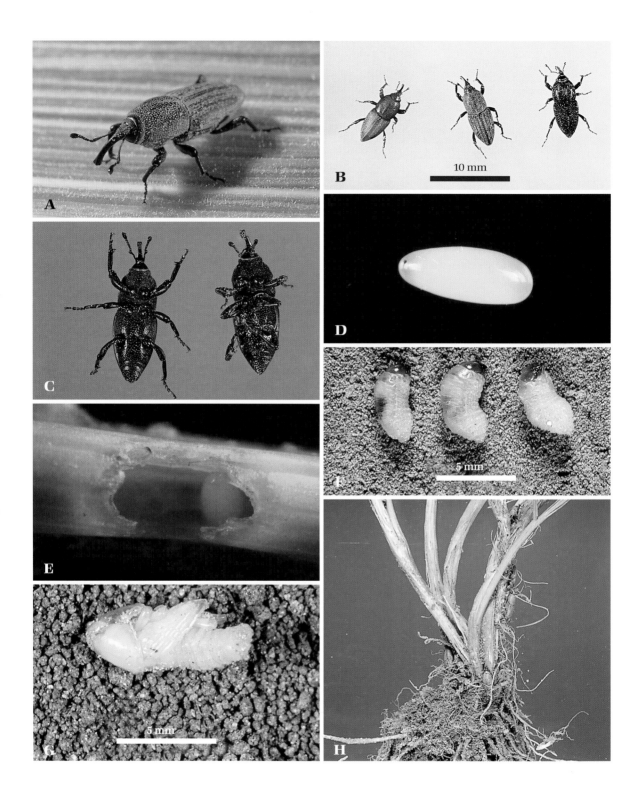

Plate 46. Bluegrass billbug, *Sphenophorus parvulus*. **A.** Adult. **B.** Teneral to mature adults. **C.** Ventral view male (left) with rounded tip of abdomen and female (right) with a more pointed tip of abdomen. **D.** Egg. **E.** Egg inside grass stem. **F.** Larvae. **G.** Pupa. **H.** Adults feeding on perennial ryegrass. (Photo A courtesy of D. J. Shetlar; photos B–H: NYSAES.)

Plate 47. Bluegrass billbug-damaged lawns. **A.** Larval frass resembling sawdust, an indication of billbug feeding. **B.** Early damage. **C.** Damage prevented (foreground) with insecticide treatment. **D.** Extensive irreversible damage. (Photos: NYSAES.)

Other important *Sphenophorus* spp. **E.** Hunting billbug. **F.** Hunting billbug adult feeding on stem. **G.** Phoenician billbug. **H.** Denver billbug. (Photos E, G, H: courtesy of D. J. Shetlar; photo F: NYSAES.)

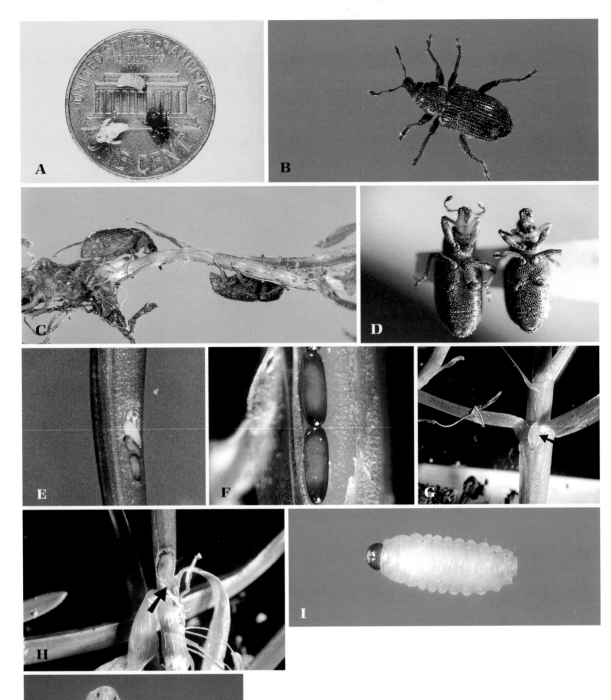

Plate 48. Annual bluegrass weevil, *Listronotus maculicollis.*
A. Larva, pupa, and adult on penny. **B.** Freshly emerged adult with full complement of scales and hairs. **C.** Teneral and mature adults. **D.** Male (left) with shallow median depression on first abdominal sternum and female (right) without such a depression. **E.** Eggs, fresh (light) and mature (dark) in stem. **F.** Mature eggs, leaf sheath removed. **G.** Second instar on stem. **H.** Third instar completely occupying stem. **I.** Mature larva. J. Pupae. (Photos A, G, H: courtesy of R. S. Cameron; photos B–F, I, J: NYSAES.)

Plate 49. Annual bluegrass weevil damage to annual bluegrass. **A.** Adults feeding on leaves and stem. **B.** Larval feeding on leaves and stems. **C.** Fifth instar feeding on crown. **D.** Early season damage predominantly at edge of fairway. **E.** Extensive destruction of annual bluegrass while a bentgrass rectangle remains healthy. **F.** Damage to collar. **G.** Damage to annual bluegrass invading a green. (Photos B, C, E, F: courtesy of R. S. Cameron; photos A, D, G: NYSAES.)

Plate 50. Annual bluegrass weevil plant relationships. **A.** White pines in rough (primary overwintering site). **B.** White pine litter containing over 400 weevils per 0.1 m² of litter. **C.** *Forsythia* in full bloom, period of weevil movement out of hibernation. **D.** Flowers. **E.** Flowering dogwood, pink. **F.** Flowering dogwood, white. **G.** White bracts, period when nearly all weevils have left hibernation sites. (Photos: NYSAES.)

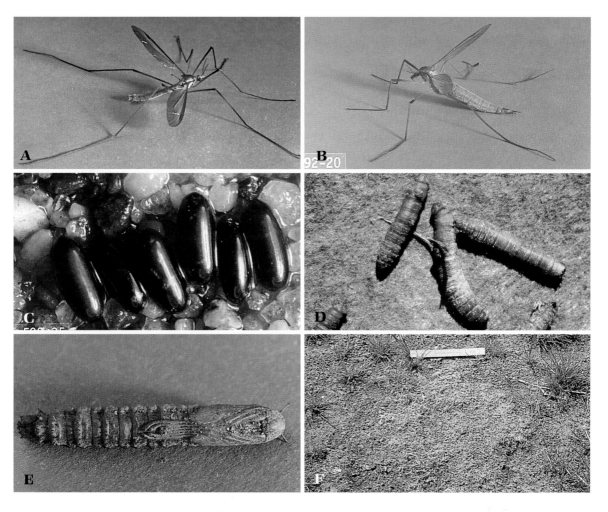

Plate 51. European crane fly, *Tipula palidosa*. **A.** Adult male. **B.** Adult female. **C.** Eggs. **D.** Larvae. **E.** Pupa. **F.** Severe lawn damage. (Photos A–C, E: courtesy of Ken Gray Collection, Department of Entomology, Oregon State University; photo D courtesy of A. L. Antonelli; photo F courtesy of S. J. Collman.)

Frit fly, *Oscinella frit*. **G.** Adult. **H.** Larva. (Photos courtesy of M. Tolley.)

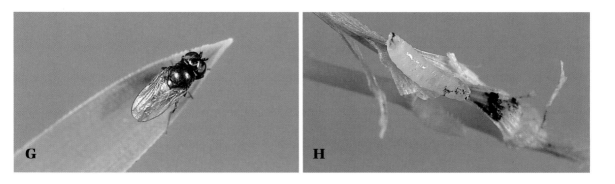

Plate 52. Ants in turf. **A.** "Anthill" in lawn. **B.** Subterranean galleries of A. **C.** Workers, larvae, and pupa from A. **D.** Anthills in golf course sand trap. (Photos: NYSAES.)

Red imported fire ant, *Solenopsis invicta.* **E.** Anthill around seedling pine. **F.** Mound in St. Augustinegrass turf. **G.** Winged female. **H.** Fire ant workers. (Photo E courtesy of P. P. Cobb; photos F–H: NYSAES.)

Plate 53. Minor turfgrass insects: Orthoptera, Hemiptera, Diptera. **A.** Northern mole cricket. **B.** Short-tailed cricket. **C.** Grasshopper. **D.** Periodical cicada. **E.** Leafhopper on Kentucky bluegrass. **F.** Leafhopper on bermuda-grass. **G.** March fly, *Bibio* sp. (Photos A, B, E, F: courtesy of D. J. Shetlar; photo G of pinned specimen from the Cornell University collection; photos C, D, G: NYSAES.)

Plate 54. Minor turfgrass insects: Hemiptera, Homoptera. Bermudagrass scale. **A.** Female. **B.** Crawlers. **C.** Bermudagrass normal (left) and scale-damaged stem (right). (Photos: NYSAES.)

Turfgrass scale. **D.** Female. **E.** Eggs in cottony mass. **F.** Infested turf. (Photos courtesy of M. K. Sears.)

Cottony grass scale. **G.** Scales on stem. **H.** Crawlers. **I.** Scale-damaged lawn. (Photos courtesy of L. M. Vasvary.)

Plate 55. Minor turfgrass insects: Lepidoptera. **A.** Burrowing sod webworm moths, female (left) and male (right). **B.** Larva. **C.** Cigarette paper sod webworm male moth. **D.** Larva. **E.** Pupal chamber liners (source of common name). **F.** Striped grassworm. **G.** Larva. **H.** Granulate cutworm moth. **I.** Lucerne moth. (Photos A–G: courtesy of D. J. Shetlar; photos H, I of pinned specimens from the Cornell University collection: NYSAES.)

Plate 56. Minor turfgrass insects: Dermaptera and Coleoptera. **A.** Earwig female (above), male (below). **B.** Corn wireworm adult. **C.** Larvae. **D.** *Polyphylla comes* male (left), female (right). **E.** Vegetable weevil females (no males occur). **F.** Vegetable weevil larva on dichondra. (Photo A courtesy of J. Kalisch; photo D courtesy of M. Klein; photo F courtesy of H. D. Niemczyk; photo E of pinned specimens from the Cornell University collection; photos B, C, E: NYSAES.)

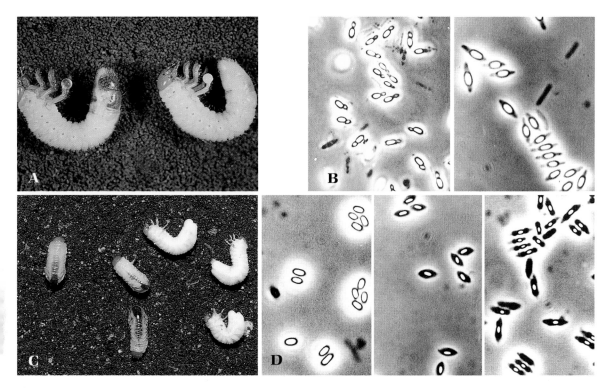

Plate 57. Entomopathogenic bacteria. **A.** Japanese beetle grub normal (left) and milky disease–infected (right) with "milky" blood exudate from snipped leg. **B.** Spore forms type A, *Bacillus popilliae* (left) and type B, *B. lentimorbus* (right). **C.** Black turfgrass ataenius normal (left) and milky diseased grubs (right). **D.** Three spore types in ataenius grubs. Only one type of spore is found in any given milky disease infection. (Photos: NYSAES.)

Entomopathogenic fungi. **E.** *Beauveria bassiana*–infected chinch bug. **F.** *Metarrhizium anisopliae*–infected tawny mole cricket. **G, H, I**: Conidial maturation of *Metarrhizium anisopliae* on Japanese beetle third instars. (Photo E courtesy of M. K. Sears; photos F–I: NYSAES.)

Plate 58. Entomopathogenic nematodes. **A.** Japanese beetle grubs healthy (left) and three infected with *Heterorhabditis bacteriophora*. **B.** Pyralid larvae healthy (left) and three infected with nematode species: *Steinernema carpocapsae, H. bacteriophora*, and *S. glaseri*, respectively. **C.** Japanese beetle grub infected with *H. bacteriophora*, with reproductive nematodes visible through cuticle. **D.** Release of infective stage *H. bacteriophora* from ruptured grub. **E.** Ambush nematode on soil particle (shiny, slender, upright). **F.** *H. bacteriophora* infectives in water column. (Photos: NYSAES.)

Sublethal effects of insect growth regulator on Japanese beetle grubs. **G.** Grub unable to shed cuticle. **H.** Cuticular growths. (Photos: NYSAES.)

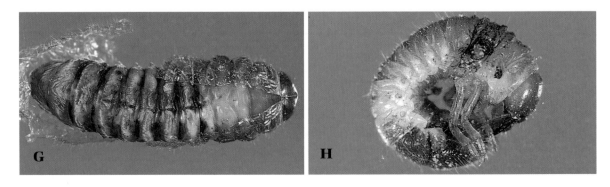

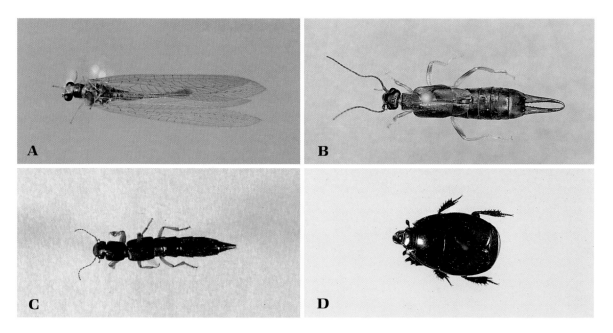

Plate 59. Arthropod predators of turfgrass insects. **A.** Green lacewing, *Chrysopa* sp. **B.** Earwig, *Labidura riparia*, predator of southern chinch bug. **C.** Rove beetle. **D.** Hister beetle.

Four common ground beetles, Carabidae. **E.** Bombardier beetle, *Brachinus* sp. **F.** Searcher, *Calosoma scrutator.* **G.** Fiery hunter; *C. calidum.* **H.** *Harpalus caliginosus.* (Photos A–H of pinned specimens from the Cornell University collection: NYSAES.)

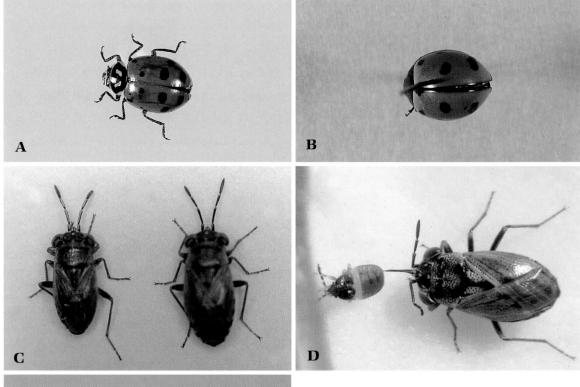

Plate 60. Arthropod predators of turfgrass insects. Two ladybird beetles. Coccinellidae. **A.** *Hippodamia convergens.* **B.** *Coccinella 9-notata.* **C.** Big-eyed bug, *Geocoris* sp. male (left) and female (right). **D.** Big-eyed bug feeding on chinch bug nymph. **E.** Assassin bug.

Three tiger beetles, Cicindellidae. **F.** *Tetracha carolina.* **G.** *T. virginica.* **H.** *Megacephala virginica.* (Photos A–H of pinned specimens from the Cornell University collection: NYSAES.)

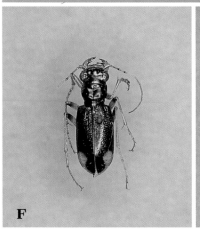

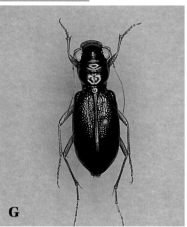

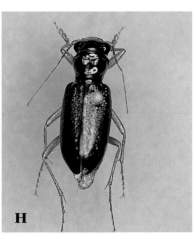

Plate 61. Arthropod predators of turfgrass insects.
A. Flower fly, *Syrphus* sp. **B.** Bee fly, *Exoprosa fasciata,*
predator of May-June beetle pupae. **C.** Flesh fly,
Sarcophaga sp. **D.** Robber fly, *Diogmites discolor,*
predator of May-June beetle pupae. **E.** Spider with egg
sac. **F.** Predatory mite. **G.** Centipede. (Photos A–D of
pinned specimens from the Cornell University
collection; Photos A–G: NYSAES.)

Plate 62. Parasitoids of turfgrass insects. **A.** Humped-back fly. **B.** Aphidiid wasp, parasitoid of aphids. **C.** Light fly, *Pyrgota undata*, parasitoid of May-June beetle adults. **D.** Tachinid, red-eyed fly, *Ormia* sp., parasitoid of mole crickets. **E.** Tachinid, "winsome fly", *Istocheta aldrichi*, egg attached to Japanese beetle adult thorax. **F.** Tachinid larva feeding on *Phyllophaga anxia* grub. (Photo E courtesy of M. Klein; photos A–D of pinned specimens from the Cornell University collection: NYSAES; photo F: NYSAES.)

Tiphiid wasps. **G.** *Myzine* sp. parasitoid of May-June beetle grubs causing permanent paralysis. **H.** *Tiphia* sp. parasitoid of May-June beetle grubs causing temporary paralysis. **I.** *Tiphia* larva attached and feeding. **J.** *Tiphia* larva nearly full grown. **K.** *Tiphia* cocoon with adult exit hole. (Photo G of pinned specimens from the Cornell University collection; photos G–K: NYSAES.)

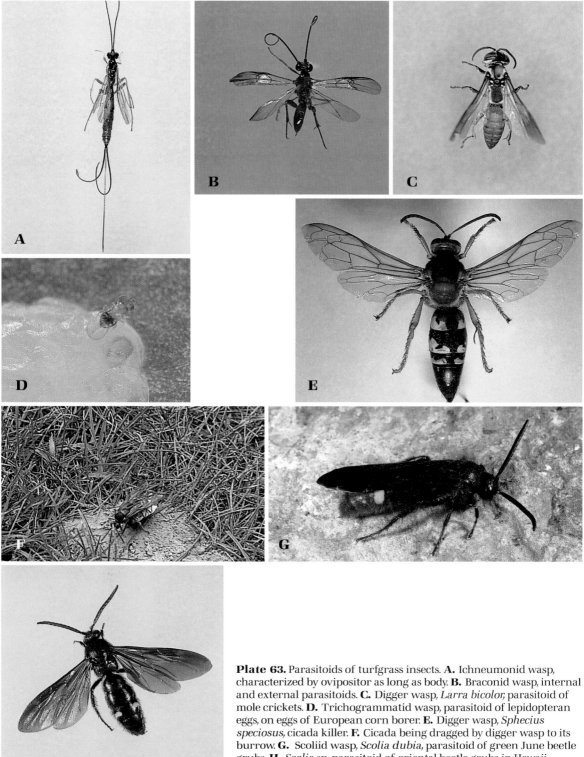

Plate 63. Parasitoids of turfgrass insects. **A.** Ichneumonid wasp, characterized by ovipositor as long as body. **B.** Braconid wasp, internal and external parasitoids. **C.** Digger wasp, *Larra bicolor*, parasitoid of mole crickets. **D.** Trichogrammatid wasp, parasitoid of lepidopteran eggs, on eggs of European corn borer. **E.** Digger wasp, *Sphecius speciosus*, cicada killer. **F.** Cicada being dragged by digger wasp to its burrow. **G.** Scoliid wasp, *Scolia dubia*, parasitoid of green June beetle grubs. **H.** *Scolia* sp. parasitoid of oriental beetle grubs in Hawaii. (Photo D courtesy of D. N. Ferro; Photo F courtesy of D. J. Shetlar; photo G by M. P. Johnson, courtesy of D. A. Potter; photos A–C, E, H of pinned specimens from the Cornell University collection: NYSAES.)

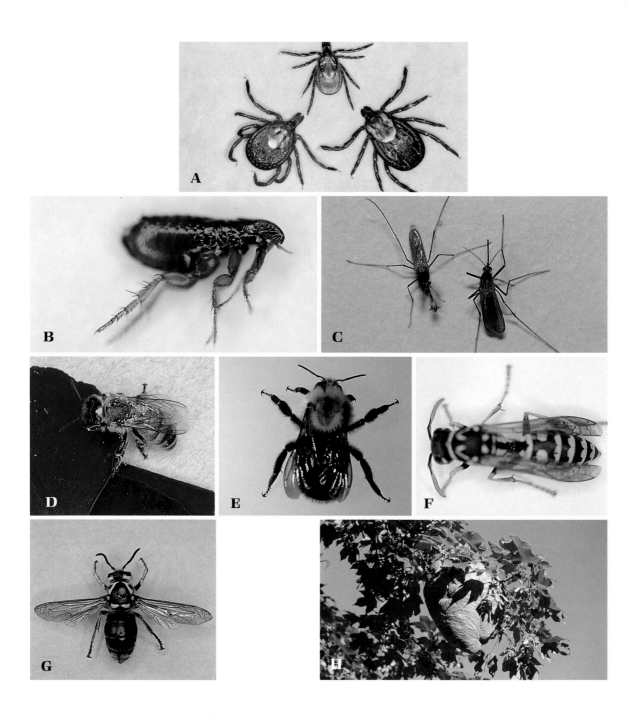

Plate 64. Biting and stinging arthropods associated with turfgrass situations. **A.** Ticks: deer tick (top), lone star tick (left), American dog tick (right). **B.** Cat flea. **C.** Mosquito, male, nonbiting (left) and female (right). **D.** Honeybee. **E.** Bumblebee. **F.** Yellow jacket. **G.** White-faced hornet. **H.** White-faced hornet nest. (Photo A courtesy of J. Kalisch; photo H by M. F. Potter; courtesy of D. A. Potter; photo G of pinned specimen from the Cornell University collection; photos B–G: NYSAES.)

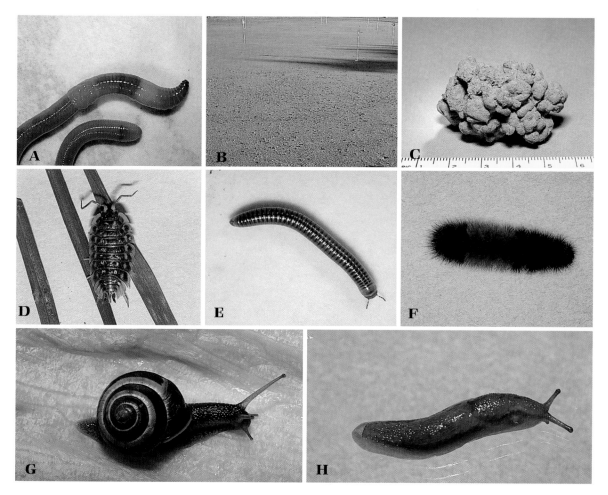

Plate 65. Miscellaneous turfgrass-associated organisms. **A.** Earthworm showing clitellum (swollen segments 31–37 having reproductive functions). **B.** Earthworm castings on golf practice green. **C.** Large dried casting adding to roughness of fine turf. **D.** Sowbug. **E.** Millipede. **F.** Woolly bear caterpillar. **G.** Snail. **H.** Slug. (Photos: NYSAES.)

Spotted pelidnota, *Pelidnota punctata*, a scarabaeid grub often found under soil surface in decaying wood of tree trunks, not injurious to turf. **I.** Mature grub. **J.** Rastral pattern. **K.** Adult. (Photo K of pinned specimen from the Cornell University collection; photos: NYSAES.)

Plate 66. Avian pests and beneficial predators. **A.** English starling. **B.** Starling peck holes from searching for grubs. **C.** Starling flock over grub-infested golf course. **D.** Starlings feeding on European chafer grubs in home lawn. **E.** Common grackle. **F.** Common crow. **G.** Cattle egret. **H.** Canada geese. (Photo A obtained from the Cornell University Laboratory of Ornithology, donated by D. P. H. Watson; E, obtained from the Cornell University Laboratory of Ornithology, donated by A. Cruickshank; F, obtained from the Cornell University Laboratory of Ornithology; G, obtained from the Cornell University Laboratory of Ornithology, donated by L. Wales; photo H by P. J. Vittum; photos B–D: NYSAES.)

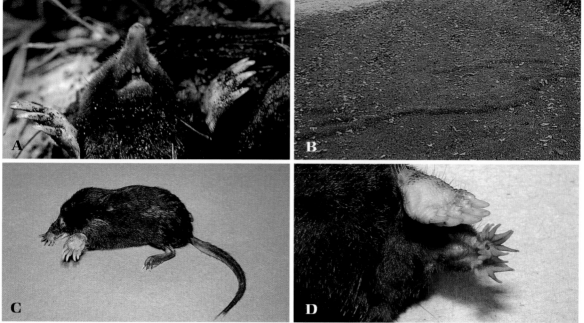

Plate 67. Moles. **A.** Eastern mole. **B.** Ridges produced by eastern mole. **C.** Star-nosed mole. **D.** Star-nosed mole anterior. **E.** Mounds of fresh soil produced by star-nosed mole. (Photo A by G. C. Hickman, obtained from the American Society of Mammalogists Slide Library; photo B courtesy of D. J. Shetlar; photos C–E: NYSAES.)

Voles. **F.** Meadow vole. **G.** Evidence of vole activity under snow. **H.** Same lawn after snow melt. (Photo F by L. L. Masters, obtained from the American Society of Mammalogists Slide Library; photos G, H: NYSAES.)

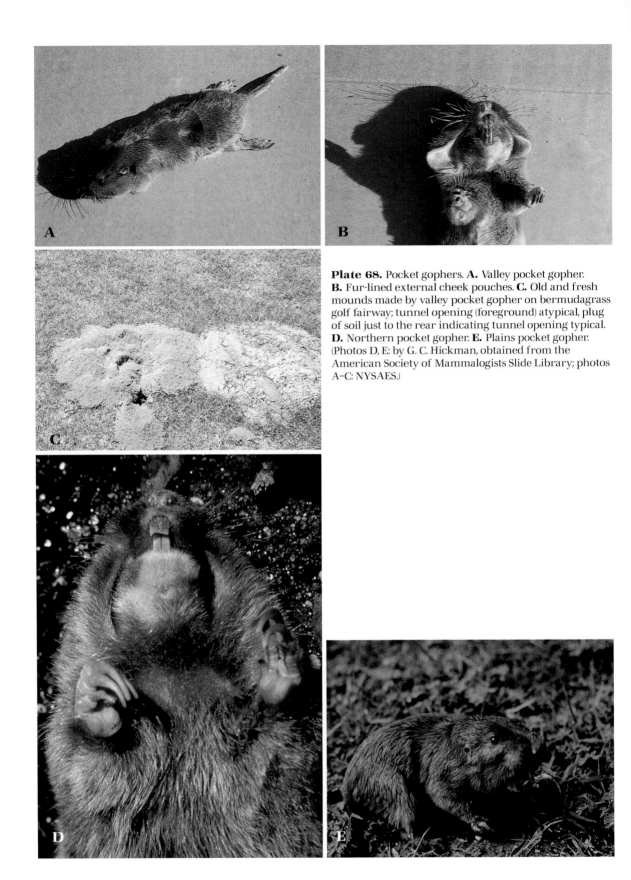

Plate 68. Pocket gophers. **A.** Valley pocket gopher. **B.** Fur-lined external cheek pouches. **C.** Old and fresh mounds made by valley pocket gopher on bermudagrass golf fairway; tunnel opening (foreground) atypical, plug of soil just to the rear indicating tunnel opening typical. **D.** Northern pocket gopher. **E.** Plains pocket gopher. (Photos D, E: by G. C. Hickman, obtained from the American Society of Mammalogists Slide Library; photos A–C: NYSAES.)

Plate 69. Skunks, raccoons, and armadillos. **A.** Striped skunk. **B.** Turf uprooted by skunks feeding on grubs. **C.** Raccoon. **D.** Turf uprooted by raccoons feeding on grubs. **E.** Nine-banded armadillo. **F.** Armadillo burrow in bermudagrass turf. (Photos A, C, E: by T. Yates, W. F. Berliner, and L. L. Master, respectively, obtained from the American Society of Mammalogists Slide Library; photos B, D, F: NYSAES.)

Plate 70. Turfgrass insect survey and sampling techniques. **A.** Lemon-scented detergent for flushing out mole crickets and cutworms. **B.** Damage rating frame, 0.6-m² grid with nine squares, used for assessing mole cricket damage. **C.** Cylinder, open top and bottom, to press into soil and fill with water for chinch bug flotation. **D.** Pitfall trap to capture insects walking on surface. **E.** Wing trap for capturing flying insects, species depending on pheromone used. **F.** Japanese beetle trap useful for capturing several species of scarabaeid adults depending on female sex pheromone or floral scent used. (Photo A by P. J. Vittum; photo B courtesy of R. Brandenburg; photos C–F: NYSAES.)

Plate 72. Turfgrass thatch. **A.** Thatch 8–10 cm thick on home lawn. **B.** Golf course fairway turf where oriental beetle grubs remained exclusively in thick thatch. (Photos: NYSAES.)

Plate 71. Turfgrass insect survey and sampling techniques. **A.** Cutting and peeling back sod (particularly for black turfgrass ataenius counts). **B.** Digging and sifting through several centimeters of soil for scarab grub counts. **C.** Golf cup cutter used for sampling soil (10.8-cm diameter and variable depth). **D.** Golf cup cutter in use. **E.** Best location of black-light trap for capturing overwintering black turfgrass ataenius adults. **F.** Blacklight trap for general capture of night-flying insects. (Photos B, E: by P. J. Vittum; photos A, C, D, F: NYSAES.)

12

Scarabaeid Pests:
Subfamily Aphodinae

Black Turfgrass Ataenius

Taxonomy

The black turfgrass ataenius (BTA), *Ataenius spretulus* (Haldeman), in the order Coleoptera, family Scarabaeidae, subfamily Aphodinae, is one of 63 *Ataenius* species recognized in the United States and Canada. The BTA was previously named *Ataenius cognatus* (Lec.). It is called the *black fairway beetle* in Canada (Cartwright 1974, Fushtey and Sears 1981, Hoffman 1935).

Importance

The BTA was considered only an incidental turfgrass pest prior to the 1970s. Since then it has become a serious pest on golf course fairways, greens, and tees over a wide area of North America. It has been reported as a damaging turf pest in at least 12 midwestern and northeastern states and in Ontario, Canada. It also occurs on golf course putting greens throughout much of California, wherever bentgrass or annual bluegrass is present (R. Cowles, Agricultural Experiment Station, Windsor, Conn., personal communication, 1998; Niemczyk and Dunbar 1976; Vittum 1995b; Weaver and Hacker 1978).

Because the BTA caused serious damage on golf courses in the 1970s, two studies were conducted and provide the most detailed information available on this insect. The work of Weaver and Hacker (1978) was conducted in southwestern West Virginia during 1976–77. The work of Wegner and Niemczyk (1981) was conducted during 1976–78 in the vicinity of Cincinnati, Ohio. These two areas are within 1° latitude of each other and are similar in elevation.

History and Distribution

The BTA first was described as a turfgrass pest in Minnesota in 1932, when it was found killing grass on golf course greens and fairways. The next report of turfgrass damage due to BTA came from New York State near the city Rochester (Monroe County) in 1969, and in Orange County in 1970, when the insect was found damaging fairway turf. During 1973 it was discovered damaging fairways on a Cincinnati, Ohio, golf course. During the next 2

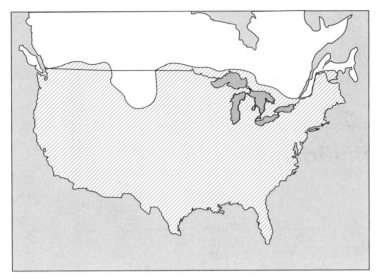

Figure 12-1. Distribution of black turfgrass ataenius. (Drawn by E. Gotham, NYSAES, adapted from Handbook of Turfgrass Insect Pests, Brandenburg and Villani 1995, Entomological Society of America.)

years, reports of fairway damage in this area became numerous. Other areas reporting damage included 12 midwestern and eastern states, Washington, D.C., and Ontario, Canada. In 1988 damage was reported from Palm Desert, California. Since then, this insect has become prevalent on California golf courses from San Diego County to San Francisco and Sacramento (Gelernter 1996, Hoffman 1935, Kawanishi et al. 1974, Niemczyk 1976).

The BTA is native to North America, and has been collected from most of the 48 contiguous states (Figure 12-1). By 1978 it was reported damaging golf course turfgrass in 24 states plus the District of Columbia and Ontario, Canada (Cartwright 1974, Niemczyk and Wegner 1979).

Host Plants and Damage

The BTA is primarily a pest of golf course turf. Fairways, greens, tees, and bentgrass nurseries are damaged as the grubs feed on the roots of annual bluegrass, Kentucky bluegrass, perennial ryegrass, and bentgrasses. Damage sometimes is observed in roughs as well. The first symptom of injury is that turf wilts despite an abundance of moisture. Wilted areas are most visible when viewed toward the sun. Under the stress caused by summer heat and low moisture, compounded by the feeding of grubs on the roots, turf dies in small, irregular patches that eventually coalesce to produce large dead turf areas (Plate 31).

Since the roots are devoured at the soil-thatch interface, the weakened and dead turf is removed easily to expose numerous mature larvae, often accompanied by prepupae, pupae, and even teneral and mature adults (Plate 31). Additional individuals often are present in the soil. As with other scarab grub infestations, birds, skunks, raccoons, or other animals can cause significant damage as they peck or dig up turf in the process of looking for beetles and grubs. Infestations of home lawns generally are considered to be only an incidental problem (Niemczyk and Wegner 1982).

Description of Stages

Adult

The adult BTA is a shiny black to reddish-brown beetle; it is thought that adults darken from red to black as they mature. BTA adults are considerably smaller than most other scarabs invading turfgrass, with a mean length of 4.9 mm and a mean width of 2.2 mm (Figure 12-2, Plate 31) (Weaver and Hacker 1978, Wegner and Niemczyk 1981).

Egg

Mature eggs average 0.7 mm in length and 0.5 mm in width. The shiny white eggs are deposited in clusters of about a dozen within a cavity formed by the female (Plate 31) (Wegner and Niemczyk 1981).

Larva

The three larval instars are scarabaeid in appearance (Plate 31). Head capsule widths increase from a mean of 0.5 mm in first instars, to 0.83 mm in second instars, to 1.3 mm in third instars. A full-grown third instar has a mean length of 8.5 mm and closely resembles (in size and appearance) late first instars of larger scarabaeids, such as the European chafer and the Japanese beetle (Figure 12-2). These species occur concurrently from July to early August in western New York. No distinct rastral pattern is present, but there are 40–45 irregularly placed hamate setae. The most distinguishing feature is the presence of two pad-like structures on the tip of the abdomen just posterior to the setae and anterior to the anal slit (Plates 28, 31) (Wegner and Niemczyk 1981).

Pupa

The mean pupal length and width are 4.7 mm and 2.5 mm, respectively (Plate 31). The sexes can be differentiated in the pupal stage by the presence of an aedeagal protuberance in males that is located on the ventral posterior region of the abdomen.

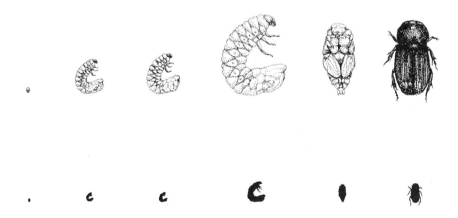

Figure 12-2. Stages of development of black turfgrass ataenius: egg; first, second, and third larval instars; pupa; and adult. Silhouettes show actual size of each stage. (Courtesy of R. Jarecke, NYSAES.)

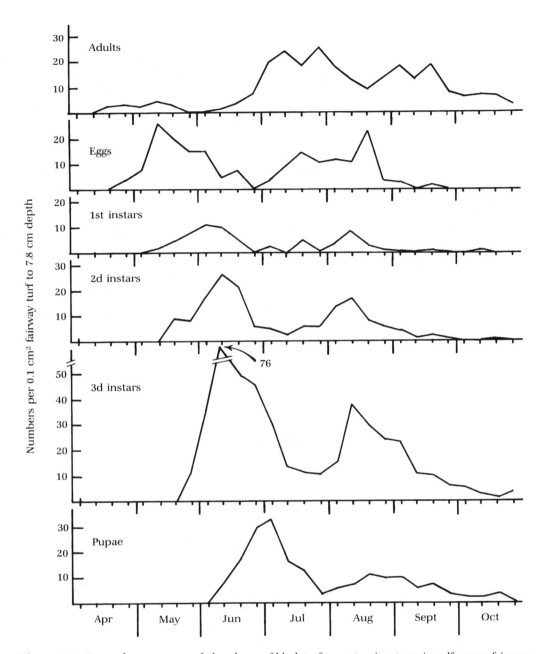

Figure 12-3. Seasonal occurrence and abundance of black turfgrass ataenius stages in golf course fairways in and near Cincinnati, Ohio, during 1977 and 1978. (Data from Wegner and Niemczyk 1981.)

Seasonal History and Habits

Seasonal Cycle and Generations

The BTA has either one or two generations each year east of the Rocky Mountains, depending on the latitude, but at least three generations often occur each year in southern California (Gelernter 1996). Two generations occur in Connecticut, northern Ohio, and southward. In western New York, Ontario, Canada, and Minnesota, the BTA is generally considered to have a single generation each year. In southern Ohio, oviposition begins in May and July. First-generation adults emerge between late June and early July, and second-generation adults emerge in August (Niemczyk and Dunbar 1976, Weaver and Hacker 1978, Wegner and Niemczyk 1981).

Seasonal Occurrence in Ohio

Wegner and Niemczyk (1981), in southern Ohio, conducted a comprehensive study on the presence of turfgrass scarabaeids in the turfgrass soil. Figure 12-3 shows the mean populations of all stages present per 0.1 m^2 in turfgrass and soil at quarter-month intervals from April through October. Data from two golf courses sampled in 1977 and a third golf course sampled in 1977 and 1978 were consolidated into Figure 12-3. The period required for the BTA to pass through one generation was 65 days ± 5 days at soil temperatures of 25° ± 6°C.

The largest population of adults occurring in the turfgrass during July in southern Ohio coincides with the capture of adults in blacklight traps in West Virginia where more than 60% of the total seasonal catch was collected between 27 June and 25 July. First-generation eggs peaked more sharply than second-generation eggs, indicating the second-generation oviposition occurred over a slightly longer period. Third instars are the most readily observed stage because of their size and position at the soil-thatch interface. Third instars of the first generation peaked in June, while third instars of the second generation peaked in August (Weaver and Hacker 1978, Wegner and Niemczyk 1981).

Adult Activities

Overwintering. In northern locations (e.g., Canada, northeastern and midwestern United States), adults leave golf course fairways during September or October and seek overwintering sites at the edge of wooded areas in the roughs and perimeter of golf courses. More than 90% of second-generation females going into hibernation are inseminated. Mature adults that go into hibernation have a high rate of survival, from 90% to 96%; larvae and pupae do not appear to be able to withstand winter temperatures. Males and females overwinter with equal success. In West Virginia, a preferred overwintering site in wooded areas is in accumulated pine needle litter and in the upper 5 cm of loose, well-drained soil. As many as 77 beetles have been found per 0.1 m^2 in such areas (Weaver and Hacker 1978, Wegner and Niemczyk 1981). In Minnesota, beetles overwintered in the upper 15 cm (moist layer) of waste piles of Milorganite and grass cuttings (Hoffman 1935, Wegner and Niemczyk 1981).

Adult Emergence and Dispersal. Adults return from hibernation sites to the golf course fairways starting in late March and continue through April and early May. On warm afternoons (4:00–6:00 PM), swarms of adults can be seen flying over the turf or walking on putting greens. Adults alight and immediately burrow into the turf (P. J. Vittum, personal observation; Wegner and Niemczyk 1981).

In West Virginia, Weaver and Hacker (1978) established two blacklight traps on a golf course and monitored them at weekly intervals. One trap was located near a known overwintering site and the second was located about 1.2 km away at a non-overwintering site. Emergence from overwintering sites began in late March and continued through mid-April.

A mass flight of beetles occurred just before dusk, and adults were captured in the traps the same night. There was a sharp increase in flight activity during late June and early July, with peak flight activity occurring during mid-July and tapering off into early August. Adult activity declined sharply in October, presumably because of movement to overwintering sites (Weaver and Hacker 1978).

An increase in flight activity before and during light precipitation on warm evenings was observed in Ohio and appears to be typical of aphodiine scarabaeids. Flight studies made in Ohio show that screen sticky traps and traps fitted with blacklight fluorescent lamps and a vacuum fan are effective devices by which to determine flights and dispersal. The largest captures of beetles occur in April and again from July through October, with minimal captures during May and June. The largest captures range from 80 to more than 800 per trap per day (Wegner and Niemczyk 1981).

Oviposition and Phenological Relations. No information is available on BTA mating habits. In addition to the high degree of insemination found in overwintering females, most females found throughout the season have been inseminated. Females examined during the months May to October were 100%, 67%, 69%, 96%, 91%, and 94% inseminated, respectively (Wegner and Niemczyk 1981).

A degree-day (DD) system of predicting BTA activity and development is based on a flight activity threshold of 13°C. The seasonal life history of the BTA is also correlated with plant phenology (Plate 32). First-generation eggs appear after 100–150 DD $_{base13C}$ and coincided with the full bloom of the Vanhoutte spirea, the horse chestnut, and the earliest bloom of black locust. Normally these plants flower during the first half of May in southern Ohio and early June in New York. Second-generation eggs appear after 650–710 DD $_{base13C}$ when the rose of Sharon begins to bloom, as it normally would during the last half of July in southern Ohio. Generation periods of 60–70 days were observed in the field where soil temperatures approached or exceeded 30°C (Wegner and Niemczyk 1981).

These phenological relationships can be useful for initiating scouting and the timing of control measures. Field trials have demonstrated that certain insecticides, applied between forsythia and dogwood full bloom, reduce adult populations significantly before females have an opportunity to oviposit. The traditional approach continues to be effective if suitable insecticides are applied when Vanhoutte spirea or horse chestnut is in bloom. Since these are very common plants growing over wide geographic areas, the relationships should be helpful in timing insecticidal applications for adult control (Niemczyk and Wegner 1979, 1981).

Larval Activities

BTA eggs are deposited in clusters of 11–12 within a cavity formed by the females in the lower 5 mm of thatch and the upper 6 mm of soil among grass roots. For this reason, it is common to find first instars clustered together, with as many as 12 per 1–8 cm³. Second instars are further dispersed, with 12 per 8–100 cm³. Third instars are dispersed even further, with 12 per 100–700 cm³, and are most common at or near the soil-thatch interface (Wegner and Niemczyk 1981).

Larvae are most abundant in closely mowed areas containing susceptible grasses. In southern California and Arizona, where bermudagrass fairways prevail, infestations are largely restricted to greens, collars, and tees. Elsewhere, larvae are more abundant on fairways, with numbers declining fivefold in the rough, even when irrigation and grass type remain constant (R. Cowles, personal communication, 1998; Rothwell 1997).

The third instar, found most abundantly during June, July, and August in northern locations, is the most destructive stage. Because of their minute sizes, first and second instars are not considered to be very detrimental to turfgrass. In the latitude of southern Ohio and West Virginia, the second-generation third instars are most prevalent during August (Figure 12-3), while in the latitude of Minneapolis and western New York, the first-generation third instars are prevalent during July and August (Hoffman 1935, Wegner and Niemczyk 1981).

Upon cessation of feeding, larvae migrate 1–8 cm deep into the soil and excavate cells in which they remain as prepupae and pupae and finally as teneral adults.

Miscellaneous Features

Threshold Populations

The most efficient method to sample grub populations is to use a cup cutter (10.8-cm diameter) to collect cores to a depth of approximately 5 cm, and carefully break up the soil. Inspect the lower thatch areas as well, for evidence of larvae or their feeding damage. An economic threshold for grubs has not been determined, but levels as low as 30 larvae per 0.1 m^2 have been suggested. In Connecticut, larval populations of 100–150 per 0.1 m^2 caused extensive damage. In Ohio, populations of 250–300 per 0.1 m^2 were frequently associated with heavy damage. A high density of 578 grubs per 0.1 m^2 is recorded (Niemczyk and Dunbar 1976, Weaver and Hacker 1978).

Damage to turf caused by BTA is determined by the vigor of the turf and by the absolute population density of larvae. For example, heat-stressed and closely mowed bentgrass during summer in the Coachella Valley (southern California) cannot support more than 20 larvae per 0.1 m^2 without showing damage, while up to 250 larvae per 0.1 m^2 can thrive in the same species of grass in moderate conditions without visible symptoms. Normal thresholds for damage vary from 20 larvae per 0.1 m^2 in the Coachella Valley in July to roughly 100 larvae per 0.1 m^2 in western New York in late August. Thus, cool-season golf course turf under average conditions often should be able to tolerate 50 larvae per 0.1 m^2. When the turf is growing vigorously, even higher thresholds can be employed. Inadequate watering, aeration, inappropriate fertility, or low mowing heights can contribute to poor root systems and increase plant stress, thus necessitating use of lower tolerance levels (Vittum 1995b).

Sampling for Adults

An irritating drench, using a lemon-scented liquid dish detergent as a disclosing agent (see Chapter 26), can be used to bring adults to the surface, where they can be counted. There appears to be no correlation between the number of adults observed and the subsequent larval population that might develop in the area, but the technique can be used to determine when adults become active.

Natural Enemies

Efforts to use biological control agents to control BTA grub populations thus far have had mixed success. Some field trials in Rhode Island indicated that *Steinernema carpocapsae* (an entomogenous nematode) could suppress grub populations. However, the trials involved application rates 10 times higher than the currently recommended rate (Alm et al. 1992, Vittum 1995b).

Microorganisms

Milky Disease. A commonly observed natural pathogen found in BTA is milky disease. In a population of third instars, prepupae, and pupae ranging from 29 to 144 per 0.1 m^2 on a golf course fairway near Rochester, New York, in 1969, about 70% of the larvae showed milky disease symptoms, and after a few days at room temperature, the balance of the grubs also became milky. Symptoms were identical to those of Japanese beetle grubs infected with *Bacillus popilliae* Dutky or *Bacillus lentimorbus* Dutky (Plate 57). Spores of the bacillus causing milky disease in BTA resemble those of *B. lentimorbus* more than those of *B. popilliae.* The bacterium is, however, entirely different and causes no infection in grubs of the Japanese beetle, the European chafer, and the northern masked chafer, which are infected by *B. popilliae* and *B. lentimorbus.*

At least three morphologically different spore forms of the bacillus infect BTA (Plate 57) (Kawanishi et al. 1974, Splittstoesser and Tashiro 1977). In Ohio, the most prevalent type of milky disease of BTA is caused by a bacillus that forms spores inside insect cells, while in Michigan the most common form of milky disease is caused by a bacillus that lyses upon spore production (Plate 57) (Rothwell 1997). A wide range in the incidence of milky disease in BTA has been reported. Incidence of milky disease in the Cincinnati, Ohio, area increased from less than 1% in May and June to 25−29% in October 1977 (Wegner and Niemczyk 1981). In New York a high incidence of disease typically is observed during July and August wherever a population is encountered (H. Tashiro, M. G. Villani, personal observations). In Michigan, milky disease was found to infect 50−60% of the live BTA larvae collected from irrigated roughs, but only 25−30% of the larvae from the adjacent fairway (Rothwell and Smitley 1998).

Invertebrate Predators

Smitley et al. (1998) studied the distribution of BTA adults and other surface-dwelling insects on golf course fairways and roughs by establishing a line of pitfall traps running 12 m into the fairway and 12 m into the rough from the rough-fairway interface. BTA adults were abundant in the fairways but became scarce immediately after crossing into the roughs. Staphylinid beetles were the most abundant predators found in the roughs. They were 3−10-fold more abundant in the roughs than in the fairways. Carabid and histerid beetles were also more abundant in the roughs, with a sharp decline in density at the rough-fairway interfaces. Ants were more active in the roughs, but the decline in density in the fairways was more gradual than that observed for staphylinid, carabid, or histerid beetles. The density of all four predator groups was inversely proportional to the number of BTA adults caught in pitfall traps at all locations (Smitley et al. 1998).

Vertebrate Predators

The most apparent vertebrate predators are crows (*Corvus* spp.) starlings (*Sturnus vulgaris* L.), and barn swallows (*Hirundo rustica* L.) that inflict considerable damage as they disrupt the turf in their search for BTA grubs (Niemczyk and Wegner 1982).

Aphodius spp.

Taxonomy

Aphodius granarius (L.), *Aphodius pardalis* Le Conte, and *Aphodius lividus*, order Coleoptera, family Scarabaeidae, subfamily Aphodinae, are occasional pests of turfgrass. Grubs of *Aphodius* often are mistaken for the BTA. More than 100 species of the genus *Aphodius* are known to exist in America north of Mexico. The adults of most species are dung feeders, and larvae feed on dung, organic matter, and live roots (Blatchley 1910).

History and Distribution

Aphodius granarius is an introduced European species that is now widespread in the United States and Canada. To date, it has been reported from 25 widely scattered states and the District of Columbia and from two provinces in Canada (Figure 12-4) (Smitley et al. 1988, Woodruff 1973).

During 1976 and 1977, *Aphodius granarius* grubs, in association with the BTA, were found damaging golf course fairways in Toronto, Ontario. In one mixed population, *Aphodius granarius* made up more than 97% of the grub population (Sears 1979).

In 1978, *Aphodius granarius* grubs were discovered in golf course fairways in Colorado and in Michigan, where the infestations first were thought to be BTA. During 1981, that same species of *Aphodius* grubs was detected in golf course fairways in Ohio for the first time.

Aphodius pardalis is a West Coast species found in the turf of lawns, golf courses, and bowling greens, where grubs cause damage by feeding on grass roots and rhizomes. Adults of another species, *Aphodius lividus*, often are attracted in large numbers to the same greens and tees as *Ataenius spretulus* (R. Cowles, personal communication, 1998; Ritcher 1966).

Host Plants and Damage

Turfgrass-damaging *Aphodius* grubs feed on the same turfgrasses as the BTA and cause similar damage. *Aphodius granarius* grubs are usually more abundant in fairways than in roughs, and are often mistaken for BTA grubs (Smitley et al. 1998).

Description of Stages

Adults

Aphodius adults closely resemble those of *Ataenius*. *Aphodius* may be distinguished from *Ataenius* by the obtuse outer apical angle of its hind tibia. When *Aphodius* adults are mounted, the hindlegs should be stretched out for easy examination (Blatchley 1910).

Aphodius granarius beetles are oblong, piceous, and shining, with reddish-brown legs and paler antennae. They are 3–5 mm long, as compared with BTA, which are 4.0–5.6 mm long (Wegner and Niemczyk 1981, Woodruff 1973).

Both *Aphodius granarius* and *Aphodius pardalis* possess transverse carinae on the tibia of the mesothoracic and metathoracic legs (Figure 12-5), while *Ataenius spretulus* does not (Woodruff 1973).

Figure 12-4. Distribution of *Aphodius granarius* in the United States and Canada. (Data from Woodruff 1973; Rothwell 1997; P. J. Vittum, personal observation; Drawn by E. Gotham, NYSAES, adapted from Handbook of Turfgrass Insect Pests, Brandenburg and Villani 1995, Entomological Society of America.)

Eggs

Eggs of *Aphodius granarius* are smooth, opaque, and oval to almost spherical. They average 0.8 mm by 0.6 mm in length and width (Wilson 1932).

Larvae

Aphodius and *Ataenius* larvae are very similar in size but can be separated by differences in the raster. Grubs of both *Aphodius granarius* and *Aphodius pardalis* have definitely V-shaped palidia, while those of *Ataenius* have teges only, with no palidia (Figure 12-5) (Jerath

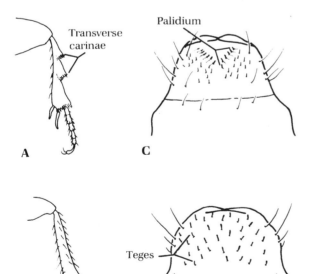

Figure 12-5. Adult and larval differences between *Aphodius* and *Ataenius* spp. **A.** Hindleg of *Aphodius granarius* with transverse carina. **B.** Absence of carina in *Ataenius spretulus*. **C.** Raster of *Aphodius pardalis* with V-shaped palidia. **D.** Absence of palidia in *Ataenius spretulus*. (Parts A and B drawn by R. McMillen-Sticht, NYSAES; parts C and D adapted from Ritcher 1966, figs. 66, 70, courtesy of Oregon State University Press.)

1960, Ritcher 1966). In Michigan, *Aphodius* and *Ataenius* grubs may be found on the same golf course but not at the same time (Smitley et al. 1998).

Larvae of the two turf-infesting *Aphodius* species show slight differences. The maximum head capsule width of *Aphodius granarius* is 1.4–1.7 mm, while that of *Aphodius pardalis* is 1.3–1.4 mm. The raster of *Aphodius granarius* has a V-shaped palidium of 14–23 caudomesally directed spinelike pali (Plate 28). The V-shaped palidium of *Aphodius pardalis* has 6–10 caudomesally directed spinelike setae. Both species have many teges scattered on both sides of the palidia (Jerath 1960).

Seasonal History and Habits

Little is known about the life history of *Aphodius granarius* and *Aphodius pardalis*. In Ontario, Canada, *Aphodius granarius* adults are present in May and again in August to September. In Ohio, adults are first collected from mid-April to early June and again in mid-November. In New Jersey, *Aphodius granarius* is reported to have a single generation that overwinters as adults. In Michigan, *Aphodius granarius* is univoltine, with overwintering adults flying in late May and laying eggs in suitable locations. Peak grub damage occurs in mid to late June, about a month earlier than damage from BTA grubs (Sears 1979, Smitley et al. 1998, Wilson 1932).

Larvae of *Aphodius granarius* are present in Ontario, Canada, in June, peaking in early July, and declining by the end of July. In Ohio, peak larval populations occur during the first week in June. They gradually diminish through June and the first week in July, indicating pupation (Sears 1979, Smitley et al. 1998).

Adult Activity

Aphodius adults as a group are among the most numerous scavengers and occur in great numbers during the first warm days of spring, seeking the fresh dung of horses and cows. The beetles burrow into dung almost as soon as it drops from animals (Blatchley 1910).

Adults of *Aphodius granarius* are the most commonly observed species within the genus in the Northeast United States. The highly polyphagous adults feed on all kinds of debris, decaying vegetation, compost, and carrion, in addition to dung. For oviposition, adults prefer dung that has dried and has formed a hard crust on its surface. The eggs are laid just beneath the hard crust (Wilson 1932). Larval infestations in turfgrass indicate that oviposition apparently occurs in turf as readily as in dung.

The limited knowledge concerning the biology of *Aphodius granarius* indicates that oviposition apparently occurs 2–3 weeks earlier than in the BTA in Ohio. The grubs of the two species, which often are collected from the same turf sites, rarely overlap temporally (Smitley et al. 1998).

Natural Enemies

Aphodius larvae, like BTA larvae, are susceptible to milky disease, caused by *B. popilliae* or a closely related bacterium. In Michigan as much as 56% of the larvae collected from a golf course fairway in 1996 were infected (Rothwell 1997). The type of milky disease found most frequently in *Aphodius* produces spores without any visible sporangia (Plate 57) (Smitley et al. 1998).

13

Scarabaeid Pests:
Subfamily Cetoniinae

Green June Beetle

Taxonomy

The green June beetle (GJB), *Cotinis nitida* L., is a member of the order Coleoptera, family Scarabaeidae, subfamily Cetoniinae (Ritcher 1966). Adults of this subfamily have the epimeron of the mesothorax visible from above (Plate 33) (Chittenden and Fink 1922). This beetle is also called the *fig eater* because it is fond of ripe figs and other thin-skinned fruits.

Importance

Historically, the GJB was considered a turfgrass pest of regional importance. During the late 1980s and early 1990s, this species increased in importance on turfgrass throughout the eastern United States. Turf damage by this insect primarily is mechanical rather than caused by feeding; the large, third-instar grubs disrupt the soil surface by burrowing in the soil and producing mounds. This activity makes mowing difficult and by exposing and damaging turf roots, can lead to turf loss due to desiccation and disease. Since the beetles are attracted to decaying organic matter, turfgrass that has been top-dressed with organic amendments is attractive for oviposition (Potter et al. 1996, Shetlar 1995a). This is a particular problem in warm-season turfgrasses where turf fertilization practices often coincide with oviposition.

The most comprehensive literature on the biology of the GJB was published during the 1920s (Chittenden and Fink 1922, Davis and Luginbill 1921).

History and Distribution

The GJB, which is native to North America, has a wide distribution in the eastern United States, extending from Connecticut and southeastern New York to Florida and westward into Texas, Oklahoma, and Kansas (Figure 13-1). Local infestations also have been reported in southern California, most probably a result of the transport of adult beetles on aircraft originating in infested areas (Hellman 1995). The beetle prefers rich sandy and loamy soils high in organic matter for oviposition (Chittenden and Fink 1922).

Figure 13-1. Distribution of the green June beetle in the United States. (Drawn by E. Gotham, NYSAES, adapted from Handbook of Turfgrass Insect Pests, Brandenburg and Villani 1995, Entomological Society of America.)

Host Plants and Damage

Adult Feeding

Beetles injure many ripening fruits, including apricots, nectarines, peaches, plums, prunes, apples, pears, grapes, figs, blackberries, and raspberries (Plate 33). Laboratory studies (Domek and Johnson 1991) determined that there is a significant positive relationship between food intake of the adult females and their longevity and fecundity. Apple-fed females lived, on average, 23 days and laid 50.8 eggs, while females receiving only water lived, on average, 14.8 days and laid 26.9 eggs.

Larval Damage

Grubs damage turf by their constant burrowing and tunneling with only limited feeding on roots. Grubs crawl to the surface at night to feed on the dead and decaying organic matter; as a result, they sometimes leave small mounds of soil 5–8 cm in diameter at the opening of their distinct vertical burrows that average 15–30 cm in depth and are approximately 2 cm in diameter. In addition to ingesting decaying organic matter, grubs also feed to some extent on the fine roots of succulent plants, especially when decayed organic matter is scarce (Davis and Luginbill 1921).

The grubs also cause similar mechanical damage on other crops, such as young corn, oats, sorghum, and alfalfa growing on heavily manured fields. Also, many vegetables, strawberries, ornamental plants, and tobacco seedbeds are damaged as the grubs disrupt the surface soil, loosen roots, and feed on roots and crowns.

Description of Stages

Adult

The beetle is velvet green dorsally with yellow-orange margins on the elytra (Plate 33). Ventrally it is a shiny metallic green mixed with orangish yellow. A hornlike process on the

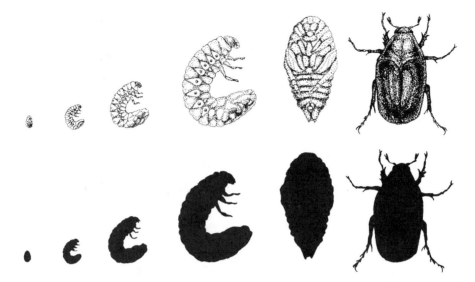

Figure 13-2. Stages of development of green June beetle: egg; first, second, and third larval instars; pupa; and adult. Silhouettes show actual size of each stage. (Courtesy of R. Jarecke, NYSAES.)

clypeus is more prominent in males. Both sexes are 2.0–2.5 cm in length and half as wide (Figure 13-2) (Davis and Luginbill 1921).

Egg
When first deposited, eggs are pearly white and 1.5 mm wide by 2.1 mm long. As the embryo develops, the eggs become nearly spherical and grow to 2.5–2.8 mm wide by 2.8–3.1 mm long (Plate 33). One or two days before hatching, the embryo is plainly visible through the chorion.

Larva
Recently hatched larvae are 7–8 mm long and 2 mm wide. They resemble older larvae except for the relatively longer hairs projecting laterally and posteriorly from the last segment (Plate 33). Full-grown third instars are 4–5 cm long by 1.3 cm in diameter and thicker toward the posterior end (Figure 13-2, Plate 33). Head capsule widths are 1.7, 3.0, and 5.5 mm for the first through third larval instars, respectively.

The three pairs of legs are very small in relation to the size of the grub and are not used for locomotion (Plate 33). The dorsal surface is transversely corrugated, with three distinct ridges per segment. Each ridge is covered with short, stiff hairs posteriorly directed that are used for its unusual dorsal locomotion. When placed on a hard surface, GJB grubs will flip onto their backs and crawl on their dorsum with their legs extended upward.

Prepupa and Pupa
As with other scarabaeids, the GJB prepupa is an active but nonfeeding stage. The pupa changes from white to light brown (Plate 34). Shortly before emergence, the metallic green and brownish tints of the adult become apparent (Plate 34). Pupae are 24–26 mm long and 13.0–14.5 mm wide through the thoracic region (Davis and Luginbill 1921).

Seasonal History and Habits

Seasonal Cycle

The GJB has one generation a year and spends about 10 months of the year as grubs, overwintering as third instars (Figure 13-3). Grubs resume feeding early in the spring, and by late May and the first week in June, form cells in which they pupate. After about a 3-week pupal period, adults emerge beginning in mid-June. In Virginia, beetles appear about mid-June, continue through July and August, and disappear by the first week in September. Oviposition occurs from mid-July through August. Eggs hatch in 10−15 days. Grubs molt twice by fall and become third instars before winter. They may be active or inactive during winter, depending on the temperature (Chittenden and Fink 1922).

Adult Activity

Daily Activity. GJB females emerge from the soil about daybreak. Shortly after sunrise, females begin to settle in the grass. Males emerge and increase in numbers rapidly. By 7:00 AM, males dominate and fly rapidly 15−45 cm above the ground, with a buzzing sound resembling that of bumble bees. Visible beetle activity gradually diminishes as the morning progresses, and by afternoon only an occasional beetle is observed (Davis and Luginbill 1921).

In Virginia, during the 4- to 6-week period when beetles are abundant, they also are active during late afternoon, flying close to the ground or soaring high among the treetops. Upon alighting, beetles gnaw into young twigs, causing them to break off. The beetles also are attracted to ripe fruit. After their daily periods of flight and mating, the males burrow just beneath the grass. Most beetles spend the night in the soil or under debris, but an occasional beetle spends the night resting in shrubbery (Chittenden and Fink 1922, Davis and Luginbill 1921).

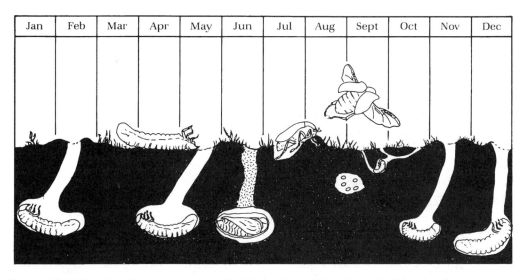

| Jan | Feb | Mar | Apr | May | Jun | Jul | Aug | Sept | Oct | Nov | Dec |

Figure 13-3. Life cycle of the green June beetle. (Adapted from Chittenden and Fink 1922, fig. 5; drawn by R. McMillen-Sticht, NYSAES.)

Adult Attractants. Both sexes of GJB adults are attracted to caproic acid baits used to trap Japanese beetles. Weekly catches of about 25–90 adults with a 1:1 sex ratio have been reported (Muma 1944). A mixture of malt extract and terpinyl acetate in Georgia peach orchards showed that the bait was attractive to GJB adults; a bait pail caught more than 400 beetles per week (Beckhan and Dupree 1952). Domek and Johnson (1988) reported evidence of a sex pheromone produced by unmated females to attract male beetles.

Oviposition. After mating and before oviposition, females fly close to the ground, buzzing like bees before entering the soil. Upon alighting, they disappear rapidly into the soil and burrow 10–20 cm deep to deposit eggs. Under golf course turf, eggs have been found within 5 cm of the surface (Hellman 1995). A sandy, moist soil containing aged manure or other decomposing vegetation is most attractive for oviposition.

The GJB places its eggs in the soil in a way that is unusual for a scarabaeid. It deposits the eggs in balls of soil about as large as a walnut and held together by a glutinous secretion (Plate 34). The number of eggs in each soil ball varies from about 10 to 30, with each egg having an individual cell within the large ball (Plate 34). Eggs in each soil ball are deposited during one continuous period of oviposition. A single beetle apparently deposits as many as 60–75 eggs during about 2 weeks under normal conditions. It is not known how the female makes the soil ball once each egg has been placed in its individual cell (Plate 34) (Davis and Luginbill 1921).

Larval Activity

Growth, Feeding, and Damage. Grubs feed on animal manure and other decomposing vegetation and grow rapidly. Eggs hatch in early August and grubs become noticeable in lawns and gardens by the middle of the month.

The mode of locomotion of GJB grubs is most distinctive, unlike that of any other turfgrass-infesting scarabaeid grub (Plate 33). Larvae crawl on their backs by alternate contraction and expansion of their body segments, making use of their transverse dorsal corrugations and posteriorly directed spines. They are unable to walk with their legs, which are vestigial. A grub enters the soil by lying on its back, bending the forepart of its body backward and downward, and burrowing in, using its head and feet (Davis and Luginbill 1921).

GJB grubs produce distinct, vertical burrows that average 15–30 cm in depth (Plate 33); in the sandy soils of Maryland golf courses, grubs have been found as deep as 0.9–1.1 m in late fall (Hellman 1995). Their burrowing activities produce mounds of earth 5–8 cm in diameter that may resemble anthills. During the day, grubs remain at the bottom of their burrows. At night, they leave their burrows and crawl about, especially on warm, wet nights. In turfgrass, examinations should be made to a depth of about 10 cm to determine infestation levels. If an average of 6–8 grubs per 0.1 m^2 is present, control should be considered (Baker 1982).

Overwintering and Spring Activity. GJB grubs burrow into the soil, and remain at the bottom of their burrows in late fall and winter, but do not have true diapause. On warm days of midwinter they may become active and briefly return to the surface. As warmer temperatures return in the spring, grubs again become active, feed for a short period, and prepare earthen cells at depths of about 20 cm in which to pupate following the prepupal period.

Cells are constructed of soil glued together by a fluid secreted by the grub (Plate 34). Cell walls are thin, about 1.0–1.5 mm, and may be crushed with moderate pressure between the thumb and forefinger. On one side of each cell is a hard protuberance filled with sand rather than with silt or clay. This protuberance apparently receives the final discharge of excrement from the grub before being sealed off and smoothed like the rest of the cell (Davis and Luginbill 1921).

Pupation occurs during late May or early June in parts of Kentucky and Tennessee, with the pupal stadium lasting approximately 16–18 days. The resulting adult remains in the cell for an additional 7–14 days before emerging from the soil (Plate 34) (Davis and Luginbill 1921).

Natural Enemies

Microorganisms

Grubs are infected by the green muscardine fungus, *Metarrhizium anisopliae*. Infection of grubs by the bacterium *Micrococcus nigrofaciens* Northrup has been reported, but there is no evidence this causes appreciable population reductions (Davis and Luginbill 1921).

An Arkansas study demonstrated that four species of entomopathogenic nematode (*Steinernema carpocapsae* [All strain], *Steinernema feltiae* [NC strain], *Steinernema glaseri*, and *Heterorhabditis bacteriophora*) caused significant mortality (65%, 45%, 64%, and 63%, respectively) when injected into the foregut of GJB larvae. However, the nematodes did not reproduce successfully in any larvae. The authors suggested that the larval habit of ingesting relatively large amounts of decaying vegetable material may have led to selection for a broad tolerance of soil microorganisms (Townsend et al. 1994). Studies testing the efficacy of entomopathogenic nematodes against the GJB under field conditions have been inconclusive.

Parasitic and Predatory Insects

The scoliid wasp, *Scolia dubia* Say, is a common parasite of GJB grubs (Plate 63). The wasp enters a burrow, stings the grub, paralyzing it, and then attaches an egg to its ventral side. Upon hatching, the parasitoid larva feeds on the paralyzed GJB grub. The wasp larva completes its growth, spins a cocoon, pupates, and remains in its cocoon until the following year (Davis and Luginbill 1921). At times these wasps may be observed in high numbers, flying low in "figure eight" patterns over grub-infested turfgrass.

Three sarcophagid flies have been reared from adults or pupae of the GJB: *Sarcophaga sarraceniae* Riley, *Sarcophaga helicis* Towns, and *Sarcophaga utilis* Aldrich (Davis and Luginbill 1921).

Vertebrate Predators

Birds are the most obvious vertebrate natural enemies of the GJB. At least a dozen species of insectivorous birds feed on adults or grubs or both. Blackbirds and robins are the most active feeders.

Mammals known to feed on the GJB include the mole, opossum, chipmunk, and skunk (Davis and Luginbill 1921).

14

Scarabaeid Pests:
Subfamily Dynastinae

Masked Chafers

Taxonomy

The northern masked chafer (NMC), *Cyclocephala borealis* Arrow, and the southern masked chafer (SMC), *Cyclocephala lurida* Bland, belong to the order Coleoptera, family Scarabaeidae, subfamily Dynastinae, tribe Cyclocephalini. The NMC is called *Ochrosidia villosa* Burm. in most of the literature prior to 1940. The SMC is called *Cyclocephala immaculata* in some literature prior to 1985. *Cyclocephala pasadenae* is the third masked chafer species of importance in turfgrass.

Importance

Masked chafer grubs are important turfgrass-infesting species, causing extensive damage to cultivated turf during late summer and early fall. They are the most injurious root-feeding pests of turfgrass throughout much of the Ohio Valley and the midwestern United States (Potter 1995a). Adult masked chafers do not feed (Potter 1981a).

History and Distribution

Masked chafers are native to North America and are distributed over a wide area east of the Rocky Mountains. The NMC and SMC occupy much of the same region (Figure 14-1). The NMC is troublesome as a turfgrass pest from Connecticut west to Ohio and Missouri, while the SMC is especially abundant in Kentucky, Indiana, Illinois, Missouri, and west to Texas (Potter 1981a). The SMC has been identified as a major, previously unrecognized, pest of turfgrass in Texas. An overall mean of 39% of the grubs collected were SMCs, with the balance consisting of *Phyllophaga crinita,* the species generally considered to be the major scarabaeid injurious to turfgrass in Texas (Crocker et al. 1982). *Cyclocephala pasadenae* is probably the most common chafer species in California and West Texas (Ritcher 1966).

Host Plants and Damage

Masked chafer grubs feed on the roots of a wide variety of pasture grasses and turfgrass (Potter 1995a). In laboratory feeding-choice studies, SMC grubs exhibited poor growth rates

Figure 14-1. Distribution of masked chafers in the United States. (Drawn by E. Gotham, NYSAES, adapted from Handbook of Turfgrass Insect Pests, Brandenburg and Villani 1995, Entomological Society of America.)

when feeding on roots of creeping bentgrass, when compared with the roots of other grass species such as perennial ryegrass, Kentucky bluegrass, endophyte-free tall fescue, and hard fescue (Potter et al. 1992). This study also determined that first-instar SMC grubs feeding on endophytic tall fescue were less fit than similar grubs in endophyte-free tall fescue, but this effect was lost when later instars were tested. In a second study, SMC grubs showed no consistent feeding preference for any of these turf species in feeding-choice experiments (Crutchfield and Potter 1995a).

In the midwestern regions of the United States, grubs of the SMC are particularly injurious to Kentucky bluegrass turf (Potter and Gordon 1984). The NMC was reported as injurious to winter wheat in Nebraska (Johnson 1941); however, in Ratcliffe's study (1991) of scarabs of Nebraska, the presence of NMCs could not be confirmed, suggesting that the damage was caused by SMCs.

The damage caused to turfgrass is typical of that caused by other scarabaeids. During late summer and early fall, third instars devour grass roots at or just below the soil-thatch interface, causing large patches of turfgrass to die through lack of roots.

Description of Stages

Masked chafers are similar in appearance and size in all stages. Because they are so similar, the general description of the stages given by Johnson (1941) for the NMC can serve as a model for all species.

Adults

The yellow-brown adult male averages 11.8 mm in length and 6.8 mm in width and the adult female, 11.0 mm in length and 6.7 mm in width (Figure 14-2). The sexes can be separated in both species by the distinctly longer lamellae of the antennae in the male, which are as long as or longer than all the other segments combined (Figure 14-3). Also, the prothoracic legs of males have heavier tarsi, one of the two claws on each front leg is distinctly

Figure 14-2. Stages of development of northern masked chafer: egg; first, second, and third larval instars; pupa; and adult. Silhouettes show actual size of each stage. (Courtesy of R. Jarecke, NYSAES.)

larger, and proximal segments 1–4 are as wide as or wider than they are long (Figure 14-3). In females these tarsal segments are longer than wide (Johnson 1941).

Both male and female NMCs can be distinguished from male and female SMCs; the male characteristics are more distinct than those in the female. Male NMCs are slightly larger, have many erect hairs on the elytra, and have much longer pygidial pubescence (Plate 35). Female NMCs have dense hairs on the metasternum and a row of stout bristles on the outer edge of the elytra (Plate 35). Both of these characteristics are lacking in the female SMC (Potter 1981b).

Both sexes of both species have dark chocolate-brown heads that shade to a lighter-brown clypeus, a characteristic that distinguishes them from other scarabaeid beetles of similar size and general coloration, including several *Phyllophaga* species and the European chafer. The head sutures of both the NMC and the SMC tend to be more pronounced than the sutures of *Phyllophaga* species.

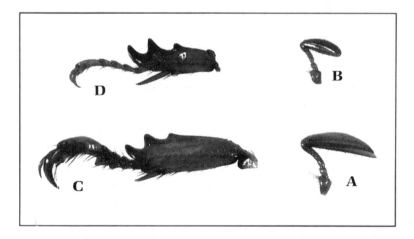

Figure 14-3. Sexual differences common to the northern and southern masked chafers. **A.** Male antenna. **B.** Female antenna. **C.** Male prothoracic leg. **D.** Female prothoracic leg. (Photo: NYSAES.)

Eggs

The pearly white, ovoid eggs of the NMC are delicately reticulate. Newly laid eggs have a mean length and breadth of 1.7 mm and 1.3 mm and expand to 1.7 mm and 1.6 mm just prior to hatching (Plate 35) (Johnson 1941).

The chorion of new eggs is only 1.25–1.60 μm thick exclusive of tubercles. By 8 days, a thick serosal cuticle with two distinct layers, an outer portion measuring 7.0–7.5 μm and an inner layer that is 12.0–12.5 μm thick, has formed. At maturity the serosal cuticle is re-absorbed; this allows the chorion to press against the egg-bursting spines on the thorax of the first-instar grub (Potter 1983).

Larvae

NMC larvae have mean head capsule widths of 1.6 mm, 2.3 mm, and 4.1 mm for the first, second, and third instars, respectively. The latter averaged 22.7 mm in total body length (Johnson 1941). NMC larvae are slightly stouter but are similar in length to European chafer larvae (Figure 14-2) (Tashiro et al. 1969). The NMC and SMC have a raster with coarse, long, hamate (hooked) spines showing no distinct arrangement. Spines become larger as they approach the anal slit, which is transverse and arcuate (Plate 28) (Johnson 1941).

Prepupae and Pupae

Soil particles that form the earthen cell used for pupation are firmly cemented together, apparently by a larval secretion (Plate 35). The prepupa in masked chafers resembles that of other scarabaeids. As in other scarabaeids, the newly formed, clearly white pupa turns reddish brown as it matures. Mean dorsal length is 16.8 mm. The sex of the pupa can be distinguished by a pair of conspicuous lobes (the aedeagal protuberance) on the venter of the ninth segment of the male. The sex of older pupae also can be determined by sex-linked differences (previously described) in the adult prothoracic tarsi, which are visible through the cuticle and described in connection with adult differences (Johnson 1941).

Seasonal History and Habits

Seasonal Cycle

Masked chafers have a 1-year life cycle, spending 14–21 days as eggs, 10–11 months as larvae, 4–5 days as prepupae, 11–16 days as pupae, and 5–25 days as adults. Masked chafers overwinter as third instars deep in the soil. They migrate upward during March and April and resume feeding until May, at which time they move downward to pupate. The nocturnal nonfeeding adults are present for nearly a month during June and July. Oviposition begins a day or two after emergence. Third instars are present in some locations by early September, but they appear later in autumn in more northern latitudes. They feed vigorously, causing most of their damage to turfgrass before they move downward in October for winter hibernation (Johnson 1941).

Adult Activity

Seasonal Presence. The most detailed study comparing the seasonal occurrence of the NMC and SMC was made by Potter (1981b). Blacklight trap catches in Kentucky revealed that flights of the NMC begin in early to mid-June, peak in late June, and end by early Au-

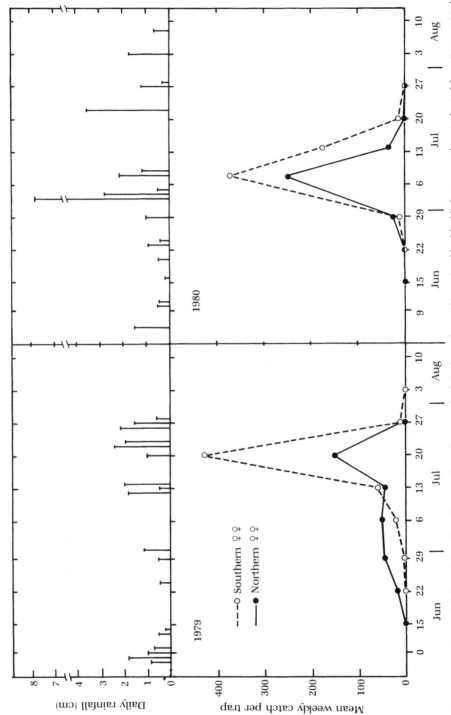

Figure 14-4. Seasonal flight activity of northern and southern masked chafer males as determined by blacklight trap catches. (Adapted from Potter 1981b, fig. 2, courtesy of the Entomological Society of America.)

gust. SMC flights begin at least a week later than those of the NMC, peak 1–2 weeks later than do those of the NMC, and terminate in early August in Kentucky. Flight activity is greater after a heavy rain (Figure 14-4).

Degree-day (DD) accumulations in air and soil with a base temperature of 10°C correlate well with the first emergence of beetles but are less useful for predicting the date of 50% and 90% flight. The first emergence of the NMC occurs after about 500 and 540 $DD_{base10C}$ in air and soil, respectively. First emergence of the SMC occurs at 585 DD (air) and 660 DD (soil) (Potter 1981b). Similar studies conducted in Indiana established that first emergence of the NMC is at 502 DD and 512 DD $_{base10C}$ for an agricultural setting and maintained turf, respectively (using air temperature). Peak flight occurs at 629 DD (NMC) and 782 DD (SMC) on maintained turf (T. J. Gibb, Purdue University, personal communication, 1998).

Adult Response to Blacklight Traps. Although the ratio of *Cyclocephala* males to females is reported to be about 1.5:1, male catches in light traps (incandescent) exceed catches of females by 7:1 (Neiswander 1938). Similarly, blacklight trap catches are 89% males for the NMC and 82% males for the SMC in Kentucky. The low proportion of females may be related to their behavior. Unlike males, which actively skim the surface of the turf in search of mates, females usually burrow beneath the sod soon after mating, and are less likely to be attracted to a blacklight trap during their limited flight (Potter 1981b).

Response to Pheromones. Sticky traps baited with crude extract from SMC females were tested on golf courses and home lawns in Kentucky. Males responded readily, confirming that females produce a powerful sex pheromone. Identification and synthesis of the sex attractant would allow season-long trapping and monitoring of adult activity (Potter and Haynes 1993). There was no correlation between numbers of beetles caught and local grub densities on golf courses, but there was a weak correlation between male captures and subsequent larval populations in home lawns (Potter and Haynes 1993). Field experiments indicate that all larval instars and pupae of both sexes of the SMC emit volatile chemicals that attract male SMC adults in a way identical to that of the sex pheromone emitted by adult female beetles; this observation has never been reported for any insect species (Haynes et al. 1992, Haynes and Potter 1995).

Nightly Activity of NMC and SMC. The nightly flights differ for the two species, with the SMC flying earlier in the evening than the NMC. Blacklight trap catches in Kentucky show that 90% of the SMC are caught between 9:00 and 11:00 PM, while more than 90% of the nightly catch of the NMC comes after 11:00 PM, with the heaviest catch between midnight and 4:00 AM (Figure 14-5).

SMC adults emerge from the soil on warm, humid evenings at about dusk. Swarms of males commonly are observed skimming over the turf in search of females. Females begin emerging and climbing up on grass blades shortly after the first male flights. Searching males generally land downwind nearby and crawl upwind toward the female. Copulation takes place almost immediately. Mating pairs often are surrounded by one to seven additional males (Potter 1980). Females produce a potent sex pheromone (Potter 1980, Haynes et al. 1995, Haynes and Potter 1995). NMC males are observed at about midnight, emerging and skimming over the sod surface in search of females. The sexual response of the two species is identical except for the temporal differences.

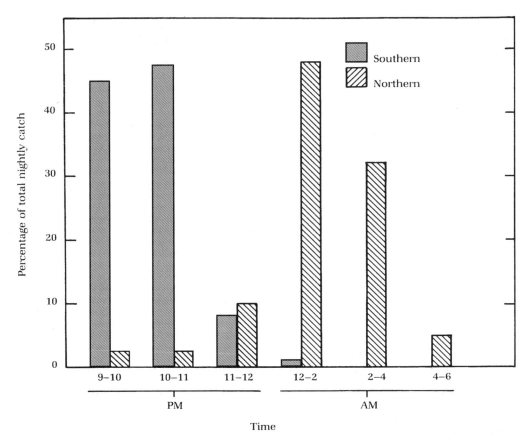

Figure 14-5. Nightly flight activity of southern and northern masked chafer males as indicated by blacklight trap catches. (Adapted from Potter 1980, fig. 2, courtesy of the Entomological Society of America.)

Traps baited with live females of either species indicate that females are cross-attractive. Males in the laboratory respond to and readily mate with females of either species. Even though there appears to be a common airborne sex attractant, the two species apparently have remained reproductively isolated (Potter 1980).

Oviposition. In captivity, NMC adults mate within a day after emergence, and oviposition begins the next day. Females deposit about 11–12 eggs each in the field, but as many as 23 eggs are laid in captivity. Hatching occurs in 14–21 days (Johnson 1941). In Kentucky, the SMC deposits most of its eggs in the upper 3 cm of soil (Potter and Gordon 1984).

Soil moisture significantly affects SMC oviposition. Eggs are not deposited in dry soils, and only a few eggs are deposited in soil of 5.0–12.5% moisture. Oviposition is greatest at 25.5% soil moisture, averaging 2.5 eggs per female per day (Potter 1983). Densities of grubs are generally higher in irrigated turf than in nonirrigated turf (Potter et al. 1996).

Eggs develop normally at soil moistures of 10.3–12.3% and above but shrivel and die in drier soils. The ability of eggs to withstand desiccation changes markedly as they age: Fully swollen 8-day-old eggs survive desiccation much better than newly laid eggs or eggs that are close to hatching (Potter 1983). In the field, no eggs survive in desiccated turf where afternoon soil temperatures exceed 40°C and where soil moisture drops to less than 8%. Egg survival under irrigated turf ranges from 55% to 75% in Kentucky (Potter and Gordon 1984).

Larval Activity

Hatching and Growth. Feeding begins almost immediately after hatching. Larvae can develop on a diet of organic matter, but their root-feeding and foraging habits make them destructive to turf. Because NMC eggs are deposited 11–15 cm deep, newly hatched grubs must migrate upward to near the soil surface for their major root-feeding activities (Johnson 1941).

Grubs grow quickly and by early fall reach their final length of about 2.3 cm. Damage to turf usually becomes noticeable in late summer and early fall. Feeding continues until mid to late October, when decreasing temperatures force them to migrate downward to spend the winter at a depth of 36–41 cm. In Connecticut, they descend as deep as 46 cm, but about 70% hibernate at depths of 13–25 cm. Winter mortality may be heavy, with as many as 50% dying in hibernation (Johnson 1941, Neiswander 1938).

Spring Activity. In the Midwest, grubs migrate upward to near the soil surface in late March and feed on grass roots until late May. Grubs move downward again to about 15 cm to become prepupae and then pupae in June (Johnson 1941, Potter 1981a).

Miscellaneous Features

Damage Thresholds

Damage thresholds for the masked chafer vary with turf species and turf vigor. Tall fescue and creeping bentgrass generally are more tolerant than Kentucky bluegrass and can withstand higher grub populations without exhibiting typical damage symptoms if vertebrate predators are not active (Potter 1995a).

Crutchfield and Potter (1995b) evaluated the tolerance of several common cool-season turfgrasses to SMC grubs by infesting field enclosures with 0–60 third-instar grubs per 0.1 m². Initial densities of at least 15–20 grubs per 0.1 m² were needed to cause any loss of aesthetic quality of most turfgrasses tested. Damage by the SMC was greater in endophyte-free tall fescue than in endophyte-infected tall fescue in some trials, and was at least equal to that caused by the same number of Japanese beetle grubs in all grasses tested.

Potter (1982b) studied the impact of soil moisture on damage thresholds for the SMC. SMC populations of 0, 3, 6, 12, 24, and 48 grubs per 0.1 m², consisting of late second and third instars, were released on Kentucky bluegrass turf within enclosures. Half the enclosures with each population were irrigated twice a week, while the remaining half received only rainfall. This test, conducted during September and October, indicated that the economic threshold for masked chafer grubs is considerably higher than the usual rule-of-thumb estimate of 6–8 grubs per 0.1 m². This study indicated that at least 9–10 grubs per 0.1 m² was required to damage moisture-stressed Kentucky bluegrass turf. Well-watered Kentucky bluegrass turf tolerated 15–20 grubs per 0.1 m² before showing noticeable damage.

Influence of Cultural Practices. Potter et al. (1996) evaluated a series of cultural practices before or during the seasonal flights of the SMC to study the effects of this manipulation on subsequent grub densities. Their findings suggest that the addition of organic amendments, the compaction of soil by heavy rollers, or aerification just prior to beetle flights has no measurable effect on subsequent grub populations. The addition of aluminum sulfate (to lower pH) resulted in significantly lower grub weights and significantly fewer third-instar grubs (in 2 years of the 3-year study) when compared to untreated controls. Increased mowing height significantly decreased SMC populations in all years of the study and affected grub weight and larval development in different years of the study. Mean weight of SMC grubs was positively correlated with soil moisture, suggesting that irrigation level impacts grub fitness.

Natural Enemies

Microorganisms

Milky Disease. Japanese beetle larval surveys for milky disease in the Northeast revealed many instances of infected *Cyclocephala* grubs in at least eight states and the District of Columbia. The bacterium infecting *Cyclocephala* grubs has been designated as the type A *Cyclocephala* strain of *Bacillus popilliae* (White 1947).

More recent studies conducted in Kentucky demonstrated that a commercial *B. popilliae* (type A) spore-talc mixture prepared with spores from milky Japanese beetle grubs are not infective to SMC grubs. Spore dusts prepared from diseased SMC grubs, however, were as infective on the SMC as commercial type A spore dusts are on the Japanese beetle. The fact that diseased grubs are common in the field would indicate that milky disease has considerable potential as a naturally occurring biological control agent for masked chafer grubs (Potter and Gordon 1984). However, commercial spore dusts containing the *Cyclocephala* strain of *B. popilliae* have not been marketed.

Masked chafers are susceptible to a number of entomopathogenic nematode species, but field efficacy has been inconsistent. This inconsistency may be attributable to suboptimal environmental conditions, nematode species, or nematode production and storage practices.

Parasitoids

In New Jersey, a native wasp of the genus *Tiphia* is reported to have a marked preference for parasitizing grubs of the SMC rather than those of the Japanese beetle in the same locality. Parasitism occurs in late August, when most Japanese beetle grubs are third instars, and while the SMC grubs still are second instars (Jaynes and Gardner 1924).

Vertebrate Predators

Predators of other, similar scarabaeid grubs are undoubtedly natural enemies. These include birds, such as starlings, grackles, and crows, and mammals, such as moles, skunks, and raccoons.

15

Scarabaeid Pests:
Subfamily Melolonthinae

Asiatic Garden Beetle

Taxonomy

The Asiatic garden beetle (AGB), *Maladera castanea* (Arrow), order Coleoptera, family Scarabaeidae, subfamily Melolonthinae, tribe Sericini, was first called the oriental garden beetle, *Ascrica castanea* Arrow (Hallock 1929). It was later named the Asiatic garden beetle, *Autocerica castanea* Arrow (Hallock and Hawley 1936). *Maladera* is the currently accepted genus. Hallock and Hawley's (1936) studies of this insect are the most complete. The following account is based on their work. Other material is cited only when it supplements theirs.

Importance

The adult AGB is a pest of ornamental plants and vegetable gardens. Although the insect breeds in greater abundance in weedy, abandoned areas than in lawns, the grubs often are destructive to turfgrass. Turf injury is most prevalent where weedy, disturbed areas and turfgrass sites exist side by side.

Because the AGB is attracted to bright lights on warm nights in July and August, the beetle is a nuisance at all types of nighttime amusement and recreational parks, open-air restaurants, and well-lit storefronts (Hallock 1936). The AGB adult can occasionally create a medical problem through its habit of entering human ears. Most cases have involved people who were sleeping on the ground in sleeping bags. Severe pain is caused when the tibial spurs pierce the delicate lining of the external auditory meatus. Medical attention generally is required for removal of the beetle, with most requests coming between 11:00 PM and 1:30 AM (Maddock and Fehn 1958, Wills et al. 1969).

History and Distribution

The AGB is a native of Japan and China. Adults were first found in the United States in Rutherford, New Jersey, in 1921, but the beetle was not identified as an introduced insect until 1926. Through 1936 it was distributed along the eastern seaboard, from Massachusetts into Maryland, and an isolated infestation appeared in South Carolina. The greatest

Figure 15-1. Distribution of the Asiatic garden beetle. (Drawn by E. Gotham, NYSAES, adapted from Handbook of Turfgrass Insect Pests, Brandenburg and Villani 1995, Entomological Society of America.)

concentrations have been in the metropolitan New Jersey–New York area, including Long Island, and in Philadelphia, but they also occur throughout southern New England (Hallock 1936; P. J. Vittum, personal observation). Since there has been no concerted effort to determine the beetle's spread, very little is known of its present distribution, but the AGB probably occurs in most northeastern states (Figure 15-1).

Host Plants and Damage

Adult Feeding

Although the AGB is considered a minor pest of turfgrass, grub populations occasionally are of sufficient densities to cause locally heavy turf loss.

Adult beetle feeding can cause significant damage to the foliage of many fruit, vegetable, and ornamental crops. Beetles may feed on more than 100 different plant species, but there are about 30 preferred food plants. Beetles begin feeding at the margins of leaves, and when feeding is heavy, the entire leaf may be eaten, leaving only the midrib. The shrubs most commonly attacked include box elder, butterfly bush, Japanese barberry, rose sumac, and viburnum. The most favored flowers include aster, chrysanthemum, dahlia, and goldenrod. Common fruits whose foliage is eaten are peach, cherry, and strawberry. Favored vegetables include carrot, beet, eggplant, pepper, and turnip (Hallock 1933, 1936).

Larval Feeding

The feeding damage by AGB larvae can be a serious turf problem. As many as 140 grubs per 0.1 m² have been found in lawns, although as few as 20 grubs per 0.1 m² can cause visible damage to lawn turf. Grubs of the AGB normally are less destructive to turf than are larvae of the Japanese beetle, European chafer, or the oriental beetle. Most larval feeding occurs at 5.0–7.6 cm below the surface, which results in less damage than the shallower feeding of other grub species.

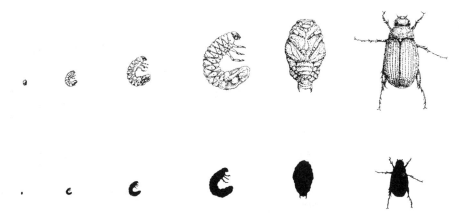

Figure 15-2. Stages of development of Asiatic garden beetle: egg; first, second, and third larval instars; pupa; and adult. Silhouettes show actual size of each stage. (Courtesy of R. Jarecke, NYSAES.)

Description of Stages

Adult

The beetle is dull chestnut brown, with a slightly iridescent velvety sheen, and 8–11 mm long (Figure 15-2, Plate 36). A few erect, irregularly arranged, backward-projecting hairs are present on the top of the head. The surface of the elytra is essentially glabrous, but the outer lateral edge of the elytra has a row of fine hairs. Ventrally, the exposed abdominal segments (5, 6, and 7) have a transverse row of spines extending the entire width of each segment (Plate 36).

Egg

As many as 19 white, ovoid AGB eggs are deposited together in a cluster. Each egg cluster is held together loosely by a glutinous material (Plate 36). The eggs increase in size (from 1.0 mm at the time of deposition) and are nearly spherical at maturity (Figure 15-2).

Larva

A newly hatched first instar is about 3.2 mm long, and a full-grown third instar is about 19 mm long (Figure 15-2). Three distinct characters distinguish the larva of the AGB from other scarab species. The raster, consisting of a single transverse row of spines in a crescent shape, is the most notable character (Plate 28). Metathoracic legs have very small claws, while the claws on the prothoracic and mesothoracic legs are more developed. Another prominent larval feature is a light-colored, enlarged bulbous stipe on the maxilla. The stipe is lateral to the mandibles (Plate 36), and is visible without further magnification. The palpi usually are in constant motion and at first sight, the larva appears to be chewing on a piece of vegetation.

Prepupa and Pupa

After larval feeding has ended, the third instar constructs an earthen cell about its body, voids accumulated excrement, and becomes first a prepupa, then a pupa that is about 8.4 mm in length (Figure 15-2, Plate 36).

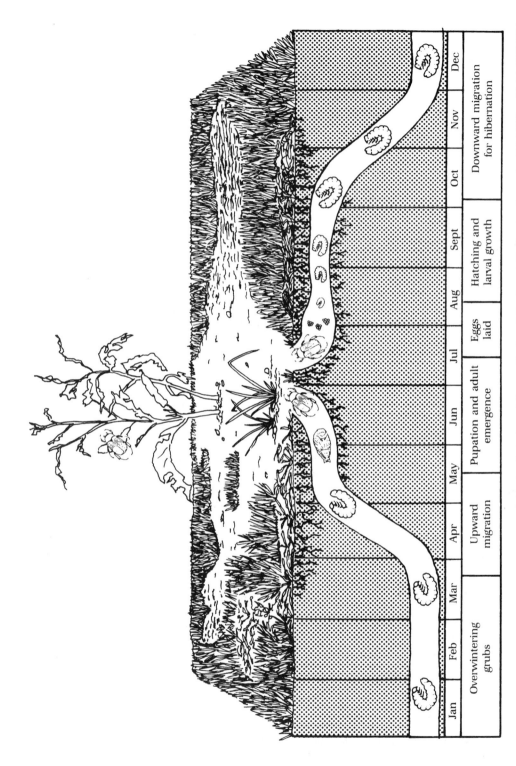

Figure 15-3. Seasonal life cycle of the Asiatic garden beetle. (Adapted from Hallock and Hawley 1936, fig. 7; redrawn by R. McMillen-Sticht, NYSAES.)

Jan	Feb	Mar	Apr	May	Jun	Jul	Aug	Sept	Oct	Nov	Dec
Overwintering grubs			Upward migration		Pupation and adult emergence		Eggs laid	Hatching and larval growth		Downward migration for hibernation	

Seasonal History and Habits

Seasonal Cycle

In the eastern United States, the AGB has a 1-year life cycle. In New York City, adults may be present from the last 10 days of June until the end of October but are most abundant between mid-July and mid-August. Eggs hatch in about 10 days. Grubs rapidly pass through the first two larval instars, with about 75% becoming third instars before winter. First instars never have been collected in the spring, suggesting that AGB grubs always complete this stage in the fall or that first instars are unable to survive the winter.

Overwintering grubs migrate upward in mid-April to resume feeding on grass roots. During late May to early June, mature larvae construct earthen cells, where they transform and remain as prepupae for about 4 days, before becoming pupae for 8–15 days. Adults remain in the earthen cells until they turn from white to chestnut brown and harden, after which they burrow upward to emerge. Figure 15-3 illustrates the seasonal life cycle.

Should the AGB spread to southern states, it may have two generations each year, since laboratory rearing has shown that it does not require a diapause. At a constant temperature of 27°C, three generations were lab-reared between September and the end of the following June (Hallock 1936).

Adult Activity

At dusk, beetles leave the ground, where they have been in hiding during the day. Large-scale emergence begins when temperatures are 18°–21°C. Beetles do not fly at temperatures below 21°C. During evenings when the temperature is above 21°C, beetles become very active and fly randomly in search of oviposition sites and favorite host plants (Allen 1944). They feed on plants near their daylight hiding areas and may remain quite unnoticed through the entire season; since adults are nocturnal, they seldom are seen on host plants.

Oviposition. Well-kept lawns and other short-grass areas near preferred host plants are the likely oviposition sites. Each female deposits a total of about 60 eggs that hatch in about 10 days. Eggs are normally deposited in the greatest numbers in overgrown, uncultivated areas where such weeds as daisy fleabane, goldenrod, plantain, orange hawkweed, ragweed, wild aster, and wild carrots abound.

Although orange hawkweed is not a food plant, the environment that it creates appears responsible for maximum oviposition. The leaves of this weed form a basal rosette that shades the ground and keeps the soil cool and moist. This creates a highly acceptable place for beetles to hide during the day, depositing their eggs in nearby turf soil. Larvae usually are 5–10 times more abundant in the vicinity of this weed than in other areas of infested sites.

Attraction to Lights. During warm nights, beetles are attracted to lights, especially to blue mercury-vapor lamps. The beetles may be found resting on the poles supporting street lamps. Flight distances range from about 180 m to 460 m when the insects are attracted by lights. A trap illuminated by a 100-watt incandescent lamp has captured as many as 2,000 beetles per hour. A trap illuminated with a G-5 mercury-vapor lamp has captured as many as 4,700 beetles per hour.

Larval Activity

AGB grubs are found in all types of sod, in weed patches, and occasionally, in cultivated gardens and nurseries (Hallock 1933). Larvae feed in the top 13 cm of soil, with major feeding at a depth of 5–8 cm. The beetles are known to burrow as far as 0.9 m horizontally through turfgrass soils. Many plants are susceptible to larval feeding, but those with succulent roots are preferred.

Winter and Spring Activity. The greatest amount of vertical movement occurs during the winter (Tashiro et al. 1969), when AGB grubs migrate down below the frost line and remain there until soils warm sufficiently in the spring. Beginning in October, as the soil temperatures drop, larvae gradually descend 15.0–30.5 cm for overwintering at usual depths of 2.3–28.1 cm. Upward migration occurs during April, and by late April through early May, the larvae have returned to feed actively on grass roots within the top 13 cm of soil. Pupation occurs at a depth of 4–10 cm during late May and June, with adult emergence beginning in late June.

Influence of Soil Moisture. Summer rainfall deficiencies have a significant influence on the survival of eggs and young larvae. Mortality may be so great during a dry summer that the number of beetles is reduced significantly the following year.

Vertical distribution is governed predominantly by moisture during the growing season. Any time that moisture is adequate, grubs occupy the upper 5 cm, where fibrous roots are most abundant. Drought drives grubs deeper into the soil in search of moisture; they may be found as deep as 20 cm. Following a heavy rainfall or irrigation that moistens the soil continuously to the depth of the grubs, the grubs will migrate to within 5 cm of the surface within 24 hr.

Natural Enemies

There have been few documented reports of natural enemies for any stage of AGB, but this may be due to the lack of research in this area rather than a true absence of predators, parasitoids, or pathogens of AGB in the field. A wasp of the genus *Tiphia* is reported to parasitize the larvae in China and Japan. Starlings and other birds feed on beetles. Hanula and Andreadis (1988) surveyed the scarab larvae on a number of turfgrass sites in Connecticut. They found AGB grubs infected with a rickettsia species (*Rickettsiella popilliae*) and a protozoan in the genus *Adelina*.

European Chafer

Taxonomy

The European chafer (EC), *Rhizotrogus majalis* (Razoumowsky), order Coleoptera, family Scarabaeidae, subfamily Melolonthinae, tribe Melolonthini, was called *Amphimallon majalis* in all publications in the United States prior to 1978 (Sutherland 1978, Tashiro et al. 1969).

Importance

Wherever the species becomes established, EC grubs often are the most serious turf pest in the region. Although population levels of the EC generally are lower than those of most other turfgrass-infesting annual scarab species, the grubs are larger, come to the soil surface to feed almost a month earlier in the spring, and feed nearly a month later in late fall than do those of the Japanese beetle and most other scarab species. The EC grubs damage all cool-season turf and pasture grasses as well as many field, forage, and nursery crops.

History and Distribution

The EC is known to occur throughout western and central Europe, excluding England or the Netherlands. The EC was first discovered in the United States in Newark, New York (near Rochester), when grubs were discovered damaging ornamentals in a large commercial nursery in 1940 (Gambrell et al. 1942).

For the first 10 years following its discovery in the United States, the insect's spread appeared to be natural and contiguous, eventually covering an area of about 997 km². During the period 1950–66, numerous infestations were discovered in Connecticut, Massachusetts, New Jersey, Ohio, Pennsylvania, and West Virginia and in adjoining and nearby states, as well as in Ontario, Canada, along the Niagara frontier. Many infestations have been traced directly to shipments of infested nursery stock from Newark. It is suspected that much of the spread of the EC can be traced to the distribution of infested nursery containers and balled and burlapped ornamental plants into uninfested regions. Other localized infestations in the same states appeared to have been the results of hitchhiking beetles, since beetles were found along major interstate highways and on railroad right-of-ways (Gambrell et al. 1942, Regnier 1939, Tashiro et al. 1969).

The results of a 1983 survey in Ontario, Canada, show that the generally infested area filled the Niagara peninsula, with the western boundary encompassing Burlington and Brantford and extending nearly to Simcoe. At least five isolated infestations were found between Toronto and Windsor (Reid 1983).

Since the federal EC quarantine was rescinded in 1971 and scouting efforts ceased, information on subsequent spread of the EC in the United States has been less defined. Populations are firmly established in eastern and northern Michigan, northern Ohio, northern/central Pennsylvania, and eastern Massachusetts. Figure 15-4 shows the beetle's currently known distribution in the United States and Canada.

Host Plants and Damage

Adult Feeding
Adults feed to such a limited extent that they are not considered to be injurious. Literally thousands congregating in a single small tree on a given night will cause only slight feeding damage along the margins of plant leaves (Plate 37).

Larval Feeding
EC grubs prefer to feed on fibrous roots, and therefore rooting habits and surrounding ground cover during the oviposition period largely determine which plants become hosts. Annual crops may escape damage because they are planted just after the grubs have completed their feeding in the spring and the crops are harvested before the next generation of grubs becomes destructive. Pastures, winter wheat, and hay crops are subject to injury. Larval feeding on ornamental and nursery plants can be severe at times because it greatly

Figure 15-4. Distribution of the European chafer. (Drawn by E. Gotham, NYSAES, adapted from Handbook of Turfgrass Insect Pests, Brandenburg and Villani 1995, Entomological Society of America.)

reduces their fibrous roots. The lining-out stock of spruce and similar conifers may be seriously damaged or killed outright where larval populations are high (Tashiro et al. 1969).

By early fall most grubs are third instars. Fall rains coupled with cooler temperatures keep the soil moisture nearly continuously at or near field capacity. Under these conditions, practically all the grubs are found feeding within 2.5–5.0 cm of the soil surface.

Unlike Japanese beetle grubs that feed predominantly in the soil-thatch interface during fall and spring, EC grubs seldom are observed in such a narrow horizon. Damage often occurs during the fall feeding period, but this sometimes passes unnoticed because the above-ground portions appear healthy due to an abundance of moisture and cool weather. As a result of such root loss, however, even a brief dry spell may cause large patches of turf to die. In the spring, however, the turf may die as the grubs resume their feeding on the previously weakened grass (Tashiro et al. 1969).

The density of EC grubs necessary to produce visible damage is difficult to predict because so much depends on the vigor of the grass and on the moisture supply. In general, 10–15 third instars per 0.1 m^2 in a high-maintenance turf, or 4–5 grubs per 0.1 m^2 in a low-maintenance turf, is sufficient to cause noticeable damage (Smitley 1995). Secondary damage may be caused by mammalian predators such as raccoons, skunks, or moles. These animals, at times, destroy turfgrass when foraging for grubs, and may be active even though grub damage is not apparent to a human observer. Thus, their foraging activity often necessitates the use of lower tolerance levels.

Action thresholds for EC grubs tend to be lower than those of Japanese beetle grubs, which are often found in the same turf sites, because EC grubs tend to be more tolerant of most common soil insecticides than are Japanese beetle grubs. In efficacy tests that included both species, chlorpyrifos, isofenphos, and bendiocarb were more toxic to Japanese beetle grubs than to EC grubs of similar stage and size (Villani et al. 1988).

Description of Stages

Adult

Adult ECs are fawn-colored beetles with a rufous-yellow head and pronotum (Plate 37). A narrow band of light-yellow hairs extends from under the caudal margin of the pronotum and slightly overlaps the proximal margin of the elytra. The heavily striated and lightly punctated elytra extend posteriorly to the middle of the penultimate abdominal segment, leaving the pygidium exposed (Gyrisco et al. 1954).

Males and females average 13.7 mm and 14.3 mm in length, respectively (Figure 15-5). Antennae have nine segments, with the club comprising three segments, and are lamellate. Males have longer lamellae than females (Plate 37). Two June beetle adults similar in size that have been mistaken for the EC are *Phyllophaga tristis* (Fabricius) and *Phyllophaga gracilis* (Burmeister). EC adults can be differentiated from *Phyllophaga* by a tooth on the tarsal claws, which is present and distinct on *Phyllophaga* but absent on the EC (Figure 15-6) (Gyrisco et al. 1954, Tashiro et al. 1969).

Egg

Newly oviposited EC eggs are about 0.5 mm wide by 0.7 mm long and become more spherical, measuring about 2.3 mm by 2.8 mm just before hatching (Figure 15-5, Plate 37). As in the case of other scarabaeid eggs, the tan mandibles of the embryo are clearly visible through the chorion prior to hatching (Tashiro et al. 1969).

Larva

The brown head capsules have average widths of 1.2 mm, 2.1 mm, and 3.0 mm for the three instars. Full-grown third instars fully stretched out are about 6 mm wide in the thoracic region (the widest area) and are about 23 mm long (Figure 15-5) (Gyrisco et al. 1954).

Grubs of the EC can be distinguished from those of all other North American scarabaeid grubs by a combination of two characters: the Y-shaped anal slit (with a stem half the length

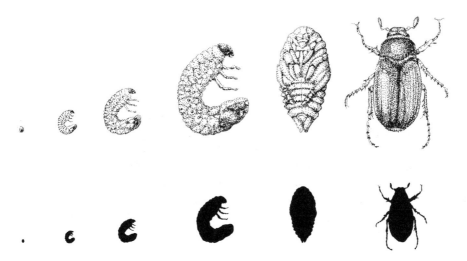

Figure 15-5. Stages of development of the European chafer: egg; first, second, and third larval instars; pupa; and adult. Silhouettes show actual size of each stage. (Courtesy of R. Jarecke, NYSAES.)

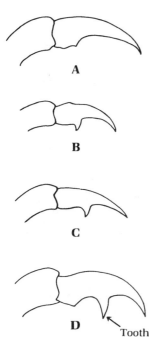

Figure 15-6. Differences in the tarsal claws between adults of the European chafer and native May or June beetles. **A.** European chafer. **B.** *Phyllophaga tristis.* **C.** *P. gracilis.* **D.** *P. hirticula.* (From Tashiro et al. 1969, fig. 10, courtesy of the NYSAES.)

of each arm), and the raster (which has two subparallel rows of palidia converging toward the anterior and outwardly curved toward the posterior) (Plate 28) (Gyrisco et al. 1954).

Prepupa

On becoming a prepupa, the larva loses its C-shaped contour, becoming nearly straight, with a slight crook on the posterior end (Plate 37). The prepupa averages 6 mm in width by about 19 mm in length and is flaccid and fragile (Figure 15-5) (Tashiro et al. 1969).

Pupa

The EC pupa resembles that of other scarabaeids (Plate 37). It turns from creamy white to the general adult color as it matures, but it remains very fragile. Unlike the Japanese and oriental beetle pupae, which are encased in the larval and prepupal exuviae, the EC sloughs off the exuvia to its posterior end. Pupae have an average width of 6.9 mm through the anterior abdominal region and an average length of 16 mm (Figure 15-5) (Tashiro et al. 1969).

Differentiation of Sexes

Adults. Males can be distinguished from females by the length of their antennal club, which is twice as long as that of females (Plate 37). Gravid females can be identified by a bright yellow-tan coloration and distention of the ventral abdominal surface caused by the presence of 40 or more eggs (Plate 37). The coloration and distention disappear as eggs are expelled, until the abdominal color resembles that of the rest of the body, as in the males (Tashiro et al. 1969).

Larvae. The sex of grubs can be determined in all three instars by the visibility of the terminal ampullae in the males, as described in Chapter 11. Although body fat sometimes obscures the ampullae, their presence is a sure sign that the insect is a male. The ampullae are more obvious in EC males than in other scarab grubs infesting turfgrass.

Pupae. The sex of pupae can be determined by differences in the ventral portion of the 10th abdominal segment (Figure 11-5). Male genital organs are much more pronounced than are female genitals. As in adults, the longer antennal club of the male is visible through the pupal integument (Tashiro et al. 1969).

Seasonal History and Habits

Seasonal Cycle

The EC normally has a 1-year life cycle, spending about 9 months of the year as a third instar, and overwintering in this stage. Less than 1% of the population requires 2 years, overwintering the first winter as a second instar, the second winter as a third instar, and emerging as an adult along with the regular 1-year-cycle beetles. Individuals requiring 2 years are found mostly in areas of tall grass and dense sod (presumably because of shading and cooler soil temperatures) or in poor soils with sparse vegetation, where the food supply is limited. Figure 15-7 represents the life cycles of both the 1- and 2-year beetles on the basis of observations made in western New York (Tashiro and Gambrell 1963).

Adult Activity

Emergence and Seasonal Occurrence. After transformation from a pupa, the teneral adult remains in the earthen pupal cell for 3–5 days before it is ready to emerge and fly. It then digs its way through the soil with its mandibles and foreleg tarsi. In western New York, EC adults start to emerge during the second or third week in June, reach peak populations during the last week in June or the first week in July, and then diminish rapidly. Flights practically are terminated by the end of July (Tashiro et al. 1969).

Phenological Relationships. The flowering period of certain common plants is correlated with the development of the EC. Temperature appears to be the controlling factor for both the beetles and the flowers. Of the phenological relationships shown in Figure 15-8, the three most useful relationships have been the full-bloom period of Vanhoutte spirea, indicating the presence of pupae; the early bloom of hybrid tea and floribunda roses, signaling early beetle flights; and the full-bloom period of common catalpa, *Catalpa bignonioides* Walt., which correlates with peak flights of adults (Plate 38) (Tashiro and Gambrell 1963).

Daily Emergence and Mating Flights. Adults begin emerging from the ground at about 8:30 PM EDT, at a light intensity of 1,400 lux (a level generally coincident with the disappearance of the sun below the horizon). At such times beetles crawl up grass stems to fly (Plate 38). They are very clumsy in their efforts to become airborne and may fall to the ground several times before they finally succeed. Flight to nearby trees or other plants becomes apparent about 8:45 PM, at a light intensity of about 323 lux.

Once airborne, the insects are strong fliers and fly to nearby or distant trees, even to the tops of trees as tall as 18–24 m. The numbers approaching a given tree increase rapidly during the next 10–15 min, and peak swarming occurs about at 9:00 PM (Plate 38). On calm evenings, beetles may be found hovering on all sides of a tree. On windy evenings

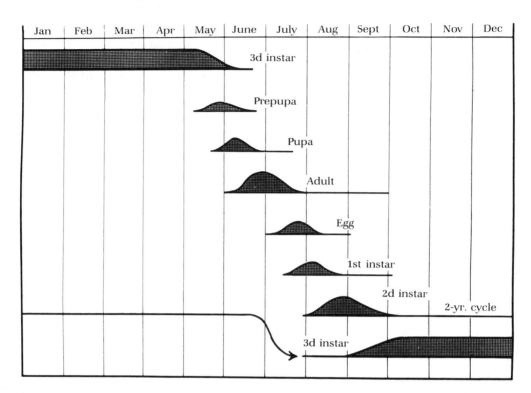

Figure 15-7. Life history of the European chafer in western New York, showing the dominant 1-year and minor 2-year life cycles. (From Tashiro and Gambrell 1963, fig. 3, courtesy of the Entomological Society of America.)

however, they hover on the leeward side. In the absence of trees or shrubs, they may swarm around such objects as light poles or chimneys. The primary attraction is a silhou-etted object. There is no evidence that a given tree is preferred in North America, but Reg-niere (1939) indicated that in France, beetles have a strong preference for poplar. When several thousand beetles are in flight around a single tree, the noise produced by their wing beats resembles that of swarming honey bees (Gyrisco et al. 1954, Tashiro et al. 1969).

Weather conditions influence flight activities. Flights are greatly reduced when tempera-tures are below 19°C or on the evenings following an afternoon rainstorm. No flight takes place at air temperatures below 11°C (Gyrisco et al. 1954).

Mating and Oviposition. Mating occurs when beetles have come to rest at aggregation sites (Plate 38). There is no evidence of either pheromones or courtship behavior by either the male or female beetle prior to mating. Shortly after 10:00 PM, mating pairs and indi-viduals start falling from the trees. Literally thousands of beetles can be dislodged by shak-ing the tree at about this time (Plate 38). Flights back to the ground continue throughout the night, and at dawn, the few remaining beetles return to the soil (Gyrisco et al. 1954, Tashiro et al. 1969).

Females deposit eggs singly in an earthen cell, generally at depths of 5–10 cm. Forma-tion of the cell and oviposition are described and illustrated in Chapter 11. The total num-

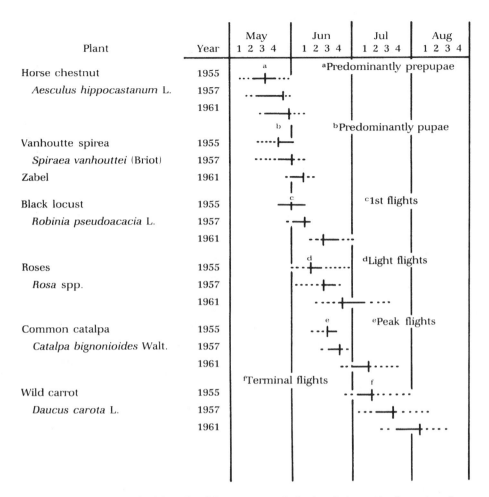

Figure 15-8. Various events in the life cycle of the European chafer in relation to the flowering of common plants during seasons that came early (1955), with average timing (1957), and late (1961) in western New York. For each year, the first broken rule indicates early bloom, the solid rule indicates full bloom, and the second broken rule indicates petal fall; the vertical line indicates chafer development. (From Tashiro and Gambrell 1963, fig. 5, courtesy of the Entomological Society of America.)

ber of eggs per female, determined through dissection, varies from 0 to 52, and in captivity, from 2 to 46 eggs are deposited, with an average of 22. Ground cover has a pronounced effect on oviposition. Soil kept fallow during the entire flight period averages only 0.3 grub per 0.1 m² in the fall, compared with 14 grubs per 0.1 m² in soil covered with turf during the flight period (Gyrisco et al. 1954).

Turf height also has a pronounced effect on oviposition. Turf that has been mowed during the entire flight season will average about 22 grubs per 0.1 m² in the fall, compared with 13 grubs per 0.1 m² in grass maintained at a height of a foot or more (Gyrisco et al. 1954).

Frequency of Flight and Longevity. Beetles feed very little, depending entirely on fat accumulated during the larval stage for their energy. Males have an average flight period of

6 days, and females 6.5 days, with an average adult life not exceeding 2 weeks. Within 2–3 days preceding natural death, abdomens become partially transparent, reflecting complete utilization of the stored fat. Many insects die on the soil surface under the trees to which they last flew (Gyrisco et al. 1954, Tashiro et al. 1969).

Response to Chemical Attractants. Exposure of more than 2,000 separate chemicals or combinations in traps originally used for the Japanese beetle revealed only a mild attraction of the EC to a few compounds. Traps baited with java citronella oil–eugenol (3:1 v/v) capture only about four times as many beetles as unbaited traps. Butyl sorbate is slightly more attractive. Both chemicals were used for scouting to a limited extent while the federal quarantine on the chafer was in force (Tashiro and Fleming 1954, Tashiro et al. 1964).

Response to Colors. Trap color has some influence on beetle response. A high-gloss finish is more attractive than a dull finish. Traps that are red, black, yellow, blue, or white (in decreasing order of effectiveness), have resulted in catches but with only a 3.6-fold difference in catches between the two extremes (Tashiro and Fleming 1954).

Response to Fluorescent Lamps. Beetles exposed at night to 15-watt fluorescent lamps ranging from peak emission of 2857–6450 Å demonstrate distinct differences in degree of attraction. From the highest to the lowest level of effectiveness, beetle catches have been made with the following types of lamps: BL black light (3,650 Å), BLB black light (,3650 Å), erythemal (3,100 Å), blue (4,400 Å), germicidal (2,857 Å), and green (5,300 Å), gold (5,900 Å), pink (6,200 Å), or red (6,450 Å). The BLB lamp filters out some of the 3,650-Å radiation at the same time as it filters out most of the visible radiation with the blue-violet tubing, so that it is significantly less attractive than a BL lamp such as normally used for attracting and capturing night-flying insects (Tashiro et al. 1967).

Unlike chemical baits that attract the beetles during their initial twilight flights, black light attracts beetles only after they have flown to a tree and have settled down. Therefore, the most effective location tends to be under a tree at which beetles congregate (Tashiro et al. 1967).

Larval Activity
Hatching and Development. Eggs hatch in about 2 weeks under normal field temperatures. Grubs are first instars for about 3 weeks, second instars for about 4–5 weeks, and third instars for about 9 months. In western New York, first instars are the dominant stage during the latter half of July, second instars during the latter half of August, and third instars from the latter half of September through the first half of May (Gyrisco et al. 1954).

Populations and Densities. After adults are discovered in a new area, the buildup of larval populations follows a predictable trend. Grubs tend not to be found in soil surveys during the first 2–3 years after discovery of the first beetles. Grubs subsequently increase steadily and reach peak populations during the next 4–6 years. Thereafter, populations decline, and a moderate level is maintained indefinitely; peak populations are never again reached (Tashiro et al. 1969).

During years of peak grub populations, a density of 22–32 third-stage grubs per 0.1 m² is common. The largest concentrations usually are found at the junction of green turf and turf that is dead as a result of larval feeding. A high-density record shows 65 third instars per 0.1 m² around a large dandelion, the only living plant, with its fibrous roots entirely consumed and its tap root more than half consumed (Gambrell 1943, Tashiro et al. 1969).

Environment Affecting Density. The highest number of grubs tends to be found out to a radius of about 25 m from the trees to which adult beetles have flown. Fewer grubs are found at locations farther from the trees. Soil moisture has a definite effect on grub density. Populations tend to be the highest where the average soil moisture during the entire larval season is below field capacity. One comparison found only 0.17 grub per 0.1 m² where the soil moisture averaged above 90% of field capacity, in contrast to 1.9–7.1 grubs per 0.1 m² where the soil moisture averaged 41% and 65% of field capacity (Shorey et al. 1960).

Larval Distribution. Eggs are deposited singly in earthen cells, but females tend to lay their eggs in batches, resulting in clumped distributions. Although the lateral movement of grubs tends to make larval distributions less aggregated, third-instar grubs still tend to be scattered in clumps and produce high and low counts (Nyrop et al. 1995). The lateral movement of third instars averages 0.3 m a day in fallow ground. Horizontal movement of this magnitude can be expected when destructive populations must continue to advance into available sources of food supply as the grass is killed (Gyrisco et al. 1954).

In study by Nyrop et al. (1995), more than 300 residential lawns were sampled intensively for EC grubs in the Rochester, New York, area, in cooperation with a local lawn care company. Data included extensive grub counts and estimates of site characteristics for each property. Golf course cup cutters were used to remove soil cores (10.8-cm diameter), which were then examined for the presence of white grubs. Cores were taken every 3.1 m in a grid pattern. Numbers and species of grubs were recorded as well as soil type, terrain, lawn age, shade, thatch, and grass species composition for both the front and the rear lawn. When comparing EC grub densities to site characteristics, the investigators found that high grub populations were strongly associated with front lawns, high proportions of Kentucky bluegrass, lawns less than 20 years old, and lawns in open (nonshady) areas.

Winter Hibernation

Vertical Distribution. During late fall or early winter, as the surface soil starts to freeze, grubs migrate downward to remain below the frost line. They may do so a month or more after cold temperatures have forced Japanese beetle grubs to retreat deeper into the soil. During early March, with 5 cm of frozen soil, 91% of the EC grubs may be found 20–36 cm deep. In areas of heavy sod and thick snow cover, however, grubs may remain within the upper 2.5–5.0 cm of the soil surface. Second instars migrate downward earlier in the fall, reach greater depths, and migrate upward later in the spring (Gyrisco et al. 1954, Tashiro et al. 1969). Winter mortality of EC grubs tends to be highly variable; 15 of 16 sites sampled in New York State in 1952–53 showed population reductions from fall to spring ranging from 6% to 53%, with an average net loss of 24% (Burrage and Gyrisco 1954).

Spring Larval Activity. During late winter or early spring, grubs migrate upward following the frost line and may even be found in ice at the soil surface during March in western New York. Grubs feed actively within 2.5–5.0 cm of the soil surface throughout April and into mid-May in western New York. Most spring damage due to grubs occurs during this period. This spring injury may be the first evidence of a grub population, as feeding commences on turf that overwintered in a weakened condition, its roots having been consumed previously in the fall. This is also a vigorous feeding period preparatory to pupation (Tashiro et al. 1969).

Upon completion of feeding, the grub changes from white and gray to cream, with the accumulation of fat. The grub moves downward about 5–10 cm in heavy soil or as deep as 15–25 cm in sandy soil. Once it has reached its ultimate depth, it forms an earthen cell. The metamorphosis of fully mature grubs to adults follows the same sequence as that for other scarabaeids, as described in Chapter 11. The prepupal period lasts 2–4 days. Afterwards, the last larval exuvia is sloughed off to the posterior (Plate 30).

During roughly a 2-week pupal period, oscillation of the body as it lies in the cell is the only movement of the pupa. Pupae are extremely susceptible to mechanical injury. Dry soil conditions are not detrimental, but saturated soils can produce high rates of pupal mortality (Tashiro et al. 1969).

Natural Enemies

Microorganisms

Milky Disease. EC grubs are susceptible both to *Bacillus popilliae* and to *Bacillus lentimorbus* (Wheeler 1946), which show the same infectivity pattern and gross symptoms as in Japanese beetle grubs (Plate 57). Infection by the former is also is known as "type A" infection and that by the latter as "type B". Bacterial strains naturally infecting EC grubs differ from those most infective to the Japanese beetle. Japanese beetles are highly resistant to the strains most infective in EC grubs (Tashiro et al. 1969).

The DeBryne strain of *B. popilliae,* selected from a farm by that name where milky disease bacterial spores were first distributed for EC grubs, consistently has been the most infective against EC. The *Amphimallon* strain of *B. lentimorbus* commonly is found in many areas where EC grub populations have persisted for 10–15 years; it is the dominant strain at the pasture (Tashiro et al. 1969, Wheeler 1946).

When temperatures are in the optimum range (20°–30°C), EC grubs are more susceptible to oral infection by milky disease organisms than are Japanese beetle grubs. In the field, however, infectivity in EC is relatively low, and milky disease has never been a major factor in providing effective biological control. The highest incidences for infection have approached 30–35%.

The need for high soil temperatures for consistent infectivity is apparent. Tashiro et al. (1969) reported that 5 weeks after inoculation, the incidence of diseased grubs was 25%, 86%, and 90% when incubated at 14°, 21°, and 32°C, respectively, for type A infections, and was 7%, 57%, and 83% when incubated at 14°, 21°, and 27°, respectively, for type B infections. North of the latitudes of Long Island, New York, EC infection rates by milky disease bacteria are low because of insufficient soil temperature during the critical feeding period when grubs are near the soil surface in early fall and late spring. Average soil temperatures of 21°C or higher are considered essential for high rates of milky disease infection. At best, during the second-instar period of mid-August to mid-September, soil temperature in those northern latitudes generally is at the minimum level for infection. During much of the third-instar period, temperatures are generally well below minimum levels (Tashiro et al. 1969).

Miscellaneous Microorganisms. Coxiella popilliae Dutky, a rickettsia which causes blue disease in Japanese beetle grubs, is infective in EC grubs under controlled laboratory conditions. Natural field infections, however, never have been observed in ECs as they have been in Japanese beetle grubs (Fleming 1962).

Laboratory studies indicate that EC grubs are highly susceptible to the green muscardine fungus, *Metarrhizium anisopliae,* and the white muscardine fungi, *Beauveria bassiana* and *Beauveria brongniartii.* EC grub mortality due to fungal infection may require several weeks or months under suboptimal soil conditions (Krueger et al. 1991). Grubs infected by these fungal pathogens are found occasionally in the field, but the incidence of infection is so low that it is not considered an important factor (Tashiro et al. 1969).

The entomopathogenic nematodes *Heterorhabditis bacteriophora* (Khan, Brooke, and Hirshmann) and *Steinernema glaseri* (Steiner) infect EC larvae in the laboratory and the field, but commercial formulations of these nematodes have proved inconsistent when used under typical turf production conditions. In paired comparisons, entomopathogenic nematodes of all species tend to be less effective against third-instar ECs than they are against Japanese beetle larvae of a similar age (Wright et al. 1988).

Insect Parasitoids and Predators

Parasitic insects contributing toward EC population reductions in Europe were introduced and released in New York during the 1950s and 1960s. Even after many releases, there has been no evidence of establishment of any parasitoid species in New York. The success of parasitoids in Europe is attributed to the presence of sufficient populations of an alternate host with a 3-year life cycle. During the period of releases, a similar alternate host in the release areas of New York varied from slightly present to practically nonexistent (Tashiro et al. 1969). Two ground beetles of the family Carabidae considered responsible for some EC reductions are *Harpalus pennsylvanicus,* a predator on grubs, and *Harpalus erraticus,* an egg predator (Gyrisco et al. 1954).

May or June Beetles

Taxonomy

May or June beetles (M-JB) belong to genus *Phyllophaga* Harris, order Coleoptera, family Scarabaeidae, subfamily Melolonthinae, tribe Melolonthini. Species that have spring reproductive flights are called May beetles and those that have summer reproductive flights are called June beetles (Crocker et al. 1995a). The most complete taxonomic work on adults and larvae of *Phyllophaga* spp. was prepared by Boving (1942); it contains larval keys for the differentiation of 61 species. Other major taxonomic contributions to this genus include the work of Ritcher (1966), which covers larvae, and Luginbill and Painter (1953), which also focuses on adults. Woodruff and Beck (1989) included a key to the *Phyllophaga* of Florida that contains excellent scanning electron micrographs of male and female genitalia of 54 species as well as micrographs of other important adult taxonomic characters. Larval keys for 23 species also are included, as well distribution maps for the United States.

Early literature refers to this group as genus *Lachnosterna.* In spite of the importance of M-JB to American agriculture, not a single species has been given a common name (Blatchley 1910, Ritcher 1966, Werner 1982).

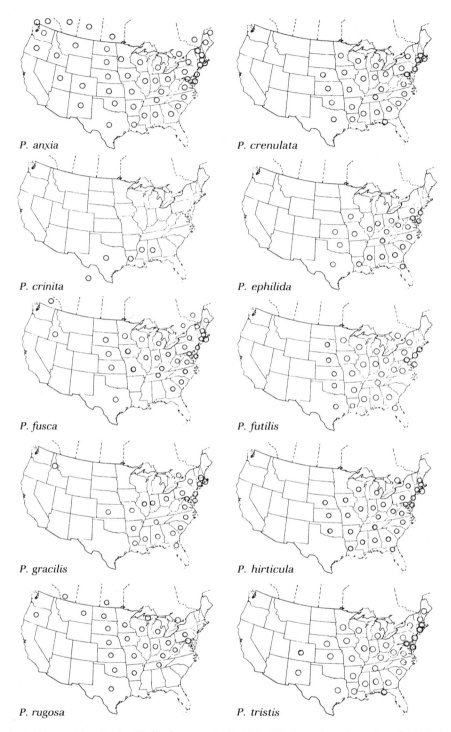

Figure 15-9. Distribution of major *Phyllophaga* spp. in the United States and southern Canada. (Adapted from Luginbill and Painter 1953; redrawn by H. Tashiro, NYSAES.)

Importance

White grubs, as the larvae commonly are called, have been recognized as major pests of agriculture for many decades in the northeastern quarter of the United States, in Texas, and in the Canadian provinces of Ontario and Quebec. Larval damage to turfgrass, whether short turf or pasture sod, resembles that caused by other scarabaeid grubs. Because individuals of many species are relatively large, a few grubs per unit area can be very destructive (Luginbill and Painter 1953).

Adults of some species are serious pests of shade trees, and where these are abundant, they can defoliate trees with young tender leaves. Favorite host trees include the oak, hickory, walnut, elm, and poplar. Adults of some species feed on herbaceous plants.

History and Distribution

More than 150 species of *Phyllophaga* have been reported in the United States and Canada. The greatest numbers of these species occur in the eastern half of the North America (Figure 15-9). Texas has almost 100 species, with *P. crinita* being the most serious turfgrass pest. A second species, *P. congrua,* also damages tall fescue stands in Texas.

The seven most common and destructive species in the north central and eastern United States include *P. anxia* (LeConte), *P. fervida* (Fabricius), *P. fusca* (Froelich), *P. hirticula* (Knoch), *P. implicita* (Horn), *P. inversa* (Horn), and *P. rugosa* (Melsheimer). The most widespread throughout the country is *P. anxia.* Even in areas where a major, economically important species occurs, relatively few studies have been conducted (Merchant and Crocker, 1995, Luginbill and Painter 1953, Ritcher 1966). Major species in several of these areas are listed below.

Northeastern United States and Canada

In the New England states and eastern Canada, the most common and injurious species is generally *P. anxia.* The principal species in southwestern Ontario, Canada, are *P. futilis* (LeConte), *P. fusca,* and *P. rugosa.* Twenty-six species are known to occur in New York. In strawberries, the more common species are reported to be *P. rugosa, P. anxia, P. fusca,* and *P. longispina* (Smith). As pests of seedbeds of a forest tree nursery near Albany, *P. gracilis* (Burmeister), *P. tristis* (Fabricius), *P. crenulata* (Froelich), *P. fusca, P. fraterna* Harris, and *P. anxia* were the most numerous (Hammond 1940, Heit and Henry 1940, Henry and Heit 1940, Kerr 1941).

Ohio and Kentucky

The most comprehensive surveys of M-JB species in the Midwest have been conducted in Ohio and Kentucky. In Ohio during the years 1935–55, the most abundant and widely distributed species, as determined by light trap collections, were *P. fusca, P. futilis, P. hirticula,* and *P. rugosa* (Neiswander 1963).

Adult collections in the bluegrass region of north central Kentucky, from host plants and larval collections during 1936–39, revealed 36 species to be present. In the inner bluegrass region, where most of the horse farms are located, *P. hirticula* constituted more than 80% of the population. Other dominant species were *P. fraterna,* which represented about 60% of the population in the intermediate bluegrass region, and *P. ephilida* (Say), which accounted for about 55% of the population in the outer bluegrass region (Ritcher 1940).

Wisconsin and Nebraska

Surveys were made in Wisconsin during 1935–37 by collecting adults from host plants. Of the 16 species collected, *P. rugosa* and *P. hirticula* were equally abundant and constituted 67% of the total M-JB population. They were followed by *P. tristis* and *P. fusca*. *Phyllophaga anxia* is the dominant species in Nebraska and occurs perennially in the sand hill areas of north central Nebraska (Chamberlin et al. 1938, Rivers et al. 1977).

Southern States

The most abundant turfgrass-infesting species in Texas, especially abundant in the lower Rio Grand Valley, is *P. crinita* Burmeister. This species is recorded in only three other states (Louisiana, Alabama, and Mississippi) and in Mexico. It resembles *P. tristis* in size but can be differentiated by its much less hairy body (Frankie et al. 1973, Luginbill and Painter 1953, Reinhard 1940). It flies in the summer and has a 1-year life cycle. *Phyllophaga congrua* is also a turf pest in Texas but is found primarily on tall fescue. This species has a 2-year life cycle and flies in the spring.

Host Plants and Damage

Adult Feeding

Feeding damage by M-JB is most common on shade and forest trees, although ornamental shrubs and even a few fruit trees are occasionally damaged. The degree of damage correlates with the emergence of beetles and the presence of young tender foliage. Oak trees are the preferred hosts for many species, and bare or partly stripped trees are a common sight in the inner bluegrass region of Kentucky. Adults of some species feed regularly on grass blades and on diverse forbs. In Kentucky, the peak emergence of *P. hirticula*, the dominant species, often comes during the first week in May, when the leaves of the pin, red, white, bur, and chinquapin oaks are young and tender. Other trees preferred and often stripped of leaves by various species of M-JB include persimmon, hickory, walnut, elm, and birch. Trees that often provide food but are never stripped include willow, plum, ash, sycamore, locust, sassafras, redbud, blackberry, and rose. Cherry and raspberry and some flower petals also provide food (Luginbill and Painter 1953, Ritcher 1940).

Larval Feeding

Damage is caused primarily during the second year of both the 3-year cycle and the 2-year cycle, when larvae are third instars. Young forest and shade trees may be damaged when the grubs eat the fibrous roots and girdle larger tap roots (Hammond 1940).

Since the early 1970s, *P. crinita* has been the most serious scarabaeid encountered in Texas turf, damaging bermudagrass, buffalograss, and St. Augustinegrass (Frankie et al. 1973, Reinhard 1940).

Description of Stages

Adults

Beetles are heavy, clumsy-looking insects, ranging in color from light or dark brown to mahogany brown and resembling several other turf-infesting scarabaeids in general appearance (Plate 39). The smallest and the largest species differ considerably in size. One of the smaller species, *P. gracilis*, varies in length from 10.5 to 13.0 mm, while *P. fusca*, one of the larger, ranges in length from 17.5 to 23.5 mm. The adult of *P. gracilis* rather closely resembles the EC, both in size and in color. Body pubescence varies considerably in the M-JB. Among the more common species, the elytra of *P. tristis* are heavily pubescent, while those of *P. anxia* are nearly glabrous. As with many scarabaeids, the sexes of many species can be distinguished by the longer lamellae of the antennae of males (Plate 40) (Luginbill and Painter 1953).

Eggs

M-JB eggs are deposited singly in earthen cells. Eggs are pearly white and elliptical when newly deposited and increase in size and become nearly spherical at maturity. *Phyllophaga hirticula*, a species of average size, newly laid eggs average 2.4 mm by 1.5 mm and increase to 3.0 mm by 2.5 mm (length by width) by maturity (Ritcher 1940).

Larvae

Newly hatched first-instar larvae are translucent white, but as soon as feeding begins, darkened areas form in the posterior region. The larvae grow from less than 6.3 mm in length as first instars to 25.0–38.0 mm as fully grown third instars for the medium-sized species (Figure 15-10, Plates 28, 40) (Hammond 1940).

The anal slit of M-JB larvae is V or Y shaped, with the stem of the Y much shorter than the arms (Plate 40). The lower anal lobe usually is divided by a sagittal cleft and sometimes is divided by a sagittal groove. The claws on the metathoracic legs are very small (Ritcher 1966).

Rastral patterns vary considerably, but the palidia of many essentially are two parallel rows of pali, or spines, and the palidia show a tendency toward convergence at both ends (Figure 15-11). In many larvae, the palidia are nearly straight, while in some they are sin-

Figure 15-10. Stages of development of June beetle (*Phyllophaga anxia*): egg; first, second, and third larval instars; pupa; and adult. Silhouettes show actual size of each stage. (Courtesy of R. Jarecke, NYSAES.)

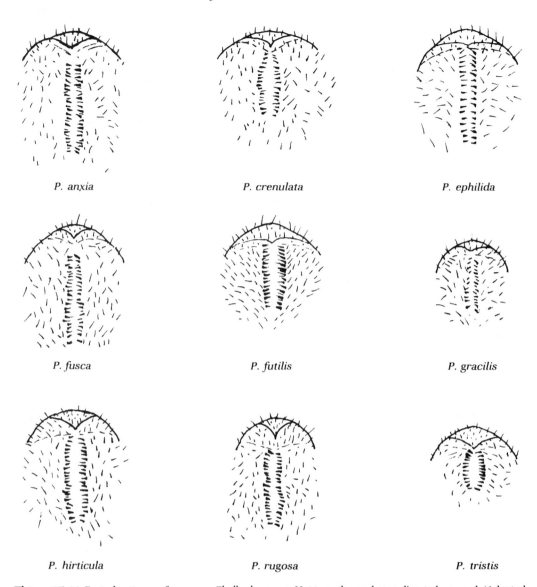

P. anxia   *P. crenulata*   *P. ephilida*

P. fusca   *P. futilis*   *P. gracilis*

P. hirticula   *P. rugosa*   *P. tristis*

Figure 15-11. Rastral patterns of common *Phyllophaga* spp. Not to scale; anal area directed upward. (Adapted from Ritcher 1949, 1966, and Boving 1942; redrawn by H. Tashiro, NYSAES.)

uate or curved (Plates 28, 40). The numbers of pali in each row vary considerably among species, from about 10 to about 30 (Boving 1942).

Pupae

As is typical of all turfgrass-infesting scarabaeids, the pupation of the M-JB occurs in an earthen cell formed by the mature third-instar. Upon transformation from a prepupa, the pupa sloughs off the larval exuvia to the posterior end, as do the EC and the AGB, which belong to the same subfamily.

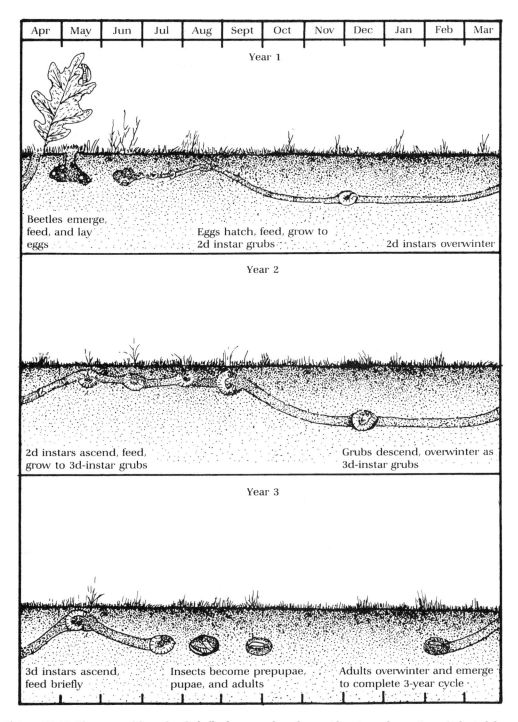

Figure 15-12. Three-year life cycle of *Phyllophaga* spp. based on midwestern observations. (Adapted from Luginbill 1938, fig. 13; redrawn by H. Tashiro, NYSAES.)

Seasonal History and Habits

Seasonal Cycles

The life cycles of the M-JB vary from 1 to 4 years for one generation, depending on the species and latitude. The most common in the geographic areas of greatest faunal presence are 3-year cycles and a few species of 2-year cycles. Those that have a 1-year life cycle are exceptions. In Kentucky, where the most comprehensive studies on the biology of the M-JB have been conducted, those with a 3-year life cycle include *P. hirticula,* the most dominant species, plus *P. rugosa, P. fusca,* and *P. crenulata,* all fairly common (Ritcher 1940).

Three-Year Life Cycle. A typical 3-year life cycle is illustrated by *P. hirticula* in Kentucky (Figure 15-12). During the first year, adults that have overwintered in the soil migrate to within 2.5 – 5.0 cm of the surface in March. They emerge as adults in late April and May to feed on foliage and deposit eggs in May and June about 8 – 10 cm deep in the soil. Eggs hatch from June into July, and larvae feed within 2.5 – 5.0 cm of the surface during July into September. They become second instars in August and September and feed into October before migrating to lower depths for overwintering (Ritcher 1940).

During the second year, overwintered second instars migrate upward in April to within 2.5 – 5.0 cm of the surface, feed, and become third instars in June, doing most of their feeding thereafter, during July and August. They migrate downward again in late September and October to overwinter.

Overwintered third instars migrate upward in late March and April of the third year and feed during April and May to complete their larval development. During June they migrate to the lower depths again and form an earthen cell, where they become prepupae in late June, transform to pupae in July and August, and become adults in August and September. These young adults remain in the same earthen cell until the next spring. Movement up from the earthen cell begins as early as February, and the first beetles arrive beneath the sod in late February and early March to complete a 3-year life cycle (Luginbill 1938, Ritcher 1940).

Life cycles studied in other areas of the Midwest, in Ohio and southern Wisconsin, indicate that with few exceptions, the M-JBs have 3-year life cycles. Most species in the Northeast and in the eastern Canadian provinces of Ontario and Quebec also have 3-year life cycles. *Phyllophaga tristis* has a 2-year life cycle in eastern New York (Chamberlin et al. 1938, Hammond 1940, Heit and Henry 1940, Neiswander 1963).

In Kentucky, some species that often have a 2-year life cycle occasionally have a 3-year life cycle in the same locality. Three of these include *P. futilis, P. bipartita* (Horn), and *P. inversa,* all present throughout the Midwest (Ritcher 1940).

Two-Year Life Cycle. Those M-JBs with a 2-year life cycle either molt twice during the first year, to overwinter as third instars, or molt once the first year, overwintering as second instars, and molt again the second year and overwinter as third instars. Kentucky species with 2-year life cycles, include *P. ephilida* and *P. tristis.* Unlike *P. hirticula, P. ephilida* overwinters at a shallow depth and pupates in June, with adult emergence the same season.

After overwintering, the third instars migrate to very near the surface in April to feed a while before returning to form their earthen cells at the depth at which they hibernate.

They become prepupae in late May, pupae in early June, and adults in late June, with emergence in July. Females deposit eggs about 8 cm deep in the soil in late July. Those eggs hatch in August, become second instars in September, and migrate downward in October to overwinter (Ritcher 1940).

During the second year, the second instars leave their overwintering depth and migrate upward, where they feed on roots from April through July. They become third instars and are the most destructive grubs until October, when they migrate to lower depths for their second winter hibernation (Ritcher 1940).

One-Year Life Cycle. The most devastating turfgrass scarabaeid in Texas is *P. crinita.* It has a 1-year life cycle typical of many other scarab species, although in northern Texas a small portion of the population is suspected of having a 2-year life cycle. They overwinter as third instars and pupate in the spring. The beetle flight period occurs mainly from June into August. The beetle is no longer considered to have both a major and a minor flight period each year. Rather, adult flights are governed by moisture, with drought inhibiting flight nearly completely. Flights occur again within a week following rainfall of 1 cm or more. Most of the damage caused by third instars occurs during the summer and fall (Frankie et al. 1973, Gaylor and Frankie 1979).

Adult Activity

Daily Flights. During the day, beetles hide in soil or sod. Emergence from the ground and flight occur at dusk or shortly before, particularly during warm cloudy days when the temperature is above 15°C. Under these conditions, beetles fly to the tops of trees to feed. They sound like a swarm of bees during heavy flights. They return to the soil a little after daybreak. Because of these nocturnal habits, most people see the beetles only when they are attracted to the light and fly into screens and windows. On cool nights adults feed near the ground (Hammond 1940; Ritcher 1940, 1949).

Feeding Habits. Beetles of most species feed only at night. Trees with young, tender leaves are the most attractive food plants although some species feed on forbes. The soft leaf tissue between veins is consumed, leaving the veins intact (Ritcher 1949). In Kentucky, *P. hirticula,* the dominant species, damages oaks heavily; the peak emergence of beetles coincides with the presence of young, tender leaves with only 2.5–5.0 cm of growth. *Phyllophaga tristis* feeds exclusively on oak. Trees that are not attacked include maple, basswood, wild cherry, and evergreens (Hammond 1940, Ritcher 1940). In Texas, *P. crinita* adults do not appear to feed.

Mating and Oviposition. Mating occurs only at night. While females are feeding, males search them out for copulation. Oviposition commences 9–10 days after mating. Each female deposits about 50 eggs during a 1–3-week period. Preferred sites for oviposition, which may occur during the day or night, include loose sod growing in sandy soil in the immediate vicinity of their food plants. Bare fallow ground is also attractive for oviposition. Most eggs are found 8–18 cm deep in small, compact balls of soil. The earthen cell is held together by a secretion of the female (Hammond 1940, Ritcher 1940).

In Texas, *P. crinita* females have a preoviposition period of just over 7 days and an oviposition period of 3–5 days. Thirty to 40 eggs are deposited by each female at a depth of 5–

10 cm in turfgrass soil. Beetles avoid heavily watered turf for oviposition (Frankie et al. 1973, Reinhard 1940).

Sex Pheromone. The sex attractant produced by females of *P. anxia* (LeConte) has been identified as a 3:1 blend of the methyl esters of L-valine and L-isoleucine. *Phyllophaga anxia* males were trapped successfully using the sex pheromone on turfgrass in New York during May and June (Zhang et al. 1997). Female-produced sex pheromones also are produced by other *Phyllophaga* species.

Seasonal Flights. Although the name *May* or *June beetles* implies the presence of adults only during these months, beetle flights may begin or end well outside either of those months. These include *P. fusca* and *P. tristis* in Kentucky. Most others begin in May; *P. fusca* is abundant through May and continues into early June (Hammond 1940, Ritcher 1940).

Attraction to Lights. M-JB adults are attracted to artificial light sources in direct relation to the brilliance of the illumination. The majority of beetles captured in light traps are males. In Kentucky, more than 85% of the M-JB adults usually are males and in Texas about 80% of the *P. crinita* adults attracted to streetlights are males (Frankie et al. 1973, Ritcher 1940).

Larval Activity

Hatching and Development. Eggs hatch in 3–4 weeks. The young larvae first subsist on organic matter in the soil but soon turn to grass roots, their preferred food. Molting to the second instars occurs during August or September for species with a 3-year life cycle. By mid-October, second instars have migrated downward for winter hibernation. Depending on species, those with a 2-year life cycle molt once or twice before overwintering (Ritcher 1940).

During the second year of a 3-year life cycle, grubs feed most ravenously and cause the most root damage. Most species feed from 4 to nearly 6 months during the second summer, compared with less than a month during the third summer. By late August of the second summer, some grubs burrow downward for overwintering, while others continue to feed until late October before hibernation (Hammond 1940).

Third-year grubs of *P. anxia* that are within the upper 5 cm are feeding actively. Grubs found below that depth are inactive and display little movement when dug from the soil. During late April and May, grubs come to the surface to feed for about a month. In June they migrate downward to form earthen cells, become prepupae, change to pupae for about a month, and become adults that remain in the ground until the following spring (Hammond 1940).

In Texas, *P. crinita* flights begin in the south and progress northward, although this pattern may be disrupted by lack of rainfall in local regions. Following oviposition in July and August, larvae emerge and spend about 3 weeks each in the first and second instars. Damage from heavy infestations can be acute within 2–3 months of egg hatch. Larvae do not necessarily migrate down in the soil profile in the winter and may feed at a shallow depth depending on soil temperatures (Crocker et al. 1996, Frankie et al. 1973).

Larval Distribution. Actively feeding grubs are found within the upper 5 cm of the soil profile. Grubs descend into the soil for winter hibernation or for completion of larval development and preparation for pupation. Depths of penetration are similar for both hibernation and pupation and vary according to species. One species that penetrates deeply, *P. hirticula*, has been found 18–58 cm deep for both purposes. A species that overwinters at a shallow depth, *P. ephilida*, has been found at 5–15 cm both in hibernation and in pupation (Ritcher 1940).

Population Densities. In Canada, a density of 63 or more grubs per m² is common and they are found most frequently on higher ground in sandy soils. The presence of 12–17 grubs on a single corn plant, sufficient to cause a complete loss of crop, is common (Hammond 1940, Luginbill 1938).

Pupation

The pupal period lasts about 1 month, depending on the species and the season, and occurs in July of the third year of a 3-year life cycle. Pupation occurs during June of a 2-year life cycle. The pupation of *P. crinita* occurs in the spring and lasts about 3 weeks in cells 8–15 cm deep (Frankie et al. 1973, Ritcher 1940).

Natural Enemies

Microorganisms

Virus. About half of all reports of viruses infecting beetles involve scarab species (Martignoni and Iwai 1986). A study in Quebec reported the first naturally occurring viral disease of *P. anxia*, but the virus appears to have little impact as a regulator of beetle populations in the field (Poprawski and Yule 1990b).

Milky Disease. *Phyllophaga* grubs infected with milky disease bacteria have been found in the field from widely scattered locations. Grubs of *P. anxia* from Clinton County, New York, in 1944, *P. fusca* grubs from Wayne County, New York, in 1945, and *P. hirticula* grubs from Kentucky, all infected with *B. popilliae*, are the original sources of *B. popilliae* strains bearing the same specific names as grubs. Grubs of *P. anxia* and *P. fusca* infected with *B. popilliae* were found on numerous occasions in western New York during 1950–82. The degree to which milky disease biologically controls M-JB grubs is not well known, but it is considered to be a relatively minor influence (Wheeler 1946). This conclusion has been corroborated in other tests of *B. popilliae* on *Phyllophaga* spp. (primarily *P. anxia*). In those tests, spores of *B. popilliae* (both a local strain and a commercial preparation) produced a maximum of 7.5% mortality when force fed to grubs (Poprawski and Yule 1990a). In this same study, an extensive survey over a 4-year period indicated that only 3% of grubs examined had died from bacterial infections. Some of the bacteria found included *Pseudomonas aeruginosa*, *Serratia marcescens*, *Micrococcus nigrofasciens*, *Bacillus cereus*, and *B. popilliae*.

Fungus. The fungus *Cordyceps ravenelii* Berk and the green muscardine fungus, *M. anisopliae*, are reported to be naturally infective to field populations of various species of grubs (Hammond 1940). In a study in southern Quebec, five species of entomogenous fungi were identified and considered to occur naturally in the *Phyllophaga* species examined. These were, in decreasing order of field incidence, *M. anisopliae*, *Beauveria bassiana*, *Fusar-*

ium sp., *Penicillium* sp., and *Aspergillus* sp. The overall incidence of fungal infections in the grub population was 7% (Poprawski and Yule 1991a).

In fungal bioassays of second-instar *P. anxia*, *M. anisopliae* yielded higher mortality (82%) in the soil inoculation than did *Beauveria bassiana* (24%). This mirrors field observations that more grubs were naturally infected with the former. A field test that used *M. anisopliae* spores against second instars resulted in a 92% rate of mycosis in the recovered grubs treated at the highest rate.

Protozoa. M-JB grubs can also be infected with protozoans. *Actinocephalus* sp. was found to be endemic (4.5%) in populations of *Phyllophaga* grubs in southern Quebec (Poprawski and Yule 1992a). The impact of this microorganism on the grub population is considered to be negligible.

Entomopathogenic Nematodes. Although many M-JB species produce larvae of large size and complete 2- or 3-year life cycles, there are relatively few studies regarding their natural associations with nematodes (Poprawski and Yule 1991b). This may be because scarab larvae in general appear to have lower susceptibility to nematode invasion than do other insects. White grubs have a relatively high defecation rate (which could expel nematodes or the xenobiotic bacteria), anaerobic hindguts, and sieve plates over the spiracular openings, which make penetration difficult. In addition, some grubs may remove nematodes by grooming, and some produce relatively low levels of carbon dioxide which may reduce the host-finding ability of searching nematode species (Villani et al. 1999).

Forschler and Gardner (1991) demonstrated that the presence of spiracular sieve plates, rapid movement of food through the digestive tract, and a thick peritrophic membrane combine to impede nematode invasions in *P. hirticula*.

Insect Parasitoids and Predators

The M-JB complex has a large complement of arthropod parasitoids and predators; these natural enemies mainly are in the orders Diptera, Hymenoptera, and Coleoptera, although other groups may be associated with them as well. The adult *Pyrgota undata* Wied. (Pyrgotidae: Diptera) alights on the dorsum of M-JB adults that are feeding, causing the latter to take flight. Flight allows the female parasitoid to insert her ovipositor into the thin integument of the abdominal dorsum. Death of the beetle occurs in 10–14 days. The parasitoid puparium is formed within the dead host (Davis 1919).

Two studies of mites found on *Phyllophaga* spp. demonstrated that they may be parasitic in some situations and phoretic in others, depending on the life stage of the host insect or species of the mite. Crocker et al. (1992) examined the relationship between the nonfeeding larval stages of a mite, *Caloglyphus phyllophaginus*, and adults of several species of *Phyllophaga*. Almost all mites were found under the wings, in the intersegmental folds of the dorsal abdominal surface. The researchers concluded that the mites were saprophytes and harmless to their beetle hosts.

In Quebec, Poprawski and Yule (1992b) examined all life stages (eggs through adults) of *P. anxia* for mite infestations and found 11 genera of mites associated with the scarab, representing 5 families and 15 species. First-, second-, and third-instar grubs were infested with mites at levels of 9.4%, 42.1%, and 54.6%, respectively, while 36.6% of adult beetles were found to be infested with mites. The authors suggested that mite infestations on grubs may be regulating populations, while mites on adults probably are phoretic only. In addition the authors suggest that parasitic and predatory natural enemies probably have a negligible impact on natural populations of *P. anxia* grubs.

Grubs of M-JB are attacked by several wasps of the family Tiphiidae. These are external parasitoids. Stings by *Myzine* females cause permanent paralysis of grubs, while stings by *Tiphia* females cause only temporary paralysis of host grubs. Both parasitoids attach an egg to the integument of the grub while it is paralyzed. After hatching, the externally attached parasitoid larva feeds on the liquid contents of the grub (Plate 62). Upon the death of the grub, when only the integument and heavily chitinized portions remain (Plate 62), the parasitoid is ready for pupation, which occurs within a fibrous cocoon (Plate 62). *Myzine quinquecincta* (Fab.) is a common parasitoid in Kentucky (Ritcher 1940).

Endoparasitic tachinid flies of several species may be found preying on *Phyllophaga* grubs. Poprawski (1994) tracked all developmental stages of *P. anxia* and found tachinid flies in 0.53% of all specimens examined.

The predators of M-JB pupae include larvae of bee flies, *Excoprosopa fasciata* Macq., and larvae of robber flies, *Diogmites discolor* Loew (Asilidae). In Kentucky, *D. discolor* was found attacking pupae of *P. hirticula, P. fusca, P. inversa, P. rugosa,* and *P. tristis,* with an estimated 12% rate of pupal destruction. Larvae of *E. fasciata* were found in pupal cells of *P. bipartita* (Ritcher 1940).

Vertebrate Predators

Vertebrate predators common to other turfgrass-infesting scarabaeids are also common to M-JB grubs. These include numerous insectivorous birds, skunks, raccoons, and armadillos that search for grubs as a source of food.

16

Scarabaeid Pests:
Subfamily Rutelinae

Japanese Beetle

Taxonomy

The Japanese beetle (JB), *Popillia japonica* Newman, order Coleoptera, family Scarabaeidae, subfamily Rutelinae (shiny leaf chafer), tribe Anomolini, is considered the single most important turfgrass-infesting scarabaeid in the United States.

Importance

JB larvae cause significant damage to turfgrass in eastern North America, and JB adults are also a major pest, feeding on foliage or flowers of nearly 300 species of plants, including fruits, vegetables, ornamentals, field and forage crops, and weeds.

History and Distribution

Introduction and Early Spread

The JB was first discovered in the United States in southern New Jersey in 1916. Prior to its accidental introduction, the insect was known to occur only on the four main islands of Japan, where it is common but not abundant. Little is known of its biology in Japan, probably because it has little importance there as a pest (Fleming 1972).

Because of quarantines established in 1920 against the Japanese beetle, accurate records have been kept on the beetles' spread in North America. From its incipient infestation of 0.5 square mile in 1916, JB spread has been rapid. By 1926, JB had spread to Pennsylvania and Delaware, infesting about 3,850 square miles. By 1946 the range had expanded to 37,500 square miles, including Virginia, West Virginia, Maryland, New York, and southern New England.

Current Distribution and Eradication Efforts. By 1972, the JB occurred in every state east of the Mississippi River except Minnesota, Wisconsin, Mississippi, and Florida. By 1974, the beetle occupied much of the United States east of the 85th meridian. The capture of numbers of adults during 2 or more years in Cedar Rapids, Iowa, and Kenosha, Wisconsin, indicates the presence of an established infestation. The insect has been observed in Minnesota and Nebraska as well (Vittum 1995c).

Figure 16-1. Distribution of the Japanese beetle. (Drawn by E. Gotham, NYSAES, adapted from Handbook of Turfgrass Insect Pests, Brandenburg and Villani 1995, Entomological Society of America.)

The insect now is well established in the Canadian province of Ontario along the entire Niagara peninsula east of Hamilton and Simcoe, in Belleville, and near Gananoque. At least six other isolated infestations are present, extending from the easternmost at Farnham (near Montreal), Quebec, to the westernmost at Windsor, Ontario (Figure 16-1) (Agriculture Canada 1983, Fleming 1976, U.S. Department of Agriculture 1972). In 1997, apparently new JB infestations were found at several sites in Texas, which are thought to have arrived via interstate shipment of nursery plants (R. Crocker, Texas A&M University, personal communication, 1998), northern Georgia, and Colorado (D. Smitley, Michigan State University, personal communication, 1998).

None of the efforts to eradicate the JB east of the Mississippi River was successful. Three infestations in California, the first in Sacramento, 1961–64, the second in San Diego, 1973–75, and the third near Sacramento, 1983–85, were successfully eradicated. Environmental conditions in California probably contributed to successful eradication. There, cities are surrounded by extensive dry land during periods of beetle flights, which inhibits oviposition and the development of eggs. The lush, irrigated turf and ornamentals inhibit the beetles from dispersing in search of food and oviposition sites (Fleming 1976, Gammon 1961).

Probable Ultimate North American Spread. The spread of the JB appears to be governed by temperature and precipitation. The beetle is adapted to a region where the mean summer soil temperature is between 17.5° and 27.5°C and winter soil temperatures are above −9.4°C. In addition, beetles thrive in areas where precipitation is rather uniformly distributed throughout the year, averaging at least 25 cm during the summer. Since eggs must absorb water before and during embryonic development (Regniere et al. 1981), soil moisture during the summer is essential for survival. The northern limits appear to be the elevated regions of the Northeast, extreme northern Michigan, and west of the Great Lakes to the Missouri River. The western limits may be about the 100th meridian, where the semi-arid region beyond is practically an insurmountable barrier. It has not become established in Florida and the Gulf Coast. The summer isotherm of 25°C appears to limit its southern spread (Fleming 1972, Ludwig 1932).

Host Plants and Damage

Adult Feeding

Adults do not injure turf but are an important pest of many other plants. Of the beetle's nearly 300 host species, those most favored include apples, cherries, grapes, peaches, plums, and blueberries in fruits; asparagus, beets, broccoli, rhubarb, and sweet corn in vegetables; Norway and Japanese maples, birch, crabapples, purple-leaf plums, roses, sassafras, mountain ash, and linden in ornamentals; soybeans, alfalfa, clovers, and corn in field and forage crops; and smartweed, crabgrass, ragweed, and cattail in weeds (Fleming 1972).

Although many of the preferred host plants are members of the family Rosaceae, more than 75 other families of plants have species that experience varying degrees of feeding injury. Since many ornamentals are favored host plants, the association of these ornamentals in landscapes makes the JB a major pest of golf courses, parks, homes, and other well-landscaped areas (Plate 42) (Fleming 1972).

Adults feed on the upper surface of the foliage of most plants, consuming soft mesophyll tissues between the veins and leaving a lacelike skeleton (Plate 42). Injured leaves eventually turn brown and die. Trees receiving extensive feeding turn brown and become partially defoliated. Often the upper canopy is defoliated first or most severely (Rowe and Potter 1996). When fruit is attacked, the fleshy tissues are eaten. Beetles feed on the maturing silk of corn, with the result that kernels often are malformed from lack of pollination.

Plants that have been fed on by JB adults subsequently attract significantly more adults than do undamaged plants (Loughrin et al. 1997). Damaged leaves produce a complex mixture of organic compounds, including terpenoids, which may facilitate host location or mate finding by the JB (Loughrin et al. 1995).

A few plants are toxic to the beetles. They include the flowers and foliage of the cultivated geranium and the castorbean, and the flowers of the bottlebrush-buckeye, which cause paralysis and eventually death (Fleming 1972). The presence of large numbers of dead beetles under food plants sometimes is interpreted mistakenly to mean that such plants are toxic to beetles. After midsummer, such accumulations generally consist of aging beetles that completed their normal life span and died during their last feeding visit (Fleming 1972).

Larval Feeding

Grubs feed on the roots of a wide variety of plants, including all cool-season grasses (Crutchfield and Potter 1995a) and most weeds (Crutchfield and Potter 1995c) that are commonly found in turfgrass sites. In laboratory feeding-choice studies, JB grubs consistently chose perennial ryegrass over other cool-season turfgrasses (Kentucky bluegrass, endophyte-free tall fescue, hard fescue, and creeping bentgrass) (Crutchfield and Potter 1995a). In a second study that tested the suitability of these grasses on JB fitness, Potter et al. (1992) determined that JB grubs exhibit poorest growth rates when they are fed the roots of Kentucky bluegrass compared with the roots of other turfgrass species (perennial ryegrass, endophyte-free tall fescue, hard fescue, and creeping bentgrass) tested. This study also determined that first-instar JB grubs feeding on endophytic tall fescue were less fit than similar grubs in endophyte-free tall fescue, but this effect was lost when later instars were tested.

Feeding by grubs on underground stems and roots often passes unnoticed until the plants are damaged severely. Injury to well-kept turfgrass is usually not apparent when there are fewer than 10 grubs per 0.1 m², but an unhealthy, poorly maintained turf may show injury, with 4 or 5 mature grubs per 0.1 m². Laboratory experiments in Kentucky determined that a density of 20–60 third-instar JB grubs per 0.1 m² did not significantly affect the foliar yield of common turfgrass species, and suggested that nonstressed turfgrasses can tolerate high levels of root herbivory without significant loss of foliar growth (Crutchfield and Potter 1995a/b).

Most severe turfgrass damage occurs in September, when third-stage grubs are feeding vigorously just before they go into hibernation. Additional damage may occur in April and May, as grubs feed for a short period of time before pupating. During these periods, most grubs are feeding within the upper 5 cm of soil or in the soil-thatch interface. As many as 130 grubs have been found in 0.1 m² of golf course fairways. Heavily damaged turf can easily be rolled back or lifted, since all of the fibrous roots have been eaten (Plate 27). Grubs may damage the roots of corn, bean, tomato, strawberry, and other field and garden plants. Nursery stock is often injured when grubs feed on and girdle the main root (Fleming 1972, 1976).

Description of Stages

Adult

Beetles are brilliant metallic green, with coppery-brown elytra that do not entirely cover the abdomen. A row of five lateral brushes of white hairs on each side of the abdomen and a pair of these brushes on the dorsal surface of the last abdominal segment distinguish this beetle from all others with similar coloration (Plate 41). Beetles vary in length from 8 to 11 mm and in width from 5 to 7 mm, with the female slightly larger than the male (Figure 16-2) (Fleming 1972, Hadley and Hawley 1934).

The sexes can be distinguished by small differences in the tibia and tarsus of the first pair of legs (Plate 41). The proximal four tarsal segments in the male are wider than they are long, while in the females they are as long as or longer than they are wide. The most pronounced difference is in the first tarsal segment. The apex of the female tibia is spatulate, while it is generally more pointed in the male. Differences in length of the lamellate club of the antennae, common in sexes of many scarabaeids, do not exist in the JB (Hadley and Hawley 1934).

Egg

Recently laid eggs are ellipsoidal and measure about 1.5 mm in diameter (Plate 41). As eggs mature, they double in size. In all other aspects they resemble the scarabaeid eggs described in Chapter 11 as they develop (Fleming 1972).

Larva

Newly hatched larvae are about 1.5 mm long and translucent creamy white. After feeding begins, the accumulation of fecal matter makes the hindgut appear gray to black. The general C-shape of JB larvae is similar to that of other common scarabaeid grubs found in turfgrass (Figure 16-2). Head capsule widths of the first, second, and third instars average 1.2 mm, 1.9 mm, and 3.1 mm, respectively.

The rastral pattern, on the venter of the last abdominal segment, has characteristics that distinguish the JB from all other North American scarabaeid grubs. Palidia are present, con-

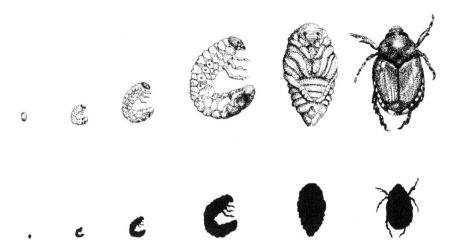

Figure 16-2. Stages of development of Japanese beetle: egg; first, second, and third larval instars; pupa; and adult. Silhouettes show actual size of each stage. (Courtesy of R. Jarecke, NYSAES.)

sisting of two conspicuous rows of six or seven pali that converge to form a V. This character, in combination with the transverse anal opening, distinguishes this grub from all others (Plate 28) (Fleming 1972).

Prepupa
At the prepupal stage the JB, like other turf-infesting scarabaeids, remains in the earthen cell constructed by the mature larva. The prepupa straightens out, leaving only a slight crook on the posterior end. Like other scarabaeids, the prepupa is very flaccid and fragile at this stage.

Pupa
The young pupa forms within the old larval and prepupal exuvia, which changes to resemble a fine, light-tan meshlike tissue (Plate 41). This covering, similar to that found in the oriental beetle, splits along the median line as the pupa matures. The pupa somewhat resembles the adult except that the wings, legs, and antennae are folded closely to the body (Figure 16-2). Its average dimensions are 7 mm wide by 14 mm long. The male pupa has a three-lobed eruption covering the genitalia on the ventral abdominal segment, which differentiates it from the female.

Seasonal History and Habits

Seasonal Cycle
The JB has a 1-year life cycle (Figure 16-3) throughout most of its range, but in the more northern latitudes or higher elevations, part of the population requires 2 years to complete a generation. These areas include southern Maine, New Hampshire, Vermont, Massachusetts, and the Adirondack regions of New York. Even in western New York, up to 5% of the overwintering third instars may be from the previous year's generation (H. Tashiro, personal observations; Vittum 1986).

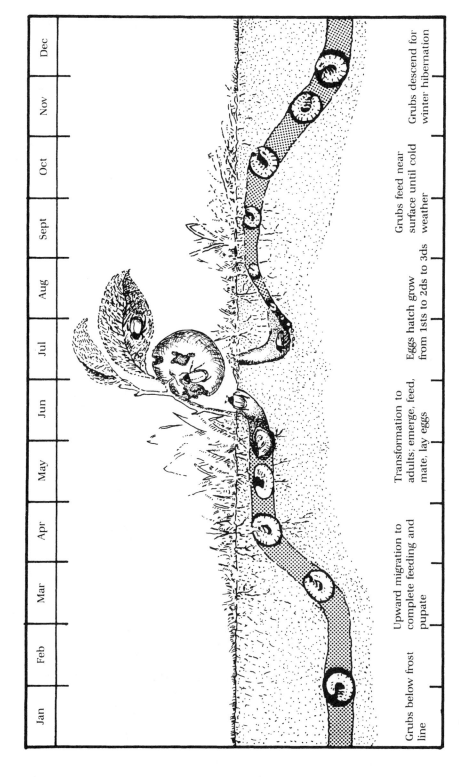

Figure 16-3. Annual life cycle of the Japanese beetle in southern New Jersey (Adapted from Fleming 1972, fig. 7; redrawn by H. Tashiro, NYSAES.)

Table 16-1. Initial Emergence and Maximum Abundance of Japanese Beetle Adults Relative to Latitude and Altitude

| | | Month and Quarter[a] | | |
	Latitude	Initial Emergence	Maximum Abundance	Nearly Gone
Central North Carolina	35	May 3	Jun 2	Jul 2–3
Mountains of North Carolina	35	Jun 4	Aug 2–3	Sept 4
Mountains of Tennessee	35.5	Jun 1	Jun 3	Sept 1
Central Virginia	38	May 4	Jul 2	Aug
Sacramento, California	38	Jun 1	Jul 1	—
Central Maryland and Delaware	39	Jun 2–3	Jul 2–3	Sept 4
Southern New Jersey, Southeastern Pennsylvania	40	Jun 3	Jul 4	Sept 3
Southeastern New York, Connecticut, Rhode Island, Southern Massachusetts	41.5	Jun 4	Jul 4	Sept 4
Southern New Hampshire and Vermont, Western New York	43	Jul 1	Jul 4	Sept 4

[a]1, 2, 3, 4 denote quarters of months.
Source: Adapted from Fleming 1972.

The typical first-beetle emergence and the peak emergence periods at increasing latitudes, represented by several states, are shown in Table 16-1 (Fleming 1962). In western New York, where mixed populations of the JB and the European chafer are common, the first emergence of the JB occurs as European chafer adult flights are ending for the season.

Individual beetles live 4–6 weeks, with females depositing eggs most of their lives. Eggs hatch in about 2 weeks, become second instars in 2–3 weeks, and become third instars about 3–4 weeks later. During September and October, most grubs feed vigorously and become nearly full grown.

Grubs begin to move downward for hibernation when air temperatures first drop below freezing in the fall and remain just below the frost line throughout the winter. Grubs move up again during April and May to complete feeding. The prepupal period occurs in May and lasts about 10 days in May. After 8–20 days as pupae, they become adults. This occurs in June in southern New Jersey and in July in western New York.

Adult Activity

Emergence and Flight. An adult remains in the same earthen cell 2–14 days while its cuticle hardens, its wings expand to normal size, and the body becomes fully pigmented. When it is ready to leave its earthen cell, the beetle uses its mandibles and fore tibiae, adapted for digging, to crawl to the soil surface, a process that may take a day or more (Fleming 1972).

After the initial emergence, most beetles leave their tunnels during the morning of clear days for feeding and mating. Beetles may return to the soil or remain on host plants in the late afternoon or evening to spend the night. Beetles also remain in the soil on cold wet days (Fleming 1972).

Evidence of Sex Attractant. Virgin females emerging from the ground are very attractive to males, especially during early summer. Males fly low over turf in the early morning of clear, warm days, searching for emerging females. As a virgin female emerges, males alight a short distance on her leeward side and crawl toward her. A single female may have as many as 25–200 males congregating around her (Plate 41). Virtually no mating takes place during this phenomenon, which ends about midday (Smith and Hadley 1926).

Sex Pheromone. The pheromone present in virgin females has been isolated and identified as (R,Z)-5-(1-decenyl)dihydro-2(3H)-furanone. A synthetic copy sex attractant has been developed; it is referred to as R,Z-furanone (Japonilure). To date, the most attractive lure is phenethyl propionate (PEP) + eugenol + geraniol, 3:7:3 (PEG) exposed jointly with as little as 5 mg of R,Z-furanone in each trap. Several commercial traps are available, most of which are baited with a dual-lure system of Japonilure and PEG. Even when this highly attractive lure is present, nearby host plants are more important in attracting beetles (Klein 1981, Ladd et al. 1981, Tumlinson et al. 1977). In addition, favored host plants may sustain more damage in the presence of such traps than in their absence, so traps must be placed and used carefully (Gordon and Potter 1985). Beetle captures are significantly higher when trapped beetles are removed daily, perhaps because decaying beetles counteract, or mask, the lure activity. Trap catches are greatest when funnel rims are 13 cm above ground, and lower when rims are 90 cm above the ground (Alm et al. 1996).

Mating and Oviposition. Copulation occurs frequently during the early morning or evening, usually on the adult food plant (Plate 41). Mating occasionally occurs on the ground. There is no uniformity in the numbers or frequency of copulation, but the females usually mate between oviposition periods (Fleming 1972).

After burrowing to a depth of 5–10 cm, the female deposits an egg in an earthen cell, presumably prepared in the same manner as that of the European chafer female (Chapter 11). There is apparently no secretion to cement and strengthen the walls of the cell. A female may enter the soil 16 or more times to deposit 40–60 eggs during her life.

Although most of the eggs are deposited in closely mown turfgrass and pastures, some are also deposited in cultivated fields of rye, corn, beans, tomatoes, and nursery stock, especially in adjacent turfgrass areas where the ground is dry and hard. During summers of deficient rainfall, females select poorly drained ground, irrigated areas, and fallow fields, where the soil is loose for oviposition (Fleming 1972, Smith and Hadley 1926).

During summers with excessive rainfall, females appear to prefer well-drained or raised ground for oviposition (M. G. Villani, personal observation). Potter et al. (1996) reported that JB grub densities increase in irrigated turfgrass when compared with unirrigated turf sites. Laboratory studies documented that females will discriminate among oviposition sites differing in soil texture, soil moisture, and percentage of organic matter (Allsopp et al. 1992). Regniere et al. (1981) reported that females did not lay eggs in air-dried soils. They further determined that as soil particle size increased, soil moisture requirements for oviposition decreased.

Most eggs are deposited during the period of maximum adult abundance, which varies by locality, depending on latitude and altitude. This oviposition occurs from early to mid-June in the more southern latitudes (e.g., Virginia and coastal North Carolina) and lasts until late July or August in the more northern latitudes (e.g., New England) (Fleming 1972).

Flights. During early morning, beetles usually rest quietly on plants. On clear days when the temperature reaches about 21°C and relative humidity is below 60%, beetles will fly. Flight usually ceases or is retarded above 35°C or when the relative humidity rises above

60%. Flights also are greatly diminished on cool windy days and cease completely on rainy days (Fleming 1972). The average outward spread from a point of introduction is about 16–24 km per year. Sustained flights of 8 km were recorded when winds were favorable (Fleming 1958).

Although beetles cause no harm to turf, they can be a nuisance on the golf greens when they are numerous, because they must be brushed aside if they block a putt. When abundant, their flight activities can also be annoying if they accidentally strike people on the head and face.

Larval Activities

Hatching and Growth. Eggs hatch in about 2 weeks, after which the grub forms a cell a little larger than its body in the soil and feeds on fine rootlets that grow into the cell. The first molt (into the second instar) occurs in 2–3 weeks. The second molt (into the third instar) occurs 3–4 weeks later for the final instar.

Eggs and first-instar grubs are sensitive to temperature and moisture extremes; under extreme environmental conditions, eggs will not hatch, and young grubs will not survive. Variation in the survival of first-instar grubs in soils of heterogeneous moistures and textures has been cited as a major cause of spatial and temporal fluctuations in population density (Fleming 1968, Hawley 1944, Ludwig 1932). Although eggs and first-instar grubs are quite sensitive to desiccation (Hawley 1944), larger grubs and pupae appear to be more resistant to desiccation and drowning (Belluci 1939, Hawley 1944, Ludwig 1936).

Regniere et al. (1981), in a series of laboratory studies, determined that no JB eggs survived prolonged contact with air-dried soil. High soil moisture causes mortality and delayed egg development in fine-textured soil but not in coarser soil textures; eggs are able to survive lower soil moistures in sandy soils when compared to clay soils. Older eggs are more resistant to soil moisture extremes than are newly laid eggs.

A JB grub shows a positive thigmotactic response to living roots, stones, and other objects in the soil. While in its earthen cell, the grub assumes a C-shaped form and does not straighten out until it is burrowing horizontally through the soil by digging with its mandibles. The cell is enlarged as the grub grows, to permit free movement of the body. Most grubs are within the top 5 cm of soil during summer if there is adequate moisture (Table 16-2). As the soil dries, many grubs will burrow into the soil to seek moisture. By fall, third-instar grubs feed and grow rapidly. With an abundance of moisture many are within the soil-thatch interface. Although moisture affects their vertical movement, soil temperature appears to have a more pronounced impact on vertical movement (Fleming 1972).

Winter Hibernation. As the surface soil cools to about 15°C, grubs begin to move downward and continue until the soil reaches 10°C. Thereafter the grub becomes inactive. By November in the latitude of southern New Jersey, the bulk of the population is at a 5–15-cm depth. Grubs remain at their hibernational depth all winter (Fleming 1972).

Most JB grubs overwinter as third instars, but those with a 2-year life cycle overwinter as second instars their first winter and as third instars during their second winter.

Winter Mortality. In sites with heavy snow and thick sod cover, grubs may remain closer to the surface, compared to sites with relatively barren soil. Lack of a snow cover on barren soil often results in increased grub mortality, particularly if the ground freezes rapidly in early winter.

Table 16-2. Vertical Distribution of Immature Japanese Beetles during a Calendar Year in Southern New Jersey and Southeastern Pennsylvania

Month	Percentage of Total at Indicated Depths (cm)			
	0−5	5−10	10−15	15−25
December–February	5	62	31	2
March	11	64	23	2
April	63	28	9	<1
May	95	5	<1	0
June	76	24	<1	0
July	84	16	<1	0
August	94	6	<1	0
September	93	7	<1	0
October	63	30	7	0
November	10	65	24	1

Note: Data cover 11 years at eight sites.
Source: After Hawley 1944.

Spring Larval Activity. Migration upward in the soil profile usually commences during the latter half of March in southern New Jersey, as soil temperatures rise above 10°C. By late April and early May, grubs primarily are within the upper 5 cm of soil. Most of the grubs feed very actively for 3−4 weeks at or very near the soil-thatch interface.

Prepupal and Pupal Activities. When feeding has been completed, the mature grub moves downward to a depth of 5−10 cm and constructs a horizontal earthen cell. Here the larva transforms to an adult in a sequence differing in only minor ways from that described and illustrated in Chapter 11 for the European chafer. Pupation occurs within the larval-prepupal exuvia, which later splits to release the pupa. As with the European chafer, the teneral adult has soft transparent elytra and still unexpanded hind wings, even after the remainder of the body has turned the metallic green characteristic of the mature adult (Plate 43).

Influence of Cultural Practices. Work conducted in Ohio during the 1950s suggests that significantly more eggs are deposited in soil of pH 4.5 or less than in soil of pH >7.6. Also, larval survival is higher in soils of pH 4.5−5.0 than in soils of pH 6.8−7.0 (Polivka 1960a, 1960b; Wessel and Polivka 1952). Laboratory studies in New York indicated, however, that there is no significant effect of soil pH on the survival of either the JB or the European chafer grubs (Vittum and Tashiro 1980). Similarly, in Massachusetts, elemental sulfur or dolomitic limestone was added to JB-infested turfgrass to adjust the soil pH to about 5, 6, 7, and 8. There was no effect on JB populations in these studies (Vittum 1984). Since most turfgrasses grow best at pH 6−7, manipulation of soil pH outside this range is not a practical method of reducing grub populations even if pH has an influence.

Potter et al. (1996) evaluated a series of cultural practices before or during the seasonal flights of JBs to study the effects of this manipulation on subsequent grub densities. Their findings suggest that the addition of organic amendments, the compaction of soil by heavy rollers, or aerification just prior to beetle flights has no measurable effect on subsequent grub populations. The addition of aluminum sulfate (to lower pH) results in fewer and smaller grubs, but not at significant levels when compared to untreated controls. Mowing height significantly impacts grub size and weight; JB grubs tend to be smaller and develop more slowly in high-mowed grass when compared to closely mowed grass, but grub densities are not affected. Irrigation appeared to have the largest impact on JB grub populations in this study: The abundance, mean weight, and developmental rate of JB grubs positively correlated with soil moisture.

Natural Enemies

JBs are attacked by a large number of microorganisms, invertebrate parasitoids and predators, and vertebrate predators. The explosive populations that developed during the years immediately following the insect's 1916 discovery no doubt reflect an initially low abundance of many natural enemies. Introduced parasitoids and predators from the Orient have become established in some locations, and native organisms have adapted to become enemies of the JB and reduced some population outbreaks.

Microorganisms

JB grubs are infected by various microorganisms, including bacteria, rickettsiae, fungi, protozoa, and nematodes. Current field and laboratory research is centered on the use of various bacteria, nematodes, and fungi to control JB grubs.

Milky Disease. A few abnormally white grubs were found in New Jersey in 1933. The hemolymph was found to be teeming with bacterial spores. Two species of the bacteria were described and named, *Bacillus popilliae* (the dominant species) and *Bacillus lentimorbus* (of relatively minor importance). The former has been called the type A milky disease organism and the latter the type B. Differences in the morphology of the spores, most readily apparent with about 400× magnification under phase contrast, distinguish the two species. Spores of *B. popilliae* have a parasporal body at the end of the sporangium, while *B. lentimorbus* does not (Plate 57).

The field efficacy of *B. popilliae* has been difficult to document in some parts of the country. A field trial in Kentucky demonstrated that commercial preparations of *B. popilliae* did not raise the incidence of milky disease in grub populations or reduce grub populations, regardless of the method of production of infective spores (Redmond and Potter 1995). However, two characteristics are certain: The bacterium works much more slowly than conventional materials and it usually remains detectable in the grub population for several growing seasons (Vittum 1995c).

Bacillus thuringiensis Variety *japonensis* (Btj), Buibui Strain. This facultative pathogen is specific to scarabs and suitable for short-term control (using high rates, or inundative releases). The Buibui strain appears to be effective against several species of white grubs including JB (Alm et al. 1997). However, the future of Btj as a commercial product is in doubt due to formulation and marketing concerns.

Entomopathogenic Nematodes. The nematode *Steinernema* (=*Neoaplectana*) *glaseri* was first isolated from JB grubs in New Jersey, and was subsequently used in a control program in the 1930s and 1940s. The success of this program was limited by a lack of economical mass-rearing techniques and by a lack of understanding of the importance of the symbiotic bacteria that are associated with these nematodes and are essential for their parasitism and reproduction in the field. Recent studies have resulted in increased knowledge of the biology and ecology of nematodes and development of suitable mass-rearing techniques.

The species that currently appear to show the most promise for field application are *S. glaseri* and *Heterorhabditis bacteriophora*. *Heterorhabditis* species may be effective against scarab larvae (Plate 58) (Klein 1995, Selvan et al. 1993, Villani and Wright 1988b). Environmental factors, especially the availability of adequate soil moisture, are critical for entomopathogenic nematode performance in the field (Georgis and Gaugler 1991, Shetlar et al. 1988).

Entomopathogenic nematodes can be used in two ways. High rates of nematodes can produce high rates of mortality quickly, whereas lower rates may produce lower initial mortality, but over longer time periods may produce high levels of control (Kaya and Gaugler 1993, Klein and Georgis 1993). This result most likely is due to the ability of nematodes to reproduce in the body of the dead insect, producing additional infective juveniles, which then disperse and can attack other insects.

Fungal Pathogens. *Metarrhizium anisopliae* (Metsch.), a widely distributed entomophthorous fungus sometimes called *green muscardine disease,* infects JB grubs (Plate 57) but the incidence of natural infection usually is low.

Microorganisms of Secondary Importance. JBs have been found in the field infected with various microorganisms. All of these natural enemies contribute toward population reductions even though they play a minor role.

Bacteria in the genus *Serratia* have long been associated with diseases of scarabs. Selected strains of *S. entomophila* are able to colonize the insect gut and cause starvation of the larvae, depletion of the fat bodies, and development of an amber color. JBs with this characteristic coloration have been found in the United States and *Serratia* species have been isolated from them (Klein 1995).

Hanula and Andreadis (1988) isolated several species of protozoans in field-collected JB grubs from multiple turf sites in Connecticut. Cephaline eugregarines were found in the guts of JB larvae from 42 locations. Over 50% of larvae collected in September were infected, and gregarines were found at every location sampled until mid-October. Several of the locations that did not have these protozoans were sampled only in November and December. Because the gregarine protozoans were nearly ubiquitous, they are presumed not to be highly pathogenic in the host grubs. A microsporidian, *Ovavesicula popilliae* Andreadis and Hanula, was found in 25% of JBs sampled throughout Connecticut. Infection rates ranged from 2% to 95%, and were not correlated with larval density (Hanula and Andreadis 1988). This organism reduced fecundity 50% in heavily infected female beetles in a laboratory test. The potential detrimental effects of *O. popilliae* apparently are modified by environmental stress (Hanula 1990).

A rickettsia, *Coxiella popilliae*, produces a greenish-blue discoloration of larval fat bodies in grubs and bears the name *blue disease.* Rickettsia are microorganisms that possess characteristics of both bacteria and viruses. They are similar to viruses in that they are obligate pathogens, not being able to grow outside of living cells (Klein 1995). Little is known about their interactions with scarab larvae.

Parasitoids

During 1920–33, some 49 species of parasitoids and predacious insects were imported from the Orient and Australia and were released into JB-infested areas to reduce beetle populations. Five species have become established as a part of the permanent fauna (Fleming 1962).

Larval Parasitoids. The spring, or Korean, tiphia, *Tiphia vernalis* Rohw. (which attacks overwintering grubs in the spring), and the Japanese tiphia, *T. popilliavora* Rohw. (which attacks young grubs in late summer), are most effective parasitoids of JB larvae in North America (Fleming 1976). A female tiphia searches out a grub in the soil, stings it to paralyze the grub temporarily, then attaches a single egg to a particular ventral body fold. Upon hatching, the maggot-like larva attaches itself externally to the grub, pierces the host's cuticle with its mouth parts, and feeds on the liquid contents of the grub. When the parasitoid larva is fully developed, it will have consumed the entire grub cadaver except for the cuticle and the head capsule. Pupation takes place in a strong, watertight, silken cocoon in the earthen cell formerly occupied by the grub. The parasitoid emerges as an adult the following year. The adults feed on the flowers of the wild carrot and on leaves of the honeydew on warm sunny days. Many new colonies have been established by collecting feeding adults and releasing them in other areas with high grub populations. During the past 20–25 years, both *Tiphia* species have been very scarce, and in many areas they have not been seen.

Adult Parasitoid. Another introduced insect, *Istocheta aldrichi,* a tachinid fly, is active in midsummer, feeding on honeydew and nectar. This female attaches an egg about 1 mm in diameter to the thorax of the beetle (Plate 62). The egg hatches in 24 hr, after which the resulting larva feeds internally on the host's vital organs as the beetle buries itself in the ground. The parasitoid kills the beetle in about 5 days, pupates, and then overwinters (Rex 1931).

During the 1930s and 1940s, *I. aldrichi* was apparently of only minor importance. During 1979, about 20% of the adult JBs in central Connecticut were found to be parasitized by this fly. It is currently found in about 10% of the early emerging adult beetles in the Connecticut River valley of Massachusetts and lower densities in eastern Massachusetts (P. J. Vittum, personal observation), but is less common later in the summer.

Predators

Invertebrate Predators. Invertebrate predators may have a significant impact on JB populations in the field. Terry et al. (1993) determined that up to 70% of artificially buried JB eggs are consumed by soil predators in Kentucky. By direct observation and videotaping through buried acrylic plates, Zenger and Gibb (unpublished data 1997) determined that ants (primarily *Solenopsis diplorhoptrum*) are responsible for at least 85% of the egg predation in Indiana.

Vertebrate Predators. Birds and mammals reduce JB grub populations considerably just as they do those of other scarabaeid grubs, but they also may cause significant turfgrass damage in the process of digging for their quarry. Starlings, grackles, robins, and other birds feed on large numbers of grubs especially during early spring and fall. When crows dig for grubs, they can cause considerable damage to weakened turf. Mammals attracted to grub-infested turfgrass as a source of food include moles, skunks, and raccoons.

Oriental Beetle

Taxonomy

The oriental beetle (OB), *Exomala* (formerly *Anomala*) *orientalis* (Waterhouse), is in the order Coleoptera, family Scarabaeidae, subfamily Rutelinae. In the United States the OB was first named the Asiatic beetle (Friend 1929); this term was replaced by the presently accepted common name in 1933 (Adams 1949a).

Importance

OB has been erroneously considered a relatively minor pest of turfgrass and ornamental shrubs, possibly because adults are relatively inconspicuous and largely unnoticed. In southeastern New York, New Jersey, and southern New England, this grub often coexists in mixed populations with the JB. Because the two species are practically indistinguishable in the grub stage without magnification, many mixed populations are often mistaken as consisting solely of JB grubs.

OB grubs are a serious pest of field-grown nursery stock and potted plants. Oviposition occurs in the soil around the plants. The third-instar grubs subsequently infest the entire soil mass of potted plants and reach soil depths exceeding 30 cm of field-grown stock (Plate 44).

Much of the information presented below comes from studies made in New Haven, Connecticut, by Friend (1929). Additional references are included to update the information.

History and Distribution

The OB, which is probably a native of Japan, was introduced to the Hawaiian island of Oahu, where it became a serious pest of sugarcane. On the mainland, adults were first collected in the United States in 1920 in a New Haven, Connecticut, nursery, having presumably been imported directly from Japan in infested balled nursery stock. Twelve years later, the OB still was limited to an area within 145 km of New York City.

The natural spread of the OB has been slow, presumably because it is not a strong flier (Bianchi 1935, Britton 1925, Hallock 1933). The recent identification and synthesis of the female sex pheromone (Leal 1993, Zhang et al. 1994) greatly facilitates monitoring and trapping of OB populations. Facundo et al. (1994) demonstrated the pheromone's potency by catching an average of 1,000 male beetles per trap per day on golf course fairways in Norwich, Connecticut. A geographic survey conducted from 1993 through 1996, using pheromone trapping of adult beetles (Alm and Villani, unpublished data, 1999), showed that the OB is present in southern New England, New York, New Jersey, Pennsylvania, Delaware, Maryland, North Carolina, Tennessee, Ohio, West Virginia (probably not established), and Hawaii (Figure 16-4).

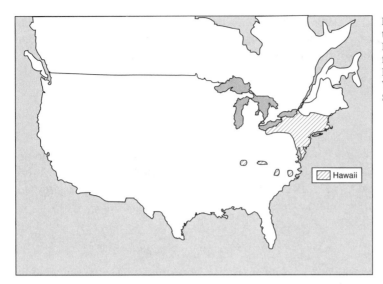

Figure 16-4. Distribution of the oriental beetle. (Drawn by E. Gotham, NYSAES, adapted from Handbook of Turfgrass Insect Pests, Brandenburg and Villani 1995, Entomological Society of America.)

Host Plants and Damage

Adults occasionally feed on flowers of such plants as roses, hollyhock, petunias, phlox, and dahlias, but they are not considered a serious pest in this life stage. Grubs, however, cause severe damage to turfgrass roots. Forty-three to 65 grubs per 0.1 m² often are encountered; these numbers are sufficient to cause complete destruction of turf. Densities as high as 590 grubs per 0.1 m² have been reported. Severe injury also has occurred in nursery stock, strawberry beds, cranberry bogs, and blueberry plots. Severe damage to sugarcane in Hawaii ceased after the introduction of a scoliid wasp, *Scolia manilae* Ashmead, which is a highly effective parasitoid of the OB under Hawaiian conditions (Britton 1925, Hallock 1930).

Description of Stages

Adult

The adult OB is a broad-bodied, spiny-legged, and convex-backed beetle. Males have a mean length of 9.0 mm; females are slightly longer, with a mean length of 10.3 mm (Figure 16-5). The sexes can be distinguished by the lamellae of the female antenna, which are distinctly shorter than the rest of the antenna. In the male, the lamellae are as long as the remainder of the antenna (Plate 44). Adult coloration ranges from black to straw, with intermediate stages having black elytral bands and thoracic markings (Plate 44). Eight variations are given (Figure 16-6).

Egg

When first deposited, the egg is white, ovoid, smooth, and about 1.2 mm wide by 1.5 mm long. At maturity it is more spherical, and about is 1.6 mm by 1.9 mm (Plate 44).

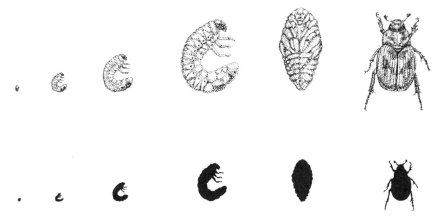

Figure 16-5. Stages of development of oriental beetle: egg; first, second, and third larval instars; pupa; and adult. Silhouettes show actual size of each stage. (Courtesy of R. Jarecke, NYSAES.)

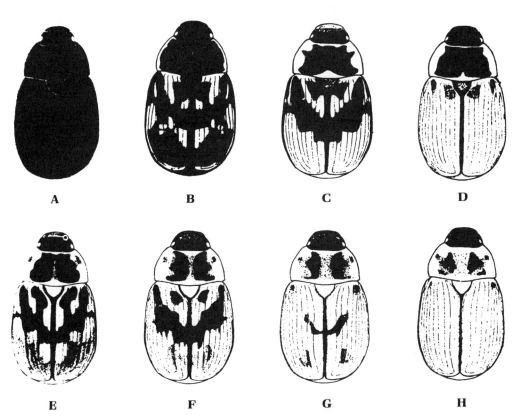

Figure 16-6. Variations in colors and patterns of the adult oriental beetle. (From Friend 1929, fig. 36, courtesy of the Connecticut Agricultural Experiment Station, New Haven.)

Larva

First instars range from about 4 mm in length on hatching to an eventual length of about 8 mm at maturity. Second instars attain a length of about 15 mm. Fully grown third-instar grubs are about 20–25 mm in length. Mean head capsule widths are 1.2 mm, 1.9 mm, and 2.9 mm for first, second, and third instars, respectively (Figure 16-5).

The characteristics of the rastral pattern on the ventral side of the 10th abdominal segment help distinguish OB larva from other turfgrass-infesting scarab larvae. Palidia are present, forming a pair of subparallel rows of medially pointed recumbent setae. The individual pali (spines) making up the palidia are much narrower and finer than in most species (Plate 28). There are usually 11–14 pali, but the number may vary from 10 to 16 in either palidium, and may differ by as many as 3 pali between palidia. This setal arrangement alone cannot be relied on completely, since other scarabaeids, such as *Phyllopertha horticola* and *Anomala (Pachystethus) lucicola*, have similar arrangements. The anal slit in the OB is transverse (Ritcher 1966).

Prepupa and Pupa

The prepupa, as in other scarabaeids, is a quiescent stage that terminates the third-instar stage. The young pupa develops much as it occurs in the Japanese beetle (Plate 44). The exuviae split longitudinally, releasing the maturing pupa. The pupa is about 10 mm long by 5 mm wide. Ventrally, the abdomen differs in the two sexes. Posterior to the ninth segment of the male there are two ventral lobes that are absent in the female.

Seasonal History and Habits

Seasonal Cycle

Normally OBs overwinter as third-instar grubs down in the soil profile, migrate to near the soil surface in April, and then feed for about 2 months before pupating in June. Adults are present during late June into August (slightly earlier than JB adults), with most eggs being laid during July and early August. First instars are dominant in August, second instars in September, and third instars in October. Downward migration for hibernation begins after the middle of October. The life cycle closely resembles that of the JB (Figure 16–3), with one generation a year typical. As many as 15% of the individuals may fail to complete their development the first year but instead overwinter as second instars. During the second year, these grubs become fully developed third instars; they then move to depths of 20–36 cm in the soil for the second winter, where they remain until June of the following year before becoming pupae and emerging at the same time as the 1-year life cycle individuals. The days required for each stage at New Haven, Connecticut, during 1926 and 1927 for individuals reared in the insectary under simulated natural conditions were as follows: adults (to oviposition), 4; eggs, 25–28; larvae, 311–317; prepupae, 5–7; and pupae, 14–15 (Hallock 1935).

Adult Activity

Adult Emergence and Flight. Newly eclosed adults remain in their earthen pupal cell for about a day, until the chitin hardens. A study by Facundo (1997) on an irrigated golf course indicated that the mating season of the OB on Long Island, New York, begins in the middle of June, peaks in the first week of July, and ends in the middle of August. He encountered differences in the emergence schedule between fairways on this golf course, as well as local differences between fairways and roughs, suggesting that microenvironmental differences in factors such as soil moisture, soil temperature, and food quality affect beetle emergence. Adult males emerged earlier than did females in the field. Both sexes were most active around sunset on shorter-cut turf (fairways, greens, and tees), and the few individuals seen during the day were males perched on vegetation at the edge of fairways. Only females devoid of eggs were observed to feed, usually in flowers adjacent to turfgrass stands. This feeding activity may be primarily in search of moisture (Hallock 1930, 1933). Beetle flights, when synthetic pheromone is not present, appear to occur primarily between 8 AM and 4 PM (Facundo 1997).

Mating and Oviposition. The OB pheromone, identified as 7-(Z)- and 7-(E)-tetradecen-2-one, was isolated, identified, and synthesized independently by Leal (1993) and Zhang et al. (1994). The male beetle locates potential mates by tracking the plume of sex pheromone released by the female. Males usually both fly and walk to reach the female. There is no overt precopulatory courtship, and the first male to reach a calling female generally mates with her (Plate 44) (Facundo 1997).

The interval between mating and oviposition may be as short as 1 day but normally is about 5 days. Oviposition occurs in 4–20 days, with the median time being about 7 days; solitary eggs are deposited both day and night at depths of 2.5–28.0 cm. Adults will oviposit in ground covered with sod or in barren soil. Females lay an average of 25 eggs, but as many as 63 eggs can be deposited by one individual. The egg stage lasts 17–25 days in the field. As with other scarabaeids, moisture seems essential for development of the embryo. Just before hatching, the brown tips of the mandibles are visible (Hallock and Hawley 1936).

Larval Activity

As soon as hatching occurs, first instars migrate upward to within 2.5 cm of the soil surface to feed on grass roots and organic matter. Larvae will live and grow on organic matter in the absence of live roots, but growth is retarded and their ultimate size is reduced.

Duration of the first instar is less than 30 days at 23°–25°C in living sod. Duration of the second instar is about the same. By the first week in September, second instars have become common, and turf damage often is apparent. By the last week in September, most of the larvae are third instars. About the middle of October, larvae begin to move deeper for hibernation.

Winter and Spring Larval Activities. Downward movement of grubs generally starts in October, when the mean soil temperature at the 8-cm depth falls to about 10°C. They spend the winter at depths of 20–43 cm, penetrating more deeply in lighter, sandier soils. Winter mortality in the field is negligible; there generally are about as many larvae in April as there were in the previous November and December (Hallock 1935).

In spring, as the soil temperature at the 8-cm level reaches a mean of about 6°C during late March or early April, grubs start to move upward. As the soil temperatures reach 10°C (usually about mid-April) grubs are at their feeding position just below the surface. They remain at this level until early June, at which time they migrate downward to pupate. Optimum conditions for larval development appear to be an open turf stand exposed to the sun and a good cover of grass on a soil of fairly rich, sandy loam. Grubs may move horizontally as far as 1.2 m (Hallock 1935).

Pupation
About the first of June, larvae dig to a depth of 8–23 cm for pupation. Some may go as deep as 30 cm. The prepupal period lasts about 6–13 days, with an average of 8 days. Pupation occurs during late June and early July in most cases and lasts about 2 weeks.

Natural Enemies

Microorganisms
OB grubs are susceptible to a milky disease, but the causal organism has not been classified. Dunbar and Beard (1975) and Hanula and Andreadis (1988) reported a low incidence of milky disease in Connecticut. Hanula and Andreadis (1988) reported a protozoan in OB grubs in a survey of scarabs collected in Connecticut.

***Bacillus thuringiensis* Variety *japonensis* (Btj), Buibui Strain.** This facultative pathogen is discussed earlier in this chapter (see page 206).

Entomopathogenic Nematodes. Several field trials have been conducted using entomopathogenic nematodes against OB grub populations. Alm et al. (1992) conducted field trials in the fall, using two strains of *Steinernema feltiae* (= *bibionis*) (Filipjev), applied at 4.9 or 49.4 billion nematodes per hectare. Half the plots were dethatched before application. The higher rate of one strain reduced grub populations significantly (but only 40–42%), with or without dethatching, while the higher rate of the second strain only reduced populations significantly in the dethatched plots. The higher rate was 5- to 10-fold higher than typically recommended for commercial applications. In a spring trial against OB, Alm et al. (1992) showed no significant difference between control plots and those treated with *Heterorhabditis bacteriophora* (= *heliothidis*) Poinar or *Steinernema carpocapsae* Weiser (All strain), even when turf was aerified prior to application to enhance nematode dispersal. Nematodes were applied at 4.9 and 19.8 billion nematodes per hectare. Cool soil and air temperatures might have reduced nematode activity. Yeh and Alm (1995) found that *Steinernema glaseri* (Steiner), applied at 1.1–2.2 billion nematodes per hectare reduced OB grub populations significantly when at least 0.6-cm irrigation was applied after treatment.

Parasitoids
In Hawaii, *Scolia manilae* Ashmead was introduced in Hawaii and was so effective in parasitizing OB grubs that the latter are no longer a serious problem there. This parasite was introduced near Philadelphia as a potential parasite of the JB but did not survive the winters (Britton 1925).

Vertebrate Predators
Vertebrates that feed on OB grubs include birds (starlings, grackles, and crows) and mammals (moles, skunks, and raccoons).

17

Coleopteran Pests:
Family Chrysomelidae

Dichondra Flea Beetle

Taxonomy

The dichondra flea beetle (DFB), *Chaetocnema repens* McCrea, family Chrysomelidae, subfamily Halticinae, is a member of a genus of minute flea beetles generally smaller than 2.5 mm and most 1.5–2.0 mm in length. A closely related species, *C. confinis* Crotch, feeds on bindweed and sweet potato plants in the morning-glory family, Convolvulaceae, of which dichondra is also a member.

Practically all the information we have on the DFB comes from McCrea (1972), supplemented by Bowen et al. (1980) and V. A. Gibeault (Cooperative Extension, University of California, Riverside, personal communication, 1985). The account below draws mainly on information from McCrea (1972) and on personal observations of H. Tashiro unless otherwise indicated.

Importance

The DFB is a very destructive beetle on lawns of dichondra, *Dichondra* spp., in California, often destroying many of them. Beetles are so small that they are difficult to detect readily. Once an infestation has become established, destruction of the dichondra lawn is very rapid, and many homeowners are unable to determine the cause until irreversible damage has taken place.

History and Distribution

Degradation of dichondra lawns in southern California by *Chaetocnema* spp. became serious during the mid-1960s and was erroneously identified as having been caused by *C. confinis* or *C. magnipunctata* Gentner. Since 1972, *C. repens* has been identified as a new species and has been determined to be the cause of dichondra devastation. This beetle is the most serious pest of dichondra throughout the southern California coastal areas from San Diego into Santa Barbara County; into Antelope Valley in the western Mojave Desert; in the interior southern California counties of Imperial, Riverside, and San Bernardino; and at least as far north as Fresno in central California. As far as is known, the flea beetle is as-

Figure 17-1. Distribution of dichondra flea beetle. (Drawn by E. Gotham, NYSAES, adapted from Handbook of Turfgrass Insect Pests, Brandenburg and Villani 1995, Entomological Society of America.)

sociated with dichondra throughout its range into the San Jose and San Francisco Bay area and near Sacramento (Figure 17-1) (V. A. Gibeault, personal communication, 1985; McCrea 1972; H. Tashiro, personal observations).

Host Plants and Damage

Until 1997, only the broad-leaved plant, dichondra, was considered to be a host plant of the DFB. During 1997, bermudagrass near Fresno, California, was damaged by this beetle (J. Kollenkark, personal communication, 1997). "Scorching" of leaves, observed on some bermudagrass, is thought to be the result of adults feeding on the epidermis. Eventually large patches of turf can be killed (Plate 45).

On dichondra, damage appears first as crescent-shaped mandibular cuts made by adults on the upper surface of the leaves (Plate 45). Soon thereafter the leaves turn brown and die (Plate 45); dying strips of dichondra appear adjacent to sidewalks. Damage is thought to start near sidewalks for any of three reasons. Hitchhiking beetles may start an infestation. Temperatures are warmer near the walks and permit earlier seasonal activity. Increased moisture due to runoff from paved areas may create a more favorable habitat for the beetle.

Injury is often assumed to be caused by lack of water or by a fertilizer burn (Bowen et al. 1980). In southern California, within 3 weeks following the first symptoms, as much as a third of a lawn of average size can be totally destroyed, and within 4–5 weeks the entire dichondra lawn is often destroyed.

If much damage by adults is evident, larvae are usually present, feeding on the roots and crown. Root damage appears to start at the tip of the small roots. Larvae work their way to the crown, leaving skeletonized roots with much frass. Sometimes the larvae burrow up roots and completely disappear from sight. McCrea (1972) is convinced that dichondra turf would survive the mandibular cuts on the leaves by adults if the larvae did not destroy the root system.

Figure 17-2. Stages of development of dichondra beetle: egg, final larval instar, and adult. Silhouettes show actual size of each stage. (Courtesy of R. Jarecke, NYSAES.)

Description of Stages

Adult

Adults are black to reddish black and slightly bronzed. Their antennae and legs are reddish yellow to reddish brown. The underside is reddish brown to reddish black and shiny. One of the more conspicuous features is the femur of the hindleg, which is enormously broad even for a flea beetle (Plate 45). Beetles are broadly oval, robust, and 1.5–1.8 mm long (Figure 17-2).

DFB adults have distinct punctures over the compound eyes and on the prothorax and can thereby be distinguished from the very similar *C. confinis.*

Egg

The minute egg is about 250 μm wide by about 480 μm long and is deposited on leaves and stems (Plate 45).

Larva and Pupa

The DFB has four larval instars. They grow from first instars with head capsule widths of 115–125 μm to fourth instars with head capsule widths of 237–265 μm. Larvae appear unusually long relative to their diameter (Plate 45). Fourth instars are about 6 mm long. Pupae are first whitish and about 1.4 mm long.

Life History and Habits

Life Cycle

Adults overwinter in the soil around the crown and in leaf trash near dichondra lawns. A warm spell in January may cause adults to surface and feed. Adults generally become active in May and continue through October in warm areas. A generation is completed in about a month.

Eggs hatch in 3 days. The four larval instars require 22–25 days, and the pupal period lasts about 5 days. Teneral adults require a day or two for maturation. A total of 32–35 days are thus required for a complete generation. Breeding is continuous through the warmer months. The last generation of pupae become adults to overwinter in November.

Adult Activity

If adults are present, moving the open palm lightly over the surface of the lawn will cause the beetles to jump and alight on the back of the hand. Beetles can also be seen on the upper surface of the leaves, with typical feeding damage (Bowen et al. 1980). Adults can be captured by sweeping the lawn with an insect net.

Mating pairs are often observed on the upper surface of a leaf (H. Tashiro, personal observation). Copulation lasts about 30 min. Eggs are laid on leaves and stems and on moist blotters in a laboratory rearing cage.

Larval Activity

Larvae live in the soil and under heavy infestations completely destroy the dichondra stand. The largest numbers are found 2.5–5.0 cm deep, near small rootlets of dichondra. Some are found 7.6–10.0 cm deep in the soil.

Young larvae begin feeding on roots soon after hatching, often burrowing completely inside the root. Upon completion of larval life, pupation occurs in the soil. Larval feeding on the roots is the most destructive activity of the DFB.

Miscellaneous Features

Larvae can be raised on damp blotters. Rearing through an entire life cycle was accomplished most readily by infesting flats of dichondra. Larvae from sod and soil samples have readily been retrieved by placing the samples in Berlese funnels.

Natural Enemies

None has been reported.

18

Coleopteran Pests: Family Curculionidae

Billbug Taxonomy

Turf-infesting billbugs belong to the order Coleoptera, family Curculionidae, subfamily Rhynchophorinae, genus *Sphenophorus*. Billbug adults can be recognized by their long snout and identified to species primarily by the markings on the pronotum and elytra. The legless larvae cannot be identified to species. Early literature placed the bluegrass billbug and other closely related species under the genus *Calendra* (Vaurie 1951), but all are classified in *Sphenophorus* now.

There are at least nine species of billbugs that attack turfgrass in the contiguous United States—the bluegrass billbug, *Sphenophorus parvulus* Gyllenhal; the lesser billbug, *S. minimus* Hart; *S. venatus* (Say); *S. inaequalis* (Say); the hunting billbug, *S. venatus vestitus* Chittenden; the Phoenician billbug, *S. phoeniciensis* Chittenden; the Denver billbug, *S. cicatristriatus* Fabraeus; *S. coesifrons* Gyllenhal; and *S. apicalis* (LeConte) (Johnson-Cicalese et al. 1990, Morrill and Suber 1976, Vaurie 1951). The bluegrass billbug and hunting billbug are the most widely recognized and serious billbug pests on turfgrass.

Bluegrass Billbug

Taxonomy

The bluegrass billbug (BGB), *S. parvulus* Gyllenhal, is thought to be the most widely distributed of the turf billbugs in North America. Three other species are similar in appearance and life cycle to the BGB. It is likely that many reports of BGBs are actually a complex of billbug species, especially in the northeastern United States, where *S. parvulus*, *S. venatus*, *S. minimus*, and *S. inaequalis* often occur together in turfgrass stands. A study in New Jersey found a nearly equal abundance of these four species, although seasonal distribution varied slightly (Johnson-Cicalese et al. 1990).

Importance

Billbugs, which may include various combinations of the four-species complex in some areas, are important pests of cool-season turf in many scattered locations. They can cause sig-

nificant damage to home lawns, athletic fields, roughs and sides of bunkers on golf courses, and commercial grounds (D. Shetlar, Ohio State University, personal communication,
1998). In Ohio, billbug injury to home lawns would rank second in importance to injury
by the hairy chinch bug. It was considered so severe a threat in Nebraska that the breeding of turfgrasses for resistance to this insect was deemed warranted (Johnson-Cicalese et
al. 1989, Kindler and Kinbacher 1975, Mahr and Kachadoorian 1983, Niemczyk 1983).

There is evidence in Ohio that billbugs developed resistance to the chlorinated cyclodiene insecticides, resulting in an upsurge of this insect in turfgrass. Prior to the development
of resistance, these insecticides, used against other turf insects such as the scarabaeid grubs,
may have held the BGB in check (Niemczyk and Frost 1978). In Rochester, New York, and
nearby areas, the BGB abruptly became a serious home lawn problem in 1967 at nearly
the same time that European chafers manifested resistance to chlorinated cyclodienes. The
eastern problem with the BGB appears to parallel the Ohio situation (H. Tashiro, personal
observations).

History and Distribution

The BGB is generally troublesome across the northern half of the United States in localized
areas (Figure 18-1). It can be found as a pest wherever Kentucky bluegrass is grown. Serious outbreaks occurred in Salt Lake County, Utah, during the early to mid-1960s. During
the late 1960s outbreaks were prevalent in Washington, Nebraska, Kansas, Wisconsin,
South Carolina, Pennsylvania, New York, and Massachusetts (Niemczyk 1983, Tashiro and
Personius 1970). According to Satterthwait (1932), the BGB has a much wider distribution,
occurring from Canada to Florida and Texas and extending from the Atlantic Coast to South
Dakota and probably throughout the United States (Figure 18-1).

Host Plants and Damage

Grasses Damaged

As its name implies, the primary host plant of the BGB is Kentucky bluegrass, but it also
attacks perennial ryegrass, tall fescue, fine fescue, and 14 other nonturf grasses (Johnson-
Cicalese and Funk 1990, Satterthwaite 1931). In a New Jersey study, the four most common adult billbugs (*S. parvulus, S. venatus, S. inaequalis,* and *S. minimus*) were found in the
field on Kentucky bluegrass, tall fescue, and perennial ryegrass. All but *S. inaequalis* were
found on Chewings fescue. Limited collections of immature billbugs revealed all four
species on Kentucky bluegrass; *S. venatus, S. inaequalis,* and *S. minimus* on tall fescue; *S. venatus* and *S. minimus* on perennial ryegrass; and *S. inaequalis* on creeping red fescue. In
related laboratory experiments, adults confined in petri dishes with either Kentucky bluegrass, perennial ryegrass, tall fescue, or bermudagrass yielded only minor differences in
survival, number of eggs laid, and amount of feeding. Bermudagrass was the least favored
but the other grasses were equally acceptable hosts (Johnson-Cicalese and Funk 1990).

Nature of Damage

Adults feed on grass stems by inserting their mouthparts, on the tip of the snout, into the
center of the stem (Plate 46). Adult feeding damage is not considered to be of major concern.

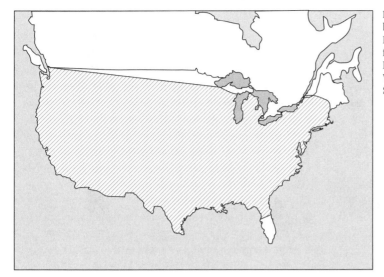

Figure 18-1. Distribution of bluegrass billbug. (Drawn by E. Gotham, NYSAES, adapted from Handbook of Turfgrass Insect Pests, Brandenburg and Villani 1995, Entomological Society of America.)

Eggs are inserted singly into feeding punctures just above the crown. Most eggs are laid on the seed stem. Cultivars with numerous thick seed stems are most likely to be damaged by BGB feeding (D. Shetlar, personal communication, 1998). As they hatch, young larvae feed within the stems and hollow them out. The larger larvae feed in the crown at the base of the stems, so that the stems can easily be pulled out by hand (Plate 47). After larvae have consumed most of the crown, the plants break off easily at that point. As the later-instar larvae approach maturity, they migrate below the stems and also feed on roots. A light-tan, sawdust-like excrement accumulates in the area of heavy feeding (Plate 47). Turf damaged by billbugs appears wilted but does not respond to watering. Feeding by scarab grubs and mole crickets makes the soil loose and soft, but turf killed by billbugs has a firm soil (Fushtey and Sears 1981, Mahr and Kachadoorian 1983, Niemczyk 1983).

Period of Damage

Damage becomes apparent in late June or July, and may continue through August, especially when the turf is under moisture and heat stress. Damage occurs as spotty brown patches, first appearing along sidewalks and driveways or near trees, where moisture or heat stress may be more apparent (Plate 47). Many infestations show brown patches scattered over the entire lawn that later coalesce to produce total turf destruction to nearly the entire lawn (Plate 47) (Niemczyk and Frost 1978; H. Tashiro, personal observations). Billbug damage often is confused with summer dormancy and misdiagnosed or overlooked.

Description of Stages

Adult

Adults of the BGB are gray, black, or brown weevils (Plate 46). Their color may be modified by the dried mud that often adheres to their bodies. The BGB has fine, even pits on the pronotum and straight rows of double pits on the elytra (Shetlar 1995a). The antennae are attached near the base of the snout in all *Sphenophorus* spp., while in *Listronotus* spp.,

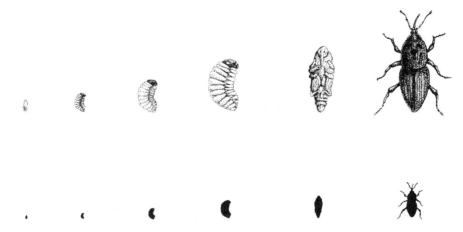

Figure 18-2. Stages of development of bluegrass billbug: egg; first, intermediate, and last larval instars; pupa; and adult. Silhouettes show actual size of each stage. (Courtesy of R. Jarecke, NYSAES.)

the only other major genus infesting turfgrass, the antennae are attached near the distal end of the snout (Figure 18-2).

The BGB is 7–8 mm long, exclusive of the length of the snout. The tarsi of the legs and most of the antennae are generally reddish brown. This same coloration often exists at the anterior edge of the pronotum, the lateral edges of the elytra, the coxal area, and the apex of the snout, and gradually fades into adjacent dark gray to black areas. Teneral adults (Plate 46) are reddish brown and are often seen with mature adults in the soil-thatch interface or in the soil. Males and females can generally be distinguished by the tip of the abdomen, which is more rounded in the male and more pointed in the female (Plate 46).

Egg
The eggs of the BGB are oblong and clear to creamy white. They are found in the stem and between leaf sheaths (Plate 46). Eggs are 0.6 mm wide by 1.5 mm long (Tashiro and Personius 1970).

Larva
The BGB larvae are white and legless, with brown head capsules (Plate 46). All instars resemble each other except in size. Mature larvae of the BGB are about 8 mm long (Niemczyk and Frost 1978).

Pupa
The pupae of the BGB are creamy white early in their development, and turn brown or tan as they mature. The bodies resemble those of adults in many ways (Plate 46).

Seasonal History and Habits

Seasonal Cycle
The BGB normally has a 1-year life cycle. Individuals overwinter as young adults produced during the summer and early fall. In the spring, adults are seen crawling over paved

sidewalks and driveways on their way to turf areas (Tashiro and Personius 1970). After a brief period of feeding, females deposit eggs in the stems. After hatching, young larvae feed in stems, but as they grow they start feeding in the crowns. By midsummer, larvae will have molted several times, and the ultimate-instar larvae produce the most damage as they continue to feed in the crowns and roots. BGB typically feed as larvae for 50–60 days, usually within 5 cm of the soil surface but they may move as deep as 10 cm during drought conditions (Kindler and Spomer 1986). Larval feeding is completed by mid to late summer, with pupation taking place in the soil. Young adults of the new generation are present during late summer (late August and early September), seeking overwintering quarters.

Occasionally summer adults will lay eggs in July or August, resulting in a partial second generation, particularly in the warmer (more southern) areas of the range (D. Shetlar, personal communication, 1998).

Adult Activities

During September and October, BGB adults are often observed climbing screens, shingles, and masonry walls to a height of 1.2–2.0 m. Adults occasionally attempt to fly, especially on windy days in late summer, but they are unable to sustain flight for more than a few inches (Shetlar 1995a, Tashiro and Personius 1970). With the onset of cooler temperatures, the adults seek overwintering sites in weeds and leaf litter, in hedgerows, in surface litter around the foundation of buildings, in worm holes, and in other protected areas. A favorite site for overwintering BGB adults is the junction of turf and sidewalk (Niemczyk 1983).

During the first warm sunny periods of late spring (when soil or thatch temperatures approach 20°C), BGB adults leave their overwintering sites and are seen crawling over paved surfaces on their way to turf areas on which they can feed. In New York this phase occurs starting in mid-May, with peak adult movement observed during June.

Oviposition

Adults feed on grass stems just above the crown by inserting their mouthparts into the stems. Shortly after feeding begins, eggs are inserted between leaf sheaths 1–2 mm above the crown. Oviposition occurs during late April and May in Ohio, or during May and June in the latitude of Wisconsin and New York (Mahr and Kachadoorian 1983, Shetlar 1995a).

Females may lay eggs over a period of 2 months under favorable conditions, which results in subsequent overlap of life stages, with early, middle, and late instars occurring concurrently. Billbugs apparently do not oviposit in dry stems, so oviposition normally ceases when reproductive tillers die in late summer (Hansen 1987).

Larval Activity

Hatching, Growth, and Maturity. Eggs hatch within the stems in about 2 weeks, and young larvae feed there a short time before burrowing downward to feed in and on the crowns. The entire crowns may be consumed by the larger larvae. Before maturity, larvae move to the root zone and feed on roots and rhizomes (Niemczyk 1983).

Wherever the BGB larvae have been actively feeding, they leave an accumulation of fine, whitish, sawdust-like frass. The presence of this excrement is positive evidence of BGB damage. Mature larvae penetrate to a soil depth of 3–10 cm to transform to the pupal stage (Niemczyk 1983).

Miscellaneous Features

Sampling Techniques

Pitfall traps are an excellent method to track the spring activity of BGB adults. A trap consists of a small plastic cup inserted in a hole made with a 10.7-cm-diameter cupcutter. The top rim of the cup should be just below the surface. A smaller cup is placed in the bottom, and a preserving agent (such as ethyl alcohol or ethylene glycol) is placed in the inner cup. (Because the latter material can be attractive to dogs, water with detergent can be used in areas where dogs or other animals might forage.) Adult BGBs will tumble into the cup as they forage on the surface. Traps should be checked two or three times a week (Shetlar 1991). When catches exceed 7 to 10 billbugs per trap, severe turfgrass injury is likely (Baxendale 1997).

The presence of BGB adults also can be confirmed when adults are observed crawling over paved areas adjacent to turf areas. When two or more adults per minute are counted during the spring, damage to adjacent turfgrass can be expected. Treating the turf to kill adults before major oviposition occurs (April through mid-May in Ohio) prevented severe turf damage by larvae during late June and July (Fushtey and Sears 1981, Niemczyk 1983).

It is almost impossible to observe adults in turfgrass, but a drench that will cause adults to surface has been developed. Turf is mowed to a height of 1.3 cm. A pyrethrin drench of 2.5 ml of a 1.2% pyrethrin and 9.6% piperonyl butoxide in 3.8 l of water is applied to a 0.1-m^2 area of turf. On turf with 0.45−1.15 cm of thatch, adults surfaced for about 20 min. During a visual search, 1.0−1.9 adults could be found, but 5.5−7.3 adults were observed using the drench. Larvae did not respond to the drench (Kindler and Kinbacher 1982).

Billbug larvae can be detected with a "tug test", which involves grasping grass stems that are turning yellow or brown and pulling upward. If the stems break off easily and the tips are filled with sawdust-like material (larval frass), billbug larvae can be presumed to be present (D. Shetlar, personal communication, 1998).

For removal of eggs and larvae from stems, 3.8-cm lengths of stem above the crown were covered with tap water and placed in a blender for three 3-sec runs. The material was passed through stacked sieve screens (U.S. Standard Sieve Series No. 10, 20, 30, and 35). Contents of the 10 and 20 sieves were washed back into a blender and were blended a second time. Extraction produced an average of 2.08 eggs and 1.15 larvae for each 25 stems (Kindler and Kinbacher 1982).

Degree-Day Accumulations

A degree-day (DD) model, using a 10°C base temperature and March 1 starting date, predicts that first adult activity will occur between 155−195 DD $_{base\ 10C}$ and 30% adult activity will occur between 311−347 DD $_{base\ 10C}$. The latter is considered to be the latest a surface insecticide, directed toward adults, should be applied (Shetlar 1995a).

Larvae begin to emerge from stems between 513−575 DD $_{base\ 10C}$. Insecticide applications can be effective from this time until damage is apparent, which normally occurs between 739−825 DD $_{base\ 10C}$ (Shetlar 1995a).

Plant Resistance

In recent years, turfgrass breeders have tried to identify cultivars that are resistant to billbugs. Among the cool-season grasses, Kentucky bluegrass usually sustains the most billbug damage. Many field trials at several locations, including Nebraska, New Jersey, and Oregon, have documented different levels of susceptibility of Kentucky bluegrass cultivars to billbugs (Ahmad and Funk 1983, Johnson-Cicalese et al. 1989, Kindler and Kinbacher 1975, Shearman et al. 1983). Several possible mechanisms for resistance have been suggested. Cultivars that were susceptible to leaf spot and melting out disease (caused by *Drechslere poae*) or had fine stems and leaves generally had less billbug damage, presumably because diseased plants or narrow stems were not as attractive to ovipositing females. In addition, cultivars that were aggressive and spread vigorously showed less billbug damage, in part because of their capacity for rapid regrowth. Recent studies suggested an association between resistance and heat and drought tolerance. The bluegrass cultivars that currently show the best resistance include 'America', 'Adelphi', 'Eagleton', 'Eclipse', 'Fylking', 'Ken-Blue', 'Wabash', and 'Washington' (Johnson-Cicalese 1997, Johnson-Cicalese et al. 1989).

Dramatic differences in susceptibility to billbugs were observed in perennial ryegrass trials in the early 1980s, and associated with the presence of the fungal endophyte *Neotyphodium* (formerly *Acremonium*) *lolii* Latch, Christensen, and Samuels, as well as additional genetic resistance (Ahmad et al. 1986). More recent work studied the effects of *Neotyphodium* endophytes on four species of billbugs (*S. parvulus*, *S. venatus*, *S. inaequalis*, and *S. minimus*). Trials included tests measuring mortality and feeding on potted tall fescue plants with and without endophyte (*N. coenophialum* Morgan-Jones and Gams); feeding preference on tall fescue tillers with and without endophyte; and no-choice feeding tests measuring mortality, oviposition, and adult feeding on either tall fescue or perennial ryegrass tillers, with or without endophyte. In each case, mortality was greater in endopyte-containing cultivars, while egg laying and amount of feeding did not differ greatly. The endophytes in both tall fescue and perennial ryegrass affected the adult survival of all four billbug species and are a useful and effective form of pest resistance in turfgrasses (Johnson-Cicalese and White 1990).

These endophytes produce toxins that are harmful to insects or induce them to avoid feeding on infected plants (Johnson-Cicalese 1997). The endophyte is transmitted by seed and vegetative propagation and gradually loses its viability when seed is stored for prolonged periods (Funk and Hurley 1984).

Natural Enemies

Microorganisms

BGB adults are susceptible to *Beauveria bassiana,* a fungus that produces copious amounts of cottony white mycelia in affected insects. *Steinernema carpocapsae* and *Steinernema glaseri,* entomopathogenic nematodes, have suppressed billbug populations in laboratory and small-scale field trials (Shetlar 1995a).

Insect Parasitoids and Predators

At least two parasitoids have been reported from several billbugs, including the BGB. These include a hymenopterous egg parasitoid, *Anaphoidea calendrae* Gab., and a dipterous adult parasitoid, *Myiophasia metallica* Tns. (Tachinidae) (Satterthwait 1931, 1932).

Vertebrate Predators

Toads, reptiles, and numerous birds, including crows, blackbirds, cardinals, and other species, are listed as predators (Satterthwait 1932).

Hunting Billbug

Taxonomy

The hunting billbug (HB), *S. venatus vestitus* Chittenden, is one of five subspecies of *S. venatus*. A second subspecies, *S. v. confluens*, is destructive to orchardgrass, *Dactylus glomeratus*, in Oregon and also feeds on bentgrass and Kentucky bluegrass (Kamm 1969, Oliver 1984, Vaurie 1951). In New Jersey, both *S. v. vestitus* and a third subspecies, *S. v. venatus*, are found, as well as specimens suggesting intergradation between the two subspecies. One or both subspecies are commonly found on tall fescue and other cool-season turfgrasses in the mid-Atlantic states (Johnson-Cicalese et al. 1990, Murphy et al. 1993).

Importance

The HB is an important pest of turfgrasses in the Southeast, especially in the Gulf States. The most severe injury occurs in zoysiagrass and in improved strains of bermudagrass. In Hawaii, the HB was second only to the lawn armyworm as a pest of lawn and turf, zoysiagrass again being the most severely damaged (LaPlante 1966b, Oliver 1984).

History and Distribution

The HB is probably the most abundant and widespread of the five subspecies of *S. venatus*. It occurs in the Atlantic coastal states from New Jersey to Florida and along the Gulf to about Brownsville, Texas, and north to Kansas and Missouri (Figure 18-3). It is also present in some of the Caribbean islands. The HB was introduced into Hawaii in about 1960 and is presently found throughout most of the state (LaPlante 1966b). In southern California the HB has repeatedly been intercepted in zoysiagrass and centipedegrass shipped from Georgia and other eastern states, and at least one golf course infestation has been found.

Host Plants and Damage

The HB damages zoysiagrass and improved strains of hybrid bermudagrasses most severely. Other turfgrasses that serve as hosts include St. Augustinegrass and centipedegrass. Other host plants include nutsedge, crabgrass, signal grass, barnyard grass, wheat, corn, sugarcane, Pensacola bahiagrass, and leatherleaf fern. In southern Louisiana, it has been the most troublesome in turfgrass nurseries, golf courses, and turf around commercial buildings (Oliver 1984, Woodruff 1966). Sod growers in Georgia, Alabama, and Florida sometimes cannot harvest billbug-infested bermudagrass or zoysiagrass sod because it falls apart. In New Jersey, *S. v. venatus* damages cool-season grasses, such as tall fescue, Kentucky bluegrass, perennial ryegrass, and Chewings fescue (Johnson-Cicalese and Funk 1990, Murphy et al. 1993).

Damage by the HB shows up as irregular elongated or rounded areas of brown and dying grass, especially in early spring and summer. On golf courses and home lawns, HB damage is often misidentified as delayed spring green-up of bermudagrass. Grass in these areas lacks sufficient roots to obtain water and nutrients or to anchor itself to the soil. Extended dry weather makes the injury more pronounced. Unlike damage by scarabaeid grubs, mole crickets, and other turf insects, which cause the turf and soil to feel soft and spongy, under grass damaged by the HB the soil remains firm underfoot.

Figure 18-3. Distribution of hunting billbug. (Drawn by E. Gotham, NYSAES, adapted from Handbook of Turfgrass Insect Pests, Brandenburg and Villani 1995, Entomological Society of America.)

Adults and young larvae feed on stolons, crowns, and new leaf buds; older larvae attack roots and runners to a depth of about 8 cm. The grass dies as a result of the feeding activity. HB-injured zoysiagrass becomes chlorotic, and tufts are lifted easily from the sod mat (Brussell and Clark 1968, Oliver 1984). Often HB damage goes undetected in summer months, when bermudagrass is growing rapidly. Populations may be high in fall as larvae damage many roots and stolons, but this activity is sometimes mistaken for normal fall dormancy.

Description of Stages

Adult

HB adults are nearly identical in appearance to BGB adults but are slightly larger, measuring 8–11 mm in length (Plate 47). The coarsely punctated pronotum, with its smooth, nonpunctated Y-shaped median area and parenthesis-like curved markings on the sides, characterizes this species and differentiates it from the BGB (Plate 47). Adults are generally black but may appear brown or gray when dried mud adheres to the punctated thorax and striated elytra. Some specimens are reddish and apparently callow. Adults usually feign death for short periods when they have been disturbed (Oliver 1984, Vaurie 1951, Woodruff 1966).

Egg

Eggs are oblong and clear to creamy white. They are inserted in stems, between leaf sheaths, and at the top of the crown (Kelsheimer 1956, Woodruff 1966).

Larva

The larva of the HB, nearly identical to that of the BGB, is a white, legless insect with a brown head capsule. All instars resemble each other except in size. The last instar is 7–10 mm long and is found in the crown or root area just below the thatch (Oliver 1984).

Seasonal History and Habits

Seasonal Cycle

Older literature indicates that HBs have one generation per year. They usually overwinter as adults or as partially mature larvae which then pupate in the spring. In most areas of their range, the adults appear to lay eggs for an extended period from May into September. Eggs laid in May produce larvae that mature and pupate by late summer. Late larvae feed and overwinter as larvae. These larvae pupate the following March to May. In many southern turf settings, HBs can be found in all stages during most of the year (Oliver 1984; D. Shetlar, personal communication, 1998). In New Jersey, *S. v. venatus* larvae were found in late fall and winter (Johnson-Cicalese et al. 1990, Murphy et al. 1993).

Adult Activity

Like the BGB, HB adults that have emerged from overwintering sites are first seen crawling over driveways and sidewalks and along curbs. Adults usually feign death for short periods when disturbed. They also cling tightly to a stem or leaf when attempts are made to collect them. Adult feeding and oviposition are most pronounced in the spring (Oliver 1984). In the southern part of its range, adults may be active and feeding all year, whenever temperatures are high enough. Most eggs are inserted in grass stems or leaf sheaths in the spring, when zoysiagrass and bermudagrass have broken winter dormancy and are growing vigorously (Shetlar 1995a).

HB adults can fly a short distance but cannot sustain flight for long. During fall, as temperatures decline, adults seek overwintering quarters in leaf litter, in weedy areas, and in other partially protected locations (LaPlante 1966b, Oliver 1984). HB adults often are active at night in July and August, perhaps to avoid the daytime heat and predation (D. Shetlar, personal communication, 1998).

Egg and Larval Activity

HB eggs take 3–10 days to hatch. Upon hatching, young larvae feed within stems, hollowing them out as well as filling them with frass. Older, larger larvae feed on the crown of the grass plants. Finally, as the ultimate-instar larvae approach maturity, they feed on the roots and stolons of plants. Larvae usually feed just below the thatch but occasionally feed as deep as 5.0–7.5 cm in the soil profile. They penetrate the soil to a maximum depth of about 8 cm. Larvae feed a total of 3–5 weeks before pupating. Because oviposition occurs over an extended period, larvae are most common from May to October, although some overwinter in the Gulf States. Pupation occurs in the soil or thatch and pupae take 3–7 days to mature (Baxendale 1997, Shetlar 1995a). Development may take longer in more northern climates (Johnson-Cicalese et al. 1990).

Natural Enemies

HB adults and larvae appear to be susceptible to two entomopathogenic nematodes, *Steinernema carpocapsae* and *Heterorhabditis bacteriophora*. The fungus *Beauveria bassiana* also occurs naturally (Shetlar 1995a).

Apparently the same insect parasitoids that attack BGB also attack HB (Satterthwait 1931, 1932). In Hawaii, giant toads of the genus *Bufo* consume large numbers of adults. Examination of their scat revealed a diet consisting, at times, nearly entirely of weevils.

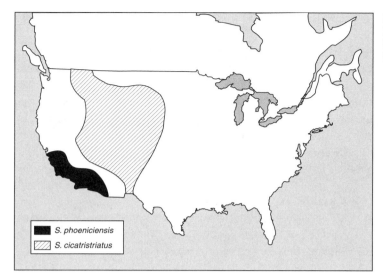

Figure 18-4. Distribution of Phoenician and Denver billbugs. (Drawn by E. Gotham, NYSAES.)

Phoenician Billbug, Denver Billbug, and Other Billbug Species

The distribution of the Phoenician billbug (PB), *S. phoeniciensis* Chittenden, is limited to southern California and Arizona, where it is the most commonly found billbug species (Figure 18-4). It has a life cycle very similar to the HB, with one generation per year. The PB feeds on warm-season grasses and the damage is similar to that caused by the HB. It commonly thins out marginally maintained bermudagrass and can kill zoysiagrass. PB adults have a broad M-shaped mark on the pronotum but otherwise are very similar in size and color to HB adults (Plate 47) (LaPlante 1966b, Morishita et al. 1971, Shetlar 1995a).

The Denver billbug (DB), *S. cicatristriatus* Fabraeus, is found in the Rocky Mountain region and northern Great Plains, from northern New Mexico to Utah, and causes severe damage in the Denver, Colorado, area (Figure 18-4) (Shetlar 1995a, Vaurie 1951). It prefers cool-season turfgrasses, primarily Kentucky bluegrass and perennial ryegrass. The life cycle of the DB is less synchronized than that of other species, because it overwinters as small and large larvae as well as adults. The DB is more likely to overwinter as a late-stage larva than is the HB, and adults typically emerge a few weeks later (early May to June) than do HB adults (Baxendale 1997). Peak adult populations may occur any time between June and mid-September, depending on altitude and weather conditions. DB populations seldom show the distinct peaks of adult activity characteristic of the BGB (Cranshaw and Ward 1996). DB adults are larger than other turf species and are shiny black with a heart-shaped double row of pits on the elytra (Plate 47) (Shetlar 1995a).

Relatively little is known about the little billbug, *S. minimus*. Its appearance, life cycle, distribution, hosts, and damage are apparently very similar to those of the BGB. It has been found on turf in New York, Pennsylvania, Ohio, and New Jersey feeding on several cool-season grasses (Cameron and Johnson 1971b, Johnson-Cicalese and Funk 1990, Johnson-Cicalese et al. 1990, Shetlar 1991). Because of its similar appearance to the BGB, it is likely that this species is often overlooked or misidentified. However, it can be distinguished by the larger, irregularly spaced punctures on its pronotum, and the irregular surface of its wing covers.

The distribution of the uneven billbug, *S. inaequalis*, is limited primarily to the eastern United States and until recently its only reported host was bermudagrass (Satterthwait 1931). It is now known to feed on Kentucky bluegrass, perennial ryegrass, and fescue as well (Johnson-Cicalese and Funk 1990). A survey of over 4,000 billbugs in New Jersey suggested it is sometimes as abundant as the BGB on turf. Its life cycle appears to be similar to that of the BGB, although adults tend to be active earlier in the spring and later in the fall (Johnson-Cicalese et al. 1990). Adult uneven billbugs are easy to recognize by their wider body shape and diamond-shaped shiny area in the center of the pronotum surrounded by four less-distinct shapes.

The relative importance of two additional species, *S. coesifrons* and *S. apicalis*, is unknown. However, both have been reported damaging turfgrasses so their possible occurrence should be recognized by turfgrass specialists. The distribution of *S. coesifrons* is throughout the southern United States (Vaurie 1951), and it has been reported damaging bahiagrass in Georgia (Morrill and Suber 1976). It is similar in appearance to the BGB but is larger and has larger punctures on the tip of its abdomen (Johnson-Cicalese et al. 1990). *Sphenophorus apicalis* occurs in the Gulf States and New Jersey (Vaurie 1951) and was reported to occur in relatively high numbers in Florida lawns (D. Shetlar, personal communication, 1998). It can be recognized by two distinct pits at the anterior medial area of the pronotum (Johnson-Cicalese et al. 1990).

Annual Bluegrass Weevil

Taxonomy

The common name, annual bluegrass weevil (ABW), has been tentatively assigned to an insect classified as *Listronotus maculicollis* (Dietz) that belongs to the order Coleoptera, family Curculionidae, subfamily Cylindrorhininae. R. E. Warner of the U.S. National Museum classified the weevil as *Hyperodes* sp. near *anthracinus* (Cameron and Johnson 1971b). Charles O'Brien (Florida A&M University, Tallahassee) has studied many of the weevils belonging to *Hyperodes* and *Listronotus*, two very closely related but separate genera. He has designated the ABW from New York State as *Listronotus maculicollis* (personal communication, 1985).

The common names *turfgrass weevil* and *annual bluegrass weevil* have been suggested. Since the insect is primarily a pest of annual bluegrass, *Poa annua* L., it appears appropriate to designate the insect the *annual bluegrass weevil*. Another common name, *Hyperodes weevil*, persists in the vernacular although it is probably inappropriate now that the generic designation has changed.

The literature lists some 48 *Hyperodes* or *Listronotus* species present in North America north of Mexico. These species are extremely difficult to classify by a novice because of their similarities in size and color. The only comprehensive key to adults indicates that *Listronotus* (formerly *Hyperodes*) *anthracinus* has sparse pubescence, a shiny body surface, and a strong median prothoracic carina, while *Listronotus* (formerly *Hyperodes*) *maculicollis* has dense pubescence, a dull body surface, and an abbreviated prothoracic carina (Blatchley and Leng 1916, Kissinger 1964, Stockton 1963).

Importance

The ABW is a turfgrass pest primarily of annual bluegrass maintained at short cutting heights of 0.6–1.3 cm or less. It is therefore a pest exclusively of golf course fairways, tees, greens, and collars and of tennis court turf. Cutting heights of 4 cm or higher preclude this insect from becoming a pest of home lawns. It has not been injurious to any other species of turfgrass associated with these two recreational facilities.

Occasionally golf course superintendents claim that minor infestations occur on pure bentgrass greens. In some cases, close examination reveals that the damage is occurring in small patches of annual bluegrass that have invaded the seemingly pure bentgrass greens. However, there have been increasing numbers of reports of damage to a few cultivars of creeping bentgrass, at least some of which have been confirmed by P. J. Vittum (personal observation).

Within a 80-km radius of New York City, major efforts often are made by golf course superintendents to eliminate annual bluegrass because it has the inherent weakness of dying during the midsummer heat. Since eradication efforts have generally been unsuccessful and automatic irrigation systems have made it easier to maintain annual bluegrass through the summer, most superintendents now make concerted efforts to maintain this grass and as a result, they must control the ABW.

Our current understanding of the ABW suggests that the major obstacle to maintaining annual bluegrass during the summer has largely been infestations of this insect and not the turf's inherent inability to withstand summer stresses. The annual "spring die-out" recognized on tennis courts and golf courses on Long Island for 40–50 years was probably caused by this insect (Cameron and Johnson 1971b, Vittum 1980).

Control is an expensive maintenance operation for golf courses; many treat the entire fairways, tees, and greens with an insecticide two to five times a year. Courses with fewer resources may treat only the areas that are most severely and repeatedly infested year after year. Whether the entire playing area is treated or spot treatments are employed, once a course has received a treatment program, it normally becomes an annual operation.

History and Distribution

The earliest reports of the ABW as a turfgrass pest came from Connecticut in 1931. On Long Island it was first reported injuring turf in Nassau County in 1957 and again in 1961. By 1965–67, additional injuries on Long Island and in Westchester County just north of New York City were being reported by golf course superintendents. In 1966–67, *Listronotus* (reported as *Hyperodes*) sp. damage occurred on an Ithaca, New York, golf course some 400 km from New York City. Turf-damaging *Listronotus* weevils were collected also in the Pocono Mountains of northeastern Pennsylvania and from Connecticut courses (Britton 1932, Cameron and Johnson 1971b, Cameron et al. 1968).

More recently, turf-damaging *Listronotus* weevils have been found throughout New York State. Several locations throughout New England have been affected, including much of Massachusetts, Rhode Island, and Connecticut. The ABW has been active in much of New Jersey and Pennsylvania. The most severe and perpetual problems with the ABW continue to occur on Long Island; Westchester County, New York; Fairfield County, Connecticut; and northern New Jersey, where there is as high a concentration of golf courses as in any part of the country (Figure 18-5).

Figure 18-5. Distribution of annual bluegrass weevil. (Drawn by E. Gotham, NYSAES, adapted from Handbook of Turfgrass Insect Pests, Brandenburg and Villani 1995, Entomological Society of America.)

Host Plants and Damage

Host Plants

In the field, annual bluegrass is the turfgrass that has suffered most ABW feeding damage and only when golf course fairways, tees, greens, and aprons of greens are close cut (0.6–2.0 cm). There has been no evidence of turf damage in roughs, presumably because of the much higher cut. Annual bluegrass maintained on tennis courts (also cut very low) has also suffered damage (Cameron and Johnson 1971a).

Adult weevils feed on the leaves and stems of annual bluegrass but cause no serious problems. Weevils occasionally feed on clover, plantain, dandelion, and even mulberry (Cameron and Johnson 1971b). In feeding preference studies with turfgrasses, adult weevils found tall fescue as attractive as annual bluegrass. Other plants fed on but found less attractive than annual bluegrass were Canada bluegrass, rough bluegrass, Kentucky bluegrass, creeping bentgrass, perennial ryegrass, and red fescue. More recently ABW adults fed and oviposited on perennial ryegrass in laboratory conditions (P. J. Vittum, unpublished data 1997).

No feeding preference tests have been performed with larvae but field observations indicate that only annual bluegrass (and occasionally creeping or velvet bentgrass) has ever shown damage. More typically, patches of bentgrass within an annual bluegrass stand remain completely healthy when the annual bluegrass is completely killed by larval feeding (Cameron and Johnson 1971a, 1971b; Vittum 1980).

Damage

Adult Feeding. Adult weevils feed mostly on leaves and stems at the base of leaf blades of annual bluegrass (Plate 49). Notches are chewed out of the edges of blades, or holes are eaten out of the centers. Adult feeding may weaken grass stems but rarely kills them (Cameron and Johnson 1971b).

Larval Feeding. Damage initially appears as small yellow-brown spots on annual bluegrass that are caused when larvae sever stems from the plant (Plate 49). One larva can kill as many as a dozen individual stems. The ultimate (fifth) instar, feeding on the crowns of the grass, causes the greatest damage. On fairways, damage is first evident near the roughs and gradually spreads toward the centers. Since hibernating adults disperse primarily by crawling, they first feed and oviposit in close-cut annual bluegrass along the edge of fairways (Plate 49). As larvae grow, the spots coalesce, causing large areas of turf to die. Concentrations of larvae as great as 500 per 0.1 m² can kill large areas of turf but may leave patches of bentgrass totally uninjured (Plate 49). Larvae severely damage the collars of greens when annual bluegrass is the dominant species (Plate 49). As previously noted, what appears to be bentgrass injury on greens is sometimes actually minute, undetected patches of annual bluegrass being killed by the larvae (Cameron and Johnson 1971c; Tashiro 1976b; H. Tashiro, personal observations).

Severe damage normally occurs during late May and early June in the metropolitan New York area, later in New England, and earlier in southern New Jersey or southeastern Pennsylvania. Larvae, pupae, and adults can be found beneath the damaged turf by mid-June in southeastern New York (Cameron 1970).

Damage during July and August by second-generation larvae is normally much less severe and more localized than first-generation damage, but in some circumstances is more severe, in part because the turf is under additional temperature and drought stresses at that time. Given areas of a particular golf course sometimes show repeated localized infestations year after year. This situation could relate directly to favorable adult overwintering sites. On fairways, while the spring damage is generally more severe along the edge, damage during midsummer is generally more widely distributed.

Description of Stages

Figure 18-6 shows all developmental stages of the ABW.

Adult

Description. ABW adults average 3.6 mm in length and resemble those of the turf-infesting billbugs but are much smaller (Plate 48). The antennae are attached near the apex of the snout, with the scape (first segment) directed back toward the base of the snout in a long groove called the *scrobe.* The scape is nearly as long as the pedicel (second segment) and flagellum (terminal, beadlike segments) combined. The snout bears three carinae (ridges) dorsally, the median carina being the most distinct. The prothorax bears a dorsal-median carina and a postocular lobe that often partially covers the compound eye (Cameron and Johnson 1971b).

Mature adult weevils are generally black and are clothed with fine hairs and yellow-brown and grayish-white scales. Scales on the elytra are in scattered groups that create a faint mottled appearance. As adults grow older, many of the hairs and scales are worn off, making the body appear more shiny and black. Callow adults, generally found only in the soil, are reddish brown to brown and gradually turn gray to black. The abdomen is the last part of the body to darken (Plate 48).

The male and female resemble each other, although the female is slightly larger. Sexes can be separated by differences in the ventral abdominal segments. In the male the third abdominal sternum (first visible) has a median depression, while in the female there is no such depression. Rather, this sternum tends to bulge slightly in the middle. The last sternum (seventh) of the male is flattened and is slightly impressed at the apex, while in the fe-

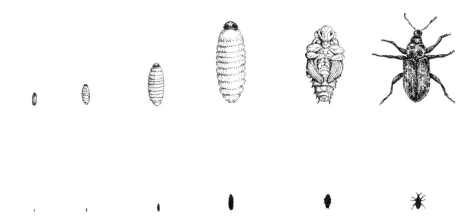

Figure 18-6. Stages of development of annual bluegrass weevil: egg; first, third, and fifth larval instars; pupa; and adult. Silhouettes show actual size of each stage. (Courtesy of R. Jarecke, NYSAES.)

male this sternum has a large, shallow depression (Plate 48). The sex ratio is nearly 1:1 (Cameron and Johnson 1971b).

Differences in Adults Revealed by Scanning Electron Microscopy. In an attempt to clarify possible species differences among ABW from different locations, adults from western New York were compared with those of southeastern New York via scanning electron microscopy (SEM). Adults from Westchester County had distinctly fewer scales per unit area immediately above coxa 1 than adults from Elmira in western New York (Figure 18-7). SEM photos of other areas of the weevils from the two locations showed only slight differences. On the basis of these studies, the weevils from Elmira were considered distinct from weevils of Westchester County (Vittum 1979). However, O'Brien's revision suggests that these are in fact two populations of the same species, *L. maculicollis* (Vittum and Tashiro 1987).

Egg

Eggs of the ABW, deposited between leaf sheaths, are approximately three times longer than they are wide and average 0.8 mm by 0.25 mm. They are rounded at both ends and are generally straight, but some may be curved longitudinally. They are also generally round in cross section, but a few may be flattened. The chorion is shiny, smooth, and usually unsculptured except for longitudinal ridges and grooves that were probably created by the ridges of the leaf sheath that press onto the eggs. Eggs are pale yellow to white, with a transparent clear chorion when they are first deposited. Within a day or two, the chorion becomes smokey black but remains transparent to reveal the developing embryo (Plate 48) (Cameron and Johnson 1971b).

Larva

The larva is creamy white, legless, and wider in the middle than either end. It grows from a length of about 1 mm in a young first instar to about 4.5 mm in a mature fifth instar (Plate 48). First instars are nearly straight but gradually become more crescent shaped as they grow. The dorsum of each segment has three fleshy lobes, and the three legless thoracic segments have ventral swellings, each with long setae. The dark-brown head capsule has

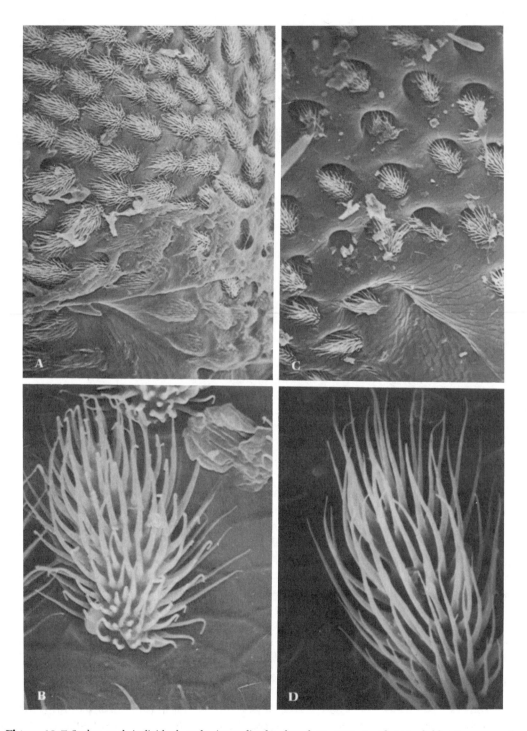

Figure 18-7. Scales and individual scale immediately dorsal to coxa 1 of annual bluegrass weevil. **A.** Weevil from Elmira (western New York). × 245. **B.** Weevil from Elmira. × 1,360. **C.** Weevil from Mamaroneck (southeastern New York). × 246. **D.** Weevil from Mamaroneck. × 1,860. (Adapted from Vittum 1979, figs. 9–12, prepared by P. J. Vittum.)

a distinct epicranial suture that makes an inverted Y. Two pairs of ocelli are present, one located just above the mandibles and the other in the genal region (Cameron and Johnson 1971b).

Prepupa

The prepupa is nearly identical to the last instar externally except that all three thoracic segments lose many of their folds, become distended, and make the prepupa slightly longer than the last instar. Prepupae do not move about but are capable of performing a circular movement of the abdomen that is often the most distinguishing characteristic (Cameron and Johnson 1971b).

Pupa

The pupa, exhibiting many adult characteristics, is first creamy white throughout (Plate 48), but parts of the body gradually darken to a reddish brown, becoming black as it matures. First the eyes, then the mandibles, the tarsal claws, the snout, and the articulations of the legs change color, in this order (Cameron and Johnson 1971b).

Seasonal History and Habits

Seasonal Cycle

The ABW overwinters as an adult primarily in the litter under trees of golf course roughs, with the largest concentrations found in white pine, *Pinus strobus*, litter (Plate 50). Large numbers also can be found in late autumn in clumps of grasses between fairways and wooded areas (P. J. Vittum, personal observation). Migration from hibernation quarters to fairways commences during early spring about the time that forsythia, *Forsythia* spp., is in full bloom (about mid-April in southeastern New York) (Plate 50). During the period when flowering dogwood, *Cornus florida* L., is in full bloom (full bract color) and when redbud, *Cercis canadensis* L., is in full bloom, all adults will have left hibernating quarters and will have moved out into fairways, tees, and greens (Plate 50) (Tashiro and Straub 1973, H. Tashiro and P. J. Vittum, personal observations).

Oviposition occurs shortly after adult feeding begins. Larval development through five instars requires about a month, and development from egg to adult normally requires about 2 months in late spring, less in midsummer. The generalized life cycle of the ABW in southeastern New York and southwestern Connecticut, found in fairways, tees, and greens during spring and summer, is shown in Figure 18-8 (Cameron 1970, Vittum and Tashiro 1987).

The first generation of the ABW shows a variation of about 2 weeks among years for any given stage. The maximum presence of small larvae extended from the last week in May to the first week in June, that of mature larvae occurred during the first half of June, that of pupae during the second half of June, and that of callow adults during the last week in June and the first week in July (Vittum 1980).

A good correlation exists between degree-day accumulations in Westchester County, New York, and maximum presence of all stages except eggs (which were not studied) during the second generation (Figure 18-9). During that generation, each stage spanned a longer period, depending on the year. Small larvae were present during the last 3 weeks of July, large larvae during the second half of July into early August, pupae during the last of July

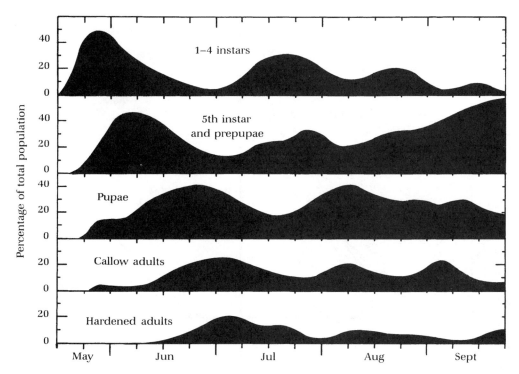

Figure 18-8. Annual bluegrass weevil fairway and tee populations exclusive of eggs, estimated on the basis of turf-soil plugs (10.8 cm in diameter) from Westchester County, New York, and Fairfield County, Connecticut, golf courses, 1976–79. (Data from Vittum 1980; drawn by R. McMillen-Sticht, NYSAES.)

and the first half of August, and callow adults during the first 3 weeks of August. However, because of extended oviposition in the spring and variable development rates of spring generation individuals, all stages (eggs; small, medium, or large larvae; pupae; adults) could be found at any given time during July and August (Vittum 1995a, Vittum and Tashiro 1987).

There was also evidence of a third generation of small larvae in 1976 and 1977 during the second half of August. These years had unusually warm springs, which apparently accelerated development throughout the entire growing season (Vittum and Tashiro 1987). Three generations per year appear to be typical from metropolitan New York south, but throughout the weevil's range, all stages can be found (in varying percentages) from late June through early September.

Adult Activity

Hibernation. The bulk of the population overwinters as adults, and migration to overwintering sites begins by early autumn. Prior to 1970, the major hibernation site of ABW adults was thought to be in tufts of fescue. Concentrations of adults varying from about 15 to a high of 42 per 0.1 m² were found in litter under hedges, in fescue under trees, and under leaves on tennis courts, but none were found in fairways, tees, greens, and aprons (Cameron and Johnson 1971b).

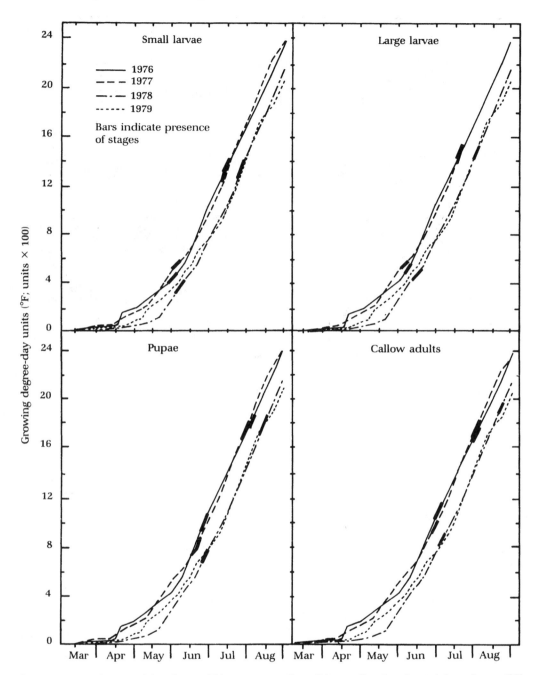

Figure 18-9. Maximum activity of annual bluegrass weevil small larvae (first four instars), large larvae (fifth instar and prepupae), pupae, and teneral adults relative to growing degree-day accumulations with a base 11°C in Westchester County, New York temperature records from Westchester County Airport, 1976–79. (Adapted from Vittum 1980, figs. 38–41.)

Table 18-1. Annual Bluegrass Weevil Adult Catches in Blacklight (15-watt) Traps at Westchester County, New York, Golf Courses, in 1976

Half-Month Period	Location				Total	Percent of Total
	1	2	3	4		
Late April	0	0	0	0	0	0
Early May	0	90	0	0	90	5
Late May	0	0	0	0	0	0
Early June	23	69	0	0	92	5
Late June	226	112	15	20	373	21
Early July	800	50	135	0	985	55
Late July	174	16	30	0	220	12
Early August	14	3	3	0	20	1
Late August	0	0	3	0	3	<1
Total	1,237	340	186	20	1,783	—
Percent of total	69	19	10	1	—	—

Note: Traps were emptied every Monday morning and catches were grouped into half-month periods.
Source: H. Tashiro, unpublished data.

Dominant Overwintering Site. During the 1970s the primary overwintering sites were determined to be in white pine needle litter. Samples of litter to the soil surface collected from the same five white pines over a year showed that as many as 443 adults were present in 0.1 m^2 of litter. There was a consistency of high populations under a given tree, but no obvious environmental conditions such as grouping of trees, thickness of litter, or tree size appeared to influence populations. Populations under the trees were highest in October and November, and began to decline in March as adults started to migrate to nearby oviposition sites. Populations increased again in September as adults returned to their overwintering sites. Soil samples removed from under white pine litter also demonstrated that roughly 40-60% of the diapausing weevils were in the upper layer of soil (Vittum 1980).

Litter samples taken from under trees of other species in the roughs of various golf courses showed weevils to be present under spruce, hickory, maple, birch, and other trees. Where white pine and other trees occurred in close proximity, weevil densities were always the highest under white pine.

Extraction of Weevils. The extraction of adult weevils from litter and soil samples is readily accomplished by submerging the samples entirely under water, which forces adults to crawl to the surface. Paper towels placed on the water allows the weevils to cling to the undersurface. The weevils can be removed readily from the paper and can be retained for further studies. About 50% of the adults surface during the first 10 min of submergence, 70% in 30 min, 90% in 60 min, and virtually 100% in 2.0-2.5 hr of submergence (H. Tashiro, personal observations).

An irritating drench (disclosing solution) can be used to force adult weevils to the surface of golf greens, tees, or fairways. One simple technique is to add 15-30 ml of lemon-scented dish detergent to 4-6 l of water and pour the solution over 0.1-0.2 m^2. Within 5 min, weevils will scramble to the surface, where they can be counted (see Chapter 26).

Emergence and Dispersal. White pine litter sampled at regular intervals during the spring of 1976 and 1977 revealed the period of weevil emergence and dispersal out of overwintering sites. All weevils were in overwintering sites throughout March. Migration into fairways began in early April both years, and by the end of the month the weevils were nearly all dispersed. Eggs were abundant in the stems during the last week in April and the first week in May. Dispersal from overwintering sites is primarily by crawling, as evidenced by turf damage first becoming apparent at the edge of fairways, greens, or tees, where oviposition first occurs (Vittum 1980, Vittum and Tashiro 1987).

Weevils also disperse by flight and are attracted to black light (Table 18-1). The highest catches occurred in July, indicating that the first-generation (summer) adults were most readily attracted. A few of the overwintering adults, however, were also attracted and were caught during early May. Weevils have also been observed flying onto fairways in full sunlight. Weevils have also been caught on 20-cm-by-30-cm rectangles of plywood painted yellow and coated with Tanglefoot or other sticky substances (Cameron and Johnson 1971b; Schread 1970b; H. Tashiro, personal observations).

Adult Feeding. Young adults are quite active during the day as they emerge from the soil. They are generally black, but some remain the reddish brown of callow adults for a week or longer after emergence. These young adults can be seen moving on the surface of putting greens and other short mowed grass. Mature adults are rather difficult to find during the day, as they tend to remain inactive in the grass and thatch. At dusk, adults begin to crawl to the tips of the grass stems to feed. During midsummer nights, adults often can be swept into a net from grass more than 1.3 cm tall but not from greens because the grass is too short. With the aid of a strong light, adults can readily be seen on the surface of greens. Adults feed mostly on leaves and at the junction of leaves and stems. They chew holes in leaves and notches along the margin. Adult feeding is relatively insignificant; adults may weaken the grass, but they rarely cause it to die (Cameron and Johnson 1971a, 1971b).

Oviposition. After a brief period of feeding, the females are ready to begin oviposition. Eggs are deposited, usually in groups of two or three between leaf sheaths. Dissections reveal the development of more than 50 eggs in each female, and the full complement is presumably deposited in the field. Laboratory rearing has resulted in an average egg production of 11.4 eggs per female. Coincidentally, when seven pairs of overwintering adults were isolated and held for as much as 10 weeks with annual bluegrass, they produced an average of 11 eggs (Cameron and Johnson 1971b; H. Tashiro, personal observations).

Larval Activity

Hatching, Growth, and Feeding. Upon hatching, most first instars burrow into the stems almost immediately, feeding both upward and downward from the point where they hatch. The youngest and most tender central portion of the stems is eaten. Larvae may emerge and reenter the same stems several times or may move to other stems nearby. Larvae are mobile and do not shun light, so they probably move from stem to stem above ground. Tunnels made by larvae are filled with frass that gradually turns brown. Feeding by small larvae seldom kills stems, but the central leaves turn yellow and gradually die (Cameron and Johnson 1971b, Cameron et al. 1968).

As the larvae increase in size, they make larger tunnels, completely hollowing out the centers and leaving only the outer leaf sheaths. As the larvae become too large for grass

Table 18-2. Location of Annual Bluegrass Larvae and Pupae in 368 Soil-Plug Samples from Golf Courses on Long Island, New York, in 1968

Stage of Development	Number	Percent of Total		
		Inside Stems	Above Soil	In Soil[a]
Small larvae	99	26	72	2
Medium larvae	214	2	84	14
Large larvae	245	0	49	51
Pupae	130	0	0	100

Note: Soil-plug samples were removed with a standard golf cup cutter (10.8-cm diameter).
[a]Varied from average of 0.3 cm for small larvae to 0.8 cm soil depth for pupae.
Source: Adapted from Cameron and Johnson 1971b.

stems, particularly the fifth (ultimate) instar, the insect begins feeding externally at the base of crowns (Table 18-2). Feeding damage occurs only on annual bluegrass and occasionally bentgrass; all other turfgrasses remain undamaged. Since annual bluegrass is likely to die under moisture stress, ABW-damaged stems in a weakened condition die out quickly with lack of moisture. Mature fifth instars are found in the thatch and surface soil among roots that are not damaged. Once in the soil, the mature larvae have completed feeding and are very close to becoming prepupae and pupae (Cameron and Johnson 1971b).

Prepupal and Pupal Activity

When development of the fifth instar is completed, the larva tunnels about 0.6 cm into the soil and forms an earthen cell. Within the cell the pupa begins to form within the cuticula of the last instar. The prepupa rotates within the earthen cell and moves its abdomen in a circular motion. This stage lasts 2−5 days.

Upon completion of the prepupal period, the fifth-instar exuvia splits dorsally just behind the head to release the young, creamy white pupa. It also rotates within the earthen cell. The pupal period requires 3−9 days. Upon completion of the pupal development, the adult breaks the pupal exuvia dorsally along the head and thorax and frees itself. The light-colored, soft adult remains in the earthen cell 3−8 days, gradually turning from a reddish-brown callow adult to a blackish mature adult. Some adults remain reddish brown for more than a week after emergence (Cameron and Johnson 1971b).

Natural Enemies

Microorganisms

A few larvae and pupae found during 1977 were flaccid with a coral to lavender hue. These were infected with a common entomogenous nematode of the genus *Steinernema* (formerly *Neoaplectana*). Several species of carabid beetles freely feed on the larvae and pupae of the ABW in the laboratory. Efforts to suppress field populations with commercial formulations of *S. carpocapsae* have thus far been unsuccessful, perhaps because spring applications are made at times when air and soil temperatures are still quite low and summer applications are made at times when nematodes are susceptible to dessication. Applications directed toward first-generation late instars (late spring) appear to be more promising, although optimum timing is difficult to accmplish (P. J. Vittum, personal observation).

Parasitoids and Predators

In studies conducted in New Zealand during 1996, *Microctonus hyperodae,* a braconid wasp that parasitizes Argentine stem weevil adults (*L. bonariensis*), attacked a low percentage (less than 10%) of ABWs in laboratory conditions. *Microctonus aethiopoides,* a less specific braconid wasp, attacked about 20% of ABWs in laboratory conditions, sterilizing them and ultimately killing them. It remains to be seen whether either species can become established in field settings (Vittum and McNeill 1999).

Vertebrates

Nearctic *Hyperodes* and *Listronotus* species, including *L. maculicollis,* have been taken from the stomachs of toads, *Bufo* spp. Blackbirds have been observed actively pecking the turf infested by the ABW (Cameron and Johnson 1971b, Johnson and Cameron 1969, Stockton 1956, Vittum 1980).

19

Dipteran Pests:
Families Tipulidae and Chloropidae

European Crane Fly

Taxonomy

The European crane fly (ECF), *Tipula paludosa* Meigen, order Diptera, family Tipulidae, subfamily Tipulinae, has elongated maxillary palpi that distinguish members of this subfamily from other subfamilies. *Leatherjacket* is the name given to the larva of the ECF (Alexander 1919, Jackson and Campbell 1975).

Importance

While the ECF has a limited distribution in North America, it is of considerable importance as a pest of lawns and golf courses in southern British Columbia, western Washington, and western Oregon (west of the Cascade Mountains). It has had an even greater impact on the dairy industry of this area by damaging pastures and hayfields. The largest populations occur in grasslands; the worst attacks are apparent usually in the spring (Jackson and Campbell 1975).

The great abundance of adults occurs for a short period during late summer and early fall. Their habit of collecting on sides of buildings creates a nuisance to the general public (Jackson and Campbell 1975).

History and Distribution

The ECF is a native of northwestern Europe, where it occurs from the lower Scandinavian area to northern Italy and from Great Britain into the USSR. In North America it was positively identified in 1955 as it infested flower beds in Cape Breton Island of Nova Scotia. Its most probable source of introduction was soil that was dumped after being used as ship ballast (Fox 1957, Jackson and Campbell 1975).

On the West Coast it was first discovered in Vancouver, British Columbia, causing severe damage to lawns in 1965. The source of this infestation is not known. In the state of Washington, it was first detected in a light trap during late summer of 1966 at Blaine, bordering Canada. During the next 7 years it gradually spread southward in localized infestations to the vicinity of Tacoma (Jackson and Campbell 1975, Wilkinson and MacCarthy 1967).

Figure 19-1. Distribution of the European crane fly. (Drawn by E. Gotham, NYSAES, adapted from Handbook of Turfgrass Insect Pests, Brandenburg and Villani 1995, Entomological Society of America.)

Presently the ECF has a continuous southward distribution to the Oregon border and has heavily infested home lawns in Spokane, Washington. There have been only isolated infestations in central Washington (G. Stahnke, Washington State University, Puyallup, Wash., personal communication, 1998). By 1992 the ECF had spread down the Oregon coast as far as Eugene (Figure 19-1) (Stahnke et al. 1993). There have been reports of infestations in pastures in northern California (A. Antonelli, Washington State University, personal communication, 1998).

The maritime climate of coastal British Columbia and western Washington of relatively mild winters and abundant rainfall is ideal for the ECF. A similar climate to the south extends the area of potential infestation as far south as northern California.

Host Plants and Damage

Larvae of the ECF feed primarily on the grasses of lawns and pastures, but strawberries, flowers, and vegetable crops have also been attacked (Wilkinson 1969).

When feeding below the surface, larvae mainly attack root hairs, roots, and crowns (Plate 51). The nightly migration of large larvae from the soil to the surface disrupts the soil surface and destroys young seedlings. They eat stems, grass blades, and leaves when they feed above ground (Wilkinson and MacCarthy 1967).

As few as 10–16 larvae per 0.1 m^2 have caused serious injury to unthrifty grassland (Antonelli and Campbell 1984). As many as 120 larvae per 0.1 m^2 have been counted in April and May in Vancouver (Wilkinson 1969). While there are varying accounts of population density relative to economic damage, counts as high as 44 larvae per 0.1 m^2 have been noted in healthy fairway turf in Washington, with no appreciably noticeable damage (G. Stahnke, personal communication, 1998).

Description of Stages

Adult

ECF adults are among the larger Tipulidae (crane flies). Males are 14–19 mm long and females 19–25 mm long. The wings are longer than the abdomen in males but shorter than the abdomen in females (Figure 19-2, Plate 51). Females have 10 clearly defined abdominal segments, with segments 8–10 modified into a functional ovipositor. The sex ratio was established as about 1.7 males to 1 female by Coulson (1962), but in Washington the male-female ratio ranged from about 1.8:1 to 1.2:1 (Jackson and Campbell 1975).

Teneral adults are pale greenish with a transparent appearance that is retained for several hours because the meconium is not discharged until after the adult emerges. Callow females have an elongated, swollen abdomen (Byers 1961, Coulson 1962).

Egg

Eggs of the ECF are typically shiny black and elongate-oval, with one side flattened and one end more pointed than the other (Plate 51). They are 1.1 mm long by 0.4 mm wide (Wilkinson and MacCarthy 1967).

Larva

Tipulid larvae have four instars, the last three of which are similar in appearance except for size (Figure 19-2, Plate 51). They are nearly cylindrical but taper slightly at both ends. Larvae are light gray to grayish, greenish brown, with irregular black specks of various sizes. The cuticle is semitransparent, revealing two longitudinal tracheal trunks and the alimentary canal (Jackson and Campbell 1975).

The truncated end of the last abdominal segment consists of the upper spiracular area and the lower anal area. The dorsal area has the spiracular disc with two rounded spiracles and six tapering anal lobes, the characteristic numbers in Tipulinae (Brindle 1960).

Pupa

The pupa is formed inside the last larval cuticle, the puparium (Plate 51). The five posterior abdominal segments have protuberances bearing caudally directed spines, so that pupae can move vertically in their burrows in response to various stimuli (Alexander 1919).

Seasonal History and Habits

Seasonal Cycle

The ECF is a univoltine species that overwinters as a third instar. The insects become fourth instars in April, feed most vigorously to maturity, and cease feeding about the middle of May. They then remain inactive for a prolonged period until pupation, which begins as early as mid-July but peaks about the end of August. During late August and September, the ECF passes in rapid succession through the pupal, adult, egg, and first-instar stages (Figure 19-3) (Coulson 1962, Wilkinson and MacCarthy 1967).

During 1972 and 1973 in Washington, a few flies emerged as early as July and a few as late as early October, but 95% emerged during about 20 days in late August and early September (Jackson and Campbell 1975).

By the time peak flights occur, most oviposition has already taken place, since eclosion,

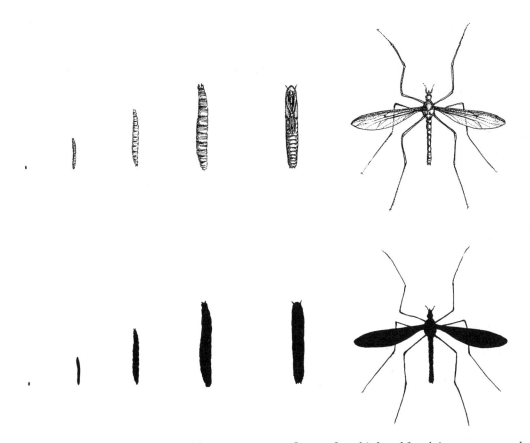

Figure 19-2. Stages of development of the European crane fly: egg; first, third, and fourth instars; pupa; and adult. Silhouettes show actual size of each stage. (Drawn by R. Jarecke, NYSAES.)

mating, and oviposition occur predominantly on the first night of adult life. Eggs hatch in 11–15 days. Larvae feed ravenously and usually complete the first two instars in less than 2 months. Typically, third instars overwinter from November into April. The ECF does not have a true winter diapause (Barnes 1937, Coulson 1962, Jackson and Campbell 1975), although it is less active in the winter months.

Adult Activity
Emergence, Mating, and Flight. Adults typically emerge shortly after sunset, finish mating by midnight, and lay most of their eggs by dawn, all during the same night. Teneral males are seen occasionally in the afternoon and can be captured on sticky board traps above 1 m in height throughout the afternoon. In Washington, the greatest male movement and mating occurred between 7:00 and 12:00 PM PDT. Males are attracted to females even while the latter are emerging from their pupal cases, and union occurs as soon as the teneral female has emerged (Jackson and Campbell 1975).

After union takes place, the female drags the male up the stem, with the male hanging downward and nearly motionless. Normally the pair remains *in copula* for about an hour,

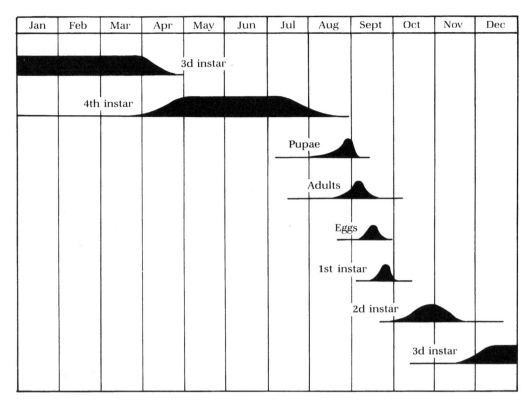

Figure 19-3. Life cycle of the European crane fly in northwestern Washington. (Adapted from Jackson and Campbell 1975, fig. 7; redrawn by R. McMillen-Sticht, NYSAES.)

but they may remain coupled for 3 hr or more. Fully gravid young females are apparently incapable of flight; they move clumsily over the soil surface, stopping occasionally to probe the soil and oviposit. Oviposition is mostly completed by dawn of the first night. Older females that have laid most of their eggs can fly. These females may take to the air while still copulating. Uncoupled females can fly 1.8 m or more in height, as evidenced by catches on sticky board traps (Coulson 1962, Jackson and Campbell 1975).

Oviposition and Fecundity. Shortly after mating ceases, oviposition begins, and it continues until nearly all eggs are laid. From 50% to 75% or more of the eggs are laid by midnight of the day of emergence and about 10% on the next day. Females insert their terminal abdominal segments just below the soil surface (no more than 5 mm) to lay eggs. Nearly 70% of the eggs are within 0−1 cm in depth, and more than 90% are within the 1−3 cm in depth. Eggs are laid in rapid succession until 200−330 have been deposited (Barnes 1937, Coulson 1962, Jackson and Campbell 1975).

Longevity. Adults take no food but do drink water, which lengthens longevity. Males live an average of 7 days, with a maximum longevity of 14 days. Females live an average of 4−5 days, with a maximum of 10 days (Barnes 1937).

Egg Development

Newly oviposited eggs of *Tipula* flies have such high moisture requirements that they will collapse within 2–4 min unless they are in a saturated environment. Fertile eggs absorb about half their weight in water when they are 2–3 days old. The optimum temperature for development and hatching is between 14° and 15°C (Barnes 1937, Jackson and Campbell 1975, Janisch 1941, Laughlin 1958).

Larval Activity

Eggs hatch in 11–15 days, and larvae feed upon hatching. First instars are about 2.7 mm long and molt to second instars about 4–5 mm long in about 2 weeks. They spend the winter as nondiapausing third instars, a stage that lasts as long as 4 months (Jackson and Campbell 1975).

ECF larvae grow rapidly in the spring, when most of the turf damage occurs. Upland organic soil holding three times its dry weight of water minimizes larval mortality. Young larvae appear to prefer green leaves to roots and grow most rapidly, with the least mortality on white clover. Larvae surface to feed at night but remain below the thatch to feed on roots and crowns during the day. On warm evenings and overcast warm days, they can be observed feeding on the surface. Otherwise they are generally found within the upper 2.5 cm of sod, often at the soil-thatch interface. After mid-May, many are found as deep as 7.6 cm below the surface (Maercks 1939, Rennie 1917, Sellke 1937, Wilkinson 1969, Wilkinson and MacCarthy 1967).

Two periods of rapid growth occur, in the fall when they grow from first to third instars and again in the spring when they transform from third to fourth instars; they reach their peak weight of 300–500 mg in June (Jackson and Campbell 1975).

Pupal Activity

Pupae migrate vertically in their earthen burrows in response to soil temperature and moisture, aided by curved spines on the abdomen. After adult emergence, the pupal cases can be found easily in the field, since half to two-thirds of their length protrudes from the soil. The sex of the individuals can be determined from their vacated genital sheaths (Jackson and Campbell 1975).

Miscellaneous Features

Factors Producing High Mortality or Survival

A very high mortality of eggs and first instars occurs when the surface soil atmosphere is less than 100% RH. Nearly 100% mortality of these stages has been observed in England, with death attributed to desiccation. Mortality from severe winter weather can be as high as 30–40% in third-instar larvae (Coulson 1962).

A 30-year average of climatic conditions of the region indicates that the wet coastal belt of British Columbia and northern Washington is particularly ideal for this insect (mild winters, cool summers, and rainfall in excess of 50 cm). More than 40 larvae per 0.1 m² may develop even in relatively warm, dry summers, which occasionally occur in this area (Wilkinson 1969). However, over a 4-year period, a 35–40% decrease in larval populations occurred from early March to late April without insecticide applications in western Washington (G. Stahnke, personal communication, 1998).

Population Survey Techniques

Adults are best sampled with a lightweight sweep net. Immatures may be sampled in the fall with a soil corer 10.16 cm in diameter to a depth of 7.6 cm. Larvae do not penetrate below this depth. The cores are examined manually (Stahnke et al. 1993). Alternatively, the cores may be dried in a Berlese funnel to drive larvae out of the thatch and soil. This method gave a more accurate count than did the application of an aqueous solution of *o*-dichlorobenzene as an irritant to force larvae to the surface of the turf (Jackson and Campbell 1975, Shaw et al. 1974). Spring sampling for larger larvae is generally more useful for the purpose of making decisions on the need for control.

Some popular journals have recommended using a drench of soapy water or some other irritant to force larvae to the surface. Stahnke et al. (1993) demonstrated that visual inspection of core samples reveals significantly more (as many as 10 times) larvae than using disclosing solutions.

Economic Threshold Population

In British Columbia, counts of 20 or more larvae per 0.1 m^2 in a lawn were considered the threshold population for a recommendation that insecticidal control be undertaken (Wilkinson and Gerber 1972).

In Washington, a survey can be made in March, or when temperatures become consistently warmer, by selecting several areas in the lawn and digging up patches of turf 0.1 m^2 and 2.5–5.0 cm in depth. If average counts exceed 25 per 0.1 m^2 in healthy turf, an insecticide treatment should be considered. Turf in poor condition can show damage with 15 larvae per 0.1 m^2 (Antonelli and Campbell 1984).

Natural Enemies

Several microorganisms, parasitoids, and predators are associated with the ECF in Europe but they appear to exert little influence on populations in North America. A tachinid parasitoid, *Siphona geniculata* (de Geer), considered to be the most promising, has been obtained from Germany and was released in Vancouver. The predation of larvae by starlings, robins, and native moles, *Scapanus* spp., exerts some influence (Wilkinson 1969, Wilkinson and MacCarthy 1967).

Field efficacy trials in Washington suggest that *Steinernema carpocapsae* "All" strain and *S. feltiae* reduced larval populations 41% and 56% respectively, when it was applied in mid-April and the turf was irrigated immediately after application. These levels of reduction are generally not considered to be acceptable in highly maintained turf, but might provide adequate levels of suppression in low-quality turf (Stahnke et al. 1993).

Frit Fly

Taxonomy

The frit fly (FF), *Oscinella frit (L.)*, order Diptera, belongs to the family Chloropidae.

Importance

The FF may be responsible for more damage to golf greens and collars (edges of greens) than is generally recognized. Damage occurs when larvae feed on the terminal shoots of

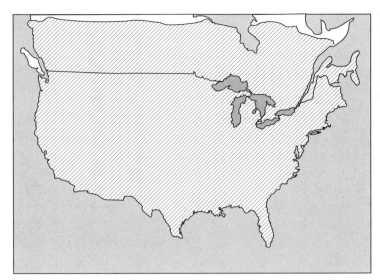

Figure 19-4. Distribution of frit fly. (Drawn by E. Gotham, NYSAES, adapted from Handbook of Turfgrass Insect Pests, Brandenburg and Villani 1995, Entomological Society of America.)

grasses. The adults annoy golfers because the flies are attracted to white balls or light-colored clothing or equipment, on which they will land. The FF is a serious pest of cereals in Europe but not in the United States. In Virginia, the FF is a serious pest of reed canary grass, *Phalaris arundinacea* (L.), a hay crop, and much knowledge of this insect in this country comes from studies on this crop (Allen and Pienkowski 1974) or from a more recent study on golf course turf that was conducted in Ohio (Tolley and Niemczyk 1988a).

History and Distribution

The FF is widely distributed in the temperate Northern Hemisphere. It is reported from northern Europe into Russia. In the United States it is commonly found at latitudes north of Virginia and Kansas, wherever grass is green for a considerable part of the year (Figure 19-4). It is most abundant in regions where winter wheat is grown, from the Great Lakes to the Ohio River and westward to Missouri. In Virginia it is more abundant at higher elevations. Infestations in lawns in Connecticut and Rhode Island have been reported several times (Aldrich 1920, Allen and Pienkowski 1974, Vance and App 1971).

Host Plants and Damage

Common cool-season turfgrasses are hosts of FF, with bluegrass and bentgrass most susceptible, especially when these grasses are irrigated and kept short by mowing or grazing. Fly density tends to be higher where tiller density is the thickest. Larvae infest stems of barley, wheat, oat, and rye in Europe and also feed on immature kernels, producing empty kernels called *fritz* in Sweden, hence the common name. In the mountains of Virginia, the FF is an important pest of reed canary grass where the yields of the second hay harvest are significantly reduced (Aldrich 1920, Allen and Pienkowski 1974, Jepson and Heard 1959).

Larval feeding in the upper primordial leaves causes general yellowing and death of the central leaf, while surrounding shoots remain green. On golf courses, damage is often first apparent on collars and approaches the center of the green. Higher elevations of greens are usually the first to show symptoms. Greens with soil high in organic matter appear more susceptible. Tunneling of the stems near ground level causes the upper portion of the plants to become brown and die (Plate 51) (Aldrich 1920, Bowen et al. 1980).

Description of Stages

Adult

The flies are 1–2 mm long and black, with a shiny dorsum, yellow halteres, and yellow markings on the tibial segments of the legs (Figure 19-5, Plate 51) (Aldrich 1920, Dahlsson 1974). They have a large triangle pattern on the head (Tolley 1995).

Egg

Eggs are pure white and 0.7–0.8 mm long, with a finely ridged surface (Figure 19-5). They are not easily detected because of their small size, and they can be jarred loose easily from leaves (Aldrich 1920, Allen and Pienkowski 1974). Eggs are placed between the leaf sheath and stem or occasionally, on leaves or outside stems (Tolley 1995).

Larva

There are three larval instars. Young larvae (maggots) are yellow white, with two conspicuous black mouth hooks and posterior spiracles (Figure 19-5). Mature larvae are 3–5 mm long and light yellow, with black curved mouth hooks and visible spiracles (Plate 51) (Aldrich 1920, Bowen et al. 1980, Tolley 1995).

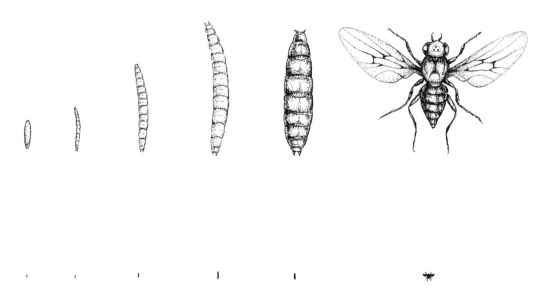

Figure 19-5. Stages of development of frit fly: egg; first, second, and third larval instars; pupa; and adult. Silhouettes show actual size of each stage. (Courtesy of R. Jarecke, NYSAES.)

Pupa
Pupae are yellow at first, turn dark brown, and are about 3 mm long (Bowen 1980, Tolley 1995).

Life History and Habits

Seasonal Cycle
Maggots overwinter in grass stems infested the previous fall. Pupation and adult emergence occur early the following spring, about March in California and mid-April at Blacksburg, Virginia. Three generations a year are reported in the higher elevations of Virginia (Figure 19-6) and four each year in Indiana. At Blacksburg, peak flights of flies occurred from April to early June from overwintering maggots. Populations of first-generation flies

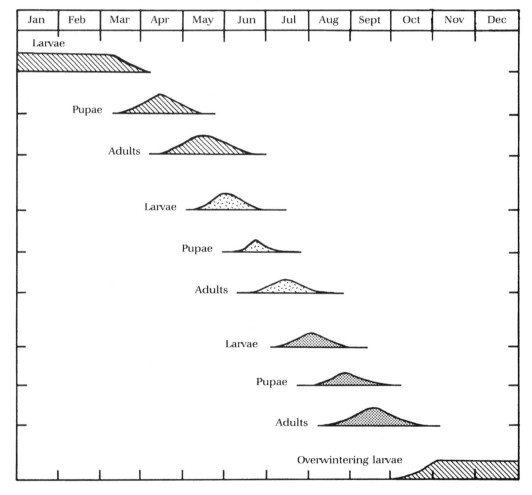

Figure 19-6. Life history of the frit fly in southwestern Virginia. (Data from Allen and Pienkowski 1974; drawn by H. Tashiro, NYSAES.)

from late June to mid-August are the lowest, and those of the last generation (during late autumn) are highest. Peaks in larval population closely followed those of the adults. First-generation larvae complete development in about 5 weeks, second-generation larvae do so in 8–10 weeks, and the third generation overwinters. The average life cycle from adult to adult is 21–58 days (Aldrich 1920, Allen and Pienkowski 1975, Bowen et al. 1980).

A study conducted in Ohio documented three full generations and a partial fourth generation in northern Ohio. Adult peak densities occurred during mid-May, late June, late July to early August, and mid-September in 1984. Peak periods were about 2 weeks earlier in 1985, apparently because of warmer spring temperatures. Throughout the study, fewer adults were collected in sweep nets following rainfall of at least 1.25 cm, because adults moved closer to the base of host plants and were not collected with the sweep net technique (Tolley and Niemczyk 1988a).

Adult Activity

Upon eclosion, adults climb from between the sheath and stem, and move to the tips of leaves and face skyward. Flies are attracted to and land on white objects, such as golf balls or light-colored clothing. They are often seen hovering over golf greens and can be collected from midmorning onward throughout the summer (Baker 1982, Niemczyk 1981). Adult flight occurs during daylight hours and does not occur at temperatures below 9°C (Tolley 1995).

Mating and Oviposition. Males mate within 6 hr of their emergence, while females do not mate until the second day of their emergence. Mating pairs have been observed on leaves and stems of host plants from about 9:30 AM until sunset. Eggs are usually laid on leaves, in leaf sheaths or axils, and in the brown stubble of grains. Females live an average of 5.5, 18.0, and 26.0 days in the spring, summer, and fall generations, respectively. Egg production averaged 17, 59, and 32 eggs for the spring, summer, and fall generations, respectively (Allen and Pienkowski 1974, Southwood et al. 1961). Egg laying does not occur below 12°C, and occurs most frequently between 18°–30°C (Tolley 1995).

Egg Survival

The incubation period of eggs in the field is 3–7 days. Rain-free periods are most important for good survival when tall grasses are infested, because eggs can be dislodged during rain. In Virginia, egg mortality is less than 5% in the spring, when rain is relatively infrequent but may exceed 50% during summer, when rain occurs more frequently. Rain (and elevated humidity), wind, and predation appear to be the primary causes of egg mortality (Tolley 1995).

Larval Activity

Upon hatching, a larva migrates downward toward the crown by moving between the leaf sheath and the stem. As the larva passes each node, rasping the internal tissue, the stem dies from the node outward. Soon the whorl leaf begins to yellow, wither, and brown. As many as six larvae may develop per tiller, with more developing per tiller in the spring than in the summer. Upon completion of larval development (a total of three instars), the puparium is formed at the base of the plant, between the leaf sheath and the stem. Overwintering larvae remain at the point of original feeding at the base of the leaf (Allen and

Pienkowski 1974). The key mortality factor for larvae is desiccation, particularly in the summer months (Tolley 1995).

Miscellaneous Features

Threshold Temperatures

When reared at constant temperatures on perennial ryegrass, the FF had lower development thresholds of 10.3°, 10.4°, and 10.8°C for eggs, larvae, and pupae, respectively. Developmental rates for all life stages increased with temperature to a maximum of 27.5°C, and decreased thereafter until the lethal threshold (34°C) was reached (Tolley and Niemczyk 1988b).

Degree-Day Developmental Relationship

Field observations determined that significant ovipositional activity begins only after at least 40% of the adults have emerged. A degree-day model using 0°C base temperature and March 1 as a start date appears to be quite accurate at predicting 40% adult flight dates for all four generations in Ohio (Tolley and Niemczyk 1988a). A separate study indicated that 49, 293, and 110 DD $_{10C\ base}$ are required for completion of egg, larval, and pupal stages, respectively, and a total of 451 DD was required for development from egg to adult in the laboratory (Tolley and Niemczyk 1988b).

Population Sampling Technique

Populations have been surveyed with a sweep net. A water trap with a wetting agent has also been used. Use of a blue- or violet-colored water trap resulted in higher catches and more balanced sex ratio catches than traps of any other color. Adult flies have been collected successfully with a DeVac machine and have been separated from grass litter using Berlese funnels (Allen and Pienkowski 1974, Oschmann 1979).

Response to Mowing

The FF shows a strong positive response to mowing, and moves in large numbers from unmowed turf to recently mowed turf. Much of the observed movement occurs within 1 hr after mowing, after which populations gradually begin to disperse again (Falk 1982).

Natural Enemies

Various species of entomopathogenic nematodes, one species of parasitic mite, at least 17 arthropod predators, and at least 39 parasitic wasps have been reported attacking the FF, but none is believed to cause high mortality in field conditions (Tolley 1995).

20

Hymenopteran Pests:
Family Formicidae

Ants—General Information

Taxonomy

Many species of ants, order Hymenoptera, family Formicidae, invade turfgrass areas throughout the country. The subfamily Formicinae includes many ants found on turf. One that has become a serious nuisance, particularly on golf courses in cool-season settings since 1985, is *Lasius neoniger*. Other species of interest in the subfamily include the cornfield ant, *Lasius alienus* (Foerster); the mound-building ant, *Formica exsectoides* Forel; and the red ant, *Formica pallidefulva* Latreille. All others mentioned herein belong to the subfamily Myrmicinae and include the pavement ant, *Tetramorium caespitum* L.; harvester ants, *Pogonomyrmex* spp.; and the complex of fire ants of the genus *Solenopsis*.

The fire ants are by far the most important and common pest ants of turfgrass in the southeastern states. There are four species of significance, two native to the United States and two introduced. The fire ant, *Solenopsis geminata* (Fabricius), is a native species found from Texas to South Carolina, primarily in the more coastal regions of its range. Mounds are usually located in open grasslands. The southern fire ant, *Solenopsis xyloni* McCook, ranges from California to North Carolina and tends to occur farther from the coast than *S. geminata*. This species has largely been replaced by imported fire ants where they occur.

The two introduced species are the black imported fire ant, *Solenopsis richteri* Forel, and the red imported fire ant (RIFA), *Solenopsis invicta* Buren. Of all the genera and species mentioned, the RIFA is by far the most important on turf (Creighton 1950, Lofgren et al. 1975, Vinson and Greenberg 1986), although *L. neoniger* is increasing in its negative impact on turf.

Importance

Ants are primarily troublesome in turfgrass areas because they build mounds as they form subterranean homes for their colonies. They seek out drier, well-drained sandy soils that have low water-holding capacity. The galleries they form, which damage roots, add to the desiccation of the soil, and the turf in the surrounding areas becomes thin and unsightly. Mounds of various sizes and shapes, formed according to the habits of the ant species, are detrimental to lawn mower blades. Some mounds are too high to mow over. Some ant species have vicious bites and stings, with the RIFA being the most serious (Baker 1982, Oliver 1982a, Schread 1964).

History and Distribution

Ants have a cosmopolitan distribution from the Arctic to the most tropical areas of the world. Most harvester ants occur west of the Mississippi River and are discussed in more detail later in this chapter. In the Northeast, *L. neoniger* is the most frequently encountered species on golf courses, although the cornfield ant is sometimes found on golf greens. Other species that inhabit turfgrass areas include the cornfield ant, the red ant, the mound-building ant, and the pavement ant (Baker 1982, Crocker et al. 1995a, Schread 1964).

Host Plants and Damage

All sunny turfgrass areas, regardless of grass species, can be invaded by ants in search of areas to start a new colony. Lawns, parks, playgrounds, golf courses, and other turfgrass areas, in addition to cultivated farm and pasture land, are likely sites as long as they are well drained (Plate 52). Ants that feed on excretions of root-infesting aphids help promote the aphid colonies, which are also detrimental to the health of grasses (Baker 1982, Schread 1964).

As the colonies grow and mature, mounds of earth are created by many species, while other species leave the surface relatively flat. On the surface the unsightly mounds smother the grass, while the galleries in the soil disturb the roots, dry the soil, and cause the grass to thin out (Plate 52). The mounds of various sizes make turf maintenance difficult, and mowing over even low mounds dulls and chips mower blades as the surface is scalped. In some habitats, ants move as much soil as earthworms, helping to reduce soil compaction (Baker 1982, Holldobler and Wilson 1990, Schread 1964, Vance and App 1971).

Mounds vary in size, shape, and appearance, depending on the ant species involved. Mounds of the RIFA may be 35 cm or more in diameter and 20–25 cm high (Plate 52). On permanent sod where ample food for the RIFA is available, as many as 125 mounds per hectare can develop. Several harvester ant species produce nests in open, sunny areas that are flattened circular areas, up to 1 m in diameter cleared of vegetation, covered with coarse gravel and with one or a few central nest openings. They also produce long (up to 7 m) foraging trails cleared of vegetation. Red ants have large colonies with several openings at the surface but do not pile soil in definite mounds. The surface soil feels spongy, and grass grows poorly. Allegheny mound ants sometimes construct mounds that are up to 0.9 m high and more than 2 m in diameter (Baker 1982, Holldobler and Wilson 1990, Oliver 1982a, Schread 1964).

When a person or animal disturbs a mound, worker ants attack the intruder by sinking their powerful jaws into the skin, afterward repeatedly thrusting their poisonous stinger into the flesh. This may be the most serious problem with ants in a turfgrass area. The RIFA has the most vicious of stings, one that produces a burning sensation, but it normally does not have the severity of a honeybee or yellow jacket sting. The area stung usually festers within a day, and the sting persists for several days. Harvester ants also inflict a painful sting.

Some species in the genus *Pogonomyrmex* produce among the most painful of all Hymenopteran stings, with intense pain lasting up to 4 hr (Evans and Schmidt 1990). People who are allergic to bee and wasp stings may also be allergic to the RIFA sting and should take appropriate precautions (Baker 1982).

Ants can be found in almost any type of turfgrass, and colonies can occur in lawns, parks, playgrounds, or golf courses, as well as agricultural land and wilderness areas. Ants feed on a wide variety of materials, although some species show distinct preferences. For example, harvester ants feed primarily on parts of seeds, while fire ants primarily consume other arthropods, including some turfgrass pests such as chinch bugs (Crocker et al. 1995, Drees et al. 1996).

Description of Stages

Adults

Ants have a very narrow, constricted connection between the thorax and head and an even more conspicuous constriction between the thorax and the abdomen. The constriction is made by the greatly modified first two segments of the abdomen. Adults may be winged males, winged females, or wingless female workers (Plate 52). The body surface may be smooth or hairy, red, brown, black, or yellow. Harvester ants are 5–6 mm long and reddish brown to yellow or black. The red ant is nearly 13 mm long, with a red head and thorax and a black abdomen (Baker 1982, Schread 1964).

Eggs

The eggs of ants are white to pale yellow and vary in shape according to species. Eggs of harvester ants are less than 0.5 mm long and elliptical (Baker 1982).

Larvae

Larvae of ants are mostly white and legless, with a body that dilates gradually from the anterior to the posterior (Plate 52). Larvae have small, distinct heads. Harvester ant larvae are shaped like a crookneck squash or a gourd (Baker 1982).

Pupae

Pupae resemble adults in size and shape, with appendages appearing distinctly as ridges on the body. Pupae of some species (e.g., fire ants) are naked while those of other species (e.g., harvester ants) occur in capsule-like cocoons. They are creamy white initially but darken to a tan or brown with time (Plate 52) (Baker 1982; B. Drees, Texas A&M University, personal communication, 1998).

Life History and Habits

Life Cycles

Male ants appear only in very large or old colonies. Winged males and females swarm from parent colonies, pair off, and mate, and the males die soon afterward. The fertile winged female, the queen, flies to a site suitable for a colony, loses her wings, finds a suitable nesting site, and begins laying eggs. She cares for and feeds the first brood of larvae that develop into workers, the wingless females. Thereafter her sole function is to lay eggs. The workers care for the eggs and the queen, construct and repair nests, gather food, feed the immatures, care for the brood, and defend the colony (Baker 1982, Schread 1964).

Red Imported Fire Ant

Importance

More public funds have been spent on this insect than on any other in history, perhaps because claims have been made that RIFA is a pest serious enough to kill livestock and wildlife and to destroy crops. Although huge sums of money have been directed for eradication and control efforts, the ant continues to spread (Oliver 1982a). For example, $10–$15 million is spent annually in Alabama alone to manage RIFA populations (P. P. Cobb, Auburn University, personal communication, 1998).

History and Distribution

Both imported fire ant species were introduced into or near Mobile, Alabama, at different times. The black imported fire ant, probably introduced from Argentina, was first recorded in 1918 and still remains confined to a small area of northeastern Mississippi and northwestern Alabama. The RIFA, a native of Brazil, was introduced about 1933–40, and has spread to infest more than 105 million hectares of land in 11 states, including all of Florida and Louisiana and portions of Alabama, Arkansas, Georgia, Mississippi, North Carolina, Oklahoma, South Carolina, Tennessee, and Texas (Figure 20-1). Because it is replacing the other three fire ants and is becoming the most common in the states from North Carolina to Texas, it is usually the insect in question whenever fire ants are discussed. The RIFA has the potential to spread west and survive in southern Arizona and along the Pacific Coast north to Washington. A narrow band passing from coastal Virginia south into the infested states and from Texas into California and up the Pacific Coast through Washington represents the area of −12°C minimum temperature. This marks the area of maximum probable spread because of the RIFA's minimum temperature tolerance (Oliver 1982a, Sparks 1995, Vinson and Sorenson 1986).

Host Plants and Damage

RIFA builds large mounds in many kinds of soils, but is most active in exposed areas such as pastures, parks, and lawns (Plate 52). The aboveground portion of a mound is normally conical and can be up to 30 cm high. The tops of mounds are less distinct in very sandy soils or where the land is disturbed frequently. Below ground, a nest may extend 1.2 m into the soil and will often be V-shaped. Sometimes colonies are produced inside rotten logs, around stumps or trees, or inside electrical switch boxes. The size of the mound depends on soil characteristics and how frequently a mound is disturbed over time (Sparks 1995).

   The RIFA is a significant predator as well as a scavenger and feeds on a variety of insects, as well as seeds, carrion, and processed foods. RIFAs can produce a painful sting and will attack anything that disturbs their mounds or food sources. Some people are highly sensi-

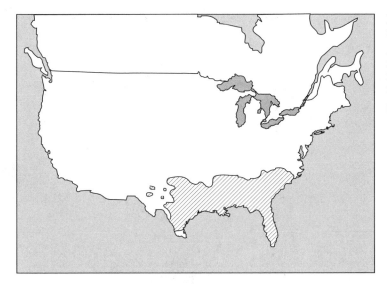

Figure 20-1. Distribution of the red imported fire ant. (Drawn by E. Gotham, NYSAES, adapted from Handbook of Turfgrass Insect Pests, Brandenburg and Villani 1995, Entomological Society of America.)

tive to the toxin injected by RIFAs and must seek medical attention after being stung. Symptoms of a RIFA sting include intense burning and itching, and may also include the eruption of a white pustule (Sparks 1995). Because the ants move aggressively and sting repeatedly, they can be a serious nuisance in turf settings such as parks, playgrounds, athletic fields, and golf courses.

The mounds that are produced are large and unsightly in a maintained turf setting. Furthermore, burrowing and foraging activity of the ants can open up the soil so that it is more vulnerable to desiccation.

Description

Adults
Reproductive females (queens) are red-brown and shed their wings after they complete a nuptial flight. Reproductive males are black and have wings, as do unmated queens. Worker ants are wingless, sterile females, and vary in length from 1.5 to 4 mm, depending on the task of the worker. Workers are reddish brown with a darker abdomen (Figure 20-2, Plate 52).

Life History and Habits

Life Cycle
Most mating flights occur from April through August, although they can occur at any time of year. Reproductive males die shortly after mating, and newly mated queens drop to the ground, shed their wings, and begin looking for an appropriate place to start the new colony. They burrow vertically into the soil to a depth of about 10 cm and construct an egg chamber near the bottom. A new colony is begun as a queen deposits 10–15 eggs. She tends the eggs as they incubate for about a week and feeds the larvae as they develop. The transition from egg to adult worker requires 3–4 weeks.

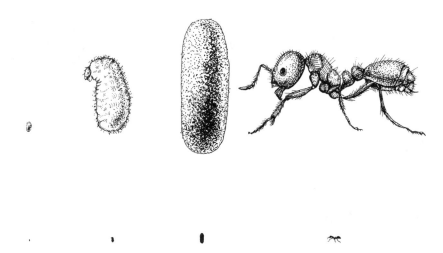

Figure 20-2. Stages of development of red imported fire ant: egg; final larval instar; pupal case; and adult. Silhouettes show actual size of each stage. (Courtesy of R. Jarecke, NYSAES.)

The first workers to develop are minims (minute, polymorphic workers and sterile females) and tend the queen, which begins to produce as many as 200 eggs per day. These eggs hatch into larvae, which are tended by the original workers. Within 2 or 3 months, minims are gradually replaced by "minor" workers, and in about 5 months to a year, the minor workers are replaced by "major" workers, which gather food for the larvae, other workers, and queens, defend the colony, and take care of the brood (maintain moisture levels, regulate temperature). Worker ants live about 8 weeks, and queens can survive for 5 years or more. A mature colony may contain as many as 500,000 workers, the brood (developing larvae), and up to several hundred winged forms and queens (B. Drees, personal communication, 1998; Sparks 1995).

Colonies are generally considered mature after 3 years. The average colony size may vary from 100,000 to 500,000 workers and includes relatively few winged or reproductive forms. Where food is abundant, as many as 50 mounds per acre may develop. The preferred food of the RIFA is other arthropods—termites, springtails, crickets, caterpillars—and other soil-inhabiting species (Baker 1982, Lofgren et al. 1975, Oliver 1982a, Sparks 1995).

Natural Enemies of RIFA

Microorganisms

A species of *Pseudomonas* bacteria is reported to be highly toxic to the RIFA. Three common entomophagous fungi are also pathogenic to the RIFA. These are *Beauveria bassiana*, *Metarrhizium anisopliae*, and *Aspergillus flavus* Link. (Lofgren et al. 1975). Laboratory experiments evaluated the effect of *B. bassiana* (Balsamo) Vuillemin strain on Brazilian and American populations of RIFAs. Responses of the two populations were similar in terms of rates of infection and mortality, which were significantly higher in the treated ants than in the untreated (control) ants during the first 3 weeks of the experiment (Stimac et al. 1993). However, several logistical considerations (e.g., application and formulation technology, effect of soil type) appear to limit the practicality of the approach, and the fungus is still being evaluated for commercial viability (Oi et al. 1994).

Some strains of entomopathogenic nematodes (*Steinernema* spp.) have induced ants in treated mounds to move away from the mound temporarily, but few colonies were actually eliminated (Drees et al. 1992). Natural enemies applied to individual fire ant mounds are generally not suitable for area-wide treatment programs (Drees et al. 1996).

Insect Predators and Parasitoids

Some native ant species compete with the RIFA for territory and resources, and are particularly effective predators on newly mated fire ant queens. The predatory straw-itch mite, *Pyemotes tritici* (Lagreze-Fossat & Montane), feeds on and paralyzes developing fire ants but is not effective when applied as directed and can be hazardous to the applicator (Thorvilson et al. 1987).

Several species of phorid flies in the genus *Pseudacteon* parasitize *Solenopsis* ants. Some species of *Pseudacteon* are native to North America and attack native species but do not recognize the imported species as suitable hosts, perhaps because the RIFA does not produce trail pheromones that the native flies can recognize (Grisham 1994).

Host specificity tests have been conducted with four South American *Pseudacteon* species, *P. litoralis* Borgmeier, *P. wasmanni* (Schmitz), *P. tricuspis* Borgmeier, and *P. curvatus* Borgmeier, to determine the degree to which these species attack the native North American fire ant, *S. geminata*. The first three species oviposited readily on the RIFA but showed little interest in ovipositing on the native species (Gilbert and Morrison 1997). While there is great hope for success from the eventual release of some of these flies, they are expected to provide only a measure of suppression over large areas, but are not expected to result in eradication of the RIFA (Drees et al. 1996, Williams and Porter 1996).

Harvester Ant

Importance

Harvester ant is a generic term applied to ants in the subfamilies Ponerinae, Myrmicinae, and Formicinae that depend on plant seeds to provide the bulk of their diet. While some of these ants (notably in the genus *Pogonomyrmex*) are primarily of concern because they can inflict painful stings when disturbed, they also can cause direct damage to turf by their nest-building activities. In contrast, Formicine ants do not sting. The most important group of harvester ants in North America are in the genus *Pogonomyrmex*. Most of the information included here is derived from Crocker et al. (1995).

History and Distribution

There are approximately 60 described species in the genus *Pogonomyrmex*, of which 22 occur in North America, primarily west of the Mississippi River in regions with relatively low amounts of annual rainfall. The others are found in Central and South America. The

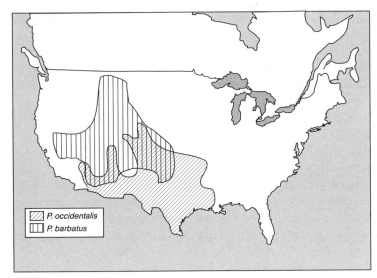

Figure 20-3. Distribution of harvester ants, *Pogonomyrmex occidentalis* and *P. barbatus*. (Drawn by E. Gotham, NYSAES, adapted from Handbook of Turfgrass Insect Pests, Brandenburg and Villani 1995, Entomological Society of America.)

four most important harvester ants in North America are the western harvester ant, *P. occidentalis* (Cresson); the red harvester ant, *P. barbatus* (F. Smith); the California harvester ant, *P. californicus* (Buckley); and the Florida harvester ant, *P. badius* (Latreille). The distribution of these species is shown in Figure 20-3 and 20-4.

Host Plants and Damage

Harvester ants construct nests in the soil, usually in areas subject to full sun. Nests often are covered by mounds varying in size and sometimes decorated with gravel. Workers of some species, notably the western harvester ant, use their powerful mandibles to remove vegetation around the nest, denuding the area and detracting considerably from the aesthetic appearance of highly maintained turf. In areas where turf seed is produced, harvester ants cause direct damage by consuming large quantities of seed.

Harvester ant venom is among the most toxic of insect venoms to mammals. Symptoms of their sting include localized swelling and inflammation, accompanied by a persistent throbbing pain. Multiple stings can result in excruciating pain, vomiting, and various systemic responses. Unlike most ants, the California harvester ant leaves its stinger in the wound.

Description

Harvester ants have two nodes on the petiole (the slender connection between the thorax and main part of the abdomen). In most harvester ant species, workers are only one size. Most harvester ant workers range from 0.5 to 0.9 mm and normally are red to black (Figure 20-5). The underside of the head has a special fringe of long hairs that assists in carrying food and soil. One exception, the Florida harvester ant, has typical minor workers, major workers with very large heads and powerful mandibles for cracking seeds, and intermediate workers of several sizes.

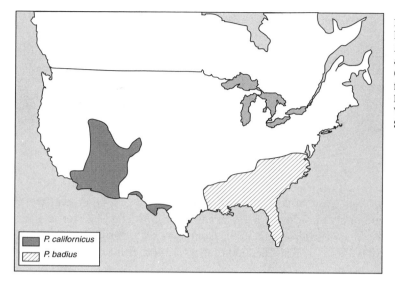

Figure 20-4. Distribution of harvester ants, *Pogonomyrmex californicus* and *P. badius.* (Drawn by E. Gotham, NYSAES, adapted from Handbook of Turfgrass Insect Pests, Brandenburg and Villani 1995, Entomological Society of America.)

Life History and Habits

Little is known about the mating habits and life cycle of harvester ants. Males of some species gather in the same site year after year in large numbers, produce pheromones to attract other reproductives of the same species, and then compete for access to females. Queens of *P. badius* and *P. owyheei* have been reported to live as long as 17 and 30 years, respectively (Holldobler and Wilson 1990).

Harvester ants collect seeds and transport them to the nest for storage and consumption. They provision nests with a wide range of seeds and may construct foraging trails as much as 7 m from the nest. They produce pheromones to recruit other ants to a new food source, but apparently rely on visual landmarks to find their way back to the nest. After wet weath-

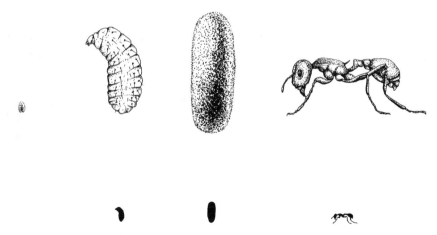

Figure 20-5. Stages of development of harvester ants: egg, final larval instar, pupal case, and adult. Silhouettes show actual size of each stage. (Courtesy of R. Jarecke, NYSAES.)

er, they sometimes move stored seeds outside to dry. Any seeds that germinate are discarded. They also prey on other insects and scavenge dead arthropods.

There is considerable variability in colony size, with some as small as a few dozen ants. Red harvester ant colonies can contain more than 12,000 ants, with tunnels extending laterally over 2 m in diameter and to a depth of 5 m.

Some species forage only during the day and block the entrance to the nest at night with a pebble. Others forage based on air and soil temperatures, with little regard for presence or absence of light.

Natural Enemies

The thief ant, *Solenopsis molesta* (Say), builds nests near harvester ants and then constructs small tunnels into brood chambers and devours the larvae, even though harvester ants are much larger than the marauders. Another small ant, *Dorymyrmex pyramicus* (Roger), often nests near harvester ant colonies and steals dead insects and other food as foragers return to the nest.

Several vertebrates, including horned lizards, toads, and various birds, prey on harvester ants.

Lasius neoniger

Importance

Lasius neoniger has become a major turf pest, particularly on golf courses throughout the Northeast. It produces small mounds, 2–4 cm in diameter, that can smother underlying grass, leaving unsightly pockmarks defacing a fairway or other highly maintained turfgrass. Most of the information currently available is based on studies conducted in grassland areas, but several turf entomologists in the United States are now studying various aspects of the ecology of the insect in fine turfgrass.

History and Distribution

Lasius neoniger is the most abundant ant species in open habitats in the northeastern United States (Wilson 1955). Preliminary data indicate population densities can exceed 100 million ants per hectare in infested turf areas (S. Werle, University of Massachusetts, personal communication, 1998). Colonies are characterized by a ring of excavated earth that forms a crater-like nest entrance (Traniello and Levings 1986). Turf managers have reported increased ant activity since 1993, particularly on fairways and other turf areas mowed at similar heights. It is unclear whether activity has actually increased or the tolerance for that activity has decreased (P. J. Vittum, personal observation).

Host Plants and Damage

Lasius neoniger produces mounds, 2–4 cm in diameter, at entrances to a colony. These mounds can smother underlying grass and leave unsightly pockmarks. Even more importantly, the gravelly or sandy mounds dull mower blades and accumulate on bed knives of mowers, resulting in uneven mowing patterns. *Lasius neoniger* also tends root-feeding aphids, presumably to extract honeydew as a sugar source.

Description

Lasius neoniger is a relatively small ant, with workers averaging 3 mm in length and queens about 10 mm in length. They range in color from light tan to dark brown. As is typical of ants in the subfamily Formicinae, *L. neoniger* does not possess a sting but instead has evolved a structure called the *acidopore,* which is an opening at the apex of the abdomen. This opening, or pore, is ringed with a crown of long hairs and is used to spray venom (mostly formic acid) at adversaries. The petiole of Formicine ants is made up of one node rather than the two nodes typical of Myrmecine ants. In *L. neoniger,* this node is very tall and wide, forming a distinctive plate-like structure separating the apparent abdomen (metasoma) from the apparent thorax (mesosoma).

    Lasius neoniger is a monomorphic species (all workers are similar in size). The only apparent exception is that newly established nests will often have some very small workers, called *nanitics,* that are analagous to the minims found in young RIFA colonies. In the northeastern United States, *L. neoniger* is the most commonly enountered small Formicine ant found infesting turfgrass. Another species that sometimes occurs in turfgrass is *L. alienus,* a very similar species in appearance and habits.

Life History and Habits

Lasius neoniger colonies often have multiple nest entrances, and each colony may be organized in "subnests". Most foraging ants show a marked preference for a single nest entrance. Once established, a colony may survive several years, and it will exhibit intense territoriality when interacting with surrounding colonies (Holldobler and Wilson 1990, Traniello and Levings 1986).

    Field observations in Massachusetts suggest that mating flights of reproductive males and females occur in late summer of each year. Inseminated queens seek a site to establish a new colony and burrow as much as 0.5 m into the soil. In late spring, soil cores reveal eggs and larvae at that depth, and these stages remain evident throughout the summer. Workers are observed foraging throughout the growing season, and are particularly active on warm, sunny days (S. Werle, personal communication, 1998).

    It is important to note that while *L. neoniger* undoubtedly can cause problems for turf managers, the beneficial aspects of the species have not been thoroughly studied. *Lasius neoniger* has been suggested as a factor limiting the northward expansion of RIFA (Bhaktar et al. 1972) and preys on various insect eggs.

21

Hymenopteran Pests:
Families Sphecidae and Vespidae

Bees and Wasps

Many ground-nesting bees and wasps become troublesome pests of turfgrass in much the same way that ants are a pest. Their nesting habits create mounds of soil as they bring subsoil to the surface in making their galleries and nests. One of the more prominent is the cicada killer, a large wasp with unusual habits. Other bees and wasps, such as yellow jackets and hornets, do not damage turf directly but can be a nuisance by virtue of their aggressive behavior and ability to inflict painful stings.

Cicada Killer

Taxonomy
The cicada killer (CK), *Sphecius speciosus* (Drury), is of the order Hymenoptera, family Sphecidae, subfamily Bembicinae. It produces major soil surface disruption while burrowing, in part because of its large size. This species is also called the *sand hornet* or *golden digger wasp*.

Importance
In size and forbidding appearance, the CK appears to be one of the most formidable of North American wasps. Its sting causes an initial sharp pain and lasts about a week. The CK has no nest-guarding instinct, however, and stings only when provoked. In addition to the possibility that it may sting people, this insect can have a slight economic importance as a pest of turfgrass because it makes unsightly mounds of earth as it digs its burrows (Damback and Good 1943).

History and Distribution
The CK is present in all states east of the Rocky Mountains and is common in areas where annual cicadas are prevalent. Its nests are made in full sun where vegetation is sparse (often preferring golf course sand traps), in light, well-drained soils (Baker 1982).

Host Plants and Damage
CK nests and the resulting mounds are not restricted to any particular kind of turfgrass. Wherever the CK makes nests in turfgrass, the mounds smother the grass (Plate 63). A very gravelly or bare area is often preferred.

Description of Stages

Adult. The wasps are large, longer than 30 mm, and resemble oversized yellow jackets or hornets. The body is black, with pale yellow bands or marks on the first three abdominal segments. The head and thorax are rusty red, and the wings are russet yellow (Plate 63) (Baker 1982, M. F. Potter 1995).

Egg. The cylindrical, cigar-shaped, translucent greenish-white eggs are 3–6 mm long and 1.0–1.5 mm wide. Each egg is placed under one femur on a middle leg of its victim, a cicada (Baker 1982).

Larva and Pupa. Larvae grow to a maximum length of 28–32 mm. When they are mature, they become somewhat shrunken and leathery, brown, and stiff. The pupa is undescribed; its case is about 32 mm long and about 11 mm wide (Baker 1982).

Life History and Habits

Life Cycle. CKs are solitary wasps and therefore do not nest together. They overwinter as late-instar larvae or prepupae within a cocoon in an earthen cell that extends 17–50 cm beneath the surface. Pupation occurs in the spring and lasts 25–30 days. The wasps emerge in about the first week in June in Arkansas and rarely before the first week in July in Ohio, but in both locations the CK closely approximates the emergence of its prey. Once emergence has begun, it continues throughout the summer. In Ohio, adults live for 60–70 days, are present from July to mid-September, and disappear as the host cicadas are on the wane. Eggs normally are deposited during late July through August. Eggs hatch within 24–48 hr, and the larvae complete their growth in 4–14 days. There is only one generation a year, closely synchronized with that of the annual cicada (Baker 1982, Damback and Good 1943).

Adult Activity. Immediately after their emergence, female wasps feed on exudates of trees and flowers. A male wasp stations himself on the top of a plant and when a female flies nearby, captures her. They mate in flight. Mating and the digging of burrows take place several weeks before the insects prey on cicadas. Each female digs her own burrows 15–25 cm deep and 1.2 cm in diameter. Broadly oval cells are then dug perpendicular to the main tunnel. The female digs by dislodging soil with her mouth and kicking the loosened particles back as a dog would dig. The excess soil thrown out of the burrow forms a U-shaped mound at the entrance, producing the wasp's primary detrimental effect on the turf (Baker 1982, Damback and Good 1943, Davis 1920).

Once a burrow is ready, a female wasp seeks adult cicadas (Plate 63) in trees, charges into a victim, and stings it between the abdominal segments to paralyze it. Straddling the cicada (Plate 63) and firmly grasping it, the female glides or flies to her burrow with her prey and places it in one of the earthen cells. She places an egg snugly under a femur of one middle leg of the cicada, then seals the cell. Occasionally two cicadas are placed in a single cell, and a larger wasp develops (Damback and Good 1943).

Danger of Stings. Female wasps are capable of inflicting a painful sting, but normally do not do so unless provoked. Mating males are aggressive (but do not sting), and females near their burrows are easily disturbed and provoked. Otherwise females are relatively docile. Nevertheless, it is wise to keep a distance from burrows to avoid the possibility of a sting (Baker 1982, M. F. Potter 1995).

Larval Activity. Eggs hatch in 24–36 hours, and the young larva immediately inserts its mouthparts into the body cavity of the cicada to feed on the liquid contents. Growth is very rapid, and the wasp larva completes its development in as short a time as 5–14 days. It then weaves a cylindrical case, the cocoon, around itself for overwintering (Damback and Good 1943).

Natural Enemies

A sarcophagid fly maggot is sometimes found feeding on the same cicada as the wasp, but whether it is predatory on the wasp is not known (Damback and Good 1943).

Yellowjacket

Taxonomy and Distribution

Wasps in the genus *Vespula* (family Vespidae) are known as "yellowjackets". The genus is worldwide in distribution, and more than 15 species occur in the United States. Among the most frequently encountered species are the German yellowjacket, *V. germanica* (F.), and the common yellowjacket, *V. vulgaris* L., which occur throughout most of the United States. The eastern and western yellowjackets, *V. maculifrons* (Buysson) and *V. pensylvanica* (Saussure), occur east and west of the Great Plains, respectively. The southern yellowjacket, *V. squamosa* Drury, is common throughout the southern states (M. F. Potter 1995).

Importance

Yellowjackets are sometimes considered to be the most dangerous stinging insects in the United States, in part because of their widespread distribution and because some people are hypersensitive to the venom. People who develop hives or dizziness or have difficulty breathing may be going into anaphylactic shock and should seek medical assistance immediately.

While yellowjackets do not cause direct damage to turfgrass, some species build nests underground or in bushes or trees, and can be a nuisance to people who are using turf for recreational activities. Nests usually are constructed in well-drained sites with sparse or moderate plant growth, but these sites may be near heavy human traffic.

Description

Workers are black and yellow or black and white. The dominant color (black, white, or yellow) varies with the species, but each has distinctive bands or other patterns on the abdomen (Plate 64). Workers are usually about 13 mm long, while queens have similar markings but are larger.

Life History and Habits

Yellowjackets are social insects with a community structure that superficially resembles that of many ants. They form annual colonies in all parts of the United States except southern Florida and parts of California. Queens mate with reproductive males in the autumn and overwinter in sheltered locations (e.g., attics of homes, under tree bark). In the spring each queen emerges and constructs a small paper nest, which consists of several individual cells, in each of which she lays an egg. The queen feeds the emerging larvae, which develop into sterile adult females ("workers") about a month later.

The colony grows rapidly and may have as many as 4,000 individuals by the end of the summer. New males and queens are produced in late summer or early autumn and mate. The colony dies, except for the inseminated queens. Abandoned nests are not reused, although a new nest may be built at or near the same site (Baker 1982, M. F. Potter 1995).

Worker Activity

Workers are attracted to human food—picnic areas, fast-food restaurants, concession areas, trash cans, refuse piles, sanitary landfills. Careful attention to sanitation can reduce foraging activity.

22

Minor Insect Pests

Turfgrass insects and mites that are troublesome on a regular basis over a wide area are naturally considered major pests. A smaller number of pests confined to limited geographic area are nevertheless regularly troublesome where they occur and are also considered major pests. Major pests are discussed in Chapters 4–20.

Another group of insects and mites become turfgrass pests only on an occasional basis, or if they occur regularly, their activities produce only minor injury. These are designated as minor turfgrass pests. Several such pests are discussed here, listed in the same sequence of orders as the major pests.

Northern Mole Cricket

Importance, Host, and Damage

The northern mole cricket, *Neocurtilla hexadactyla* (Perty) (Orthoptera: Gryllotalpidae), is found occasionally on turfgrass from Georgia to New Jersey and as far west as Illinois. It is not an agricultural pest and typically occurs near wetland habitats. While the northern mole cricket does not feed on turfgrass roots directly, its burrowing activity can be a nuisance in finely maintained turf.

Description

The northern mole cricket superficially resembles the southern and tawny mole crickets discussed in Chapter 5 except it has six tibial dactyls (southern and tawny mole crickets have four). Females are slightly larger than males (Plate 53).

Life Cycle and Habits

Little is known about the life cycle of the northern mole cricket. The description that follows is taken from a South Carolina study (Semlitsch 1986), which found that young nymphs first emerge in June, overwinter as nymphs, and continue growing the following spring. During that second summer or autumn (14–16 months after emergence), individuals mature to adults.

While mating has not been observed, it is assumed to occur in autumn when adult males call intensively (see Chapter 5). Females produce fertile eggs the following spring. Eggs have not been found in the field, but females produce 30–35 eggs in small chambers in the soil in the laboratory. Oviposition apparently occurs in May or June. The proportion of adult males remains constant (30%) from October to March.

The northern mole cricket appears to take 2 years to complete development from Georgia to New Jersey, but is univoltine in parts of Florida (Semlitsch 1986).

Short-tailed Cricket

Importance, Host, and Damage

The short-tailed cricket, *Anurogryllus muticus* (De Geer) and other species (Orthoptera: Gryllidae), occurs along the Atlantic Coast from New Jersey to Florida and west to Texas and eastern Oklahoma. It feeds on grasses, weeds, pinecones, and pine seedlings. Burrows constructed by nymphs and adults produce unsightly mounds of small soil pellets that are most abundant after a rain, when the soil is moist and the weather warm. Damage caused by night feeding on grass blades is negligible.

Description

Adults are brown, with fully developed brown to black wings, and resemble field crickets except for the short ovipositor (Plate 53). The body length varies from 14 to 17 mm. The off-white glabrous, oblong eggs are 1.8–2.8 mm long and 0.6–1.2 mm wide. All nymphal instars are light brown.

Life Cycle and Habits

This cricket overwinters in the penultimate instar. Individuals become adults in early spring, mate, and oviposit during late spring or early summer in a multichambered burrow constructed by the adult. Both eggs and young nymphs occupy the burrow, but the fourth- to sixth-instar nymphs leave to construct their own burrows. At first the burrows are 2–3 mm in diameter and 5–10 cm deep. As the cricket matures, it enlarges burrows, which eventually are 30–50 cm deep. Only one mature cricket occupies a burrow, but females with their new brood of young can be found in a single burrow (Baker 1982, Weaver and Sommers 1969).

Grasshoppers

Importance, Host, and Damage

Grasshoppers (Orthoptera: Acrididae) have a worldwide distribution and occur throughout the United States but are most troublesome in the semiarid regions from Montana to Minnesota south into New Mexico and Texas. Grasshoppers feed on a wide range of forage legumes and grasses but do not feed on turfgrasses unless other forage is scarce or the weather is particularly dry.

Description

Grasshoppers are about 19–38 mm long when full grown. Their color varies from reddish brown to dull yellow to greenish, with other markings (Plate 53). Egg pods are oval-elongate and curved, with eggs that are white, yellow-green, tan, or various shades of brown and about the size of rice kernels. Nymphs resemble small wingless adults and are pale green when they first emerge after molting. Later they assume the colors and markings of adults.

Life History and Habits

Many species overwinter as eggs and hatch from April to June. Nymphs molt five or six times over 35–40 days to become adults. Oviposition begins about 2 weeks after mating. Egg pods placed in the soil may contain 15–150 eggs, and each female may produce 300 eggs. Some species swarm to feed and to deposit eggs on sunny, warm days. Generally one or two generations occur each year (Baker 1982).

Periodical Cicadas

Importance, Hosts, and Damage

Periodical cicadas, *Magicicada* spp. (Homoptera: Cicadidae), occurring in the eastern United States, are known best for their very long nymphal life under trees and because they emerge from the soil rather suddenly as full-grown nymphs. The numbers of exit holes may be quite striking, with as little as 5 cm separating each hole and marring the surface of the turf. Twigs and small branches are injured by the female as she slashes these tissues with her ovipositor when depositing eggs.

Description

The periodical cicada *Magicicada septendecim* (L.), also called the *17-year cicada*, is the best known of the periodical cicadas. It has reddish eyes and wing veins. Its body is 27–33 mm long but is about 40 mm long to the tips of the folded wings (Plate 53).

Life History and Habits

The presence of cicadas is obvious from the day-long shrill "singing" of males on summer days. Females lay eggs in the twigs of various trees and shrubs. Eggs hatch in about a month. Nymphs drop to the ground, burrow into the soil, and feed on roots of trees and shrubs. After spending 17 years as nymphs in the soil, the last nymphal instars emerge, climb some object, usually a tree trunk, and fasten themselves to the bark, from which point adults escape for a short existence of about a month (Borror et al. 1989, Shurtleff and Randell 1974). The emergence of adults is coordinated so that it occurs the same year in a given location.

Buffalograss Mealybugs

Importance, Host, and Damage

Two species of mealybugs have been collected from heavily damaged buffalograss stands in Nebraska. These species, *Tridiscus sporoboli* (Cockerell) and *Trionymus* sp., are in the family Pseudococcidae and can only be distinguished by clearing and staining specimens.

The mealybugs cause a general thinning and yellowing of foliage that resembles drought stress. In severe infestations, plants turn straw-brown and enter dormancy or die. Damage appears to be restricted to buffalograss. Mealybugs are found behind leaf axils enclosing pistillate spikelets or within leaf sheaths in the crown of the plant. Turf decline often becomes apparent 2–3 years after establishment of a new planting and is most severe in mid to late summer (Baxendale et al. 1994).

Description

Nymphs and adult females of both species are dark pink to purple-gray and covered with a white waxy substance. Early instars are 0.2 mm long, while adult females are almost 3 mm long. Pale pink eggs are deposited in a cottony mass within leaf sheaths. Adult males have one pair of wings and three pairs of red eyes (Plate 15) (Baxendale et al. 1994).

Life History and Habits

Little is known about the life cycle of buffalograss mealybugs. There is little correlation between mealybug population density and the extent of associated plant damage. This may suggest a tolerance of some buffalograss selections to mealybug feeding or reflect a difference in the length of time plots had been established before sampling (Baxendale et al. 1994).

Natural Enemies

Two parasitic wasps, *Pseudaphycus* sp. and *Rhopus nigroclavatus* (Ashmead), family Encyrtidae, have been documented parasitizing buffalograss mealybugs (Heng-Moss et al. 1998). Research demonstrated the effectiveness of *R. nigroclavatus* as a biological control agent for buffalograss mealybugs under greenhouse conditions. The release of this parasitoid in greenhouses when mealybug populations are high may be an effective technique for reducing pest populations and eliminating potential detrimental side effects associated with pesticides.

Leafhoppers

Importance, Hosts, and Damage

Many species of leafhoppers (Homoptera: Cicadellidae) are common throughout the eastern United States. All turfgrasses may be attacked. At least three of the species known to feed on turfgrasses are the painted leafhopper, *Endria inimica* (Say), on bluegrass; *Draecu-*

lacephala minerva Ball; and *Deltacephalus sonorus* Ball. Both adults and nymphs suck plant sap from leaves and stems, causing a bleaching or drying out of the grass, which eventually turns brown. Leafhoppers rarely damage turf in the northern and eastern states but can cause serious damage in the Prairie States and in some Gulf States. Newly seeded lawns can be killed outright by leafhoppers.

Description

Adults of various species average about 4 – 7 mm in length, have triangular heads, and have a general body color of yellow, green, or gray (Plate 53). Eggs are white, elongate, and 1 mm or less in length. Nymphs may be various colors but are usually lighter colored than adults. They are wingless at first but develop wing stubs as they molt and have the general wedge shape of adults except that they are smaller. Nymphs have a habit of moving sideways or backward when disturbed or may run to the other side of a leaf to avoid detection.

Life History and Habits

Leafhoppers may overwinter as eggs or as adults. Some species overwinter only in the Deep South and migrate annually into the northern latitudes to establish an infestation. In the spring after feeding and mating, females insert from less than a hundred to several hundred eggs singly beneath the epidermis of leaves and stems. Eggs hatch in about 10 days and develop through five instars to adults in about 12 – 30 days. From one to four generations may be produced, depending on the latitude and on the species. Populations often are very high by late summer (Baker 1982, Bowen et al. 1980, Shetlar 1995, Shetlar et al. 1983).

Leafbugs and Fleahoppers

Importance, Hosts, and Damage

One common species on California turf is the white-marked fleahopper, *Spanogonicus albofasciatus* (Reuter) (Heteroptera: Miridae). Adults and nymphs suck juices from leaves and stems of all grasses and dichondra. When infestations are heavy, growth is retarded, and grass may die in spots. A hand or foot rubbed lightly along the surface of the grass will induce fleahoppers to hop about if they are present.

Description

Adults of the above species are about 3 mm long and blackish or grayish, with white markings on the wings, which are folded flat over the back. They can be differentiated from flea beetles by their long, thin antennae. Although the tarnished plant bug, *Lygus lineolaris* (Palisot de Beauvois), is not a common turf-infesting species, it is included as a representative of this group.

Life History and Habits

Little is known about their life history. Several generations are thought to occur each season. Fleahoppers jump readily and fly short distances when they are disturbed (Bowen et al. 1980).

Bermudagrass Scale

Importance, Host, and Damage

Bermudagrass scale, *Odonaspis ruthae* (Homoptera: Diaspididae), sometimes called *Ruth's scale,* is a pest wherever bermudagrass is grown. It seldom causes serious damage but is most injurious when the grass is under stress. Heavily infested turf has a dry appearance, with severely retarded new growth. The turf becomes weakened, thin, and susceptible to other stresses (Plate 54). The scale thrives in heavily thatched turf and does its greatest damage in heavy shade.

Description

Adults are about 1.6 mm in length and are covered with a hard, white circular or clam-shaped covering. They are found on stems, clustered around the nodes, and occasionally on the leaves (Plate 54).

Life History and Habits

Females lay eggs under their waxy coverings. Upon hatching, the crawlers move about the grass plant and soon settle down, insert their piercing mouthparts into the grass, and become sessile (Plate 54). In molting they lose their antennae and legs, exude their waxy covering, and remain sessile for many months before laying eggs to repeat the life cycle (Converse 1982).

Turfgrass Scale

Importance, Host, and Damage

Turfgrass scale, *Lecanopsis formicarum* Newstead (Homoptera: Coccidae), was found during 1983 and 1984 on turfgrasses of home lawns, parks, and sod farms in the province of Ontario, Canada. It occurred on Kentucky bluegrass, red fescue, and creeping bentgrass but has not been seen on annual bluegrass. Sodded lawns 3–5 years old were the most commonly infested. In Poland, red fescue was the most commonly infested turfgrass. Nymphs occur on the tillers of host grasses, usually at the base of leaves and stems and on rhizomes but not on roots (Plate 54).

Description

The oblong adult females are yellowish with broad brown lateral stripes on each side of the median area to a uniformly orange-brown body (Plate 54). They are about 1.5 mm wide by 2.5 mm long. Females produce a cottony mass of silk containing eggs. The minute elongate-oval crawlers are reddish. Several nymphal instars are present.

Life History and Habits

In Ontario, mature nymphs are present throughout the year and are the only stage that overwinters. These become adults in May and June and produce the cottony masses of silk containing eggs. Crawlers are present in late June and July. At the peak of their activity, crawlers move to the tips of the grass blades, causing the turf to become reddish. This stage is dispersed by the wind. First-stage nymphs are present in July and August. Intermediate-stage nymphs are present from August into October (Boratynski et al. 1982; Sears and Maitland 1983, 1984).

Cottony Grass Scale

The cottony grass scale, *Eriopeltis festucae* (Fonscolombe) (Homoptera: Coccidae), has been found infesting and killing home lawns in New Jersey (Plate 54). Cottony masses are produced on grass stems. Very little is known about this scale. It was found in large numbers in Nova Scotia in 1889 and thereafter also in New Brunswick. It is said to occur in the Dakotas, Indiana, and Illinois on timothy and redtop, *Agrostis alba* L. (King 1901; L. M. Vasvary, Rutgers University, personal communication, 1985).

Lucerne Moth

Importance, Host, and Damage

The larvae of the lucerne moth, *Nomophilia noctuella* D. and S. (Lepidoptera: Pyralidae), normally feed on clover and similar legumes but will feed on grass leaves and stems. If damage appears, it is usually late in the summer, and control measures are seldom necessary (Bowen et al. 1980, Shurtleff and Randall 1974).

Description

Adults are mottled gray-brown with two pairs of indistinct dark spots on the forewing and a wingspan of 25–30 mm (Plate 55). Larvae are slender, about 25 mm long, and spotted like those of sod webworms but larger. When disturbed, they become very active (Bowen et al. 1980, Shurtleff and Randall 1974).

Life Cycle and Habits

Moths fly at night and lay eggs on clover, other legumes, dichondra, and turfgrasses. Larvae live in silken tubes near the base of the plant (Bowen et al. 1980, Shurtleff and Randell 1974).

Burrowing Sod Webworms

Importance, Host, and Damage

Burrowing sod webworms (Lepidoptera: Acrolophidae) are primarily tropical insects but occur from Pennsylvania and New Jersey south through North Carolina into Florida and westward through Texas and Nebraska and into Arizona. *Acrolophus plumifrontellus* (Clemens), *Acrolophus popaenellus* (Clemens), and *Acrolophus arcanellus* (Clemens) are the species most commonly encountered on turf. They infest the roots of most lawn grasses and other plants, severing the grass near the thatch line and feeding on leaves within its burrows. These burrows are vertical and deep, unlike the burrows near the surface made by the temperate-region webworms. The cigarette-paper sod webworm, *A. plumifrontellus*, has injured turf in Illinois (Plate 55). Larvae spin silken tubes about 0.5 cm in diameter and 3.8–5.0 cm long in which to pupate. Robins pull the tubes up to feed on the insects. The empty tubes superficially resemble cigarette butts.

Description

Adult
Adults of the burrowing sod webworm, *A. popeanellus*, with a wingspan of 25–38 mm, have predominantly yellowish- or grayish-brown forewings with irregular dark-brown to black spots (Plate 55). Hindwings are yellowish to bronzy brown. Males of this genus have large, hairy labial palpi recurved over the head (Shetlar 1995a).

Egg
Eggs of *A. popaenellus* are spherical and 0.5 mm in diameter. They are cream colored when first laid, but darken to gray-brown after a few hours. Each egg has 18–20 longitudinal ridges (Shetlar 1995a).

Larva
Larvae of most species have a velvety gray or brown body with a chestnut-brown head capsule. Under magnification, the head is seen to have several raised ridges. The crochets on the prolegs are in a pear-shaped row of large hooks surrounded by several smaller hooks. Late-instar larvae are up to 20–30 mm long and have no dark spots on the body (Plate 55) (Shetlar 1995a).

Life History and Habits

Moths may emerge as early as May but normally do so in June or July. Adults fly for a few hours after sunset. Females drop eggs to the ground rather than placing them on plant material. Eggs take 1–2 weeks to hatch. Larvae construct vertical burrows the size of a pencil and 15–60 cm deep into the soil. The tubular webs mixed with frass and soil extend from the lower leaf blades into the burrow to the point where larvae retreat when they are disturbed. Most species overwinter as larvae in burrows. Pupation occurs in late spring in the tunnel near the ground surface (Baker 1982, Banerjee 1967c, Shetlar 1995a).

Granulate Cutworm

Importance, Host, and Damage

The granulate cutworm, *Feltia subterranea* (Fabricius) (Lepidoptera: Noctuidae), is a widely distributed pest throughout most of the United States and feeds on a large number of plants, including bermudagrass and dichondra. It cuts off plants near the soil surface, climbs plants, and feeds on foliage or bores into fruits or vegetables touching the ground.

Description

Moths are various shades of black, brown, and gray, appearing inconspicuous (Plate 55). They have a wingspan of 35–40 mm. Their eggs are about 0.63 mm in diameter. Larvae develop through six or seven instars, with head capsule widths of about 3.3 mm in mature larvae.

Life History and Habits

Both adults and larvae are nocturnal. Peak activity occurs a few hours after sunset. The insects go into hiding before sunrise. Eggs are deposited singly or a few together on the upper surface of plants. Each female lays from 800 to more than 1,500 eggs. Pupation takes place in the soil after the larva burrows 5–15 cm deep and forms a cell. Light trap catches show the granulate cutworm to be active throughout the year in the Gulf States (Snow and Callahan 1968).

Striped Grassworm

Importance, Host, and Damage

The striped grassworm, *Mocis latipes* (Guenee) (Lepidoptera: Noctuidae), is an occasional pest of turfgrass but is a perennial major pest of pasturegrasses in Florida. Turfgrasses infested include bahiagrass, bermudagrass, and St. Augustinegrass. In heavy, unchecked infestations of 30–75 larvae per 0.1 m², the larvae eat the grass so closely that only bare stolons remain. Large populations that build up on pasturegrasses often migrate to adjacent parks and residential lawns.

Description

The moth is mottled gray to brown, with a dark-brown stripe near and parallel to the apical margin of the forewing. It has a wingspan of 30–45 mm (Plate 55). Mature larvae are about 0.5 cm in length, striped with brown and yellow that blend with the vegetation. There are six or seven larval instars. Their cocoons are spindle shaped, with leaf clippings incorporated on the surface.

Life History and Habits

The striped grassworm has several generations a year in south Florida and is present throughout the year, with peak adult populations in the fall. Females begin laying eggs on the third day after emergence and deposit more than 300 eggs each during the next several days.

Newly hatched larvae feed on the upper cell layers and start to eat the margins of leaves in the fourth instar. Larvae passing through six instars require a mean of 30.5 days, and those passing through seven require a mean of 32.4 days, to develop from eggs to adults at 27°C and about 80% RH. Pupation occurs after the larva has folded a grass blade about it. Pupae are enclosed in a silken cocoon covered with a whitish powder (Reinert 1975).

Wireworms

Importance, Hosts, and Damage

Wireworms are the larvae of click beetles (Coleoptera: Elateridae) and are strictly soil insects. Most wireworms that are active in agricultural crops feed on seeds, but some feed on roots of grasses and bore into fleshy underground stems. The gradual loss of turf roots causes patchy, withered spots and dead grass. The larval period may require 2–6 years for completion.

Description

Adults are click beetles and are generally recognized by their elongate body, usually parallel sided and rounded at both ends. The posterior corners of the pronotum come to sharp points or spines. The body is slightly compressed dorsoventrally. Most species are 12–30 mm in length and are light brown, dark brown, or black, depending on the species. The larvae are slender and are usually compressed dorsoventrally, shiny, and heavily chitinized. They are generally rust to brown and 1.3–3.8 cm long (Plate 56).

Life History and Habits

After larvae mature (2–6 years), they pupate during late summer, and adults overwinter. Females lay eggs in the soil during spring to early summer on roots of grasses and live for about 1 year.

Adults have a flexible union between the prothorax and mesothorax. When they are upside down on a hard surface, they can arch at this union and, with a sudden jerk and a clicking sound, throw their bodies into the air to turn over (Borror et al. 1989, Shetlar et al. 1983, Shurtleff and Randell 1974, Vance and App 1971).

Polyphylla spp. Grubs

Importance, Hosts, and Damage

Some 23 species of the genus *Polyphylla* (Coleoptera: Scarabaeidae) occur in the United States. *Polyphylla decimlineata* (Say) is probably the most widely distributed, but *P. variolosa* Hentz and *P. comes* Csy are considered the more injurious species. Members of the genus rather closely resemble the May or June beetles, *Phyllophaga* spp., in life history, habits, and economic importance. Grubs feed on the roots of a wide variety of plants including grasses, coniferous nursery plants, strawberries, corn, and so forth. Adults of at least one species feed on the needles of conifers.

Description

Polyphylla adults are somewhat larger than *Phyllophaga* adults, mostly striped with brown and white, and easily separated from the latter by the presence of six or seven segments in the lamellae of their antennae (Plate 56). *Phyllophaga* spp. have only three segments. *Polyphylla* larvae have transverse anal slits. The raster has two short, longitudinal parallel palidia, with fewer than 16 pali per palidium. In the *Phyllophaga* the anal slit is Y shaped, and the palidia are much longer, with generally more than 20 pali per palidium.

Life History and Habits

Like the *Phyllophaga*, *Polyphylla* spp. have life cycles of 3 or more years. *Polyphylla* larvae have a greater tendency to coil and remain coiled when disturbed. Their coiled bodies are more rigid than those of *Phyllophaga* grubs (Cazier 1940, Heit and Henry 1940, Ritcher 1966, Yeager 1949).

Vegetable Weevil

Importance, Hosts, and Damage

The vegetable weevil, *Listroderes costirostris obliquus* (Klug) (Coleoptera: Curculionidae), occurs in the Gulf Coast states and in California, where it damages dichondra lawns. It does not infest turfgrasses. It infests such vegetables as turnips and carrots. Both adults and larvae cut small holes in the leaves of dichondra, completely skeletonize leaves, or remove leaves completely, leaving only the base of stems. Damage is usually localized or spotty, because adults do not fly.

Description

There are no males. Females, grayish-brown to dull black "snout" beetles about 9.5 mm long, have very rough or punctate wing covers with sparse, short setae (Plate 56). Each elytrum has a pointed protuberance on top near the rear and a gray mark that when paired with the mark on the adjacent elytrum, forms a V. Larvae are greenish, legless grubs about 9.5 mm long when fully grown (Plate 56).

Life Cycle and Habits

Since males do not exist, reproduction occurs without mating. Eggs are laid on plants or in soil during late summer and fall. Larvae develop during the winter. By early spring they pupate and become adults. There is one generation a year. Both adults and larvae are slow and sluggish in their movements. Normally they feed at night and hide in the soil around plants during the day. Occasionally adults can be seen during the day hiding under plant foliage. They often crawl to nearby lights at night. In a close-cut dichondra lawn, larvae respond to a pyrethrin drench, which can be used for population survey (Bowen et al. 1980).

March Flies (Bibionids)

Importance, Host, and Damage

Several species of bibionid flies have larvae that inhabit soils in the root zone. The genera *Bibio* and *Dilophus* (Diptera, family Bibionidae) are most commonly encountered in turf. Bibionids occur throughout the world but are known to damage roots of various crops and grasses in North America, Europe, and Asia. Larvae normally feed on decaying organic matter but also are found feeding on roots of Kentucky bluegrass, annual bluegrass, bentgrass, and fescue lawns.

Damage often appears early in spring, shortly after snow cover has disappeared. Dead patches, 5–25 cm in diameter, sometimes resemble snow mold. When dead turf is removed, light-tan larvae can be seen in the soil (Shetlar 1995a).

Description

Adult
Adults are black or dark-red flies with brown or yellow legs. They are clumsy walkers. Wings are folded flat over the abdomen, and have a dark spot midway down the front margin. Antennae are short and beadlike (Shetlar 1995a).

Egg
Eggs are cylindrical with rounded ends, 0.5 mm long and 0.2 mm wide. They are laid in clusters of 200 or more and are cream colored before darkening to reddish brown (Shetlar 1995a).

Larva
Larvae are torpedo shaped, widest at the head and tapering to the posterior. Young larvae (1.5 mm long) are nearly white and covered with long bristles. Later instars are yellow-brown with a distinct black head capsule, and are 6–20 mm long depending on species. A row of short fleshy protuberances develops on each segment, replacing the bristles (Shetlar 1995a).

Pupa
Wing pads, legs, and antennae are readily apparent on the pupa, which is yellowish white and 8–15 mm long (Shetlar 1995a).

Life Cycle and Habits

Adults emerge in late March to May, sometimes in very large numbers. Adults are active on warm sunny days, and fly close to the ground for short distances. They often visit early spring flowers or fruit trees. After mating, a female digs a burrow in moist soil, showing a distinct preference for high organic matter content. Within a day, a female can produce a chamber 2–4 cm into the soil, where she lays 100–400 eggs. She then retreats to the surface and dies.

Eggs hatch in 3–6 weeks, depending on species and soil temperature. The legless larvae use the long bristles to aid their movement through soil. Larvae apparently do not feed directly on healthy turf roots, but do feed on decaying plant material or on plants recently damaged by other insects.

Larvae complete several molts, feeding actively in cool wet conditions and remaining dormant when soils are hot or dry. During the winter, larvae may congregate in areas damaged by white grubs or snow mold. In early spring (early March) mature larvae burrow 1–2 cm into the soil and pupate.

23

Turfgrass-Associated Arthropods and Near Relatives

Another group of arthropods causes no injury to turfgrass but is present in the turf grass ecosystem for a variety of reasons. Some are attracted to the usually damp soil surface with a protective cover that makes an ideal habitat for feeding and hiding. Others use turfgrass areas mainly as a resting place when they are not on the wing. Still others that are parasitic on warm-blooded animals are present in lawns because pets rest there. These turfgrass-associated arthropods and near relatives are listed in their normal phylogenetic sequence. We do not include in this section the beneficial parasitic and predatory insects and mites that are host specific on major turfgrass pests. Normally these beneficial insects are only present in significant numbers when their hosts are numerous, and are mentioned in the appropriate individual chapters. Also see Chapter 27.

Turfgrass-Associated Invertebrates

Earthworms

Earthworms (phylum Annelida: class Oligochaeta) are found in almost any moist soil, feeding mainly at night on the organic matter in or on the soil. They do not feed on living plants.

The soil eaten with organic matter is deposited on the soil surface as "castings" that can be detrimental on highly maintained turf. Castings on putting greens and fairways have become a problem in the 1990s, perhaps because lower mowing heights have made earthworm activity more apparent. Many insecticides and fungicides currently used on turfgrass do not have the detrimental effects that were so characteristic of the organochlorine insecticides and benzimidazole fungicides used in the 1970s and 1980s, perhaps enabling earthworm populations to recover. The castings of larger earthworms can make a smooth, level lawn very bumpy and irregular (Plate 65). This nuisance effect, balanced with their beneficial "tilling" activities, makes them a neutral soil turf organism except where turfgrass is very highly manicured (e.g., golf courses).

Earthworms typically have many similar-looking segments and range in length from 2.5 to 15.0 cm or more. A glandular organ, the clitellum, used during reproduction, appears as a slightly enlarged section from segments 31 or 32 through 37 (Plate 65).

Earthworms are hermaphroditic (having both sexes in each individual). Each is capable of functioning as a male and a female, but only one sex is functional at any given time. Self-fertilization does not take place. In mating, two individuals come together and acting as males, exchange sperm. Then the worms separate. The clitellum secretes a bandlike tube that is pushed forward. A worm's own eggs and sperm from the mating partner are discharged into this band. As the band is forced off the anterior end, it becomes a closed cocoon, or "egg", containing the fertilized egg, which develops within the hardened cocoon into a minute worm. Upon hatching, the young worm feeds on the contents of the cocoon before emerging as a tiny replica of its parents to start its independent life (Baker 1982, Hegner 1942).

Snails and Slugs

Snails and slugs (phylum Mollusca: class Gastropoda). These and earthworms are the only organisms discussed in this section that are not in the phylum Arthropoda. Snails and slugs are most active under moist conditions, hiding under vegetation and coming out mainly at night to feed. They occasionally eat dichondra. They never bother well-maintained turfgrasses, but weedy lawns may have large numbers of these slimy pests.

Snails differ from slugs in having a protective shell in which they can enclose their entire body, while slugs are completely naked (Plate 65). They both move on a slime trail with wavelike contractions of the foot (Bowen et al. 1980, Judge 1972).

Turfgrass-Associated Arthropods

Pillbugs and Sowbugs

Pillbugs and sowbugs (phylum Arthropoda: class Crustacea: order Isopoda) are common in damp places, feeding on decaying organic matter such as mulch, grass clippings, and manure. Some may attack roots and succulent stems. Most of the feeding occurs at night.

Adults are 5–10 mm in length, with a slightly flattened, segmented body. Pillbugs can roll into a ball when disturbed, but sowbugs remain flat. Both have seven pairs of legs, long antennae, and well-developed eyes. They are blackish, gray, or brownish (Plate 65).

Both require about a year to become full grown. One to three broods occur each year, with 25–200 young per brood. Eggs are carried by the female in a ventral brood pouch, where the young remain 1 or 2 months after hatching. Each individual may live 2 or 3 years (Baker 1982, Paris 1963).

Millipedes

Millipedes (class Diplopoda) usually live in damp places and feed on decaying vegetation, but some attack growing plants. Occasionally they are found in large numbers in mulch at the base of ornamental plantings that border turf sites. They do no significant damage to turf but become a nuisance when they leave this habitat and wander into buildings.

Several species may occur in turf. Some are dark brown to almost black and have a generally cylindrical or tubular body (Plate 65). Others are pinkish, orange, or cream colored with brown markings and have hard, somewhat flattened bodies. Most of these millipedes are 13–38 mm long and have a distinct head with short antennae. The anterior segments of many millipedes are modified for copulation and have only one pair of legs per segment. The rest of the segments have two pairs of legs each.

Millipedes overwinter as adults in protected areas and lay eggs during summer in nest-like cavities in the soil or in a dark, damp place (e.g., in mulch, in basements). Each egg, usually white, hatches into a six-legged millipede which adds more legs in subsequent molts. Many species give off an unpleasant-smelling fluid that in some cases is reported to be hydrogen cyanide. They do not bite humans. They walk slowly, their legs moving in a wave-like motion. When disturbed, they will curl up tightly (Baker 1982, Borror et al. 1989, Matheson 1951).

Centipedes

Centipedes (class Chilopoda) are terrestrial and predacious, feeding on insects and spiders. They are generally found in protected places in the soil, under bark, or in rotting logs. Unlike millipedes, they can move very quickly. All possess poison jaws for paralyzing their prey. Those in the northern parts of the United States and Canada are harmless to humans, while those in the South or Tropics can inflict a painful bite. Generally, centipedes more than 4 cm long can pierce human skin.

The centipede body is dorso-ventrally flattened, with a distinct head bearing a pair of long antennae, a pair of mandibles, and two pairs of maxillae. The first body segment behind the head has a pair of clawlike appendages functioning as poison jaws. All other segments except the last two bear a single pair of legs (Plate 61).

Centipedes overwinter as adults in protected places and lay eggs during summer. The sticky eggs, laid singly, become covered with soil. In some species this covering protects them from being eaten by the males (Baker 1982, Borror et al. 1989, Matheson 1951).

Spiders

The spiders (class Arachnida: order Araneae) are a large, distinct, and widespread group occurring in many habitats, including turfgrass areas. Spiders generally are considered beneficial as predators of pest insects and mites in the turfgrass system (see Chapter 27 for a discussion of spiders as natural enemies). Nearly all spiders have venom glands, but only a few are dangerous. Chief among these are the black widow spider, *Latrodectus mactans* (Fabricius), and the brown recluse spider, *Loxosceles reclusa* Gertsch and Mulaik.

The black widow spider, the most venomous, is common in the southern states. Females are coal black, with a red hour-glass figure on the ventral abdominal area. Appropriately named, the female often kills the male soon after mating. The brown recluse spider, second only to the black widow in its venomous nature, occurs east of the Rocky Mountains, mainly in Arkansas, Missouri, and Texas. Known also as the *fiddle back*, it is not hairy and has a violin-shaped mark on the cephalothorax.

Spiders have two body regions, the cephalothorax, bearing the eyes, mouthparts, and four pairs of legs, and the abdomen, bearing the genital structures, spiracles, anus, and spinnerets for weaving (Plate 61).

Usually spiders lay eggs in a silken sac that is attached to leaves, bark, or crevices or is carried by the female. Baby spiders undergo very little metamorphosis and look like miniature adults when hatched. They molt 4–12 times to become adults. All spiders are predacious, feeding mainly on insects, killing their prey by biting and injecting a poison into it. Their habit of spinning silk into webs of various shapes and sizes is one of their most characteristic features (Borror et al. 1989).

Several small spiders occur in turfgrass in surprisingly large numbers and prey on various soft-bodied insects in the thatch and upper part of the soil. Some of the more common families observed in turf are line weaving spiders (Linyphyiidae), wolf spiders (Lycosidae), and crab spiders (Thomisidae) (P. J. Vittum, personal observation). Such spiders generally are considered beneficial in the turfgrass ecosystem.

Mites

Many different kinds of mites inhabit the soil and thatch in the turfgrass environment. They superficially resemble spiders, in that they have eight legs and two main body regions. Oribatid mites are usually the most abundant group, with as many as 100 mites being recovered from a single 10.8-cm-diameter core (P. J. Vittum, personal observation). Oribatid mites are tiny, 0.2–0.5 mm in diameter, rounded or pear shaped, and usually brown or black. They are active decomposers, feeding on fungi and decaying organic matter and playing a key role in decomposition of thatch and plant litter (Potter 1995b).

Predatory mites also are active in the thatch and help reduce populations of plant-feeding mites and insects (e.g., winter grain mite, aphids).

Scorpions

Scorpions (class Arachnida: order Scorpiones) are well-known animals in the southern and western United States. They are found on the ground, whether barren or covered with low vegetation, and often in lawns. Scorpions will sting if disturbed. Their painful sting is followed by local swelling and discoloration but is generally not dangerous. Of some 40 species, only 1, *Centruroides sculpturatus* Ewing, is dangerously venomous. As far as is known, it occurs only in Arizona. In areas where scorpions occur, objects on the ground should be picked up with care. Scorpions on the body should be brushed off rather than swatted.

The scorpion body is divided into the prosoma, which bears the eyes, pedipalps, and four pairs of legs, and the broadly joined opisthosoma. The opisthosoma is differentiated into two sections, a broad, seven-segmented anterior mesosoma and a narrow, five-segmented posterior metasoma terminating in a sting. Most scorpions are small to medium and seldom exceed 2.5 cm in length. The largest are about 12.5 cm long.

Scorpions are largely nocturnal and carnivorous, feeding on insects and spiders that they catch with their pedipalps and sometimes sting. The young are born alive and for some time after birth are carried on the mother's back. When scorpions run, the pedipalps are held outstretched and forward and the metasoma is usually curved upward (Borror et al. 1989).

Ticks

Ticks (class Arachnida: order Acari) have two families in the United States, the Ixodidae, or hard ticks, and the Argasidae, or soft ticks. Both are present on turfgrass and other areas covered by vegetation. The most common species are the American dog tick, *Dermacentor variabilis* (Say), abundant in the eastern two-thirds of the country; the brown dog tick, of worldwide distribution; and the lone star tick, *Amblyomma americanum* (L.), which occurs from Texas and Oklahoma eastward to the Atlantic. Ticks attach themselves to their hosts to feed on blood, and some inject toxic saliva that produces paralysis. Tick-borne diseases transmitted in the feeding process include, among others, Lyme disease, spotted fever, relapsing fever, tularemia, and Texas cattle fever.

Adult ticks are brown to reddish brown and 3–7 mm long, according to species, and have four pairs of legs (Plate 64). Eggs are about 0.5 mm long by 0.4 mm wide, shiny, oval, and yellowish to pale yellow. Larvae, 0.5–1.0 mm and brownish, have three pairs of legs. Nymphs are 1.5–2.5 mm long and brownish, with four pairs of legs.

Ticks lay their eggs in various places but not on their hosts, which are warm-blooded animals. After hatching, immature ticks climb vegetation and attach themselves to their host as they pass by. Hard ticks take only one blood meal in each of their three instars and drop off their host to molt. Soft ticks feed several times before molting (Baker 1982, Borror et al. 1989).

The deer tick, *Ixodes scapularis* (formerly *I. dammini*), is considered a pest, particularly in the northeastern states, because it is a vector of Lyme disease. These ticks are about 2 mm long. Adult females are light to orange-brown with a dark spot near the head. The etiology of the disease, caused by *Borrelia burgdorferi*, a spirochete, is highly complicated. Incidence of the disease is rising, in part because the spread of suburban communities increases the probability that people will come in contact with the tick.

Chiggers

Chiggers (class Arachnida: order Acari) are a worldwide pest. Larvae of the chigger mite, *Trombicula irritans* (Riley), are often abundant in grass and waste lands. Humans walking into infested areas can find larval chiggers clinging to the clothing or bare skin, where they insert their mouthparts, if possible in a hair follicle, and suck blood. This bite causes severe irritation, itching, and sometimes intense pain. When immature mites have completed feeding, they drop to the ground to molt. The nymphs and adults are free-living predators and scavengers. No diseases are transmitted by chiggers in the United States.

Both adults and nymphs are usually bright red and about 1.25 mm long, with four pairs of legs. Eggs are minute and globular, changing from light to dark with age. Larvae are orange-yellow to light red and hairy, with three pairs of legs.

Larvae, the parasitic stage, attach themselves to the vertebrate host until fully engorged, then drop to the ground to molt first to the nymphal stage, then to the adult stage. An entire life cycle requires about 50 days, and there is usually one generation a year in North Carolina (Baker 1982, Matheson 1951).

Turfgrass-Associated Insects

Springtails

Springtails (order Collembola: families Entomobryidae, Isotomidae, and Sminthuridae) are tiny (1–4 mm), relatively primitive insects that often occur in very large numbers in thatch and soil. As many as 1,000 individuals have been recovered from a single 10.8-cm-diameter core (P. J. Vittum, personal observation). These insects derive their name from a hinged-forked structure on the ventral surface of the abdomen which, when "sprung" suddenly, propels the springtail into the air, away from a predator. Collembola feed on decaying organic matter, fungi, bacteria, and other microorganisms in the thatch and soil and appear to be important decomposers of plant litter in a variety of habitats.

Earwigs

Earwigs (order Dermaptera) occur throughout North America and are found wherever there is moisture and cover, so that turfgrass areas are a natural habitat. Golf course superintendents sometimes find earwigs when changing cups on putting greens. The European earwig, *Forficula auricularia* L., is most common. Its diet is variable, with incidental feeding on foliage, but it is more of a nuisance because of its presence. Other species are beneficial, feeding on other insects and on decaying organic matter.

The common species are brown to black, elongate insects, 19–25 mm long, with a pair of large forceps at the end of the abdomen (Plate 59). Those of the female are straight, but those of the males are noticeably curved. Adults may be winged or wingless. Forewings, if present, are short, thick, and veinless, covering a membranous hindwing. Earwigs rarely fly. White, ovate eggs about 1 mm long are deposited in clusters. Grayish nymphs resemble adults but are smaller, and forceps of both sexes are straight.

Few earwigs can overwinter successfully outdoors in the northern states. In warmer areas all stages can overwinter. Upon hatching, nymphs feed and develop through five instars, taking about 45 days to do so in summer or more than 150 days in winter. There is only one generation a year (Baker 1982, Borror et al. 1989).

Ground Beetles

Ground beetles (order Coleoptera: family Carabidae) have a worldwide distribution. There are more than 2,500 species in North America alone. They are most commonly seen in the grass, in the soil, and under bark or debris. Some are often seen running over golf greens during the summer. Some species may feed on seed or pollen, but most prey on other insects and are considered beneficial.

Adults are variable in shape, color, and size, ranging from 3 to 25 mm in length. Most are black or brown but often appear shiny or metallic. They are broad and have hard wing covers with many parallel longitudinal ridges (Plate 59). Their eggs are oval and cream colored. The larvae of ground beetles are elongate, 10–45 mm long, and slightly flattened and taper toward the rear. Their heads are large, with sickle-shaped mandibles directed forward.

Most ground beetles complete a generation within a year, overwintering as larvae or adults. Eggs are deposited singly in the soil. Larvae develop in or on the soil, generally hidden under stones, boards, or other debris. Adults are usually nocturnal. Both adults and larvae are predacious and are therefore beneficial (Baker 1982, Matheson 1951).

Rove Beetles

Rove beetles (order Coleoptera: family Staphylinidae) are fast-moving, elongated, dark-colored beetles with short elytra that leave much of the abdomen uncovered (Plate 59). They are distributed worldwide and range in size from 2 to 12 mm. Both the larvae and adults are predaceous on a range of insects and are considered beneficial.

Fleas

Fleas (order Siphonoptera) have a worldwide distribution and are commonly found in bedding or near areas where host animals sleep. During the warmer months lawns can become heavily infested with fleas when infested pets rest and sleep on the turf. Most commonly encountered are the cat flea, *Ctenocephalides felis* (Bouche), and the dog flea, *Ctenocephalides canis* (Curtis). Adult fleas hop onto humans and suck blood, causing irritation, itching, and sometimes pustules.

Adults are 1.0–2.5 mm long, brown, wingless, tough spiny insects compressed laterally, with legs modified for jumping (Plate 64). White oval eggs are about 0.5 mm long. Larvae are 4–6 mm long, slender, white, and legless. Pupae are dingy white, oval, and about 4 mm long in a silken cocoon.

Adult females require a blood meal to lay eggs, which are deposited on the host but later fall off and hatch. Outdoors in lawns or in pet quarters, the larvae feed on the excrement of adult fleas, rodents, domestic pets, and general decaying organic matter. Pupation also occurs on the ground or in bedding. A life cycle may be completed in as little as 2 or 3 weeks, several months, or as much as 2 years. Adults are capable of living for as long as 2 years without a blood meal (Baker 1982).

Mosquitoes and Biting Midges

Adult mosquitoes and midges (order Diptera) rest in turfgrass and adjoining shrubs during the day. They emerge in the evening to feed on blood and cause a nuisance with which everyone is familiar during the summer. Like fleas, females require a blood meal to produce eggs, and only the females bite.

Mosquitoes (family Culicidae), represented by more than 100 species in North America, have as their most common genera *Culex, Aedes,* and *Anopheles.* Adults are slender, long-legged flies 5–10 mm long, with an abundance of scales and hairs on the wings and appendages (Plate 64). The biting midges (family Ceratopogonidae), often called *no-see-ums* or *punkies,* are so small that they are often not seen, even when they are biting. Adults are less than 2.0 mm long. Most of the bloodsuckers attacking humans belong to two genera, *Culicoides* and *Leptoconops.*

Both groups of insects require water for reproduction. Mosquitoes deposit eggs on the surface of the water or on soil that will be submerged later. Both larvae and pupae are strictly aquatic. Biting midges are also aquatic, with larvae occurring in mud and ooze, among algae in ponds, and in decaying vegetation. Adults that are not bloodsuckers are predacious on other insects.

Mosquitoes and biting midges transmit serious human diseases, making them much more important than their bites alone would indicate. Most of these diseases, such as malaria, are restricted to tropical or subtropical zones (Borror et al. 1989, Matheson 1951).

Wild Bees

Wild bees (order Hymenoptera: family Andrenidae) do little, if any, damage to turf in their ground nesting activities. They select well-drained sites of moderate to sparse plant growth with little organic matter. Like yellowjackets (discussed in Chapter 21), they can inflict painful stings and are important because of this habit. They are small to medium-sized insects, 8–17 mm in length, in a wide range of colors of metallic red, black, blue, green, or copper.

Wild bees overwinter as adults in their burrows in the soil. They emerge in April and dig new burrows, provision them with pollen balls 3–5 mm in diameter, and deposit a single egg on each pollen ball. The larvae develop within the burrows, become adults, and remain there until the following spring (Baker 1982).

Effect of Insecticides on Nontarget Arthropods

Many of the insects and other arthropods discussed in this chapter are beneficial to the turf environment because they prey directly on pest species. In fact, even though pest insects often have enormous reproductive capabilities (e.g., greenbugs can reach outbreak densities in less than a week), low-maintenance turf usually is *not* overwhelmed by pest insects. This is in part a result of the activity of predatory arthropods, such as ants, ground beetles, rove beetles, tiger beetles, spiders, and mites. It appears that many turf insect pest populations are suppressed, at least to some extent, by natural enemies (Potter 1995b).

Unfortunately, many of the broad-spectrum insecticides used to manage pest populations also have a detrimental effect on some of these beneficial arthropods. Sometimes the pest population rebounds to levels even higher than those observed before an insecticide application. This phenomenon ("pest resurgence") usually is presumed to occur because natural enemies were destroyed and are no longer able to suppress pest populations through predation or parasitization. An example is the resurgence of the winter grain mite in New Jersey following applications of carbaryl (see Chapter 4).

24

Vertebrate Pests

Vertebrates become destructive to turfgrass in most cases simply because an attractive food supply exists in the turfgrass or soil below. Most often that food is insects or earthworms. The vertebrate pests are either birds or mammals. Information in this chapter on the biology and distribution of these pests is primarily from Peterson (1980) on avian pests and from Burt and Grossenheider (1964) and Timm (1983) on mammalian pests.

To control turf-damaging birds it is necessary merely to eliminate the attractive food supply. The control of mammalian pests, apart from elimination of the attractive food supply of turfgrass insects, is discussed briefly, since the various methods of control are long-standing and are not subject to frequent changes. Recommendations for the control of mammalian pests are derived primarily from Timm (1983).

Birds

Starlings

The European starling, *Sturnus vulgaris* L. (order Passeriformes: family Sturnidae), often simply referred to as *starling*, was introduced from Eurasia but now is distributed throughout the United States and southern Canada. It is a gregarious, garrulous, short-tailed blackbird, 19–22 cm long from the tip of its bill to the tip of the tail (Plate 66). In winter it has a heavily speckled body and a dark bill. In spring it becomes iridescent, with a long pointed yellow bill.

Starlings often nest in large trees or in small cavities in buildings. Thousands of birds may gather in "roosts", after the nesting season. Once established, these large congregations are difficult to drive away (Shetlar 1995a). Starlings feed on insects, seeds, fruits, and berries. Webworm larvae in the thatch and scarabaeid grubs in the soil are favorite foods. Once food has been found, flocks of starlings descend to feed by inserting their beaks into the thatch or soil to obtain the insects. They leave peck holes the diameter of a pencil, which can be interpreted as minor bird damage. More often than not, such holes serve as a warning signal that destructive turf insects may be present and that further examination is in order.

Grackles

The common grackle, *Quiscalus guiscula* (L.) (order Passeriformes: family Icteridae), ranges over the southern half of Canada and the United States east of the Rocky Mountains. It is a large, iridescent blackbird, 28–34 cm long with a long wedge-shaped or keel-shaped tail (Plate 66). The males have iridescent purple on their heads with deep bronze or dull purple on their backs. Females have little or no iridescence. Grackle food habits are similar to those of the starling, and the two are often seen together, feeding on grubs and other turf-inhabiting insects. Grackles are generally not as abundant as starlings but because of their larger size can disrupt a weakened lawn or athletic field while scratching for and feeding on insects.

Crows

The American crow, *Corvus brachyrhynchos* (order Passeriformes: family Corvidae), ranges from the southern half of Canada south throughout the eastern two-thirds of the United States to the arid Southwest and north from Baja California. Almost everyone is familiar with this large, chunky, ebony-colored bird with a strong bill and feet (Plate 66). In strong sunlight its plumage shows a purplish gloss. Its body length is 43–53 cm. Crows are often gregarious. Their diet includes almost anything edible, including insects such as grasshoppers, scarabaeid beetles, ground beetles, and caterpillars. Annual bluegrass weevil–infested fairways and tees have been further damaged by crows tearing and uprooting the weakened turf as they scratch in search of the weevil larvae.

Cattle Egret (White Ibis)

The cattle egret, *Bubulcus ibis*, is found in the southeastern United States, as well as Europe, Asia, and Africa. It is a relatively short-legged and short-necked heron. The cattle egret is almost entirely white throughout its development, and is 47–52 cm long, with a wingspan of 90–95 cm (Plate 66). The head has pale yellow plumelike feathers along the forehead to the nape. The bill is pale yellow or orange-brown. The legs and feet usually are black.

The cattle egret is much less shy but more silent than most North American herons. It is usually found in close association with cattle, often perched on the backs of cattle or other grazing animals, or walking alongside the animal. It feeds on insects (especially crickets, grasshoppers, and fly larvae) as well as ticks, frogs, and lizards. When the bird is searching for prey, the neck sways or wiggles in a characteristic manner, probably facilitating the sighting of insects and other prey. Cattle egrets scatter when feeding, but congregate when flying. They beat their wings rapidly, in a small arc, unlike most herons. They feed on moist or dry ground and avoid standing in water (Palmer 1962).

While cattle egrets are occasionally sighted in New England, few breed as far north as Maryland. Most breeding populations in the Western Hemisphere occur in Florida, Cuba, and the Caribbean islands.

Canada Goose

The Canada goose, *Branta canadensis*, is widespread throughout North America, particularly in open country with lakes and rivers. It is a gregarious bird, with as many as 12 distinct races, each differing slightly in size and appearance and often interbreeding with neighboring races. All populations have brown body plumage, black head and neck, and an extensive white facial patch (Plate 66). Adults are 55–110 cm long and weigh 2.3–5.2 kg.

Canada geese usually migrate from summer breeding grounds (throughout much of Canada and the northern United States) to warmer locations during the winter. They often return to the place of their birth each spring. Adults usually maintain lifelong pair bonds. Many populations breed near lakes and rivers in open country, with nests on dry ground but near open water. Geese swim readily, but spend most of their waking hours grazing, usually on grains and grasses. Large numbers of geese are seen flying in characteristic V formations, particularly at twilight.

About 40% of Canada geese are found in the Atlantic Flyway—from the tundra of northeastern Canada to Maryland and the Carolinas. Recently many geese have begun to overwinter in Maryland (which holds more than half the total population of the Atlantic Flyway), Delaware, and New Jersey. These areas contain large acreages of corn fields (and residues of grain), as well as large expanses of aquatic sanctuaries (Addy and Heyland 1968).

Geese can be a major nuisance in turf, particularly on golf courses and athletic fields. Their aggressive behavior, especially noticeable when nesting, can be disruptive or intimidating. The substantial (2–4 cm long, 0.7–1.2-cm diameter) fecal deposits are unsightly and can be numerous enough to disrupt play on putting greens or on athletic fields. The presence of geese on expanses of turfgrass can make scheduling of pesticide applications difficult, because geese are sensitive to several pesticides, particularly organophosphorous and carbamate insecticides, and often are reluctant to leave their territory.

Mammals

Moles

Moles (order Insectivora: family Talpidae) live most of their lives underground. The presence of one species or another can be detected by the low, snaking ridges pushed up in the turf as they move just under the surface or by surface mounds of 2–8 l of soil. Moles leave no trace of the entrance to their burrow (Plate 67). Deeper tunnels may be marked with cone-shaped earthen mounds. Turf may die along tunnel surfaces, especially in very dry conditions, as roots are disturbed or destroyed. Damage is most frequent and extensive during spring and fall during cool, moist periods, especially along ditches.

Moles consume from 70% to 100% of their own weight each day and need access to large amounts of food. The stomach contents show what is eaten. Of 100 moles examined, 67 had beetles in their stomach; 64, white grubs; 40, earthworms; 44, other beetle larvae; and 43, seedpods and husks. Other food items in descending order included centipedes, spiders, ants, and crickets (Henderson 1983). As soon as insects and earthworms disappear from a turfgrass area, disruption of turf generally ceases, and the moles move on to another food supply.

Moles have broad front feet, with palms usually facing outward. Their eyes are pinhead size. They have no external ears, and their body is covered with soft, thick fur that has a sheen and varies from golden, to brown, to slate, and to nearly black.

At least four species in the United States invade turfgrass areas, consuming insects and worms but in the process damaging lawns, golf courses, and other turf as they burrow through soils of moist, sandy loam. These are the star-nosed mole, *Condylura cristata,* with an unusual finger-like projection of 22 tentacles surrounding the end of the nose (Plate 67); the eastern mole, *Scalopus aquaticus* (Plate 67); the hairy-tailed mole, *Parascalops breweri;* and the California mole, *Scapanus latimanus.* Moles range in size from 11 to 16 cm in body length and from 2.5 to 9.0 cm in tail length.

The first three species occur east of the Rocky Mountains, and their ranges overlap such that one or more can be found in any part of this region. The California mole occupies much of California and southern Oregon. Certain differences exist in the habits of the three eastern species. Damage by the eastern mole is recognized by tunneling just 2.5–5.0 cm below the surface that creates a long, winding ridge. It prefers well-drained, loose soils in fields and lawns, but occasionally burrows along the edge of wooded areas. The star-nosed mole makes similar ridges during cool moist weather when the food supply is very near the surface, but ordinarily it tunnels at depths of 10–15 cm, raising numerous and frequent mounds rather than ridges. It is semiaquatic and is often found in wetlands or burrowing near streams or ponds (Shetlar 1995a). Meadow mice and shrews frequently invade and use mole tunnels.

Control of moles can be accomplished by direct killing, trapping, or fumigation or by insecticidal applications that eliminate the food supply and force the animals to seek another source of food. Moles burrow at night but continue until shortly after dawn. When the ridges are being pushed up, a sharp blow with the back of a shovel just behind the leading edge of the ridge can kill the eastern mole.

When trapping, baiting, or fumigating moles, it is essential to select active tunnels for placement. To confirm activity, flatten the ridges and mounds. Ridges that are pushed back up within 12–24 hr or mounds that are pushed back up in 24–48 hr indicate active habitats (Dudderar 1983). Trapping is the most effective and practical way of eliminating moles because it capitalizes on the mole's natural habits. Excavation of a mole tunnel is the first requisite for trapping. Capturing a mole live is possible with the use of a large jar and a board to exclude light. Mechanical traps are sprung by the animal's natural instinct to reopen obstructed passageways. Of the three types of traps available, the Victor trap is the simplest to set for moles making the surface ridges, because no portion of the trap need be placed in the tunnel. If a small section of an active tunnel is flattened and the trigger pan is set in contact with the firmed turf, the trap will be activated when the mole pushes up the ridge again (Henderson 1983).

Shrews

Shrews (order Insectivora: family Soricidae) are mouse size, with beadlike eyes not covered with skin. They can be further separated from mice and voles by the fact that they have no external ears. Although shrews are primarily insectivorous, they also eat roots, seeds, and other vegetable matter occasionally. They are not harmful to turfgrass. The short-tailed shrew, *Blarina brevicauda,* occupies the entire eastern half of the United States and southern Canada. Its saliva is poisonous. This shrew is often found occupying the tunnels of the star-nosed mole.

Voles

Throughout Canada and the United States, wherever there is good grass cover, one or more species of voles (order Rodentia: family Cricetidae) are likely to be found. Mice and voles belong to the same family and are similar in size and general appearance. Mice have large ears, large eyes, and long tails, however, while voles have small ears, small eyes, and short tails (Plate 67). Voles have mostly brownish-gray fur. They range in body length from 10 to 15 cm, with tails 3.0–7.5 cm long.

Voles are active both day and night. Since they are mainly vegetarians, insects need not be present before voles will damage turf. They make narrow runways 2.5–5.0 cm wide through the grass. During winter snow, their round openings to the surface reveal their presence (Plate 67). After the snow is gone, their trails, which are completely devoid of grass, make an unsightly lawn that will not fill in completely for several months (Plate 67). In addition to the turfgrass damage, they can severely harm trees by removing bark from roots.

Rodenticides, usually formulated as baits, have been the mainstay in vole control. Weeds, ground cover, and litter provide food and cover for moles, and so eliminating these around lawns and turfgrass areas reduces damage. Trapping, fumigation, and shooting generally are not effective methods of control (O'Brien 1983).

Chipmunks

Chipmunks spend most of their time on the ground but occasionally climb trees. They feed on plants, snails, and insects, often digging for scarabaeid grubs in the lawn. They disrupt soil much less than skunks but still leave excavations. The eastern chipmunk, *Tamias striatus* (order Rodentia: family Sciuridae), with a head and body 13–15 cm long and a tail 7.6–10.0 cm long, is squirrel-like and runs with its bushy tail straight up. It ranges east of the 100th meridian throughout southern Canada and most of eastern United States except the southern half of the Gulf Coast states, Florida, and eastern North and South Carolina. It feeds on scarabaeid grubs in the Northeast (H. Tashiro, personal observations).

Since chipmunks damage turfgrass in the course of seeking grubs or other insects as a source of food, elimination of the food supply generally eliminates the chipmunk problem.

Pocket Gophers

Pocket gophers (order Rodentia: family Geomyidae) do not seek turfgrass insects as a source of food but burrow in moist, friable soil to depths of 10–45 cm below the surface, making large fan-shaped surface mounds on turfgrass areas. A plug of soil placed in the entrance hole is plainly visible (Plate 68). Many golf courses with sandy loam soils have numerous unsightly mounds on fairways, roughs, and even on greens. Home lawns, parks, cemeteries, and other turfgrass areas are also affected.

Gophers are considered harmful wherever they occur. The surface mounds, in addition to being unsightly, interfere with normal turfgrass mowing operations. Gophers can destroy underground utility cables and plastic irrigation pipes in addition to girdling stems and feeding on roots, tubers, and surface vegetation.

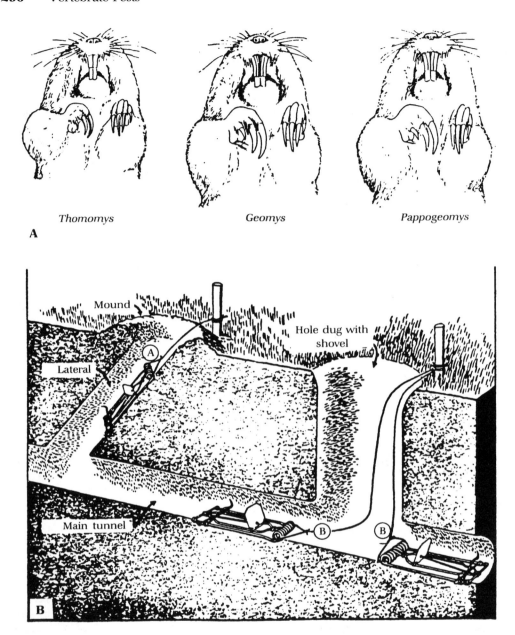

Figure 24-1. Pocket gopher genera and trapping techniques. **A.** Differentiation of three genera by upper-incisor characteristics and relative size of forefeet. **B.** Tunneling system and trap locations. (Part A courtesy of the Colorado Agricultural Experiment Station, Fort Collins; part B from leaflet attached to trap at purchase, no manufacturer's identification.)

There are at least 10 species of pocket gophers in North America, scattered mostly throughout the West. One of the most widespread in the West is the valley pocket gopher, *Thomomys bottae,* occupying most of California eastward through Utah and south into western Texas and Mexico. Another is the northern pocket gopher, *Thomomys talpoides,* which occupies an area east of the Sierra Nevada–Cascade Range eastward into North and South Dakota, into southern Canada, and southward into New Mexico. The plains pocket gopher, *Geomys bursarius,* occupies a large area from southern Texas into northern Minnesota and from Wyoming, Colorado, and New Mexico eastward into Indiana (Plate 68). One species, the southeastern pocket gopher, *Geomys pinetis,* is found in Florida, Georgia, and Alabama.

Members of this family are small to medium-sized rodents with a body length of 13–36 cm, varying in color from nearly white or mostly brown to nearly black. Their two most distinguishing features are their yellowish incisor teeth, always exposed in front even when the mouth is closed, and their external cheek pouches, which are fur lined, reversible, and open on either side of the mouth (Plate 68). Characteristics on the front of upper incisors and forefeet distinguish certain genera (Figure 24-1) (Case 1983).

Whether traps or toxicants are to be employed to control pocket gophers, it is first necessary to locate their tunnel system. The main burrow normally is found 30–45 cm from the plug on the fan-shaped mounds. Trapping is an effective means of controlling gophers in small areas and can eliminate the remaining animals following a poison control program over a large area. One trap is placed in a lateral tunnel, or two traps facing in opposite directions are placed in the main tunnel, to capture an animal coming from either direction (Figure 24-1). If rodenticides are used, they should be placed as baits in the gopher's tunnel system (Case 1983).

Skunks

Skunks (order Carnivora: family Mustelidae) are chiefly nocturnal and omnivorous, feeding on insects, mice, eggs, carrion, berries, roots, and other vegetation. In their search for insects, primarily scarabaeid grubs, skunks destroy turf, tearing up and uprooting the already weakened grass (Plate 69). Their foraging activities are most common or most visible during the fall and early spring, when larvae are feeding near the surface. They may leave golf ball–sized pits when digging out individual insects.

Four species of skunks occur in the United States. The striped skunk, *Mephitis mephitis* (Plate 69), ranges over the entire country, and the spotted skunk, *Spilogale putorius,* ranges over most of the country except the Northeast, the southern Great Lakes region, and Montana. The hooded skunk, *Mephitis macroura,* and the hog-nosed skunk, *Conepatus leuconotus,* range over the Southwest into Mexico. Skunks range in size from the spotted skunk, the smallest, 23–33 cm in body length and 11–23 cm in length of tail, to the hog-nosed skunk, the largest, with a 36–48-cm body length and tail of 18–30 cm.

Eliminating or greatly reducing populations of grubs, one of the skunk's favorite foods, often reduces its turf-damaging activities. Skunks' beneficial habits in eating other insects, rodents, and other pestiferous small animals outweigh their occasional destruction in damaging beehives, killing fowl, and eating eggs (Knight 1983).

Raccoons

The raccoon, *Procyon lotor* (order Carnivora: family Procyonidae), also called *coon* and *ringtail,* is chiefly nocturnal and omnivorous, feeding on fruits and nuts, insects, frogs, bird eggs, and whatever else it can obtain. When feeding on scarabaeid grubs during the fall or spring, it tears up and turns over large chunks of sod, so that the grubbing activities of skunks appear relatively insignificant by comparison (Plate 69).

Raccoons range over the entire United States except the Rocky Mountains and parts of the Great Basin area. They are 46–70 cm long, with a tail 20–30 cm long. The pepper-and-salt body, the black mask over the eyes, and the yellowish rings alternating with black on the tail identify this animal (Plate 69).

In addition to grub-infested turf, raccoons may destroy poultry, bird nests, and sweet corn or other vegetables; they also raid garbage cans. By eliminating or reducing grub populations in turf, turf damage can be limited (Boggess 1983). However, because their diet is so varied, once raccoons become familiar with a territory, they may revisit turf sites, even though insect populations have been reduced (P. J. Vittum, personal observation).

Armadillos

The nine-banded Texas armadillo, *Dasypus novemcinctus taxanus* Bailey (order Xenarthra: family Dasypodidae), a peculiar-looking animal, is chiefly a tropical mammal about the size of a house cat (Plate 69). It ranges from most of Florida across the Gulf Coast and northward from Mississippi into eastern Kansas and southward through most of Texas. The armadillo's body and tail and the top of the head are covered with heavy, leathery plates. The head is covered with small scaly plates and has a distinctly pointed snout. Wide scaly plates on the shoulders and pelvis are connected by nine smaller jointed plates. These joints enable the armadillo to roll into a compact and well-protected ball when disturbed (Shetlar 1995a). It is the only mammal in the United States with such protective armor.

The armadillo is mainly nocturnal during the summer but may be diurnal during the winter. It digs burrows that are usually 18–20 cm in diameter and as much as 4.5 m in length. The burrows are located in rock piles and around stumps, brush piles, and dense woodlands or occasionally along stream banks. After mating in the fall, females usually produce four offspring, which are mobile within hours after birth. Armadillos can run surprisingly rapidly, but often are dazed by headlights of vehicles.

More than 90% of the armadillo's diet is insects (adults and larvae), but the animals also feed on lizards, frogs, snakes, eggs of birds, berries, fruits, and roots. Armadillo feeding activity in turf normally is an indication that insects or earthworms are present, and further inspection is warranted. An armadillo pursuing turfgrass insects will tear and uproot the turf much as skunks and raccoons do but at somewhat deeper depths of 2.5–7.6 cm (Plate 69) (Hawthorne 1983).

Since the armadillo prefers to have burrows in areas with cover, the elimination of such habitats discourages it from becoming established. Removal of the insect food source from turfgrass areas largely eliminates its turf-destroying habits (Hawthorne 1983).

25

Principles of Integrated Pest Management

General Introduction

Integrated pest management (IPM) often is defined as a program that in the context of the environment and the population dynamics of pests, uses many different techniques and strategies in as compatible a manner as possible to maintain pest population levels below those causing economic injury (Dent 1991). The concept of IPM was initially developed in traditional agriculture, where the success of a crop was measured in economic yield (quantity and quality of produce). Pest activity could be measured against the eventual yield—would the loss of yield as a result of pest damage exceed the cost of controlling that pest? Numerous studies were conducted to determine how much yield would be reduced if a pest insect were present at various densities, and field trials provided good estimates of pest population reductions that occurred when pesticides were applied.

The key to such agricultural IPM programs has always been establishing consistent and reliable "economic thresholds"—pest populations at which the cost of expected crop damage exceeds the cost of implementing control. The IPM concept involves a shift from expectations of pest eradication to pest management—managing entire pest populations, not just localized ones (Dent 1991). Most IPM definitions also include a statement indicating that traditional pesticides should be used only when all other management strategies (e.g., renovation, resistant varieties, cultural control, or biological control) have been tried or proved inadequate.

In turf, the expected benefit from reducing a pest population usually is much more difficult to measure than in agriculture. It is difficult, if not impossible, to determine the economic value of suppressing pest insects. How many grubs per square meter will a golf membership tolerate? How many chinch bugs does a homeowner find objectionable on a home lawn? How much is the turf "worth"? The answers depend on many factors, not the least of which is individual expectations or preferences. As a result, "economic thresholds" in turf IPM usually are more accurately described as "tolerance levels", or "action thresholds". In a turf IPM program, the turf manager must determine what pest populations can be tolerated without incurring unacceptable damage. The economic part of the equation often is difficult to calculate, but the concepts of pest management are the same. Instead of "crisis intervention", an IPM-based program is a process that works with the surrounding environment and takes advantage of some of the natural conditions that help keep turf healthy (Schumann et al. 1997).

There are several keys to an IPM program. While the examples cited here primarily refer to insect pests, the concepts hold true for diseases, weeds, and other pests as well. The basic components of an IPM approach include site assessment, monitoring and predicting pest activity, setting thresholds, stress management, identifying and optimizing management options, and evaluating the results (Grant 1995). Much of the following is derived from Schumann et al. (1997).

Site Assessment

The first step in establishing an IPM program is to assess the current condition of the site. Often the easiest approach is to draw a map or sketch of the area, including surface features (slopes, shade patterns, location and kind of trees, location of other strategic plants, surface water). The site profile should also include the location of drainage lines and irrigation systems, as well as a sketch of the actual delivery pattern of irrigation. Soil texture and percolation rates, traffic patterns, and areas scheduled for renovation should also be noted. Finally the map should mark areas where insect (or other pest) activity or agronomic problems have occurred in recent years.

Mapping problem areas is critical, as it enables the turf manager to schedule monitoring activity to concentrate in the "trouble spots". Treating only the heavily infested areas reduces application costs and minimizes environmental disruption. In addition, insects often return to the same areas year after year. By mapping and identifying these areas, one may identify pest "reservoirs"—locations where insects overwinter before reinfesting the turf or areas where they congregate before moving into prime turfgrass sites.

The assessment should also include a realistic appraisal of the resources available—for example, technical support for identifying potential pests, the potential for improving irrigation or drainage equipment, or the support of the client (perhaps a homeowner, golf course owner, or a board of directors of a park).

Site assessment also includes developing *pest profiles* for the most commonly encountered pests. These profiles collect the critical information that is needed when determining action thresholds and management options. The profiles should include the common name and scientific name of the pest, typical symptoms (description of the appearance, what kinds of turf are affected, time of year usually encountered), favorable environmental conditions, biology (identification, life cycle, which stages are vulnerable to management), appropriate monitoring techniques, possible action thresholds, predictive models, cultural control practices that can put the pest at a disadvantage, biological control options, and chemical control options. A list or a photocopy of references is also very helpful.

Finally, a thorough evaluation of the current situation can prove helpful. Determine which pests have been the most difficult to manage in past years, identify areas that might be renovated with turfgrass cultivars or other plants that are resistant to insects or diseases, and note the areas where implementation of IPM strategies might make the most difference (where pest activity has been persistent but action thresholds are also relatively high). It is also helpful to identify areas where small-scale trials can be conducted to determine how well a new management technique will work in local conditions (a new formulation of a biological control agent, a new pesticide rate, or a tank mix).

Scouting or Monitoring

A key element of any IPM program is scouting or monitoring the turf regularly. Scouting is discussed in detail in Chapter 26, but a few points are stressed here. Inspecting the turf regularly enables a turf manager to monitor pest presence and development throughout the year, leading to well-founded decisions regarding the implementation of pest management strategies. Monitoring should also include an assessment of agronomic conditions, and helps identify some cultural stresses before they become detrimental to turf vigor—for example, localized dry spots, leaky irrigation heads, compacted soil, or dull mower blades. While many turf managers already scout the turf for which they are responsible, they do not always keep good records of the conditions they find. Good records are critical in the monitoring process, and greatly enhance the ability to make informed decisions in subsequent years (Potter 1998, Villani 1996, Vittum 1997e).

A turf scout should describe and identify damage as it occurs (e.g., symptoms, where the damage occurs, the time of year and recent weather patterns, soil texture and condition, air circulation).

Setting Thresholds

Establishing action thresholds, or tolerance levels, may be the most difficult aspect of an IPM program (Schumann et al. 1997). As mentioned previously, action thresholds in turf are more difficult to establish because the economic value of the turf is virtually impossible to determine. Thresholds are site specific and will vary widely within a given turf area. Several factors, including turf use, replacement value, safety issues (particularly on athletic fields), agronomic conditions, and cultural practices, must be taken into consideration when setting thresholds (Potter 1998).

A golf course may have several different thresholds for the same insect pest, depending on the mowing height, irrigation pattern, location (e.g., proximity to the clubhouse), time of year (and stage of development of the insect), tournament schedule, expectations of the membership, and the course budget (Schumann et al. 1997). Setting thresholds is discussed in more detail in Chapter 26.

Predicting Pest Activity

Part of setting thresholds is understanding the likelihood that an insect population will increase over time or knowing when certain stages of development will be present (Brandenburg 1995b). Predictive models have been developed for several insects, based primarily on air temperatures and the accumulation of degree-days. Each insect has a base temperature below which it does not develop physiologically and above which it does develop. The base temperature must be determined for each insect species, but many insects found in cool-season turfgrasses have base temperatures around 10°C. The warmer the temperature (above the base temperature), the faster the insect develops. Determination of degree-day values provides a measure of the rate at which an insect develops.

One simple method of calculating degree-days is to determine the daily average temperature (obtained by adding the minimum and the maximum temperature and dividing by 2), and subtract the base temperature appropriate for the insect. For example, if the high and low temperatures are 20° and 10°C, and the base temperature is 10°C, 5 DD would be accumulated that day (15° − 10° = 5°). The process is repeated daily and the number of degree days is added up to maintain a cumulative total. (If the average temperature for a given day is less than the base temperature, no degree-days are accumulated for that day, and none are subtracted from the cumulative total.)

Degree-day models use the information from several years of field observations and collections of temperature data to correlate degree-day accumulations and insect activity. For example, the peak in annual bluegrass weevil feeding activity may vary as much as 4 weeks on a calendar basis from one year to another, but occurs at about the same number of degree-days each year (Chapter 18). Degree-day models also have been developed for various stages of development for chinch bugs (Chapter 6), webworms (Chapter 8), masked chafers (Chapter 14), and billbugs (Chapter 18), among others. However, degree-day models usually are developed within a given climatic region and may not necessarily be accurate when used in a different climatic area. A turf manager should always check with a local turf specialist before relying on degree-day models to predict insect activity.

Plant phenology is another technique that can be used to predict insect development (Potter 1998). Plants, like other living organisms, develop more rapidly in warmer conditions. Flowering plants bloom earlier in warm springs than in cool seasons. In some cases the progression of blooming for certain key plants can provide indicators of pest activity. For example, black turfgrass ataenius adults begin to lay eggs when Vanhouette spirea or horse chestnuts bloom (Chapter 12), and annual bluegrass weevil adults become active in the spring between the times when forsythia and dogwood are in full bloom (Chapter 18).

In some cases, turf conditions can be used to predict the likelihood of insect infestation. For example, European chafers are more likely to infest home lawns that are less than 20 years old with low shade and relatively high amounts of Kentucky bluegrass (Villani 1994).

Stress Management

Much of an IPM program revolves around minimizing agronomic stress. In fact, Schumann et al. (1997) suggest that IPM should stand for "intelligent plant management", or commonsense turf management. Cultural strategies usually have two primary goals: to alter the turf microenvironment so it is less favorable for insect development and to provide optimum growing conditions for the turfgrass to make it less susceptible to attack by insects and other pests (Schumann et al. 1997).

Mowing

Mowing is a critical aspect of turf management, and often a client insists that mowing heights be lowered for aesthetic reasons. However, whenever turfgrass is maintained below the recommended height for an extended period of time, the shoot and root growth rates are slowed and plants do not recover as fast when insects attack. Turfgrass mowed consistently at putting green height (3–5 mm) has shallower and less fibrous roots than the same grass mowed at fairway (1.2–3.0 cm) or lawn (2.5–6.0 cm) height. In addition, mowing creates wounds that provide points of entry for pathogens. Mowing in the same pattern often leads to compaction, particularly along the collars of putting greens or at key turning points on a home property.

Many insects are affected by mowing heights. For example, black cutworms generally only damage creeping bentgrass mowed at putting green heights or fairway heights, even though eggs and small larvae are found in areas of higher heights (Chapter 9). The annual bluegrass weevil only damages annual bluegrass on greens, tees, and fairways. Even though larvae and adults can be found in the higher cuts typical of golf course roughs, damage is never observed in these areas (Chapter 18).

Thatch Management

Thatch, the accumulation of organic matter between the leaf tips and the crown, is a haven for many turf insects. Heavy thatch accumulation favors the development of several insects, including chinch bugs (Chapter 6), mealybugs and two-lined spittlebugs (Chapter 7), and masked chafers (Chapter 14). Regular thatch reduction strategies can put such insects at a disadvantage. Furthermore, the efficacy of many insecticides may be enhanced because they can penetrate farther into the thatch and achieve better contact with the target insect.

Water

Turfgrass needs water to grow, and turf managers often use irrigation to supplement natural rainfall. The amount of water and timing of application are important factors in reducing insect damage and minimizing water stress. Excess soil moisture leaches nutrients from the soil, increases the likelihood for disease development, and limits the amount of oxygen available to the root system. Water deficiency, often first noticed when plants begin to wilt, reduces photosynthesis and slows plant recovery after injury and allows soil insects to move deeper into the soil profile.

Many turfgrass insects are directly affected by moisture deficiencies. For example, bermudagrass mites thrive in warm, dry conditions, and regular irrigation can minimize damage caused by mite activity (Chapter 4). Hairy chinch bugs cause more damage in turfgrass growing in well-drained soils with moisture deficiencies (Chapter 6). Sod webworms often cause more injury when temperatures are high and soil moisture levels are low (Chapter 8). White grubs often move deeper in the soil profile when surface soil moisture declines (Chapters 11–16).

Fertilization Practices

All turfgrasses need a base level of nutrition to grow and thrive. Nitrogen affects the overall rate of photosynthesis, while potassium plays an important role in several physiological functions, including photosynthesis, respiration, carbohydrate metabolism, and water regulation. In addition, several nutrients must be present in minute amounts to allow optimal conditions for growth. When fertilizers are applied incorrectly (deficiencies, excesses, or improper balance), plant growth, stress tolerance, and recovery from injury are all reduced (Schumann et al. 1997).

While fertilizer products should be used according to the nutritional needs of the turf-grass, there are some secondary aspects of their use that should be mentioned. Southern chinch bug (Chapter 6) and two-lined spittlebug (Chapter 7) populations often increase after nitrogen has been applied to turfgrass. Fall armyworms are attracted to overfertilized grasses (Chapter 9). Insect damage sometimes appears more severe in turfgrasses that are underfertilized, and differentiation between nutrition deficiency and insect feeding can be difficult.

Soil Conditions

Soil pH

Soil pH (a measure of the acidity or alkalinity of a substance) affects the availability of many nutrients, particularly the micronutrients. In addition, each turfgrass has a range of soil pH within which it grows best (usually between 5.0 and 7.5, depending on the grass species). If soil pH is unusually high or low, turfgrass will be stressed and less able to withstand insect activity.

Soil Texture

Each soil type has a characteristic texture, as defined by the relative amounts of sand, silt, and clay particles in a given volume of soil. Sandy soils have relatively large particles and air spaces, and water normally percolates through these soils readily. Sandy soils are more likely to experience drought stress, and are more vulnerable to damage from insects such as chinch bugs (Chapter 6). Soils with higher proportions of clay and silt (smaller particles) retain moisture and nutrients more readily, but are also more vulnerable to excess moisture and the accompanying problems.

Some insects, including mole crickets (Chapter 5) and white grubs (Chapters 11–16), seem to thrive in "light" soils and do not do as well in soils with considerable clay content. In addition, the activity of some insecticides can be directly affected by soil properties. For example, as soil pH increases, the mortality of Japanese grubs exposed to carbaryl decreases. Similarly, as the percentage of organic matter increases from 3% to 10%, mortality of grubs exposed to bendiocarb decreases from 96% to 75% (Villani 1995).

Cation Exchange Capacity

The cation exchange capacity (CEC) is a measure of the number of "exchangeable" (i.e., available) cations the soil can absorb. Soils with high CECs are able to retain more nutrients than soils with low CECs, and often can recover from pest stress more readily.

Compaction

When a turfgrass area is subjected to heavy traffic over a repeating pattern, the underlying soil often is compacted. Soil particles are pushed closer together and soil pore spaces are compressed. As a result, oxygen supplies are restricted and drainage of water occurs more slowly. In such conditions, roots have more difficulty penetrating through the soil profile. Because turfgrass on compacted soils is stressed agronomically and growth is stunted, insect damage often is much more noticeable than on similar soils that are not compacted.

Selection of Well-Adapted Turfgrass Species and Cultivars

Each turfgrass species has certain climatic conditions to which it is best adapted, and other areas in which its survival is marginal (Chapter 1). Grasses are adapted to specific temperature ranges, soil moisture and rainfall patterns, soil salinity and pH, and soil textures. Some grasses are well adapted to direct sunlight, while others thrive in shady conditions. Finally, each turfgrass species has mowing heights at which it can thrive (and limits on mowing heights below which it will die), particular nutrient requirements, wear tolerance, and ability to recuperate following heavy use (Schumann et al. 1997).

A well-designed IPM program is centered on using turfgrass species that are best suited to local conditions. In addition, there are many specially bred cultivars that provide specific levels of resistance to insect activity. One example is cultivars of perennial ryegrasses and tall fescues that contain endophytes (fungi that grow inside the grass plant and produce toxins detrimental to certain insects). These endophytic cultivars provide resistance against chinch bugs (Chapter 6), greenbugs (Chapter 7), webworms (Chapter 8), fall armyworms (Chapter 9), and some billbugs (Chapter 18).

Several other examples of turfgrass cultivars have been developed with physiological resistance that does not involve endophytes, for example, in bermudagrass against bermudagrass mites (Chapter 4) and mole crickets (Chapter 5); St. Augustinegrass against southern chinch bugs (Chapter 6); and Kentucky bluegrass against hairy chinch bug (Chapter 6), sod webworms (Chapter 8), and bluegrass billbugs (Chapter 18).

Biological Control Strategies

Biological control refers to the use of living organisms or by-products of living organisms to suppress or manage insect (or other pest) populations (Schumann et al. 1997). Turfgrass insect pests live in a complex environment that includes numerous natural enemies and pathogens. An IPM program takes advantage of natural enemies that are already present or introduces biological control agents that might suppress the pest population.

Biological control strategies are discussed in detail in Chapter 27, but a few concepts are presented here briefly. Many turf insect pests (e.g., southern and tawny mole crickets, Japanese beetles, oriental beetles, European chafers, imported fire ants) are not native to North America, but were imported accidentally. When these "imports" arrived on the new continent, their natural enemies usually did not arrive with them, and the pests were able to expand rapidly because they encountered little resistance. *Classical biological control* occurs when researchers return to the land of origin of an organism, identify some of the natural enemies there, and release those natural enemies in the newly infested sites. While there are several controversial issues in such an approach (Chapter 27), there are a few examples of such releases that have been directed toward turf insects in North America, including *Ormia depleta* (a tachinid fly) from South America to suppress mole cricket populations (Chapters 5 and 27) and *Istocheta aldrichii,* another tachinid fly, from Japan to suppress Japanese beetles (Chapters 16 and 27).

There are several naturally occurring predatory insects and parasitoids (insects that parasitize other insects) (Chapter 27). Unfortunately these insects are sensitive to many of the traditional insecticides so when a chemical insecticide is applied to control a pest population, many of the natural enemies are killed at that same time. An IPM program attempts to conserve natural enemies that are already present by avoiding insecticide applications when those beneficial insects are active or by avoiding the use of broad-spectrum insecticides whenever possible.

Finally, several insect pathogens have been identified and are available commercially. These pathogens can be considered "biopesticides" because they are applied in a manner similar to traditional insecticides—through typical application equipment, timed to coincide with the presence of susceptible stages. Although most of these biopesticides target certain kinds of insects and are less likely than traditional insecticides to be detrimental to the beneficial insects that abound, some can be harmful.

An integrated insect management program attempts to identify the biological control alternatives that are viable and reliable, and incorporates as many of those techniques into the management scheme as possible (Schumann et al. 1997).

Chemical Control Strategies

The backbone of most turf insect management programs continues to be traditional insecticides, particularly when cultural practices and biological control efforts cannot maintain pest populations at tolerable levels (Villani 1996). However, a turf manager committed to IPM must consider several characteristics of an insecticide before deciding which ones might be appropriate. Insecticides are discussed in more detail in Chapter 28, but a few concepts are mentioned here.

The environmental profile of a pesticide is very important. Some materials are more likely to leach or run off, thereby increasing the probability that they will end up in surface water or groundwater. Some insecticides are more sensitive to degradation by microbes, alkaline water, or ultraviolet rays. Some insecticides are highly toxic to birds, fish, or other vertebrates, and should be avoided unless conditions can ensure minimal potential for unintended exposures (Balogh and Anderson 1992, Brown and Hock 1989).

Each insecticide has specific chemical characteristics that dictate how quickly it becomes effective against the target insects and how long it will remain active in the turf environment (Schumann et al. 1997, Vittum 1997f). Meanwhile each insect has a period during its development when it is most susceptible to insecticides. This "window of opportunity" may be a few days or several weeks long, depending on the insect. If an insecticide is to be applied late in the "window", the insect probably has already emerged as a damaging stage and the selected material should work fairly quickly on the target insect. However, if an insecticide is applied early in the "window", it is likely that not all the insects will have emerged, and a slower-acting but longer-lasting material would be a preferable choice.

Most IPM programs, predicated on scouting and subsequent implementation of control strategies only if tolerance levels have been (or will be) exceeded, promote the use of curative insecticide applications. These fast-acting insecticides can be used after an insect has reached a damaging stage and yet will reduce populations significantly. However, several relatively long-lasting insecticides are also available and many turf managers prefer to use these materials preventively. Such a pattern normally does not fit in an IPM program because these insecticides must be applied well before an insect population approaches damaging numbers, and sometimes are applied unnecessarily.

Finding Compatible Strategies

An IPM program tries to incorporate as many cultural and biological control strategies as possible before resorting to the use of traditional insecticides (Schumann et al. 1997). Several cultural practices are compatible with other strategies, particularly the use of well-adapted species and insect-resistant cultivars. Most other cultural practices simply involve common sense.

Some of the biological control alternatives that are available are not compatible with traditional insecticides. For example, entomogenous nematodes can be applied to suppress populations of mole crickets (Chapter 5), cutworms (Chapter 9), billbugs (Chapter 18), and some other insects. However, the nematodes are sensitive to several of the traditional insecticides that are also used on turf. Therefore, applications of the latter should not be made within a couple weeks of a nematode application. Other insecticides are directly compatible with nematodes and can even be used in a tank mix.

The development of compatible strategies takes time, experience, and some experimenting. Turf managers should begin to conduct their own small trials, using biological control agents and traditional insecticides in a range of combinations. Careful observation and assessment of insect activity will demonstrate which combinations can be used successfully and which ones must be avoided.

Evaluation

One commonly overlooked aspect of an integrated insect management program is evaluation. This process should occur every time an effort is made to manage an insect population (Schumann et al. 1997). The first step is to assess the current pest population; that is, how many insects were present per 0.1 m^2 before the control strategy was implemented? Make notes about the procedure. If the approach involved a biopesticide or an insecticide, what was the active ingredient? At what rate was it applied? Was it watered in? What were the weather conditions at the time? Following an appropriate period of time (often a week or two), assess the pest population again. Has the number of pests decreased? If so, was the reduction as great as expected (compared to the initial population)? If not, what steps could be taken next time to ensure better results?

Such a procedure provides invaluable information that can be incorporated into subsequent decisions, either the same growing season or the next year. Without assessing the initial population, a turf manager cannot be certain whether a control effort had any effect.

Advantages

Turf managers who practice integrated insect management find the approach provides several advantages. While the cost of insect management may increase in some situations (e.g., when regular monitoring documents the presence of a previously unrecognized pest), in many cases the use of insecticides is reduced substantially. Regular monitoring often enables a turf manager to treat only infested parts of the turf area, and leave other areas untreated where insect populations do not warrant control. An understanding of pest biology and interactions with weather conditions often results in delaying insecticide applications until weather conditions are conducive to insect development. Sometimes this delay allows for elimination of some insecticide applications. In any case, an IPM program puts a mechanism in place that serves as a cost-benefit analysis, and helps a turf manager determine the most economical approaches to pest management (Schumann et al. 1997).

The use of pesticides has come under increasing scrutiny, particularly in turf and ornamental settings. Overuse of pesticides in the past has led to the development of resistance in some pests, for example, southern chinch bug resistance to chlorpyrifos (Chapter 6), or white grub resistance to chlordane (Chapter 28). Sometimes an application of an insecticide destroys natural enemies and enables other insects to develop as pests. Many pesticides used in turf have detrimental effects on a variety of nontarget organisms in the turf microenvironment (Chapter 28). IPM minimizes disruption of the environment by implementing a management program that is based on the needs of the turf ecosystem. Although there are exceptions, IPM programs often lead to a reduction in the total amount of pesticides used during a growing season (Schumann et al. 1997).

Many people do not understand the actual environmental impact of pesticides applied to turf. Healthy turfgrass often serves as a sponge that absorbs pesticides and excess water, in the form of either rainfall or irrigation. This reduces the rate of lateral runoff or leaching through the soil profile (Brown and Hock 1989). Carefully documented IPM programs can demonstrate to the general public the vast array of research-based studies that are considered when making pest management decisions. Emphasis on minimizing disruption to the environment is a concept the general public can understand.

26

Sampling Techniques and Setting Thresholds

T he cornerstone of any integrated insect management regimen is accurate identification of pest insects (including assessments of population density) and development of appropriate action thresholds (Hellman 1994, Schumann et al. 1997, Vittum 1997e). While a true IPM program includes management of insects, diseases, weeds, and other pests, the following discussion is limited to turf insect pests.

Accurate Diagnosis

Many insects that are found in turf settings are innocuous or beneficial to the turf or to other aspects of the environment. For example, springtails (Collembola) are tiny insects that are found in huge numbers in the thatch, sometimes as many as 1,000 insects in one 10.8-cm-diameter plug. They are saprophytes and play a key role in decomposing organic matter in the thatch and upper root zone (Vittum 1997d). Other insects that are found commonly in turf are active predators of turfgrass pests. For example, ground beetles (family Carabidae) are very mobile and feed on eggs and small larvae of turf pests such as white grubs and cutworms (Cockfield and Potter 1984, Potter 1998). Many such arthropods associated with turfgrass are discussed in more detail in Chapter 23.

While many turf insect pests tend to occur in the same general locations from one year to the next, infestations may expand into new areas. Large numbers of insects can move in with storm fronts or high-pressure systems. For example, black cutworms typically overwinter as moths in the Carolinas and mid-Atlantic states and are carried into the northeastern United States and southern Canada on storm fronts each spring. Sometimes insects change their behavior and begin to feed on host plants that previously were not thought to be susceptible.

Some turf insect pests look very similar to other innocuous insects and may be misidentified. In some cases, inaccurate diagnosis leads to attempts to control an insect population that is in fact beneficial (e.g., ground beetles). In some areas of the country, mites damage turf, but because of their small size, they may be mistaken for aphids or greenbugs or other tiny insects. However, because mites are different from insects in significant ways, the insecticides that are most familiar to turf managers often are not very effective against mites

(Vittum 1997a). Thus it is imperative that a turf manager obtain an accurate diagnosis of any insect or mite activity that is suspected to be causing damage to the turf.

Scouting or Monitoring

Early Detection and Diagnosis

Insects and mites that damage turfgrass sometimes are so small that their invasion or the early development of potentially damaging populations can be difficult to detect (Schumann et al. 1997). Early symptoms of damage are frequently evident before the actual pest is observed. Once the earliest symptoms of damage appear, the progression of deterioriating turf can be very rapid, simply because of the sheer numbers of insects present. Therefore any adverse appearance of the turf should be investigated immediately to determine the cause.

The best way to find the cause of turf deterioration is to examine the leaves and stems closely, manually separating the thatch to search for insects in this layer. Insects often move actively and willingly, and can be seen in this region of the turf. In addition, the turf roots must be inspected to determine the presence (or absence) of insects and mites, and to look for evidence of feeding damage on the roots. To accomplish this, remove a section of sod and soil to a depth of at least 10 cm, break it apart, and inspect the soil closely. Insects are much more apparent in moist, friable soil than in dry soil, where dust particles adhere to the hairs of the body and make insects more difficult to see.

Relating Symptoms to Cause

The first symptom of damage caused by leaf- and stem-feeding insects usually is yellowing of the leaves in small isolated patches. When strictly root-feeding insects are involved, the earliest symptom may be a gradual thinning of the turf resembling drought stress.

Turfgrass deterioration caused by insects, pathogenic fungi, improper soil pH, inappropriate fertility practices, or something else may have symptoms that are similar in appearance. Positive correlation of the symptoms with a pest requires confirmation that the pest is present in a turf-injuring stage and at sufficient densities to cause the observed damage. Turf that is already dead may no longer be harboring the pest responsible for the damage, since the insects may have moved on to a fresher food supply, entered the soil to pupate, or dispersed as adults. However, insects often leave evidence characteristic of their presence, for example, the dried sawdust-like frass of billbugs at the crowns of plants or the dried fecal pellets of webworms, cutworms, or armyworms.

Seriously damaged turfgrass may eventually be heavily infested with saprophytic insects attracted to this habitat by decaying vegetation. In addition, many insects and mites that are found in turfgrass are beneficial organisms, often preying on pest species. Thus it is critical to identify turf-damaging pests accurately.

Collecting and Labeling Specimens for Identification

The vast numbers of insect and mite species present in turf make proper identification difficult, even for those familiar with turfgrass arthropods. A magnifying lens (10×–20× is ideal) is often needed to aid in insect identification in the field. Examination of the smaller

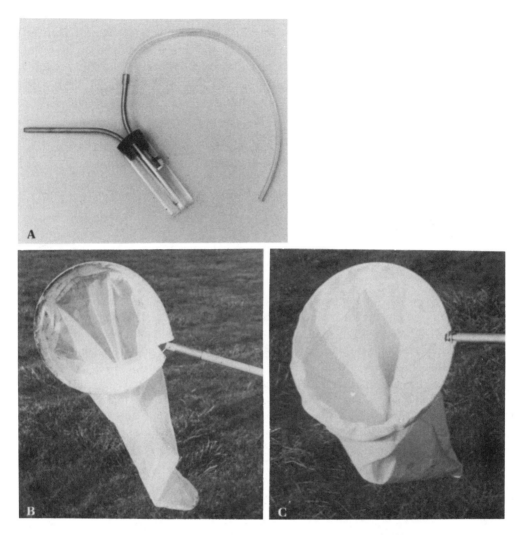

Figure 26-1. A. Aspirator used to collect insects. **B.** Aerial net for capturing insects in flight. **C.** Sweeping-beating net for capturing insects on vegetation. (Photos by H. Tashiro, NYSAES.)

details of an insect, such as the arrangement of spines or the presence or absence of certain markings, is usually necessary. If a specimen cannot be identified readily, help can be obtained from a specialist. It is necessary to submit good, representative specimens, properly preserved so the body and all appendages are intact. Both adult and immature forms should be submitted if possible.

The most reliable way to confirm an insect identification is to collect several specimens and deliver them to an entomologist who specializes in identifying turf insects. Small insects that move rapidly in the thatch or on foliage can be captured with an aspirator (Figure 26-1). An aerial insect net (i.e., a "butterfly net") is a very practical tool to capture flying insects. Other kinds of turfgrass insects (e.g., mole crickets, white grubs, billbugs) are captured most efficiently directly by hand.

With the exception of butterflies and moths, most insects can be placed directly in ethyl or isopropyl alcohol (30–70%). "Rubbing alcohol" is a serviceable and convenient preservative for a few days. Some soft-bodied insects (e.g., white grubs, caterpillars) will maintain their shape and color longer if they are first blanched. This is accomplished by bringing water to a boil, removing it from the heat, and dropping the insects in the water for 30–90 sec before storing them in alcohol.

Each specimen that is submitted for identification should include, as a minimum, the following information:

Name, address, and telephone number of the collector
Location where specimen was collected (including a description of the habitat)
Host plant (cultivar and species, mowing height, typical cultural practices)
Thorough description of damage (when first noticed)
List of pesticides used in the area in previous 6–12 months
Photographs of damage, which can be very helpful

Population Survey Techniques

Various methods have evolved by which the populations of insects present in the turf environment can be determined relatively rapidly and efficiently. These methods have become standard techniques for reporting insect densities. Hellman (1994) differentiated between passive sampling techniques (use of traps) and active visual inspection techniques (actual quantification of insect populations).

Furthermore, turf insects can be classified by the part of the turf they attack. Shetlar (1995a) described three zones: foliar and stem, stem and thatch, and thatch and root zones (Chapter 3). Some insects are limited to the *foliar and stem zone* and feed on the upper leaves and stems of turfgrass plants. Many of these insects hide in the thatch when not foraging, while others may hide beneath leaf sheaths or between nodes. Other insects are limited to the *stem and thatch zone,* which is composed of stems, stolons, and other organic matter. Many of the insects that are active in this zone (e.g., chinch bugs, mealybugs, billbugs) escape detection until significant damage has occurred. In settings where there is a history of activity of these insects, the turf should be monitored regularly using soap flushes or water flotation. Finally, some insects dwell primarily in the soil and are active in the *root and lower thatch zone.* The insects that inhabit this zone can be some of the most damaging insects because their feeding destroys roots, crowns, stolons, and rhizomes, and because their feeding often goes undetected until the insects have reached late instars.

An effective monitoring program must be accurate and affordable (McCarty and Elliott 1994, Vittum 1997e). Efficiency of monitoring can be enhanced by concentrating on *key pests* (those responsible for major turf losses in a given setting) and *key locations* (areas that the insects habitually select and damage from year to year) (Hellman 1994). Examples of key locations abound: Billbugs tend to lay eggs adjacent to sidewalks and driveways and the first symptoms of damage from larval feeding occur in these areas (Chapter 18); some white grub species survive better in well-watered soil that drains well but may also occur in low-lying areas in unusually dry years (Chapters 12–16); chinch bugs prefer turf that is predominantly fine fescue or ryegrass in sunny exposed sites (Chapter 6); mole cricket activity in the summer often is worst where adults were active in the spring (Chapter 5); and annual bluegrass weevils prefer annual bluegrass maintained at low mowing heights on golf courses (Chapter 18). Much of the following description of sampling techniques is derived from Hellman (1994) and Schumann et al. (1997).

Visual Inspection Techniques (Active Sampling)

Visual inspection is the quickest way to diagnose turf insect problems and to measure pest populations. The characterization of damage can provide important information. Are there many discrete patches, all similar in size and color? Do some patches appear to coalesce? Are individual plants affected and does the pattern appear to be more generalized? Where the damage occurs can provide additional critical information, such things as species or cultivars affected, mowing height, slope, underlying soil type, soil moisture and condition, and slope.

Spot Sampling

Spot sampling may be the simplest form of monitoring insect activity, and involves inspecting a small area of turf for insect activity. For example, a 30-sec spot count per 0.1 m² in 20 or more locations would provide information on the scope of damage, stages of pest present, and population density for a given home lawn. However, environmental conditions may affect insect activity and therefore, the number of insects observed, so comparing spot counts can be misleading. For example, most insects are markedly more active on warm, sunny days than during cloudy or rainy conditions. With practice, most turf insects can be detected with spot sampling but visual inspections only provide information about insect activity during the conditions under which the area was scouted. Generally, samples taken under extremes of temperature or moisture underestimate the actual population (Hellman 1994).

Food Plants

Populations of some scarabaeid species (e.g., green June beetle, May or June beetles, and Japanese beetle) can be estimated by observing the density of adults on their favorite host plants. Once the early beetles of the season are observed, Japanese beetle populations build up rapidly and beetles congregate on their favorite food plants during warm, sunny days to feed and mate. They feed on many plants, but one of the most favored is grape leaves. Other favorites in a landscaped area include ornamental roses, cherries, peaches, and plums. Some of the most favored trees include mountain ash, linden, and sassafras. When most of the leaves on these trees turn brown as a result of the skeletonization of leaves, adult populations may be high enough for many eggs to have been deposited in nearby turf. While a direct correlation between adult feeding and subsequent larval activity does not always exist, larval damage often can be expected later in the year when substantial beetle feeding is observed (Chapter 16).

May or June beetles feed on the young tender leaves of oaks as these trees are starting to leaf out in the spring. Other favored trees include persimmon, hickory, walnut, elm, and birch. Defoliation of these trees by large adult populations indicates potential turf damage for three seasons, with the most severe damage being incurred during the second year (Chapter 15).

Green June beetle adults feed most heavily on ripening thin-skinned fruits, such as figs, peaches, and grapes. Heavy feeding damage sometimes indicates larval populations will be sufficient to cause significant turf damage (Chapter 13).

Mating Flights

European chafer beetles leave the ground shortly after sunset on warm sunny days and congregate around silhouetted trees for mating. During nights of peak flight, so many beetles may congregate around a single tree that their beating wings sound like a swarm of honeybees. Peak flight activity occurs within half an hour of sunset. Flights of this intensity usually result in such massive oviposition in adjacent and nearby turf areas that damage is observed later in the fall and in the following spring (Chapter 15).

Preoviposition Adults

Twice during the year, billbug adults are often observed on sidewalks and driveways adjacent to home lawns or industrial parks and appear to be wandering aimlessly. The heaviest populations of wandering adults are seen on warm, sunny days of late September and October as adults search for overwintering quarters. On warm sunny days in May and June in cool-season turfgrass, bluegrass billbugs sometimes are seen wandering about, usually in smaller numbers than in the fall and apparently in search of grass for feeding and oviposition. When adults are so numerous that at least one is observed every minute, turfgrass damage usually can be expected in adjacent lawns (Chapter 18).

Black turfgrass ataenius adults are often present on golf greens during early spring when overwintering adults begin to fly or walk in from overwintering sites and again during midsummer as the first-generation larvae develop into adults. Adults also may be seen on the surface of swimming pools and around lights (Chapter 12). Similarly, annual bluegrass weevil adults often are seen in large numbers on golf greens and fairways during early summer, after first-generation larvae have pupated and emerged as young adults. Large numbers often are collected in the baskets of the mowers.

Surveys of Hibernation Quarters

Annual bluegrass weevil adults start moving from golf course fairways into roughs during the fall to seek hibernation quarters. The most favored quarters are the needle litter under white pine trees (Plate 50) and tall tufts of bunch-type grasses near the margins of woods (P. J. Vittum, personal observation). The density of weevil populations can be determined by collecting white pine litter (or grass clumps) and surface soil and completely submerging the sample in lukewarm water. Weevils will start to surface in a few minutes and will have nearly finished doing so within an hour. Hibernating populations of more than 100 adults per 0.1 m² often cause extensive damage to adjacent annual bluegrass fairways, greens, or tees the following spring and summer (Chapter 18).

Flotation Sampling

A flotation (or flooding) sample is used primarily to determine the presence or absence of chinch bugs in a turfgrass area (Hellman 1994, Vittum 1997e). Select a large can with a diameter of at least 15 cm, from which both ends have been removed. Where extensive studies on chinch bugs have been undertaken, cylinders made of heavy sheet metal or heavier material were specifically designed for this purpose. One end (preferably with a sharp edge) is forced through the thatch and into the surface soil at least 5 cm. The cylinder is then filled with water, and more water is added as the level recedes (Plate 70) (Mailloux and Streu 1979, Niemczyk 1981). Chinch bug adults and nymphs in the thatch soon float to the surface, and within 5–10 min, the entire population present within the cylinder will have surfaced. The insects are removed and counted as they surface. Both adults and nymphs are detected easily with this method, since they will be moving constantly on the surface of the water (Plate 70). Big-eyed bugs also float to the surface and can be counted.

In areas where thick or dense thatch makes it difficult to insert a metal can or cylinder, a 10.8-cm-diameter core or a 10- to 15-cm square of turf, cut to a depth of at least 2.5 cm, can be removed from the turf and placed in a bucket. When the bucket is filled with water, chinch bugs and big-eyed bugs will float to the surface.

A further modification of this technique makes it possible to count eggs as well. Water pressure is used to loosen the thatch forcibly and to wash it, along with the dislodged insects, out of the cylinder (or bucket) into a 100-mesh sieve. All material is collected in 40% ethyl alcohol and then centrifuged to separate the insects, including eggs, from the thatch. Further centrifugation in a sugar solution results in recovery of 95–100% of the chinch bug eggs, in addition to adults, nymphs, and other arthropods present in the cylinder (Chapter 6) (Mailloux and Streu 1979).

Although the flotation method is very accurate, it is also time-consuming. The method is probably best suited to confirm chinch bug presence in areas where the insect is suspected, and where populations must be measured relatively accurately.

Irritant Sampling

An irritating drench (or soap flush) is most effective when sampling relatively mobile insects that are active in the thatch and leaf zones, but is also highly effective at forcing mole crickets to the surface (Schumann et al. 1997). It involves spreading water that contains an irritant over an area of turf. Contact with the irritant forces thatch-inhabiting insects, primarily larvae of moths and butterflies, to move to the surface. Other insects that can also be forced to the surface include bluegrass billbug and annual bluegrass weevil adults, black turfgrass ataenius adults, mole crickets, and chinch bugs. Accuracy of the technique varies with thatch thickness, soil temperature, soil moisture, and depth of the insect at the time of sampling (Hellman 1994).

Several flushing agents (or "disclosing solutions") can be used in turfgrass IPM scouting programs (Potter 1998). Perhaps the most common is 15–30 ml of lemon-scented liquid dish detergent diluted in 4–8 l of water, subsequently applied to an area 0.3–0.6 m on a side. Another flushing agent is pyrethrin, diluted to 0.002% in water. One or two drops of permethrin or other synthetic pyrethroid diluted in a gallon of water is also effective. Higher concentrations of synthetic pyrethroids may kill younger larvae before they reach the surface, resulting in an underestimation of the population density (Hellman 1994).

A more thorough study of the role of the various components of this technique was made using the grass webworm infesting 'Sunturf' bermudagrass, *Cynodon magennisii* Hurcombe, and 'Seashore' paspalum, *Paspalum vaginatum* Sw. (Tashiro et al. 1983). Five concentrations of liquid detergent (0.063–1.000%) and five of pyrethrins (0.0004–0.0060%) were tested. The 0.25% detergent (10 ml/4 l water) and 0.0015% pyrethrin (1 ml/4 l of water) applied to 0.19 m² of turf within a metal frame were most efficient at forcing grass webworms to the surface. The study also found that a dilution rate comparable to applying 0.44 l per 0.1 m² was not sufficient to make all the larvae surface, but that a greater volume of water (1.75 l/0.1 m²) was needed. The detergent forced larvae to the surface more quickly than did pyrethrin. Continuous observation was essential for accurate larval counts because significant numbers of larvae reentered the thatch within 5–10 min of their emergence (Tashiro et al. 1983).

The irritant method requires a thorough soaking of the thatch layer and the soil surface. Furthermore, the soaked area must be watched very closely for 5–10 min because insects normally emerge from the thatch during that period and may also move out of the sample area quickly. Tashiro et al. (1983) suggested that a sample area no larger than 0.4 m² should be monitored at a time to obtain accurate counts.

Rating System

Cobb and Mack (1989) devised a rating system to monitor mole cricket activity (Chapter 5). A frame with 1.3–2.5-cm-diameter PVC pipe, 0.6–1.0 m on a side, is constructed. The frame is subdivided into nine equal squares by stringing line from one side of the frame to the opposite side, two lines in each direction (Plate 70). The frame is placed on the area to be monitored and each small square is inspected visually and manually for evidence of mole cricket activity. If any activity is noted within a small square, it is assigned a score of 1, while if no activity is noted, the square receives a score of 0. The "damage rating" is the total of all the scores of the nine squares.

Although this system does not provide absolute insect counts, it does provide a method of assessing mole cricket (or cutworm) damage and comparing damage from different sites or at different times of year. In addition, there is a direct correlation between damage ratings and number of mole crickets recovered with an irritating drench (Cobb and Mack 1989).

Soil Sampling

The techniques required to sample the soil underlying turfgrass areas are the most arduous and the most disruptive to the turf. The population of soil insects such as white grubs usually is distributed unevenly, so soil samples often must be as large as 0.1 m². Because some turf-damaging insects remain strictly in the soil, disruption of the soil is necessary to obtain accurate counts. When samples are taken with a spade, the depth of sampling is often variable because of the rooting habits, soil moisture, and soil texture. Most often sampling is much deeper than necessary in order to ensure that specimens of all the insects present have been obtained. The least disruptive method of examining samples of sod and soil is to cut three sides of a square to a depth of 7.5–10.0 cm and turn back the cut area to expose the soil (Plate 71). This procedure allows many of the plants to keep their root system intact. Samples that involve 0.1 m² (0.3 m on a side) sometimes heal very slowly, leaving dead or dying patches of turf. Smaller samples (e.g., 15 cm on a side) often provide as accurate an assessment of the grub population but recover much more quickly.

The quickest and least destructive method of collecting a soil sample is to use a cup cutter, 10.8 cm in diameter, to collect samples to a depth of 5–10 cm (Plate 71). Samples can be taken and inspected very quickly, enabling a scout to check several different locations and to provide more accurate information about the spatial distribution of white grubs. The technique is also useful for black turfgrass ataenius and annual bluegrass weevil larvae. To convert insect counts to individuals per 0.1 m², multiply each sample by 10.15.

Figure 26-2. Jacobson sod cutter with a cut 30.5 cm wide, used for spring and fall sampling of sod at predetermined depths for scarabaeid grub surveys. (Photo by H. Tashiro, NYSAES.)

Another method to collect soil samples is to use a sod cutter, which cuts a strip of sod 30 cm wide (Figure 26-2). When grubs are collected from a large number of turfgrass plots, one pass of the sod cutter is made through a complete row of plots, but only a small portion of the cut strip in each plot is lifted for examination. The depth of cut can be adjusted to penetrate only deep enough to include most of the larvae present. During the most active feeding period of scarabaeid grubs in the spring and fall, most of the population is in the upper 5 cm of soil. However, during other times of the year or when one is sampling grubs that move up and down in the soil rapidly in response to soil moisture or temperature (e.g., European chafer, green June beetle, and oriental beetle), some grubs will be missed if only the top 5 cm is sampled. When a sod cutter is used, the portion not lifted for larval counts reroots immediately, showing little or no evidence of disturbance (H. Tashiro, personal observation). Furthermore, soil is not examined to an excessive depth, so rerooting in sample sections occurs readily in most situations.

Thatch and Foliar Sampling

Suction or vacuum devices can be effective for general sampling of insects on foliage. Such devices are not used commonly for sampling insects from turfgrass but have the potential for large-scale sampling of insects in thatch and foliage. A similar unit is used to collect large numbers of Argentine stem weevils (closely related to annual bluegrass weevils) in ryegrass pastures in New Zealand (P. J. Vittum, personal observation).

The Berlese funnel is an effective apparatus that uses heat to drive insects out of thatch and foliage samples (Figure 26-3). Samples taken from the field are placed on a 0.6-cm mesh screen near the top of a large funnel. A 25-watt lamp is mounted in a cover reflector placed on the top of the funnel. Insects move away from the heat and fall through the funnel into a receptacle placed beneath it. Within 24–48 hr, depending on the moisture content of the sample and the insect species, the receptacle (filled with 50–70% ethyl alcohol) will con-

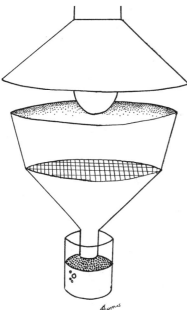

Figure 26-3. Schematic diagram of Berlese funnel, which uses heat to drive live specimens out of foliage, thatch, and soil samples into collection jar. (Drawn by S. Thomas, University of Massachusetts, Amherst, Mass.)

tain insects that were driven from the turf sample (Vittum 1997e). Examination and identification can be undertaken at any later time, since the alcohol will preserve the insects. This technique is particularly useful for sampling populations of billbug adults, chinch bugs, webworms, cutworms, and mites (Niemczyk 1981).

Passive Sampling Techniques

Light traps, mechanical traps, and pheromone traps are most effective when used in large-scale sampling programs (e.g., for a golf course, municipality, or large park). Such techniques usually will not provide accurate information about population density, but are valuable in monitoring the presence of a pest, the date of arrival or date of emergence of adults, and seasonal population fluctuations. They can help determine when the frequency of active sampling techniques should be increased (Hellman 1994, Potter 1998).

Blacklight Traps

Blacklight traps, generally fitted with a 15-watt blacklight lamp with a peak emission of 3,650 Å, are standard equipment for sampling many night-flying insects (Plate 71) (Hellman 1994, Potter 1998). Insects attracted to the radiation strike the metal baffles that are at right angles to the lamp and are then deflected into the funnel and into a receptable below the light source. The latter can be charged with a volatile insecticide to kill the insects quickly and thus preserve their natural appearance and prevent escape. When an AC circuit is employed, the location of the trap is somewhat restricted, but the lamp is generally operated continuously, since it consumes very little power. If the unit is powered with a storage battery employing a photoelectric switch, it becomes much more expensive, but its placement is much more flexible. During sampling for certain insects, it is a disadvantage to trap large numbers of extraneous insects, since much time must then be spent sorting out the captured insects (Tashiro et al. 1967).

None of the turf-damaging stages are caught in a blacklight trap, but the adults of several damaging species are attracted to the traps. All of the turfgrass-infesting moths, including webworms, cutworms, and armyworms, are attracted, as are several scarabaeid beetle species, including the black turfgrass ataenius, the northern and southern masked chafers, the Asiatic garden beetle, the European chafer, and May or June beetles. Annual bluegrass weevil adults are also collected from traps, particularly in the spring (Chapter 18).

While blacklight trap catches do not estimate damage potential, the technique is helpful in monitoring the first occurrence of pests or delineating the species distribution over a large geographic area. When adults of a pest species are captured in significant numbers, a turf manager then should initiate appropriate active sampling methods. Trap catches also help determine the relative abundance of species and the risk of damage from one year to another (Hellman 1994).

Pheromone or Floral Lure Traps

Many insects produce pheromones, or chemicals that enable individuals within a species to exchange information (Chapter 27). Often adult females produce pheromones that indicate to males of the species that the female is ready to mate. Perhaps the most widely used trap in turf IPM programs is the Japanese beetle trap, which uses a combination of a female sex pheromone (which attracts males) and a floral scent (which attracts both sexes) (Potter 1998). A study in Kentucky (Gordon and Potter 1985) indicated that beetle activity and damage to ornamental plantings increased when a trap was placed within 10 m of indicator plants. Pheromone traps normally cannot reduce insect populations but are invaluable as a monitoring tool, detecting population buildup over time, date of first emergence, and date of peak activity. Daily trap catches are strongly influenced by temperature, rain, location (proximity to host plants), soil type, groundwater levels, surface water, and presence of natural enemies (Hellman 1994).

Pheromone traps are available for several species of white grubs, including masked chafers and oriental beetles (Zhang et al. 1994), and may be on the market within a few years (Chapters 14 and 16). Lures for black cutworms, true armyworms, and fall armyworms are currently available and used in agricultural settings.

Pitfall Traps

The conventional type of pitfall trap can be made with three plastic cups, with the upper interior cup modified to act as a funnel to direct captured specimens into the collection cup below (Figure 26-4) (Hellman 1994). Alcohol or ethylene glycol is placed in the collection cup to kill and preserve captured insects. The lip of the exterior cup is placed at or just below the soil-thatch level. Arthropods crawling through the turf fall into the outer cup, through the funnel, and into the collection cup, where they are captured. These traps can be used to monitor the activity of many turf pests (e.g., chinch bugs, adult billbugs) (Niemczyk 1981), as well as other arthropods in the turf, such as springtails, predatory mites, spiders, and ants (P. J. Vittum, personal observation).

A linear pitfall trap, designed primarily for collecting mole cricket nymphs, captures the insects at all stages plus many other arthropods (Figure 26-5) (Hellman 1994). Major components of the trap are a PVC pipe 2.5 m long with a slot 2.5 cm wide cut to run nearly its entire length. A 20-l plastic pail, a 3.8-l plastic jug, a screen, and an end cap complete the assembly. The PVC pipe is placed in a trench with the open slot up, at the soil level or slightly below, and the pail is embedded in the ground. About 1 cm of soil is placed inside the

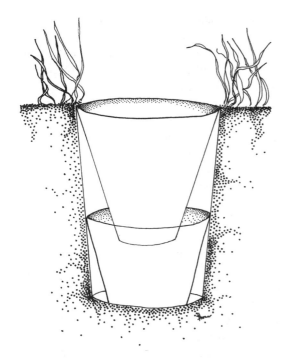

Figure 26-4. Conventional pitfall trap. (Adapted from Niemczyk 1981. Drawn by S. Thomas, University of Massachusetts, Amherst, Mass.)

PVC pipe and the jug to prevent cannibalism among captured mole crickets. Insects that fall into the pipe eventually move to the open end and fall into the jug. These traps work equally well on sod and bare ground and catch large numbers of mole crickets at all stages as well as other turf-inhabiting insects (Lawrence 1982).

Another modification of the pitfall trap is used to monitor mole crickets. A wading pool 1.5 m in diameter, partially filled with water, and an electric calling device is suspended over the pool. Males of each mole cricket species produce species-specific calls, which are tape-recorded for later use. When the calls are played back, attracted females fly to the synthetic call and collect on the surface of the wading pool. Daily counts during the flight period determine first occurrence and peak activity, and allow comparison with data from previous years (Walker 1982).

Setting Thresholds

IPM can be defined as using several different kinds of management strategies to maintain a pest population below a "threshold". In agriculture, the threshold is based primarily on economics: Will the anticipated gain in yield (either quantitative or qualitative) justify the cost of the management strategy? In turf, however, the expected benefit from reducing a pest population usually is measured in the aesthetic quality of the turf (i.e., an improvement in the appearance of that turf). It is difficult, if not impossible, to determine the economic value of such an improvement, so thresholds in turf IPM usually are more accurately considered to be "tolerance levels", or "action thresholds". In a turf IPM program, the turf manager must determine what pest populations can be tolerated without incurring unacceptable damage.

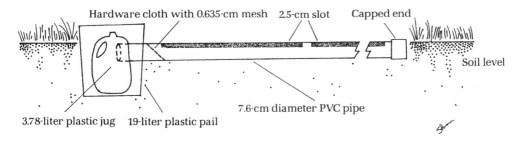

Figure 26-5. Linear pitfall trap. (Adapted from Lawrence 1982. Drawn by S. Thomas, University of Massachusetts, Amherst, Mass.)

The tolerance level, or action threshold, for turfgrass insects is site specific and depends on many factors, such as pest species or complex, turfgrass species and cultivar, turf use, turf vigor, time of year, expectations, availability of curative control options, and budgets (Potter 1998, Schumann et al. 1997). Turf managers often are less concerned with insect infestations that are found in autumn on cool-season grasses, because most insects are noticeably less active at cooler temperatures and the turf is more able to recover from any damage that occurs.

Pest Species or Pest Complex

In the case of white grubs, several species damage turf (Chapters 12–16). Some species are inherently more damaging than others so fewer grubs per 0.1 m² will cause visible damage. For example, many species of May or June beetle larvae are much larger than other scarabaeid species and cause significantly more damage per grub (Chapter 15). Therefore, the tolerance level for May or June beetle larvae usually is lower than that for other species of white grubs. Masked chafers feed more on decaying organic matter and less on turf roots than do other white grub species (Chapter 14), so the tolerance level for masked chafers often is higher than that for similar-sized species. The European chafer feeds longer in autumn and returns to the root zone earlier in the spring than do other univoltine species of grubs (Chapter 15). In addition compared to many other species, it is less susceptible to most insecticides, so the tolerance level usually is lower than that for similar-sized species.

While southern and tawny mole crickets cause severe damage on bermudagrass and other warm-season grasses, their feeding behavior differs (Chapter 5). Tawny mole crickets feed on turf roots and other vegetation as well as small insects, while southern mole crickets feed primarily on other arthropods. Thus an individual tawny mole cricket often will cause more damage than a southern mole cricket that is similar in size, and the tolerance level for tawny mole crickets will be lower.

Some insect pests can be damaging in turfgrass but only at relatively high densities. For example, if turf receives adequate moisture, a chinch bug population must exceed 80–100 chinch bugs per 0.1 m² to cause visible damage, but damage may be apparent at lower densities when the turf is under moisture stress (Chapter 6).

Turf Species and Cultivar

Most turfgrass species are adapted to particular regions of North America (Chapter 1). Cool-season grasses thrive in cooler climates, while warm-season grasses do better in warmer climates. When grasses are grown in areas outside their normal range, they are often more stressed (too much or too little moisture, high or low temperature extremes) and do not grow as vigorously. As a result, the additional stress of insect feeding can result in visible damage, even though comparable insect populations in less-stressed turfgrass do not damage the plants (Schumann et al. 1997). For example, creeping bentgrass, a cool-season turfgrass, is grown on putting greens on some golf courses in Georgia and Alabama, but the tolerance level for white grubs and other insects is much lower on these golf courses than it is on comparable greens in the northeastern United States, because the bentgrass is less able to recover as a result of the agronomic stresses.

Some turfgrass species are resistant to insect activity, while others are highly susceptible. Annual bluegrass is highly vulnerable to feeding by the annual bluegrass weevil (Chapter 18), so the threshold for that insect is much lower on annual bluegrass than it is on creeping bentgrass. Several endophytic cultivars of tall fescue and perennial ryegrass are resistant to chinch bugs (Chapter 6), webworms (Chapter 8), and billbugs (Chapter 18), so these cultivars will tolerate the presence of more insects than will nonendophytic cultivars.

Turf Use

Some turfgrass is highly maintained or subjected to heavy use (golf course greens, tees, fairways, athletic fields, some home lawns), while other sites are less intensively managed (some industrial parks, rights-of-way). Customers or clients whose turf is highly maintained often expect pristine conditions and will not tolerate much, if any, insect damage. Heavily used athletic fields, with high traffic patterns, will be more vulnerable to insect damage because the compaction and disruption caused by athletes digging into the turf add stress to the plant and it is less able to overcome insect-feeding damage.

Turf Vigor

Turf plants often can tolerate one or two forms of stress but when additional stresses occur, plant vigor suffers. Often the initial stresses are agronomic or abiotic (e.g., lack of water, inadequate drainage, mowing height, lightning damage), while the added stress may be a result of pest activity. The result is that a turf manager blames the pest for the damage, but in fact the situation is much more complicated. Stress management becomes the key: If agronomic and abiotic stresses are minimized, pest thresholds often can be much higher (Schumann et al. 1997).

Water

Soil moisture and rainfall patterns play a significant role in turf insect activity. In general if soil moisture is too low (plants are wilting) or too high (saturated soils, inadequate air in the root zone), the turf is more stressed and less able to withstand insect activity. White grub damage often is more severe on golf course roughs than fairways, in part because roughs are less likely to receive supplemental irrigation (P. J. Vittum, personal observation). Chinch bug populations on turfgrass that receives adequate moisture are less likely to cause visible damage and therefore thresholds can be higher (Chapter 6).

Mowing Height

Turfgrass is unusual in the plant kingdom because the meristematic tissue is near the base of the plant. It is this feature that enables grasses to withstand frequent mowing. However, excessively low mowing heights eventually result in stunting of the entire plant, including the roots. Turfgrass mowed consistently at putting green height (3–5 mm) has shallower and less fibrous roots than does the same grass mowed at fairway (1.2–3.0 cm) or lawn (2.5–6.0 cm) height. Thus insect thresholds are much lower on lower mown grasses.

The annual bluegrass weevil only damages annual bluegrass on fairways, tees, and greens (Chapter 18). The insect is seen in the roughs but has never been reported damaging turfgrass at that height. Black cutworms cause damage on greens and occasionally on fairways. While eggs of cutworms often are laid in higher mown areas (Chapter 9), damage is much more severe in the nearby areas with lower mowing heights.

Fertility

Fertilization practices have a direct effect on turf vigor and, as a result, on the ability of the plant to withstand insect pressure. If plant nutrition is unbalanced, the plant is stressed. Inadequate fertility may stunt the turfgrass or reduce growth rate (and recuperative potential) while excess nutrients may induce unnaturally lush shoot growth that may be particularly attractive to insects. For example, southern chinch bug infestations often increase rapidly after heavy fertilization of St. Augustinegrass (Chapter 6).

Recuperative Potential

Some turfgrasses can recover from insect damage simply by virtue of their growth habits. Species that produce many stolons or rhizomes often can tolerate more feeding activity because the side shoots can regenerate tissue nearly as quickly as it is lost through insect feeding.

Time of Year

The amount of damage incurred by turfgrass depends in part on the seasonal timing of insect activity. Insects that feed during the summer months on cool-season turfgrasses often cause more damage than similar insects feeding in the spring or late summer, because cool-season grasses are under the greatest agronomic stresses in July and August. For example, as few as 10–20 annual bluegrass weevil larvae per 0.1 m² have been observed damaging annual bluegrass on a golf course rough in early August, while 30–50 larvae per 0.1 m² often are present in late spring without signs of visible damage (P. J. Vittum, personal observation).

The reproductive rate of the insect also affects thresholds. Populations of insects that have only one generation per year often increase more slowly than do other species, and once an initial assessment of the population has been made, a turf manager can predict what to expect with some accuracy. But insects that complete several generations per year may increase in numbers very rapidly. The threshold for greenbugs will be lower in July than it will be in September, because the July greenbugs may complete two or three additional generations, each resulting in an increase in the total population.

Expectations

Most turf managers maintain turf for someone else (e.g., members of a golf course, a home-owner, an athletic director, a cemetery owner, or a parks and recreation director). Sometimes a turf manager does not have direct control over decisions made regarding pest management. The client or customer may intervene and dictate that little or no pest activity will be tolerated, even if damage from that pest is insignificant or has no effect on turf vigor. Some homeowners expect (and demand) 100% control of insects, even though such demands are unreasonable. Members at an exclusive private golf course often will demand a higher level of maintenance and pest management (i.e., lower tolerance level or threshold) than golfers at many public golf courses.

The reality is that expectations or limitations imposed by the decision-maker will directly affect a turf manager's working action threshold. If a golf course membership demands that no grub damage be incurred anywhere on the golf course, a turf manager is forced to employ a lower threshold than what otherwise might be necessary. If a university has a football team that appears on national television regularly, the action threshold for insect activity might be lower. If a homeowner is caught up in the "competitive lawn syndrome", a lawn care company might be forced to use insecticides preventively to ensure that no insect damage occurs.

Availability of Curative Control Options

The establishment of an action threshold depends in part on the kinds of control strategies that can be employed against the insect pest. Some biological and chemical control options work only when applied preventively (before an insect population begins to develop to a damaging stage), while others can be used curatively (applied after pest insects have reached their most damaging stage). Some insects are very difficult to control once they have developed to their most damaging stages. For example, late-instar billbugs are relatively difficult to kill, even with fast-acting insecticides. In cases such as this, the action threshold (for earlier instars) may be somewhat lower, and preventive strategies may be employed more quickly to ensure that fewer insects survive to the damaging stage. In contrast, hairy chinch bugs are sensitive to traditional insecticides throughout most of their development, so curative applications usually work quite well. Thresholds can be higher, as a result, because even if populations increase or damage becomes more apparent throughout the summer months, a curative material can be used to stop the progression.

Budget

Most turf managers have specific budget limitations and must use biological or chemical control strategies judiciously. Some control strategies are highly effective but very expensive and cannot be considered by managers with budget limits. In these instances, tolerance levels may be higher than those for managers with less restricted budgets.

Getting Started

Tolerance levels, or action thresholds, must be established for each different turf condition. Thresholds within a single golf course will vary, depending on the time of year, whether the tees or fairways are affected, proximity to the clubhouse, tournament schedules, and other factors. Athletic field managers must consider practice and game schedules, availability of water, and mowing heights, as well as other factors, when setting thresholds. Lawn care professionals often base thresholds, at least in part, on customer demands and expectations.

Many turf managers begin by setting conservative thresholds for some of the most critical areas under their management, and then expand the concept as they become familiar with the techniques and the variables. They concentrate on key pests and key locations initially and include additional pests and locations as they gain confidence.

27

Biological Control Strategies

*B*iological control refers to the suppression of pest populations through the activity of living organisms or their by-products. Although a majority of this book is devoted to understanding turfgrass pests, most organisms associated with turfgrass are not pests but instead may be considered beneficial because they reduce thatch, help recycle soil nutrients, or are natural enemies of pest species. Pest outbreaks can sometimes be traced to the absence of natural control agents in the turf environment. Vertebrate and invertebrate predators, insect parasitoids, and microbial pathogens may act as natural enemies of turfgrass pests. Although the impact of one species of natural enemy may be minor, the combined impact of predators, parasitoids, and pathogens can cause considerable reductions in pest populations (Hoffmann and Frodsham 1993).

Under our broad definition of biological control, additional agents can be considered biological control agents. These include fungal endophytes, botanicals, and synthetic compounds that mimic the activity of insect-produced bioactive compounds (such as hormones and pheromones). It is our intention to review the importance of the range of biological control agents that can reduce turfgrass insect pest populations. Specific biological agent interactions with pest species are discussed within the appropriate pest-related chapters.

Use of Natural Enemies in Turfgrass Management

In traditional biological control programs natural enemies are released, managed, or manipulated by turfgrass managers to avoid or reduce damaging pest populations (Weinzierl and Henn 1994a). Classical introduction, augmentation, and conservation of natural enemies are the three primary methods for implementing or enhancing biological control with natural enemies:

1. Classical introduction. Classical introduction refers to the importation, release, and establishment of introduced natural enemies, usually to control introduced or exotic pest species. Many introduced turf pests arrive in new locations without their associated predators, parasitoids, and pathogens. Freed from their natural control agents, these pests may spread unchecked, causing significant feeding damage to plants. Over time, native beneficial organisms often stabilize pest populations, but in other situations, researchers must travel to the place of origin of the pests to find and import specific predators, parasitoids, or pathogens to control these introduced pests. When successful, classical introduction is a long lasting and inexpensive approach to biological control of turfgrass pests. However, failures are not unusual. The reasons for these failures include the poor adaptation of the introduced natural enemy to environmental conditions at the release location, the absence of alternate hosts, and the lack of synchrony between the life cycle of the natural enemy and the pest (Hoffmann and Frodsham 1993).

2. Augmentation. Pest managers can release natural enemies into a site in an effort to augment (increase) the activity of existing biological control agents. There are two basic approaches to augmentative releases of natural enemies. Inoculative augmentation involves the release of small numbers of natural enemies with the hope that these agents will establish, persist, and multiply at the site. One expects relatively slow, but long-term, suppression of pests with successful inoculative release programs. Alternatively, one can release a massive quantity of natural enemies in an effort to inundate, or overwhelm, the pest population. With inundative releases, the natural enemy can be considered a natural, short-residual insecticide that confers immediate but usually temporary control of the pest population.

3. Conservation. Conservation of existing or introduced natural enemies is often the most important factor ensuring long-term suppression of turfgrass pests. Potter (1998) noted that outbreaks of turfgrass insects are uncommon in golf course roughs and low-maintenance lawns, and suggested this may be because most pests are held in check by natural enemies (pathogens, predators, and parasitoids). In many instances the importance of existing, resident natural enemies has not been adequately studied, nor does it become apparent until insecticide use is stopped or reduced.

Among the practices that conserve and favor increases in populations of natural enemies are recognizing natural enemies to avoid mistaking (and treating) a high population of a natural enemy for a damaging population of a pest species; minimizing broad-spectrum insecticide applications to reduce unintended harmful effects on nontarget organisms; and maintaining complex plantings to provide alternative food sources (pollen, nectar, other hosts) and cover to provide overwintering sites and protection from enemies (Glenister 1994, Weinzierl and Henn 1994a).

Predators

Predators are animals that find, attack, and consume many prey items during their lifetime. They are usually as large as, or larger than the prey organisms that they attack. Predators are often generalist feeders that attack a variety of pest and nonpest species in turf. They are usually quite mobile so they can seek out prey, and feed on many individuals during their lifetime. Some predators can survive only on animal prey (insects and mites), but others will supplement animal prey with other types of food including plant roots, foliage, and seeds. Following are some of the predatory groups that are reported to help reduce pest populations in turfgrass systems.

Green Lacewings (Family Chrysopidae)

Adults in the family Chrysopidae are common insects found in grassy or weedy sites, or resting on the foliage of trees and shrubs. Green lacewings have large, lacy wings, and threadlike antennae (Plate 59). Most adult lacewings feed on nectar and pollen but several species have been reported as feeding on other insects. Larval lacewings are flattened and elongated with sickle-shaped mandibles that pierce their prey and suck body fluids. The larvae of most species are predaceous, chiefly on aphids and are commonly called *aphid lions* (Borror et al. 1989).

Earwigs (Order Dermaptera)

Earwigs occur throughout North America and are found wherever there is moisture and cover, so that turfgrass areas are a natural habitat. The common species are brown to black, elongate insects, 19–25 mm long, with a pair of large forceps at the end of the abdomen (Plate 59). Predatory earwig species are nocturnal, generalist predators. The name *earwig* is derived from an old superstition that these insects enter people's ears. This belief is without foundation (Borror et al. 1989).

Ground Beetles (Family Carabidae)

Adult ground beetles are variable in shape, color, and size, ranging from 3 to 25 mm in length (Plate 59). Most are black or brown and are shiny or metallic in appearance. The larvae of ground beetles are elongate, and slightly flattened and taper toward the rear. Their heads are large, with sickle-shaped mandibles directed forward. Most ground beetles complete a generation within a year, overwintering as larvae or adults. Eggs are deposited singly in the soil. Larvae develop in or on the soil, generally hidden under stones, boards, or other debris. Adults are usually nocturnal. Both adults and larvae are predacious, although a few species feed on seeds (Baker 1982).

Rove Beetles (Family Staphylinidae)

Rove beetles are fast-moving, elongated dark-colored beetles with short elytra that leave much of the abdomen uncovered (Plate 59). They are distributed worldwide and range in size from 2 to 12 mm. Most rove beetle species overwinter as adults. Both the larvae and adults are predaceous on a range of insects and are considered beneficial.

Hister Beetles (Family Histeridae)

Hister beetle adults are small, shiny black oval-shaped insects (Plate 59). They are usually found living in or near decaying material and are reported to be predaceous on small insects living in these habitats (Borror et al. 1989).

Ladybird Beetles (Family Coccinellidae)

Adult ladybird beetles are common and easily identified insects. They are oval, convex, and often brightly colored, usually with black spots (Plate 60). Larvae are dark colored, with orange, or yellow stripes across their bodies. Ladybird beetle adults and larvae feed on a number of small important insect pests, most notably aphids. They also feed on insect eggs, particularly moth eggs (Borror et al. 1989).

Tiger Beetles (Family Cicindellidae)

Adult beetles have long, cylindrical, brightly colored, metallic bodies with large mandibles for grasping prey (Plate 60). They are often found hunting in sandy, open areas. Larvae live in vertical burrows dug in sandy soils and wait at the mouth of the tunnel to ambush any suitable prey that wanders by (Arnett 1985).

Big-eyed Bugs (Family Lygaeidae)

Big-eyed bugs are important predators of chinch bugs and other small insects living in the thatch and leaf zones of turfgrass (Plate 60). They possess piercing sucking mouthparts with which they feed on the liquid contents of captured prey. Big-eyed bugs resemble chinch bugs, but are more robust and have large eyes that protrude at the sides of the head. They also move more rapidly among the grass blades and stolons than chinch bugs. Common big-eyed bug species have been observed feeding on turfgrasses, but this activity is presumed to be inconsequential (Crocker and Whitcomb 1980).

Assassin Bugs (Family Reduviidae)

Assassin bugs belong to a diverse family of predatory bugs possessing piercing sucking mouthparts with which they feed on the liquid contents of captured prey. Most members of the family are black or brown, but brightly colored species are not uncommon (Plate 60).

Flower Flies (Family Syrphidae) and Bee Flies (Family Bombyliidae)

Adults of these families can often be found hovering around flowers that provide the flies with pollen and nectar. Many species are bee or wasp mimics both in coloration and also in the buzzing sound associated with bee and wasp species (Plate 61). Predatory flower flies often attack aphids, while the larvae of bee flies feed on the eggs or immature stages of beetles, grasshoppers, caterpillars, and wasps (Arnett 1985).

Flesh Flies (Family Sarcophagidae)

Flesh fly adults are black or gray nonmetallic flies whose larvae feed on a variety of animals including many insect pest species (Plate 61). Some flesh fly species are predacious, while other species are insect parasitoids. Many species lay their eggs in the nests of bees or wasps, where their larvae feed on the materials with which these nests are provisioned (Borror et al. 1989).

Robber Flies (Family Asilidae)

Adult robber flies are large, beelike insects with bulging eyes (Plate 61). Larvae are usually found in rotting wood, under bark or fallen leaves, or in loose soil. Both the adult and larval robber flies are predacious (Arnett 1985).

Ants (Family Formicidae)

All ant species are social insects meaning that they live in organized colonies. Adult ants have a very narrow, constricted connection between the thorax and head and an even more conspicuous constriction between the thorax and the abdomen (Plate 52). Larvae of ants are mostly white and legless, with small, distinct heads, and bodies that dilate gradually from the anterior to the posterior. Consumption of eggs and immature stages of scarab

beetles, webworms, phytophagous ants, and mole crickets has been implicated in significant population regulation of pest species in turfgrass (Chapter 20).

Spiders (Order Araneae)

Spiders are a large, distinct, and widespread group occurring in many habitats, including turfgrass areas (Plate 61). All spiders are predacious, feeding mainly on insects, killing
their prey by injecting a poison through their fangs. Their habit of spinning silk into webs
of various shapes and sizes is one of their most characteristic features (Borror et al. 1989).
Several small spiders occur in turfgrass in surprisingly large numbers and prey on various
soft-bodied insects in the thatch and upper part of the soil. Some of the more common families observed in turf are line weaving spiders (Linyphyiidae), wolf spiders (Lycosidae), and
crab spiders (Thomisidae).

Predatory Mites (Order Acari)

Predatory mites in the genera *Phytoseiulus* and *Amblyseius* are pear-shaped predators
with short life cycles and high reproductive capacities (Plate 61) (Weinzierl and Henn
1994a). Predatory species of mites important in turfgrass sites are active in the thatch and
help reduce populations of plant-feeding mites and insects (e.g., winter grain mite, aphids).

Centipedes (Class Chilopoda)

Centipedes are terrestrial and predacious, feeding on insects and spiders (Plate 61). They
are found in protected places in the soil, under bark, or in rotting logs. Unlike millipedes,
they can move very quickly. All possess poison jaws for paralyzing their prey. Centipedes
overwinter as adults in protected places and lay eggs during summer. The sticky eggs, laid
singly, become covered with soil. In some species this covering protects them from being
eaten by the males (Baker 1982, Borror et al. 1989).

Vertebrates

Birds, mammals, amphibians, and reptiles all reduce insect pest populations in turfgrass
populations (Plates 66–69). Starlings, grackles, robins, and other birds feed on large numbers of soil-dwelling stages of turf pests in thatch and can cause considerable damage to
weakened turf. Mammals attracted to insects in infested turfgrass as a source of food include moles, skunks, raccoons, red foxes, armadillos, coyotes, and black bears. However,
they also may cause significant turfgrass damage in the process of searching for their prey.
Toads in the genus *Bufo* are reported to consume larval webworm and armyworm and
the adult stages of several weevil species including billbugs.

Parasitoids

Parasitoids are insects that lay their eggs in or on an insect host. When these eggs hatch,
the parasitoid larvae feed internally or externally on their hosts, usually killing them shortly before the parasitoid pupates and emerges as an adult from the cadaver of the host. In
contrast to predators, parasitoids are often highly specialized to attack a particular host
genus or species and a specific stage (e.g., egg or adult). Unlike parasites, parasitoids commonly kill the host during their development. Following is a discussion of some of the parasitoids known to help reduce pest populations in turfgrass systems.

Humped-backed Flies (Family Phoridae)

Adult Phorid flies are usually black, with a deflected head that gives them a humped-back appearance (Plate 62). Adults are common around decaying vegetation. Parasitoid species of this family attack wasps, beetles, and caterpillars (Arnett 1985).

Scelionid Wasps (Family Scelionidae)

Scelionid wasps are egg parasitoids, and are reported to attack chinch bug eggs (Mailloux and Streu 1981, Reinert 1972). Upon finding a suitable host, the female wasp thrusts her ovipositor into a host's egg and deposits her own egg inside. The developing parasitoid usually consumes the entire content of the host egg before pupation. Unlike a normal chinch bug egg, which turns reddish when it is about 3 days old, a parasitized egg remains tan until the parasitoid pupates, then becomes blackish, with the parasitic pupa clearly visible inside.

Tachinid Flies (Family Tachinidae)

Tachinid adults are highly mobile, strong flying individuals, that typically feed on honeydew and nectar. Females of most species deposit their eggs directly on the body of their hosts. Upon hatching, the larvae burrow into and feed internally on the host (Plate 62) (Borror et al. 1989). The larvae of tachinid species are parasitoids of mid to late instars of beetles, lepidopterans, true bugs, and grasshoppers. Some tachinid species also parasitize the adults of these groups (Arnett 1985).

Light Flies (Family Pyrgotidae)

Light flies are specialist parasitoids of scarab beetle larvae. The adult fly lands on the dorsum of a scarab adult that is feeding, causing the beetle to take flight. Flight allows the parasitoid to insert her ovipositor into the thin integument of the beetle's abdominal dorsum (Plate 62). The beetle dies in 10–14 days, and the parasitoid puparium is formed within the dead host (Davis 1919).

Aphidiid Wasps (Family Aphidiidae)

These tiny wasps are internal parasitoids of aphids. Before pupating, the aphid parasitoid makes a hole in the underside of its host and briefly emerges from it to fasten the skin of the aphid to the grass blade (Plate 62) (Arnett 1985).

Tiphiid Wasps (Family Tiphiidae)

Tiphiid wasps are external parasitoids of beetle larvae, particularly scarab grubs (Plate 62). A female tiphia searches out a grub in the soil, stings it to paralyze the grub temporarily, then attaches a single egg to a particular ventral body fold. Upon hatching, the maggot-like larva attaches itself externally to the grub, pierces the host's cuticle with its mouthparts, and feeds on the grub's liquid contents. When the parasitoid larva is fully developed, it will have consumed the entire grub cadaver except for the cuticle and head capsule. Pupation takes place in a strong, watertight, silken cocoon in the earthen cell formerly occupied by the grub. The parasitoid emerges as an adult the following year.

Braconid Wasps (Family Braconidae)

Braconids are small dark wasps with long curved antennae (Plate 63). This family contains examples of both internal and external parasitoids, and includes species that attack most insect pests found in turfgrass. All pest life stages, from egg to adult, may be attacked by braconid wasps (Borror et al. 1989). Braconids that are internal parasitoids often will emerge from their host to pupate. The presence of silken cocoons on or near dead host larvae is characteristic of mortality due to braconid wasp parasitism.

Ichneumonid Wasps (Family Ichneumonidae)

Adult ichneumonids are dark, slender wasps with long antennae. Females possess ovipositors that are typically as long as, or longer than their bodies; ovipositors are permanently extended and cannot be retracted into the body as they are in many wasp families (Plate 63). All ichneumonids are parasitoids of insects with complete metamorphosis. Both internal and external parasitoid species are represented in this large family. Some are host specific but it is more likely that they are habitat specific, parasitizing several kinds of larvae that share a specific environment. For example, leaf-feeding caterpillars in warm, moist environments may be parasitized by a specific ichneumonid (Arnett 1985).

Trichogrammatid Wasps (Family Trichogrammatidae)

These minute wasps are egg parasitoids of lepidopterans and occasionally other insect species (Plate 63). Parasitized eggs often turn black as the wasp larvae develop inside.

Digger Wasps (Family Sphecidae)

Digger wasps are large, dark-colored, solitary parasitoids of insects and spiders (Plate 63), and are common parasitoids of scarab grubs. The wasp locates and stings the grub, paralyzing it, and then attaches an egg to its ventral side. Upon hatching, a parasitoid larva feeds on the paralyzed grub. The wasp larva completes its growth, spins a cocoon, pupates, and remains in its cocoon until it emerges as an adult wasp (Davis and Luginbill 1921).

Scoliid Wasps (Family Scoliidae)

Scoliid wasps are large, hairy, hard-bodied insects whose larvae are parasitoids of scarab beetle grubs, especially species in the genus *Phyllophaga*. Scoliids are usually black with bright patterns of red, yellow, or white that are often characteristic of the species (Plate 63). The female wasp flies low over the ground seeking scarab grubs. Upon locating such a site, she will land and dig with her mandibles until she uncovers a suitable host. She then deposits an egg on the cuticle of the grub and the wasp larva that hatches from it parasitizes and eventually kills the grub (Arnett 1985).

Pathogens

Pathogens are microorganisms that cause disease in pest organisms. Insect pathogens that have been observed in nature to cause pest population crashes can be enhanced by manipulating the environment or by releasing large numbers of infective stages that were produced in a laboratory. An advantage of entomopathogens for turfgrass protection is the specificity of their effects. None of the insect pathogens considered for commercial development as biopesticides has been demonstrated to have serious effects on humans, mammals, or other vertebrates (Cranshaw and Klein 1994). Following is a discussion of some of the pathogen groups known to help reduce pest populations in turfgrass systems.

Fungi

Every major group of turfgrass insect pests, including scarab grubs, chinch bugs, billbugs, annual bluegrass weevils, sod webworms, cutworms, and mole crickets, is susceptible to a wide variety of fungal pathogens. Outbreaks of the pathogens *Beauveria bassiana* and *Metarrhizium anisopliae* have been reported to decimate damaging populations of hairy chinch bugs and bluegrass billbugs under favorable environmental conditions in the field (Plate 57). Other genera of fungi that have been reported as infecting insects in turfgrass sites include *Cordyceps, Fusarium, Penicillium,* and *Aspergillus.*

Fungi differ from most other microorganisms because they do not have to be ingested to be effective. Infection is initiated by the adhesion of a fungal spore to the body of an insect. If conditions are correct, the spore will germinate and a tube will grow from the spore into the insect and penetrate the circulatory system. After penetration, the fungus reproduces within the insect and produces toxins that quickly kill the insect. After death of the host, hyphae emerge from the insect and develop into structures that produce more infective spores. This new generation of spores can then be spread through the environment, infecting additional insects.

Although often considered a broad-spectrum microbiological agent, individual fungal isolates can display considerable specialization with respect to host range. Specificity studies have reported differences in virulence among fungal isolates when tested against a specific insect species (Milner 1992). In some cases a fungal strain is most virulent in the insect species from which it was isolated; in other cases, a fungal strain isolated from a distant taxon proved to be highly virulent in screening studies (Ferron 1978, Latch 1976).

Bacteria

Bacteria are small, single-celled organisms that may be shaped like rods (bacilli), spheres (cocci), or spirals (spirilla and spirochetes). Bacteria are the most common microorganisms associated with insects, with a significant number behaving as entomopathogens (Tanada and Kaya 1993). Entomopathogenic bacteria may cause host mortality in three distinct ways: They may invade the hemocoel, multiply, and produce toxins that kill the host (septicemia); they may be confined to the host's gut lumen and produce toxins that kill the insect (toxemia); or they may multiply in the host's hemolymph without toxin production (bacteremia). Many bacteria species have been isolated from field-collected insect cadavers. However, interest in bacterial pathogens for turfgrass insect control has focused on three groups.

Bacillus thuringiensis. *Bacillus thuringiensis* (*Bt*) is a bacterium common in soils that was first discovered and identified in Japan in 1901. Over 30 subspecies and varieties have been identified (Tanada and Kaya 1993). This family of bacteria produces one or more toxic protein crystals that are active against a fairly narrow group of insect species. An important feature of these toxins is this specificity; specific toxins usually only kill the larvae of certain species within a single order of insects (Knowles and Dow 1993). Various strains or varieties have been identified with activity against caterpillars, fly larvae, and beetle larvae. Commercial *Bt* products have been used to control insects for many years; *Bt* varieties that are currently being commercialized include *Bt kurstaki* and *Bt aizawai* (caterpillars), *Bt israelensis* (mosquitoes), and *Bt tenebrionis* (leaf beetles).

Bt bacteria can be produced readily in great quantities on artificial media. Therefore, there is great interest in the commercialization of this bacteria for insect control. *Bt* typically has been used as a microbial insecticide for short-term control because the bacteria do not normally reproduce in the insect host, persist in the environment, or spread from the treatment site.

The *Bt* bacterium produces protein crystals (endotoxins) that are toxic to susceptible insect species; this protein crystal must be ingested by an insect and then dissolve within its gut to become active. Gut pH determines the solubility of these protein crystals. In some cases, additional products produced by the living bacteria must be consumed for maximal activity. Proteins released from the dissolved crystal bind to specific sites in the gut lining of susceptible insect species, causing rapid paralysis of the gut. The insect stops feeding almost immediately, as the gut wall deteriorates. In most cases a susceptible insect will die 2–7 days after ingestion of the *Bt* toxin. If an insect is not susceptible, it is usually because improper gut pH does not allow the toxic protein crystal to dissolve, or the toxic proteins are unable to bind to the insect's gut (Knowles and Dow 1993).

Although *Bt* products are registered for use against several turf-feeding caterpillars including cutworms and sod webworms, these products have not been widely recommended or accepted in the turf industry. Likely reasons for this lack of interest for *Bt* products currently available include low residual activity, short shelf-life, limited systemic activity, low potency, narrow host range, slow action, and the need for viable spores rather than toxins for maximum efficacy in some isolates (Lambert and Peferoen 1992).

Until recently there were no *Bt* varieties known to cause significant mortality in scarab grubs inhabiting turfgrass. In 1991, *Bt japonensis* strain *BuiBui* was discovered in Japan. *Bt japonensis Buibui* has a spherical protein crystal that is toxic to certain kinds of scarab grubs. Unlike most commercial *Bt* products currently on the market, maximum activity of *Bt japonensis BuiBui* against scarab grubs occurs when formulations include both the toxic protein crystals and live spores produced by the bacteria (Alm et al. 1997).

Milky Disease (Bacillus popilliae). This bacterial disease was discovered when a few abnormally white Japanese beetle grubs were found in New Jersey in 1933. The hemolymph of those grubs, normally translucent, was milky white and teeming with bacterial spores. The origin of these native organisms is unknown, but they are thought to be obligate parasites of native white grubs of the family Scarabaeidae.

The spores normally are found in the soil and remain viable for several years. When ingested by grubs (along with roots, organic matter, and soil), the spores lodge in the epithelial tissues of the midgut. Spores germinate in the gut and then pass through the gut wall into the coelomic cavity, where multiplication and sporulation occur. The normally clear hemolymph becomes turbid as vegetative bacterial cells multiply. The hemolymph eventually turns milky white as sporulation of the bacteria occurs (Plate 57). Billions of spores are produced, causing the grubs to die. Histological studies on milky disease infections clearly demonstrated that vegetative cells penetrate the midgut tissues where vegetative growth, multiplication, and some sporulation occur (Splittstoesser et al. 1973, Splittstoesser et al. 1978).

Bacillus popilliae kills by bacteremia, a massive production of bacterial cells (spores) in the insect host. Grubs infected with milky disease may live for weeks or months, and continue to feed as they gradually change in external appearance, become weaker, and eventually die. Infection is possible in all three instars, but infected grubs never transform to the next stage. As maximum sporulation occurs and grubs approach death, spore contents can vary from 2 billion to 5 billion per grub. A decaying cadaver releases a high concentration of spores into the soil, and any grub subsequently ingesting the spores is subject to infection. Spores remain viable after passing through the digestive tracts of birds and mammals, which is believed to be the primary method of dispersal.

The temperature range for growth and development of *B. popilliae* in a grub is about 16°–36°C, and about 21°C is required for a rapid buildup of spores. In order to reduce feeding damage, soil temperatures must be sustained at levels greater than 20°C during the critical larval feeding period in late summer and early fall. This minimum temperature requirement slows the rate of disease progression in scarab populations in northern regions.

The field efficacy of *B. popilliae* has been difficult to document in some parts of the country. A field trial in Kentucky demonstrated that commercial preparations of *B. popilliae* did not raise the incidence of milky disease in grub populations or reduce grub populations, regardless of the method of production of infective spores (Redmond and Potter 1995). Klein (1995) suggested that although the value of individual applications has often been questionable, the primary value of "milky disease" has been its contribution to general population suppressions and not as a biological insecticide.

Amber Disease. This bacterial disease is caused by bacteria in the genus *Serratia*. Pathogenic species of this genus infect over 70 insect species in the orders Orthoptera, Coleoptera, Hymenoptera, Lepidoptera, and Diptera (Tanada and Kaya 1993). *Serratia* strains are able to colonize the foregut of susceptible species, leading to starvation, depletion of fat bodies, and development of the characteristic amber coloration associated with the late stages of the disease (Jackson et al. 1992). Investigations of *Serratia entomophila* determined that endemic populations of this bacteria were associated with significant reductions of the grass grub (*Costelytra zealandica*) in New Zealand pastures (Jackson 1984). Extensive efforts by Jackson and his collaborators (reviewed in Jackson et al. 1992) to optimize production, distribution, and application technology have resulted in development of a commercial *Serratia* product for grass grub control. In the United States, Japanese beetle and masked chafer larvae with similar amber coloration have been observed, and *Serratia* species have been isolated from them (Klein 1995).

Entopathogenic Nematodes

Entomopathogenic nematodes are tiny worms that attack insects. These nematodes commonly enter the host through natural openings such as the mouth, spiracles, or anus. Entomopathogenic nematodes in the genera *Heterorhabditis* and *Steinernema* carry a pathogenic bacterium (Plate 58). Once inside the insect, these bacteria are released by the nematode. The bacteria multiply within the insect and produce toxins that kill the infected host rapidly. Nematodes feed on the bacteria and host tissues, and reproduce inside the dead insect, producing several thousand new nematodes. Under ideal conditions these new infective nematodes escape from the dead insect in as little as 10 days and begin to search for new hosts to infect.

Entomopathogenic nematodes have recently received attention as alternatives to insecticides for turf insect control. Many factors make nematodes an ideal microbial control agent: They have a relatively broad host range, will not attack plants or vertebrates, are easy to mass produce, and can be applied with most standard insecticide application equipment.

Additionally, nematodes can search for hosts and kill them rapidly, and can release thousands of mobile progeny able to locate and infect new insect hosts within weeks of the initial infection. Because they are considered predators, entomopathogenic nematodes are exempt from registration by the U.S. Environmental Protection Agency (USEPA). This exemption from long-term safety and water-quality studies has greatly reduced the costs and risks typically associated with registering a new insecticide.

Although there have been many successful field applications of entomopathogenic nematodes for turf insect control, problems with product quality, persistence, and host specificity have led to some unsatisfactory results. Entomopathogenic nematodes have fairly broad host ranges in laboratory studies, but different strains and species of nematodes vary in activity against different insect species in the field. Overall, the nematodes *Heterorhabditis bacteriophora* and *Steinernema glaseri* are more effective against white grubs, because they move down in the soil actively and search for insects. *Steinernema carpocapsae* has been effective for control of billbugs and caterpillars such as cutworms, webworms, and armyworms.

Many unsuccessful field applications of entomopathogenic nematodes for scarab grub control can be traced to inappropriate conditions at the time of application and for several weeks after application. Environmental factors, especially the availability of adequate soil moisture, are critical for entomopathogenic nematode performance in the field (Georgis and Gaugler 1991). Nematodes are extremely sensitive to exposure to ultraviolet light and will last only a few minutes when exposed to full sunlight. They are also quite prone to desiccation and therefore require high relative humidity and a film of moisture on leaf and soil surfaces to survive and move. Therefore nematodes should be applied very early in the morning or late in the day and be irrigated immediately with at least 1 cm of water. However, excessive irrigation should be avoided since nematodes cannot search in saturated soils.

Entomopathogenic nematodes can be applied through inundative or inoculative releases. High rates of nematodes (inundative releases) can produce high rates of mortality quickly. On the other hand, lower inoculative application rates may produce lower initial mortality, but significant pest population suppression may occur over time (Kaya and Gaugler 1993, Klein and Georgis 1993). This is most likely because the nematodes can reproduce in the insect's cadaver and disseminate new infective juveniles, which can then disperse and attack other insects.

Rickettsias

Rickettsia are "bacteria-like" organisms, similar to viruses in that they are obligate pathogens that multiply in the cytoplasm of infected cells (Tanada and Kaya 1993). They resemble bacteria in that they contain both RNA and DNA and possess true cell walls (Jackson and Glare 1992). One such rickettsia, *Coxiella popilliae*, produces a greenish-blue discoloration of larval fat bodies in grubs and bears the name *blue disease* (Klein 1995).

Ingestion of the infective stage of rickettsia is required for the disease to progress. The period of infection, from onset to death, may take 6 months or more, and is directly related to the quantity of infective propagules consumed (Jackson and Glare 1992). Although naturally occurring rickettsias may cause significant mortality of scarab grub populations in the field, the lack of an artificial diet for their production and concerns about mammalian pathogenicity have hampered commercial interest (Klein 1995).

Viruses

The word *virus* is derived from Latin and means slimy liquid, poison, or stench (Tanada and Kaya 1993). Viruses are simple organisms that consist of nucleic acids packaged with a coat of protein, glycoprotein, lipoprotein, or a combination of all three (Glare and Crawford 1992). More than 1,600 viruses, in nine families have been recorded as infecting insects (Martignoni and Iwai 1986). Baculoviruses, also called *nuclear polyhedrosis viruses* or NPV, are among the most important viruses, and cause the characteristic "wilt disease" of many caterpillar species (Cranshaw and Klein 1994). Baculoviruses usually infect their insect hosts when larvae eat the virus particles that adhere to food plants. Young larvae are usually much more susceptible to viral infections than are older insects.

Protozoans

Protozoans are free-living one-celled organisms that can be important in the natural suppression of scarab grubs and other turfgrass pests in the field. Microsporidia are the most important group of insect pathogenic protozoa (Hanula and Andreadis 1992) and *Nosema locusta,* a microsporidian, is marketed for grasshopper control in the United States. Virulent spores are usually ingested and germinate in the midgut cells, causing infection that leads to cell death. Microsporidia often cause significant mortality to highly stressed pest populations over extended periods of time. Therefore, a commercial formulation of a microsporidian protozoan would be most appropriate in low-maintenance turfgrass (Klein 1995). Many scarabs are infected by coccididian protozoan in the genus *Adelina.* Adult scarab beetles infected with an *Adelina* protozoan have reduced fat reserves when compared with uninfected beetles, and most infected females die before ovipositing (Klein 1995).

Endophytes

Endophytes are fungi associated with certain plant species that grow entirely within the plant (Fraser and Breen 1994). The endophyte is transmitted primarily by seed and vegetative propagation and gradually loses its viability when seed is stored for prolonged periods (Funk and Hurley 1984).

Endophytes produce alkaloids that may protect infected plants from insect feeding damage. These alkaloids, which are commonly found in highest concentrations in the aboveground parts, may act as a direct toxin to susceptible pest species or serve as feeding deterrents. Feeding on endophyte-infected plant tissues causes reduction in growth or mortality in some insect species. Other insects can detect endophytic grass and will avoid feeding on infected plants (Johnson-Cicalese 1997). Still other turf pest species (e.g., black cutworms) do not appear to be affected when they feed on grasses containing endophytes (Williamson and Potter 1997d). Alkaloid levels are typically lower in the roots of endophytic turfgrasses than in other plant tissues, suggesting that damage by root-feeding pests such as scarab grubs is possible in endophyte-enhanced turfgrass varieties (Murphy et al. 1993, Potter et al. 1992). At the present time commercial varieties of endophytic turfgrasses include tall fescues (endophyte species, *Neotyphodium coenophialum*), fine fescues (endophyte species, *Epichtoe typhinum*), and perennial ryegrass (endophyte species, *Neotyphodium lolii*).

Pheromones

Pheromones are chemical signals that are released by one individual and stimulate a reaction in other individuals of the same species. Insects of the same species can communicate with one another by emitting small quantities of chemical substances from their bodies into the air. The released pheromone travels through the air until it is intercepted by antennae (or possibly other body parts) containing sensory receptors that recognize the pheromone.

Pheromones have been used in a number of strategies for managing turfgrass pests. Most often pheromones aid in the detection and monitoring of exotic (foreign and introduced) or endemic (native and established) pests. There is also great interest in using pheromones to manage rather than monitor pests by either trapping a large portion of the pest population (mass trapping) or flooding the landscape with synthetic pheromone to confuse or disrupt the normal mating behavior of a pest population (mating disruption).

Mass Trapping

Anyone who has seen Japanese beetle traps overflowing with insects might believe that this trapping will translate into less adult feeding damage to ornamental plants in the area, and fewer Japanese beetle grubs infesting the surrounding turf the following fall. Although the pheromone-floral lure baits will attract large numbers of beetles, many beetles will be intercepted by attractive host plants in the landscape (Gordon and Potter 1985). Mass trapping may be a useful control tactic for an established pest species if large geographic regions are targeted and if a great number of highly efficient traps are used. Mass trapping may also be useful if an introduced pest that has not become established in an area is targeted. For example, it may be possible to mass trap oriental beetle adults in a nursery or interiorscape in an area with no previously established beetle population.

Mating Disruption

This approach involves saturation of the environment with massive amounts of synthetic pheromones so that males and females cannot locate each other to mate. Mating disruption has the best chance for success when the initial pest population is small; there is little chance for immigration of individuals into the target site; the synthetic lure will outcompete calling females; and males emerge before calling females, thereby reducing the competition of the lure with calling females.

Insect Growth Regulators

As part of normal growth and development, insects molt regularly. This process is governed by several natural hormones that regulate the production of new chitin, destruction of old chitin, and the development from immature to adult. Insect growth regulators (IGRs) are insecticides that mimic natural insect hormones, and interfere with the normal molting process. For example, some IGRs accelerate the molting process, while others signal the insect to remain in an immature stage (Plate 58). IGRs tend to be much less toxic to mammals or other vertebrates than traditional insecticides. Many IGRs are quite specific and only affect specific insect groups (Cowles and Villani 1996). The speed of efficacy often is quite slow, and most IGRs must be used when susceptible immature stages are still early in their development (Shetlar 1995a, Vittum 1997f).

High doses of these IGR products typically cause rapid insect mortality, while sublethal effects include rapid maturation to the adult stage, larvae showing deformities, and larvae undergoing additional larval molts instead of changing to pupae. Affected larvae usually stop feeding. Specific IGR products have shown activity against scarab grubs, cutworms, and sod webworms.

It is important that the target insect is feeding actively when IGRs are applied. Therefore, the use of an IGR on scarab grub populations in late fall or late spring is ill advised. Laboratory and field studies indicated that early larval stages of insects are more susceptible to IGRs than are later instars of the same species. These studies also suggested that there may be a wide range of susceptibility to a specific IGR among closely related insect species (Cowles and Villani 1996).

Botanicals

Botanicals are natural products, derived from plants, with insecticidal properties. The use of botanical teas and dusts with insecticidal properties has been reported in most primitive societies for control of insects. These crude preparations usually contain mixtures of chemicals with toxic, deterrent, or repellent activity against insect pests, and are probably produced by a plant to protect itself from insect attack in the field. Even today, many botanical insecticides are mixtures of natural compounds that interact to reduce insect feeding damage. Other botanical products are highly purified compounds with activity that closely resembles that of synthetic insecticides. While there are many botanical products with insecticidal activity, turfgrass managers will most likely be exposed to the following three.

Neem

The neem tree (*Azadirachta indica*) is a tropical evergreen related to mahogany. Native to East India and Burma, it grows in much of Southeast Asia and West Africa. The principal active compound in neem is azadirachtin, which is both a feeding deterrent and a growth regulator. Minor compounds in neem products may act as repellents, growth regulators, oviposition suppressants, sterilants, or toxins (Weinzierl and Henn 1994b)

Pyrethrum

Pyrethrum is a natural insecticide derived from the dried flower head of the African chrysanthemum, *Chrysanthemum cinerariaefolium*. The active compounds contained in pyrethrums are known as pyrethrins. Pyrethrum has long been used to control flying insects because of its fast-acting knockdown effect. Because insects can recover rapidly, natural pyrethrins are usually combined with synergistic compounds such as piperonyl butoxide, which reduces the rate at which the target insect is able to metabolize pyrethrins. (Synthetic pyrethroids, discussed in Chapter 28, are insecticides whose molecular structure is based on the naturally occurring pyrethrins.)

Nicotine

Nicotine is a naturally occurring alkaloid obtained from the dried leaves of tobacco. Commercial preparations of nicotine are often used as a fumigant for insect control in greenhouses. Nicotine teas are sometimes concocted by home gardeners as a home remedy for insect pests under the mistaken belief that natural products are inherently safer than synthetic insecticides. Nicotine in its pure form is an extremely toxic compound, with a mammalian toxicity that places it in the most dangerous class of all insecticides (Weinzierl and Henn 1994b).

28

Chemical Control Strategies

Despite the promising developments that have occurred recently in insect-resistant cultivars, biopesticides, and improved cultural practices, turf managers often find that circumstances dictate the use of a traditional insecticide. The pressure to use insecticides to manage turf insect populations often comes from clients (e.g., homeowners) or golfers who have come to expect and demand virtually pest-free turf and assume that the use of an insecticide will produce the desired result. When there is a sudden and unexpected heavy infestation of a pest insect, there often is no alternative but to depend on a traditional insecticide to reduce the population in a timely manner.

Accurate Diagnosis

Accurate diagnosis of an insect or mite pest is critical in any pest management program. The insect (or mite) has several distinguishing characteristics that can help identify it, including type of metamorphosis, whether it has wings or legs or both, where in the turf it occurs, how large it is, or what color it is. Several sampling and diagnostic techniques are discussed in more detail in Chapter 26. The damage associated with the insect (e.g., discrete off-color patches, general droughtlike stress, or destroyed roots) provides additional clues. The location of pest damage (e.g., height of cut, grass species or cultivars attacked, soil conditions, or exposure to sunlight) provides even more information. Daily and seasonal activity of the insect also help in identification—some insects are active only at night and most insects (and mites) have fairly predictable life cycles such that certain stages are present at specific times of year (Schumann et al. 1997).

Threshold Populations

When an insect (or mite) infestation has been identified and confirmed as a potential pest, it is necessary to determine whether that infestation is severe enough to require action. In general, if the turf is growing well or has the ability to recover from various agronomic stresses, it can tolerate more pest insects than turf that is already under stress.

Action thresholds, or tolerance levels, are discussed in more detail in Chapter 25, but one concept is reinforced here. Some insecticides are most effective when applied before an insect population has reached its damaging stage (preventive application). Several insecticides (e.g., fipronil, halofenozide, imidacloprid) developed in the 1990s are best used in this manner. In addition some of the insect growth regulators (e.g., azadirachtin, spinosad) must target small larvae or nymphs. Other insecticides should be applied after a population has developed to the damaging stage (curative application). Many of the insecticides developed or used in the 1970s and 1980s are most effective when used this way. If there are no effective curative insecticides for a given insect pest, threshold numbers typically are lower because a turf manager must be prepared to treat the affected area early, before the insects reach damaging levels.

Influence of Thatch

Thick, dense thatch greatly limits the effectiveness of insecticides, particularly in the control of soil-inhabiting turfgrass insects. Insecticides applied to the surface of turfgrass often are adsorbed by the thatch, preventing their movement to the upper root zone (Niemczyk et al. 1988). When long residual chlorinated hydrocarbon insecticides, such as chlordane and DDT, were used for turf insect control in the 1950s and 1960s, these materials eventually moved to and penetrated the soil as a result of rainfall and other actions of weather (e.g., expansion and contraction of soil from freezing and thawing).

Turfgrass insecticides that have been developed and used since 1970 tend to be much less persistent. To be effective, organophosphate, carbamate, and synthetic pyrethroid insecticides must move into or through the thatch rapidly, and come in contact with the target insect quickly, because of their relatively short residual activity. Some of these insecticides are degraded completely in less than a month. The most effective medium for percolation through the thatch and into the soil is water, either natural rainfall or irrigation. Chlorpyrifos, a commonly used organophosphate insecticide, is tightly bound by thatch and has one of the lowest water solubilities. Trichlorfon, another organophosphate compound, is highly soluble in water and is not tightly bound by thatch. Soils with high organic matter content may also reduce the effectiveness of insecticides, even in the absence of an established thatch layer (Cowles and Villani 1994). These relationships of solubility and thatch absorption hold true for several turf insecticides, with bendiocarb a notable exception (Table 28-1).

Isazofos, an insecticide that is relatively mobile in surface water, was applied to turfgrass and plots were irrigated 0, 8, 24, or 36 hr after treatment in an Ohio study. Subsequent analysis showed that 96–99% of the residues that were recovered were found in the thatch as much as 2 weeks after application. In many cases, thatch residues were significantly lower in unirrigated plots than in those irrigated immediately after application. While the authors (Niemczyk and Krueger 1987) suggested that this may have been because posttreatment irrigation washed the residue off the grass blades (which were not sampled) into the thatch, it is also possible that some of the material in unirrigated plots volatilized more rapidly than in irrigated plots.

In another study, ethoprop, isazofos, or isofenphos was applied to a golf course fairway and irrigated. A soil probe was used to collect samples of thatch, the top inch of soil, and the second inch of soil, 2, 5, 15, 29, 57, and 91 days after application. At 2, 5, 15, and 29

Table 28-1. Water Solubility of Insecticides and Their Binding
Characteristics on Turfgrass Thatch

Insecticide	Water Solubility[a] (ppm)	Units of Thatch Required to Bind 50% of Insecticide Applied
Chlorpyrifos	3.5	4
Diazinon	50	75
Trichlorfon	135,000	>500
Bendiocarb	260	>640

[a]Water solubility figures from Material Safety Data Sheets.
Source: Adapted from Niemczyk 1977 and Tashiro 1983.

days, more than 97% of recoverable residues from all treatments were found in the thatch. At 57 and 91 days after application, more than 90% of recoverable residues were still in the thatch. These results suggest that thatch traps several insecticides (Niemcyzk et al. 1988).

Thatch management can be a critical component of turf management. Several turf insect pests (e.g., chinch bugs, webworms, cutworms) thrive in thick thatch, so reducing or thinning thatch may help reduce those populations. In addition, reducing thatch enables insecticides to penetrate more completely.

Timing of Application—Seasonal

Effective control of a given turfgrass pest with an insecticide depends on several factors. Proper timing of application, directed at the most vulnerable stage, is one of the most important factors. In many cases the most vulnerable stage is the young larva (e.g., webworms or white grubs) or nymph (e.g., mole crickets or chinch bugs) just after emergence from the egg. In some instances, the adult is susceptible to insecticides just prior to egg laying (e.g., annual bluegrass weevil, bluegrass billbug, or black turfgrass ataenius). These variables suggest the need to have a good working knowledge of the biology and behavior of potential insect pests. Figure 28-1 indicates the period of seasonal occurrence of the destructive stage of the major groups of injurious insects and adds to an understanding of proper timing of insecticides relative to the biology of the insect. Note that the most vulnerable stage rarely is present when damage becomes apparent. Normally damage occurs when most of the population is in large larval or nymphal stages and not as susceptible to insecticides as are younger stages.

Winter Damage

The winter grain mite is the only major pest damaging turf during the winter. Streu and Gingrich (1972) noted that mite infestations are often greater in turf that was treated with a carbamate the previous summer. Autumn treatments with certain organophosphates have not proved effective, but several organophosphates have been more effective in early spring applications (Chapter 4). March flies also may be active in late winter or early spring, particularly in soils with high organic matter content (Potter 1998) (Chapter 23).

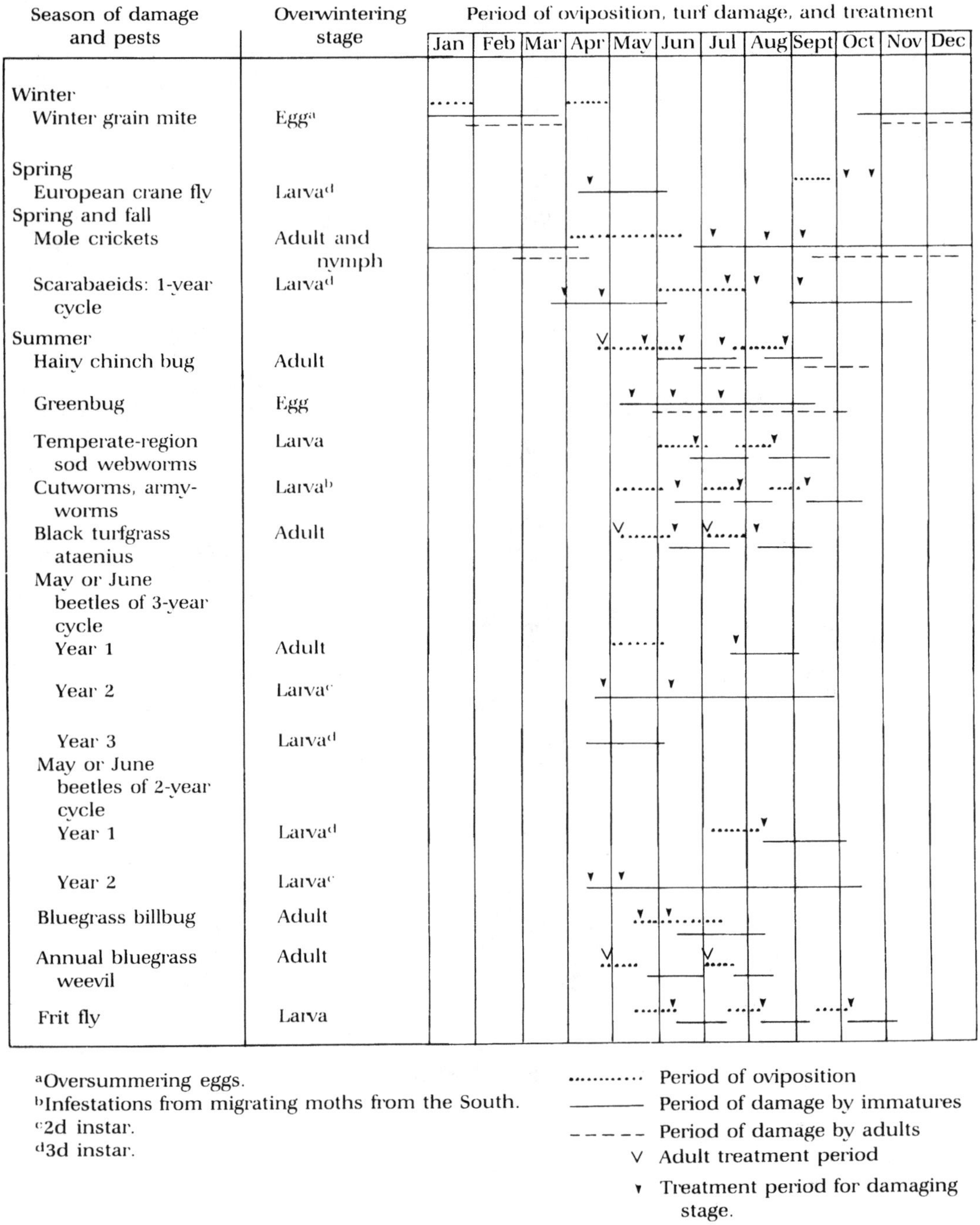

Season of damage and pests	Overwintering stage	Period of oviposition, turf damage, and treatment
		Jan | Feb | Mar | Apr | May | Jun | Jul | Aug | Sept | Oct | Nov | Dec
Winter		
Winter grain mite	Egg[a]	
Spring		
European crane fly	Larva[d]	
Spring and fall		
Mole crickets	Adult and nymph	
Scarabaeids: 1-year cycle	Larva[d]	
Summer		
Hairy chinch bug	Adult	
Greenbug	Egg	
Temperate-region sod webworms	Larva	
Cutworms, armyworms	Larva[b]	
Black turfgrass ataenius	Adult	
May or June beetles of 3-year cycle		
Year 1	Adult	
Year 2	Larva[c]	
Year 3	Larva[d]	
May or June beetles of 2-year cycle		
Year 1	Larva[d]	
Year 2	Larva[c]	
Bluegrass billbug	Adult	
Annual bluegrass weevil	Adult	
Frit fly	Larva	

[a]Oversummering eggs.
[b]Infestations from migrating moths from the South.
[c]2d instar.
[d]3d instar.

```
..........   Period of oviposition
_____    Period of damage by immatures
_ _ _ _ _    Period of damage by adults
   V         Adult treatment period
   ▾         Treatment period for damaging
             stage.
```

Figure 28-1. Seasonal occurrence of destructive stages of turfgrass insects and timing of insecticide applications. (Drawn by H. Tashiro, NYSAES.)

Spring Damage Only

European crane fly larvae cause damage in the spring only. Major turf damage occurs during March and April as the overwintering third instars become fourth (ultimate) instars. Normally, insecticides should be applied during early April. Preventive fall applications during October, after most of the eggs have hatched and when the larvae are small and vulnerable, have also been effective (Chapter 19). Bronzed cutworms occasionally are active in late winter or early spring (Chapter 9).

Spring and Autumn Damage

Southern and tawny mole crickets complete their egg laying in May to July. During late summer or early autumn, the majority of the population is large nymphs capable of causing severe damage. The most effective time for control with insecticides is during May through mid-July throughout much of the range where mole crickets are active (Chapter 5).

Scarabaeid species that normally have a 1-year life cycle cause most damage during autumn as a new generation of grubs become third instars. Midsummer (mid-July) to early autumn (early September) is the most effective period for insecticide applications to kill the first or second instars before turf damage becomes apparent, provided that the soil is moist and the insecticide is watered in. If the grubs are not treated during the fall, they will move down in the soil to overwinter and migrate back to the soil-thatch interface in late March and early April (depending on soil temperatures and species) to resume feeding. Insecticide applications should be made as larvae return to the surface. In spite of the generally cool weather, insecticide applications can be effective. Apparently, the frequent spring rains encourage grubs to remain in the upper root zone and lower thatch, and promote relatively rapid movement of the material through the thatch into the surface soil zone where grubs are feeding (Chapters 13–16).

Summer Damage

All remaining turfgrass insects damage the turf primarily during the summer, although some may start early in the spring or continue into autumn. Treatments should be applied for the hairy chinch bug, greenbug, temperate-region webworms, cutworms, and armyworms when nymphal or larval populations become established but before turf damage becomes noticeable (Chapters 6–9). Insecticides are most effective when directed toward early nymphal or larval stages, which are usually more vulnerable to insecticides than are later stages.

Two approaches to controlling the black turfgrass ataenius are available. Phenological correlations indicate that overwintering adults start laying eggs when Vanhoutte spirea and horse chestnut are in full bloom and when black locust is in early bloom. This is generally the first half of May in southern Ohio and early June in New York and southern New England. Traditional grub insecticides applied to the turf during or shortly after this period will kill small larvae as they hatch. The deposition of second-generation eggs coincides with the early bloom of rose of Sharon. In southern Ohio this is normally during the second half of July. These phenological correlations also coincide with 100–150 DD $_{13C\ base}$

for first-generation egg laying and $650-710$ DD $_{13C\ base}$ for second-generation egg laying. The second approach targets overwintering adults as they are migrating from protected overwintering sites to oviposition sites on fairways, greens, or tees. Insecticides that are readily adsorbed in thatch can be applied between the times of forsythia and dogwood (*Cornus florida* L.) full bloom. In New York and southern New England, this is normally in late April or early May (Chapter 12).

May or June beetles with a 3-year life cycle lay eggs in May or June of the first year. Young grubs, as first and second instars, cause only minor damage in July and August of the first year, and this is the most opportune treatment period to prevent serious damage the following year. Major damage occurs during the second year as overwintering second instars migrate to near the surface and third instars feed all summer into September. A spring or early summer treatment during the second year can provide satisfactory results, particularly if a fast-acting insecticide is watered in heavily. During the third year, damage again is minor, as third instars feed for a short period in May before moving down in the soil profile to pupate, and insecticide applications normally are not appropriate or effective (Chapter 15).

May or June beetles with a 2-year life cycle deposit eggs during midsummer (normally during late July). Treatment during August and early September, directed at young first and second instars, is normally most effective. Missing this opportunity, treatment the following spring during late April through May would be directed at second or third instars that have migrated to near the soil surface to feed throughout the summer (Chapter 15).

Phyllophaga crinita in Texas is univoltine (one generation per year), although some individuals may require 2 years for a generation in the northern part of their range. Treatment periods (targeting first instars shortly after they emerge from eggs) vary from June to early July in the extreme south to late July to mid-August in the extreme north (Chapter 15).

Damage caused by bluegrass billbugs appears beginning in mid-June. Since the most effective way to prevent an infestation is to target adults before oviposition, insecticide applications made to the turf in late April to mid-May (depending on soil temperature and species) should kill adults before major oviposition occurs during May and June (Chapter 18). Shetlar (1995a) indicated that some soil insecticides can be effective against bluegrass billbug larvae when they move to the soil to feed on crowns in mid-June to mid-July.

An identical principle holds for the annual bluegrass weevil, which produces its major damage during late spring into early summer. Normally damage is less severe with the second and third generations in mid to late summer. Treatment during the spring to kill overwintering adults before or during egg laying provides effective control. Adults move out of overwintering quarters normally from mid-April to early May, timing that coincides with the blooming of forsythia and the full brack color of flowering dogwood. Treatments during this 2-3-week period are effective, particularly when insecticides that will remain in the thatch are used. The greatest efficacy normally is achieved during the period when the dogwood bracts are in full color. Spring applications usually can be restricted to the outer edges of fairways, greens, and tees. Midsummer treatment periods are more variable—problem areas are more localized and often involve the same areas from one summer to the next. Generally only localized areas (i.e., areas that incurred loss of turf from the first generation) need attention for the second generation (Chapter 18).

Frit flies have as many as three generations a year, normally from the summer to early fall, with larval damage occurring shortly after each adult appearance. Seldom do all three generations create a threat of damage. However, if damage appears imminent, treatments are recommended when the flies or larvae are present, with a second treatment 10–14 days later (Chapter 19).

Timing of Application—Daily

Many turf insects are nocturnal and remain hidden in burrows during the day. Recent studies (Williamson and Potter 1997b, Williamson and Shetlar 1995) showed that applications of insecticides as late in the day as possible are significantly more effective against black cutworms than the same materials applied in morning or midday. However, some insecticides registered by the Environmental Protection Agency may have "reentry" restrictions, which would further dictate when they could be applied.

Some insecticides are sensitive to breakdown in ultraviolet light, so these materials should be applied in early morning or late afternoon to avoid the most intense ultraviolet rays from sunlight.

Windows of Opportunity

Most turf insects are susceptible to insecticides at some point during their development, normally as young larvae or nymphs. Turf managers are not always able to apply insecticides at the optimum time so they must learn to recognize the "window of opportunity" (period when insects are susceptible) and select the insecticide best suited to the circumstances (Schumann et al. 1997). Specifically, some insecticides begin to kill insects almost immediately after application, but break down quickly. Such materials should not be used early in the treatment period, because the insecticide may degrade before all the insects have hatched. Instead these insecticides should be used curatively only. Conversely, some insecticides take 2 or 3 weeks after application before they begin to kill insects, but remain active for several weeks. These materials should not be used late in the treatment period, because the insecticide will not become active soon enough to affect the damaging stages. Normally these kinds of insecticides should be used preventively (before damaging stages develop), and only in areas with a history of pest insect activity.

Names of Insecticides

All insecticides have a chemical name, a common (or generic) name, and a trade name. Chemical names are extremely complex and provide enough information so an organic chemist would be able to sketch the structure of the molecule. Common names often are derived in part from the chemical name and may provide clues about the chemical nature of the insecticide. Regardless of crop use or manufacturer, a common name refers to only one compound and is spelled in lowercase letters. Trade names (names by which a chemical is sold) are often confusing, because a given compound may have two or more trade names or formulators. For instance, chlorpyrifos prepared for turf use is called Dursban, but the preparation for other agricultural crop uses is Lorsban. The carbamate bendiocarb is sold as Turcam in the turf market, Dycarb in the greenhouse and ornamentals market, and Ficam for structural pest control. A single compound may have different trade names, depending on the company packaging and selling the product. Trichlorfon as it is market-ed by Bayer Corporation is called Dylox, while that sold by AgrEvo is called Proxol. We use common names of insecticides here to reduce confusion.

Chemical Properties of Insecticides

Each insecticide is a relatively complex organic compound that consists of carbon, hydro-gen, and usually oxygen. Each compound has several inherent characteristics, based in part on the physical structure of the molecules that make up the compound.

Water Solubility

Water solubility is a measure of how readily an insecticide will "dissolve", or go into solu-tion, in water. Generally, a more soluble insecticide will penetrate thatch more readily than other compounds. However, soluble compounds are also more likely to leach or run off from the point of application (Balogh and Anderson 1992). The solubility of some com-monly used turfgrass insecticides is given in Table 28-1.

Adsorption Coefficient

The adsorption coefficient (K_d) indicates how readily an insecticide is adsorbed (bound) to soil particles. K_d is influenced by soil characteristics (e.g., percentage of organic matter or clay) as well as chemical properties of the insecticide. Chemicals that are highly adsorbed (high K_d value) are less likely to move through thatch or to move beyond the soil profile. These compounds are much less likely to leach or runoff than are less highly adsorbed compounds (Balogh and Anderson 1992). K_{oc} takes into account the effect of organic mat-ter on K_d, and is determined by dividing K_d by the percentage of organic matter in the soil. Some references provide K_d values, while others provide K_{oc} values. In both cases high val-ues indicate a high rate of adsorption and less potential to leach or run off. The adsorption coefficient of some turfgrass insecticides is given in Table 28-2.

Vapor Pressure

The vapor pressure of an insecticide indicates how readily it will volatilize (transform from the solid or liquid state to a gaseous phase, or "evaporate"). Insecticides that have high va-por pressures will volatilize readily and may be lost from the point of application. Volatiliza-tion occurs most rapidly in high temperatures, relatively low humidity, and light air movement (Balogh and Anderson 1992). The vapor pressure of some turfgrass insecticides is given in Table 28-2.

Persistence and Speed of Efficacy

Some insecticides are extremely stable and remain in their original form for several months or even years. DDT and chlordane, which were used as turfgrass pesticides in the 1950s

Table 28-2. Water Solubility, Adsorption Coefficient, Vapor Pressure, and Half-life of Several Turf Insecticides

Insecticide	Water Solubility (ppm)	Soil Adsorption K_{oc} (mg/ml)	Vapor Pressure (mm Hg) ($\times 10^{-4}$)	Soil Half-life (days)
Acephate	818,000	2	NA	3–6
Bendiocarb	260	570	0.00000052	3–28
Carbaryl	36	300	1.3	6–110
Chlorpyrifos	3.5	7,500[a]	0.9	6–139
Cyfluthrin	0.005	NA	0.00033	56–60
Diazinon	50	350[a]	1.4	7–103
Ethoprop	725	75[a]	3.8	14–63
Fipronil	2.4	5,100[a]	0.000028	90–220
Fluvalinate	0.002	1,000,000	0.001	30
Fonofos	13	NA	2.1	46–90
Imidacloprid	510	350[a]	0.000015	61–150
Isofenphos	22	265[a]	0.04	30–365
Lambda-cyhalothrin	0.004	NA	0.000015	NA
Trichlorfon	135,000	4	0.016	3–27

[a]Range of values for K_{oc} as follows: chlorpyrifos, 2,500–14,800 mg/ml; diazinon, 40–570 mg/ml; ethoprop, 26–120 mg/ml; fipronil, 2,670–7,820 mg/ml; isofenphos, 17–535 mg/ml; imidacloprid, 292–410 mg/ml. NA, not available.
Sources: Manufacturer's Material Safety Data Sheets.

and 1960s, often remained active against insects for 2 or more years after they were applied. However, such highly persistent compounds often have detrimental effects in the environment (e.g., some buildup in fat tissue of insects and insect predators) and so are no longer registered for use in the United States or Canada.

The persistence of each insecticide can be measured as "half-life"—the amount of time it takes for half of the insecticide to break down from the original molecule to a degradation product, which may be more or less toxic than the original compound. Most insecticides used in turfgrass in the 1990s have half-lives of 4–60 days, although a few have even longer half-lives (Balogh and Anderson 1992). Insecticides that have long half-lives remain active against the target insect longer than insecticides with short half-lives. Half-lives of some turfgrass insecticides are shown in Table 28-2.

Ordinarily short-lived insecticides should be applied just after most of the individuals have hatched from eggs, when the immatures are most vulnerable to insecticides, or just after adults emerge from overwintering sites. Such materials are normally used in a "curative" manner, targeting insects after they have developed to damaging stages or numbers, but should not be applied too early because the material may degrade before the susceptible stage occurs. Conversely, longer-lasting insecticides often can be used "preventively" (applied before insects develop to damaging stages), targeting insects that have been present in past years and have proved difficult to control with curative measures (Schumann et al. 1997).

Some insecticides are toxic to target insects almost immediately after application, while others are much slower acting and may take 2–3 weeks to begin to kill the target insects. The "speed of efficacy" (the length of time for an insecticide to begin to kill target insects) is an important feature to consider when determining which insecticide would be an appropriate choice. In some cases, an insect population is already well established and the lar-

Table 28-3. Common Name, Trade Name, Chemical Class, and Oral and Dermal LD$_{50}$ Values for Several Turf Insecticides

Common Name	Trade Name	Chemical Class	LD$_{50}$ Oral	Dermal
Acephate	Orthene	Organophosphate	1,030–1,477	>10,250
Bendiocarb	Turcam	Carbamate	50–179	>1,000
Bifenthrin	Talstar	Synthetic pyrethroid	375	>2,000
Carbaryl	Sevin, Sevimol	Carbamate	100–850	>2,000
Chlorpyrifos	Dursban	Organophosphate	60–1,000	>2,000
Cyfluthrin	Tempo	Synthetic pyrethroid	500	5,000
Diazinon		Organophosphate	450–1,250	>2,000
Ethoprop	Mocap	Organophosphate	61	<10
Fipronil	Choice	Phenylpyrazole	97	>2,000
Fluvalinate	Mavrik	Synthetic pyrethroid	261–1,100	>20,000
Fonofos	Crusade, Mainstay	Organophosphate	8–18	25
Halofenozide	Mach 2	Insect growth regulator	435–2,850	>2,000
Imidacloprid	Merit	Chloronicotinyl	450	450
Isofenphos	Oftanol	Organophosphate	20–150	70
Lambda-cyhalothrin	Battle, Scimitar	Synthetic pyrethroid	56	696
Spinosad	Conserve	Spinosyn	3,780	>2,000
Trichlorfon	Dylox, Proxol	Organophosphate	150–649	1,500

Sources: Manufacturer's Material Safety Data Sheets.

vae or nymphs are relatively large. In such a situation, an insecticide that takes a week or more to kill the insects may not have a significant effect on the population because the immature insects are able to complete their feeding before they are killed by the insecticide (Schumann et al. 1997).

In most cases, insecticides that are fast acting tend to have relatively short half-lives, while slow-acting insecticides tend to have relatively long half-lives. Table 28-3 gives estimates of persistence for several turfgrass insecticides.

Chemical Classes of Insecticides

Insecticides can be classified by their chemical structure. There are several classes of insecticides, each with its own general characteristics.

Organochlorines (Chlorinated Hydrocarbons)

The first class of organic insecticides to be developed commercially was the organochlorines or chlorinated hydrocarbons, which contain carbon, hydrogen, and chlorine. DDT was used in agricultural settings, in forestry, and in mosquito control programs for many years, beginning in the 1940s. However, many organochlorines have long residual activity and buildup in the food chain or have detrimental effects on nontarget organisms. Chlordane, which was used widely for white grub and crabgrass control, was banned from use on turf in the United States in 1975, and was banned from all uses in 1982 (Vittum 1997a). The pattern of use of chlordane against white grubs had already declined because in many locations, grubs had developed resistance to chlordane (Tashiro and Neuhauser 1973).

There are no organochlorine insecticides currently labeled for use in turf in the United States or Canada.

Organophosphates

Organophosphates (OPs) contain carbon and hydrogen, but also contain at least one atom of phosphorus and either oxygen or sulfur attached to that phosphorus atom. The first OPs were developed in the late 1940s and early 1950s, in part because of concerns about the persistence of DDT and its relatives. There is a wide range of characteristics within the class: Some OPs remain active against target insects for several weeks while others degrade within a few days. Some OPs (e.g., chlorpyrifos, isofenphos) are virtually insoluble in water while others (e.g., trichlorfon) are very soluble. There is a wide range of toxicity to vertebrates and other nontarget organisms. Most OPs are "broad-spectrum" materials, effective against many different kinds of arthropods, including beneficial insects and spiders. OPs are cholinesterase inhibitors (neurotoxins), and normally do not build up in the food chain (Vittum 1997a). Table 28-3 lists several examples of OP insecticides. Although new chemistry is being developed, many of the current insecticides registered for use on turf are OPs.

Carbamates

Carbamate insecticides are derived from carbamic acid, and contain at least one atom of nitrogen and two oxygen atoms. Like OPs, there is tremendous variation in the characteristics of carbamates. However, only two carbamate insecticides are available for use on turf in the United States or Canada—carbaryl and bendiocarb. Both of these insecticides are broad spectrum, and tend to be quite toxic to honey bees, earthworms, and some other nontarget arthropods. They are intermediate in solubility, persistence, and toxicity (Vittum 1997a), and often are used to control armyworms, cutworms, and webworms.

Synthetic Pyrethroids

Pyrethroids are synthetic chemicals that resemble (in structure and activity) the botanical compound pyrethrin. Most pyrethroids are relatively fast acting and recently developed pyrethroids have residuals of a few weeks. Some pyrethrins are highly toxic to fish and are very sensitive to photodegradation (breakdown in sunlight). Most are less toxic to mammals and birds than are other classes of insecticides. Whereas OPs and carbamates are typically applied at 2−8 lb active ingredient per acre, pyrethroids are applied in fractions of pounds per acre (Vittum 1997a). Table 28-3 includes some examples of synthetic pyrethroids.

Chloronicotinyls

This new class of insecticide chemistry is currently represented by only one insecticide, imidacloprid. The chemical structure vaguely resembles nicotine in that it contains several nitrogen atoms and two ring structures. Imidacloprid is relatively low in acute toxicity to mammals and most other vertebrates, is quite persistent in the soil, and is relatively soluble but degrades quickly in surface water. It is much slower acting but also longer lasting than most other insecticides available for use in turf. It appears to be less detrimental to many beneficial arthropods and earthworms than most other turf insecticides (Vittum 1997a).

Phenylpyrazoles

This new class of insecticide chemistry is currently represented by only one insecticide, fipronil. The compound is effective against mole crickets at very low rates of application, and provides longer residual activity than most insecticides. Fipronil is virtually immobile in soil, so most applications are made directly to the soil (see "Subsurface Placement", later in this chapter). Fipronil is moderately toxic to mammals and other vertebrates, but apparently is not as disruptive to beneficial insects as some other turf insecticides (Vittum 1997a).

Insecticide Resistance

History

Turfgrass insects have developed resistance to some of the commonly used modern synthetic insecticides. Most insects have the ability to break down or detoxify many of the chemicals with which they come in contact. Some insects in a given population are inherently adapted to break down certain insecticides before the material can kill the insect. When such insecticides are used, these insects will survive, and pass on the genes that enable them to degrade the insecticide. If the insecticide is used repeatedly, natural selection occurs and eventually most of the insects in the population will carry the "resistance gene" (Vittum 1997a).

The resistance of the scarabaeid larvae to the chlorinated hydrocarbons, such as chlordane and dieldrin, that developed after 10–20 years' use, effectively terminated some of the most efficient insect control practices ever known. A single application provided nearly complete control for 5–10 years with chlordane and for as long as 20 years with dieldrin (Tashiro and Neuhauser 1973).

In the 1960s and 1970s, various turfgrass insects developed resistance to several of the OP insecticides. Some of the most severe cases included the southern chinch bugs in southeastern coastal areas of Florida, which developed resistance to diazinon and chlorpyrifos after 11–20 years of use (Chapter 6). The mechanisms that an insect uses to detoxify a given insecticide are often virtually identical for detoxifying other insecticides within the same chemical class. Thus, once an insect develops resistance to one insecticide, it usually becomes resistant to others in the same class, a phenomenon known as *cross-resistance*.

Resistance Management

Several procedures can be followed to delay the development of resistance (Potter 1998, Schumann et al. 1997). Turf managers should avoid using the same insecticide, or insecticides in the same chemical class, repeatedly in the same location or on the same pest generation. To delay the inevitable, treat only when absolutely necessary, when monitoring has confirmed the presence of a pest population that is or will be above the threshold level. Finally, avoid treating every section of turfgrass; spot treat the most heavily infested areas and leave other areas, with lower populations, untreated.

Effect of Insecticides on Nontarget Organisms

Many insecticides used in the turf industry are "broad-spectrum" materials that affect many kinds of arthropods, including beneficial insects (honey bees and other pollinators, natural insect predators and parasitoids, and saprophytes). Many studies have demonstrated varying degrees of reduction of nontarget insect populations. Cockfield and Potter (1983) showed that chlorpyrifos or isofenphos, when applied to Kentucky bluegrass, reduced populations of predaceous arthropods for up to 6 weeks, while effects of bendiocarb and trichlorfon were generally less severe and more transient.

In another study, Arnold and Potter (1987) maintained Kentucky bluegrass plots for 4 years on a schedule of fertilizer, herbicide, and insecticide treatments (high maintenance) or with no chemical input. Trap catches of predaceous arthropods (notably some spiders and certain predatory beetles) were significantly reduced by insecticides, particularly late-summer applications of diazinon, but predators repopulated treated plots the following spring. Soil and thatch pH decreased significantly and thatch accumulation was three times greater in the high-maintenance program, but earthworms were not affected significantly and oribatid mite populations normally were higher in high-maintenance plots. The authors suggested that thatch accumulation may have been a consequence of more vigorous turf growth rather than decreased decomposition. They added that the effects of insecticides are variable and in some cases less severe than had been anticipated.

One group of organisms that plays a critical role in turf growth and vigor is earthworms (Oligochaeta: Lumbricidae). They aid in the natural decomposition of turf by fragmenting plant debris in their digestive tract before excreting the material, which is ultimately broken down by soil microbes (Curry 1987). In addition earthworms mix organic matter into subsurface soil layers. Their burrowing activity increases aeration and water infiltration (Randell et al. 1972). In a Kentucky study, the toxicity of 17 commonly used turfgrass pesticides to earthworms was determined. Of the insecticides tested, a single application of bendiocarb, carbaryl, or ethoprop at labeled rates reduced earthworm populations by 60–99%, with significant effects lasting for at least 20 weeks. Other insecticides (including chlorpyrifos, diazinon, isazofos, isofenphos, and trichlorfon) resulted in somewhat more variable results, but caused significant earthworm mortality in some tests (Potter et al. 1990).

Insecticide Formulations

Insecticides consist of some amount of active ingredient (the actual insecticidal product), usually 1–80% of the total material, and the remainder as inert ingredients. The inert ingredients often dictate the "formulation", or final form, that an insecticide takes. Some examples of inert ingredients are talc (used to incorporate the active ingredient on a dust for dry application), chalklike powder (used to incorporate the active ingredient so that it can subsequently be dissolved in water for spray application), corn cob or bentonite clay (used to form granules on which the active ingredient is incorporated), or petroleum-based solvents (used to get the active ingredient into emulsion in water so it can be sprayed). Formulations are devised to make the product safer and more convenient to use. Most of the information that follows is derived from Gaussoin (1995) and Vittum (1997b).

Granular Formulations

Granular formulations have a mix of coarse, porous, free-flowing particles on which the active ingredient is incorporated (Potter 1998). The base may be corn cob (relatively light weight), vermiculite, clay, or a synthetic gel. Particle size varies but is usually quite uniform within a given product. The concentration of active ingredient usually ranges between 0.5% and 10%, although occasionally a formulation may contain even more active ingredient.

Granular products have several advantages over other formulations. They are ready to use without further mixing. Most granular products produce minimal drift. Most granules are less than 10% active ingredient, so they are less concentrated and therefore less hazardous to the handler.

However, some granular formulations release fine dusts during application, resulting in inhalation hazards. If granules differ in size or density, some may settle out during application, leading to an uneven distribution pattern (P. J. Vittum, personal observation). The active ingredient often is not released from the granular carrier until an irrigation or rainfall event (Potter 1998). Finally, shipping and storage costs tend to be higher for granular products because there is more bulk involved.

Sprayable Formulations

Many formulations are added to water and applied through standard hydraulic sprayers. Each of these has advantages and disadvantages. Some of the more common formulations are discussed here.

The inert ingredient in *soluble* or *wettable powders* (SPs or WPs) is a fine clay-like material. When the powder is added to water, it goes into solution (or suspension). The powder usually is not readily absorbed through the skin of the handler. Spills are often easier to clean up than spills involving liquid carriers. SPs and WPs are less likely to cause phytotoxicity than are many other formulations. Shipping and storage costs usually are much lower than for granular formulations.

SPs and WPs can generate fine dust when added to a sprayer, leading to the potential for inhalation hazard. Powders may clog nozzles or lead to abrasion if used repeatedly, and they tend to settle out in the tank unless they are agitated constantly. Powder formulations may leave a visible residue on the leaf surface, which is usually visible only for a few days.

Dry flowables and *dispersible granules* or *water dispersible granules* (DFs and DGs or WDGs) are clay-based formulations that have been reconfigured into small pellets or beads. They are much less dusty than WPs or SPs, so inhalation hazard during mixing is greatly reduced. However, they require even more agitation during application and can clog or abrade nozzles.

Many insecticide active ingredients are not soluble in water, and therefore cannot be formulated as WPs or WDGs or other solution-based products. An emulsifiable concentrate (EC) is a petroleum-based product which contains emulsifying agents that enable the active ingredient to go into suspension in water. The product is clear, but the final spray solution is opaque.

ECs are usually quite concentrated, so shipping and storage costs are minimized. Small nozzle orifices can be used because the formulation is less likely to abrade nozzles. However, it tends to be more corrosive to metal components of the sprayer. Applications of emulsifiable concentrates normally will not wash off the leaf surface as readily as most other sprayable formulations, and will remain active on the surface of plants longer than most other formulations.

However, ECs have some handling hazards of which turf managers must be aware. For example, ECs tend to be hazardous during the mixing process, because the oil-based carrier penetrates skin much more readily than any other common formulation, so risk of dermal exposure is considerable. Spills can be difficult to clean up. Some ECs have a low flash point, so they must be stored in relatively cool places. Finally, ECs are more likely to be phytotoxic than most other sprayable formulations.

Flowable formulations consist of finely ground particles of the active ingredient which are suspended in water. These products go into suspension in water, and mix well in a sprayer. Flowables have numerous favorable traits: They seldom leave visible residues on the plant surface, are much less likely to burn plants than are ECs, and are less hazardous during mixing than are ECs. They need some agitation but not as much as WPs or SPs. Soluble concentrates (SCs or CSs) are similar in physical appearance (although they tend to be a bit thicker and more subject to settling in storage) and handling characteristics.

A *microencapsulated suspension* (ME) is a suspension of the active ingredient in microscopic capsules, which produces an extended release of the active ingredient. MEs usually are less hazardous than comparable ECs. Handling and storage characteristics of MEs are similar to those of ECs, but they usually provide even better residual activity. Because the technology is still developing, MEs tend to be more expensive than other formulations.

Selecting a formulation sometimes appears to be a daunting task, but it can be simplified. Many field trials suggest that generally there is no difference in efficacy between granular and sprayable formulations for soil insects (e.g., white grubs). While granulars sometimes take a day or two longer to become active than sprayable formulations of the same active ingredient, they usually remain active just a little longer as well. The resulting mortality to the soil insect population is seldom significantly different between formulations of the same active ingredient. The evidence regarding efficacy of different formulations against surface-feeding insects is less conclusive, and some differences in performance may be encountered.

Toxicity of Insecticides

Acute toxicity refers to the harmful action of a material following one or a few relatively high doses. It is measured by exposing rats, mice, or rabbits to a range of doses of the compound and determining the survival rate of the test animals. Such a test establishes the LD_{50}, or lethal dose 50%, the amount of material needed to kill 50% of the test population. The value is expressed in terms of milligrams of actual toxicant per kilogram of total body weight of the test animal (mg/kg), which is equivalent to parts per million (ppm). The tests may be administered orally (oral LD_{50}) or applied to the skin (dermal LD_{50}). The larger the LD_{50}, the less toxic the chemical (Potter 1998, Vittum 1997c). Table 28-3 lists some common insecticides that are or have been recommended for use on turfgrass, and the toxicity of the technical ("pure") chemical. In actual use these are not the toxicities encountered, since most formulations lower toxicities by reducing the percentage (or availability) of the active ingredient.

There is generally no correlation between high toxicity (low LD_{50}) and effectiveness of an insecticide, because many other factors, such as solubility, timing, weather, insect activity, and insecticide degradation, are involved. All necessary precautions relating to toxicity and use are stated on the label. Ordinarily, if all other considerations are equal, the least toxic pesticide that provides acceptable control should be chosen. This minimizes accidents when label directions are not followed strictly. The danger (misapplications, misuses) is greatest with less-experienced users such as homeowners.

Application Technology

Subsurface Placement

There have been several developments in insecticide application technology in the 1980s and 1990s. One involves the direct placement of an insecticide beneath the surface of the turfgrass. This is accomplished by using a slicing unit (a coulter or other device slices the turf and the insecticide is delivered directly into the slit, either as a liquid or as a granular formulation) or using high-pressure liquid injection (pressures of up to 5,000 psi delivered through nozzles placed no more than 1 cm above the ground). Other variations (low-pressure injection systems for liquids, high-pressure systems for granules) are being developed as well.

Subsurface placement enables an applicator to deliver the insecticide directly to the soil, and is particularly effective when used against soil insects such as mole crickets and white grubs. The application rates of some insecticides can be reduced without loss of efficacy (Vittum 1994), the amount of insecticide on the surface can be reduced (P. J. Vittum, unpublished data 1991), and the material appears to be less sensitive to photodegradation. Materials applied below the surface apparently are less likely to run off.

Subsurface appplication equipment is very specialized and therefore may not be cost-efficient for some turf settings. While most high-pressure liquid injection systems have numerous safeguards, there can be an inherent hazard when using very high pressures. Materials applied below the surface might be more likely to leach, although there are no published data to confirm this supposition. Furthermore, not all insecticide labels or state interpretations support the use of this technology.

Environmental Issues

When an insecticide is applied to turfgrass, it may adsorb to soil particles, move away from the point of application, or degrade to various by-products. Each of these actions can have a direct effect on the environment. Much of the following discussion is derived from Brown and Hock (1989) and Vittum (1997d).

Pesticide Adsorption

The process of adsorption binds pesticides to soil particles. Chemical characteristics of the insecticide and various soil factors influence pesticide adsorption. For example, soils that are high in organic matter or clay content adsorb insecticides more readily than do coarse, sandy soils. Pesticides that are tightly adsorbed to soil are much less likely to move through the soil profile, and thus are not apt to run off or leach.

Pesticide Transfer

Volatilization is the transformation of a solid or liquid into a gas, similar to evaporation. Once an insecticide has volatilized, it may move on air currents away from the original point of application. Insecticides with a high vapor pressure are more volatile. Environmental conditions in which volatilization is most likely are high temperature, low relative humidity, and light wind. An insecticide that is tightly adsorbed to soil is less likely to volatilize.

Runoff refers to the movement of water over a sloping surface, which occurs when water is applied (in rainfall or irrigation) to the soil at a rate faster than it can enter the soil. The water may carry insecticides directly or bound to soil particles that are moved through erosion. Highly soluble insecticides are more likely to dissolve in and move with runoff, while short residual materials may degrade before they reach surface water. Insecticides that are tightly adsorbed to soil particles are less likely to run off (unless the entire soil particle is carried in an erosion event). Runoff is most severe in areas with steep slopes, compacted or already saturated soils, or thin turf cover.

Leaching refers to the vertical movement of water through the soil profile and, ultimately, to groundwater. Highly soluble insecticides are more likely to leach, while insecticides that break down rapidly are less likely to leach because they remain in the soil a short time. Insecticides that are highly adsorptive are less likely to leach. Leaching is most severe in soils that are highly permeable or low in organic matter, and in areas with thin turf cover.

Absorption (uptake) is the movement of pesticides into plants. Once an insecticide has been absorbed by a plant, it may be broken down directly or remain inside the plant for a period of time. Systemic insecticides, such as acephate and to some degree imidacloprid, are absorbed into plant tissue readily, while some other insecticides primarily remain on the plant surface. The degree to which an insecticide may be absorbed depends on the chemical properties of the insecticide as well as physical characteristics of the plant (e.g., presence and density of leaf hairs, waxiness of cuticle).

Collection of clippings can be of concern to turf managers because each time clippings are collected and removed to another location, some pesticide residue may be moved with the clippings. While golf course clippings normally are not used for mulch, homeowners sometimes do incorporate clippings into mulch or compost. To minimize risks from clipping removal, insecticides with short residual properties should be used in such circumstances.

Insecticide Degradation

Insecticide degradation (breakdown) is usually beneficial, because most reactions that destroy insecticides transform the residues into less toxic compounds. Degradation can pose a problem, however, when the insecticide is broken down before it has controlled the target insect (Brown and Hock 1989).

Microbial degradation is the breakdown of insecticides by microorganisms, usually in the soil, that use insecticides as a primary food source. Soil conditions (moisture, texture, organic matter content, temperature, pH) have a direct effect on the rate of microbial degradation because of their effect on the microorganisms. In addition, microbial degradation is more likely to occur when the same insecticide is used in a given location several times in quick succession. Repeated applications stimulate the buildup of populations of the microorganism, which in turn lead to accelerated degradation of the insecticide (Brown and Hock 1989).

An example of microbial degradation in turf involved isofenphos, a soil insecticide first used to control corn rootworms in the Midwest. In some areas where isofenphos was used against white grubs several consecutive years in the early 1980s, the effectiveness of the material declined over time. Niemczyk and Chapman (1987) sampled soil from golf courses where spring applications of isofenphos failed to control summer-generation white grubs, and found that isofenphos disappeared (91–99%) in 3 days, compared to no degradation from samples from an untreated site. While the phenomenon of accelerated (enhanced) degradation was blamed for many of the instances in which isofenphos did not meet expectations, some of the "failures" could be attributed to poorly calibrated equipment, inaccurate or poorly timed applications, inadequate watering, or unreasonable expectations (P. J. Vittum, personal observation).

Chemical degradation occurs when an insecticide is degraded by processes that do not involve living organisms. These processes are governed by soil and thatch temperature, moisture, and pH, as well as by chemical properties of the insecticide. One of the most common chemical reactions is hydrolysis, in which an insecticide reacts with water and is broken into two or more smaller molecules. Many OPs and carbamates (notably, acephate, isazofos, and trichlorfon) are very susceptible to hydrolysis under alkaline (basic, or opposite of acidic) conditions (Cowles and Villani 1994). This sensitivity to alkaline hydrolysis (breakdown in water with pH >7.0) can also lead to incompatibility with other pesticides or fertilizers.

Photodegradation refers to changes in chemical bonds of insecticides (and other pesticides) following exposure to ultraviolet radiation from sunlight. Photodegradation can affect insecticides on leaf blades, in the thatch, and in the air, but insecticides that have been applied below the surface appear to be less subject to it. Photodegradation is most severe when sunlight is most intense (summer months, midday hours, and lower latitudes).

Selecting an Insecticide

If an insect population has reached damaging numbers or other circumstances dictate that insect management strategies must be employed, and there are no viable cultural or biological control alternatives available, several factors should be considered before determining which insecticide should be used.

Identify environmental conditions (nearby ponds or streams, sandy soils, compacted soils, shallow water table) that may be impacted by an application. Note weather conditions (prevailing winds, high temperatures, anticipated heavy rains) that might affect the efficacy of an application, or might lead to increased movement of the material. Select materials that are least likely to leach, run off, volatilize, or drift. Consider the application equipment available and your ability to apply specific products or formulations accurately (Vittum 1997b,d).

Identify toxicity issues that may be unusual to a given location. Consideration of game and practice schedules must be included before insecticides are applied to athletic fields. If children play in the area regularly, further consideration might be given to biological or cultural control strategies. Choose insecticides that have the lowest levels of toxicity to humans and other nontarget organisms, and have the shortest reentry intervals (Potter 1998, Vittum 1997c).

The use of irrigation before or after an insecticide application often is critical to its efficacy. If irrigation is not available and rain events are unreliable, some insecticides should be avoided. Never use an insecticide that must be watered in unless able to do so.

Identify any beneficial insects that must be protected, including insects protected under the Endangered Species Act and such common insects as honey bees and other pollinators. Avoid using broad-spectrum insecticides whenever possible, or apply these products when beneficial insects are not present or active.

Determine whether the "window of opportunity" dictates use of a curative material (insects already present and causing damage) or a preventive material (before insects reach damaging stages or exceed threshold levels).

Avoid using the same insecticide or insecticides in the same chemical class more than 2 or 3 consecutive years if at all possible. Avoid treating the same pest generation repeatedly with insecticides in the same chemical class. Only treat selected areas and use insecticides only if absolutely necessary.

When used carefully, and at the right time in the insect's life cycle, insecticides are an important tool in insect pest management.

APPENDIX 1

English and Metric Units
of Measure and Conversions

The International System of Units (SI), which was adopted and endorsed in 1960 by the International Bureau of Weights and Measures, forms the basis for the modernized metric system. In the United States, as in the rest of the world, the scientific community has been using the metric system for a long time. The U.S. Congress enacted the Metric Conversion Act of 1975, but this has done little to promote general public acceptance of the metric system. Today, however, with few exceptions, the entire world is using the metric system or is shifting to it, although the process has admittedly been slow in this country.

The metric system is a decimal-based system, with each unit of measure related to all other units by factors of 10. For example, 10 mm = 1 cm = 0.1 dm = 0.01 m. Measurements in the text are given in metric units only. Conversion factors for units of length, area, volume, and weight are provided here. The following information is adapted primarily from Beard (1982) and U.S. Department of Commerce (1981).

Units of Measure

Length

English Units	Metric Units
1 ft = 12 in	1 cm = 10 mm = 10,000 μm
1 yd = 3 ft = 36 in	1 m = 100 cm = 1,000 mm
1 mi = 1,760 yd = 5,280 ft	1 km = 1,000 m

Conversions:

1 m = 3.3 ft = 1.1 yd   1 ft = 0.305 m

Examples:

1.0 m	=	3.3 ft	7.5 m = 24.6 ft	
2.5 m	=	8.2 ft	10.0 m = 32.8 ft	
5.0 m	=	16.4 ft		

1 km = 0.6214 mile   1.0 mile = 1.61 km

Examples:

1.0 km	=	0.6 mi	30 km	= 18.6 mi
2.5 km	=	1.6 mi	40 km	= 24.8 mi

5.0 km = 3.1 mi	50 km = 31.1 mi	
10.0 km = 6.2 mi	80 km = 49.7 mi	
25.0 km = 15.5 mi	100 km = 62.1 mi	

10 mm = 1 cm = 0.39 in   1.0 in = 2.54 cm = 25.4 mm

Examples:

0.10 mm = 0.0039 in	1.0 mm = 0.039 in	1.0 cm = 0.39 in
0.25 mm = 0.0098 in	2.5 mm = 0.098 in	2.5 cm = 0.98 in
0.50 mm = 0.0197 in	5.0 mm = 0.197 in	5.0 cm = 1.97 in
0.75 mm = 0.0295 in	7.5 mm = 0.295 in	7.5 cm = 2.95 in
1.00 mm = 0.039 in	10.0 mm = 0.39 in	10.0 cm = 3.94 in

Typical mowing heights

$3/32$ in = 2.38 mm	$1/4$ in = 6.35 mm	1.0 in = 2.54 cm
$7/64$ in = 2.78 mm	$3/8$ in = 9.53 mm	1.5 in = 3.81 cm
$1/8$ in = 3.18 mm	$1/2$ in = 12.70 mm	2.0 in = 5.08 cm
$9/64$ in = 3.57 mm	$5/8$ in = 15.88 mm	3.0 in = 7.62 cm
$5/32$ in = 3.97 mm	$3/4$ in = 19.05 mm	4.0 in = 10.16 cm
$3/16$ in = 4.76 mm	$7/8$ in = 22.23 mm	

Area

English Units

1 ft^2 = 144 in^2
1 acre = 4,840 yd^2
1 mi^2 = 1 section = 640 acres

Metric Units

1 are = 100 m^2
1 hectare = 100 ares = 10,000 m^2
1 km^2 = 1,000,000 m^2

Conversions

1 cm^2 = 0.155 in^2 1 in^2 = 6.45 cm^2

Examples:

1.0 cm^2 = 0.16 in^2	25.0 cm^2 = 3.88 in^2
2.5 cm^2 = 0.39 in^2	50.0 cm^2 = 7.75 in^2
5.0 cm^2 = 0.78 in^2	75.0 cm^2 = 11.6 in^2
7.5 cm^2 = 1.16 in^2	100.0 cm^2 = 15.5 in^2
10.0 cm^2 = 1.55 in^2	200.0 cm^2 = 31.0 in^2

1 m^2 = 10.76 ft^2 1 ft^2 = 0.093 m^2

Examples:

0.10 m^2 = 1.1 ft^2	1.0 m^2 = 10.8 ft^2	10.0 m^2 = 107.6 ft^2
0.25 m^2 = 2.7 ft^2	2.5 m^2 = 26.9 ft^2	25.0 m^2 = 269.1 ft^2
0.50 m^2 = 5.4 ft^2	5.0 m^2 = 53.8 ft^2	50.0 m^2 = 538.2 ft^2
0.75 m^2 = 8.1 ft^2	7.5 m^2 = 80.7 ft^2	75.0 m^2 = 807.0 ft^2

1 hectare (ha) = 2.47 acres (A)   1 acre (A) = 0.40 hectare (ha)

Examples:

0.10 ha = 0.25 acre	1 ha = 2.47 acre
0.25 ha = 0.62 acre	5 ha = 12.4 acre
0.50 ha = 1.24 acre	10 ha = 24.7 acre
0.75 ha = 1.85 acre	25 ha = 61.8 acre
1.00 ha = 2.47 acre	50 ha = 123.6 acre

Volume

| English Units | Metric Units |

1 cup = 16 tbsp = 48 tsp   1 ml = 1,000 μl
1 pt   = 2 cups = 16 fl oz   1 l   = 1,000 ml = 0.001 m^3
1 qt   = 2 pt = 4 cups
1 gal   = 4 qt = 8 pt
1 gal   = 231 in^3 = 128 fl oz

Conversions

1 ml = 0.203 tsp   **1 tsp = 4.93 ml**
1 ml = 0.034 fl oz   **1 fl oz = 29.6 ml**
Examples:
  1.0 ml = 0.2 tsp   10 ml = 2.0 tsp
  2.5 ml = 0.5 tsp   25 ml = 5.5 tsp
  5.0 ml = 1.0 tsp   50 ml = 10.1 tsp
  7.5 ml = 1.5 tsp   75 ml = 15.2 tsp

1 liter (l) = 1.056 quart (qt)   **2 pints = 1 quart = 0.946 liter**
Examples:
  1.0 l = 1.06 qt
  2.5 l = 2.64 qt
  5.0 l = 5.28 qt = 1.3 gal
  7.5 l = 7.92 qt = 2.0 gal
 10.0 l = 10.57 qt = 2.6 gal

Mass or Weight

| English Units | Metric Units |

1 lb   = 16 oz   1 g = 1,000 mg
1 ton = 2,000 lb   1 kg = 1,000 g
                          1 metric ton (mt) = 1,000 kg

Conversions

1 g = 0.0353 oz   **1 oz = 28.35 g**
Examples:
  1.0 g = 0.035 oz   10 g = 0.35 oz
  2.5 g = 0.088 oz   25 g = 0.88 oz
  5.0 g = 0.176 oz   50 g = 1.76 oz
  7.5 g = 0.264 oz   75 g = 2.64 oz
 10.0 g = 0.353 oz   100 g = 3.53 oz

1 kg = 2 lb 3.27 oz   **1 lb = 0.454 kg**
Examples:
  0.5 kg = 1 lb 1.64 oz   10 kg = 22 lb 0.74 oz
  1.0 kg = 2 lb 3.27 oz   15 kg = 33 lb 1.11 oz
  3.0 kg = 6 lb 9.82 oz   20 kg = 44 lb 1.48 oz
  5.0 kg = 11 lb 0.37 oz

Temperature

English units: °Fahrenheit (°F), freezing = 32°F, boiling = 212°F
Metric units: °Celsius (°C), freezing = 0°C, boiling = 100°C
Examples:

−30°C = −22°F	20°C = 68°F
−20°C = −4°F	25°C = 77°F
−10°C = 14°F	30°C = 86°F
0°C = 32°F	35°C = 95°F
5°C = 41°F	40°C = 104°F
10°C = 50°F	100°C = 212°F
15°C = 59°F	

APPENDIX 2

Abbreviations Used in the Text

Units of Measure

English		Metric	
°F	Fahrenheit	Å	angstrom
fl oz	fluid ounce	°C	Celsius
ft	foot	cm	centimeter
ft-c	footcandle	a	are
ft^2	square foot	dm	decimeter
gal	gallon	g	gram
hr	hour	ha	hectare
in	inch	kg	kilogram
lb	pound	km	kilometer
mi	mile	l	liter
mph	miles per hour	m	meter
pt	pint	m^2	square meter
qt	quart	m^3	cubic meter
tbsp	tablespoon	mg	milligram
tsp	teaspoon	ml	milliliter
yd	yard	mm	millimeter
		mt	metric ton
		μm	micrometer

Other Abbreviations and Symbols

$>$, $<$—greater than, less than

AC/DC—alternating current/direct current

BL—black light; electromagnetic radiation peaking at 3,650 Å and highly attractive to night-flying insects; generally a fluorescent lamp

BLB—blacklight fluorescent lamp with a blue-violet tube that filters out most of the visible radiation

DD—degree-days

EPA—Environmental Protection Agency, federal agency responsible for registering pesticides

ESA—Entomological Society of America (headquarters in Lanham, Maryland)

LC$_{50}$—lethal concentration of a toxic compound killing 50% of the test animals

LD$_{50}$—lethal dose of a toxic compound killing 50% of the test animals

pH—symbol used to indicate acidity or alkalinity, with pH 7 defined as neutral, pH >7 as alkaline
   (basic), and pH <7 as acidic; logarithm of the reciprocal of the hydrogen ion concentration in
   gram atoms per liter of solution
ppb—parts per billion
ppm—parts per million
RH—relative humidity
SEM—scanning electron microscopy

Glossary

abdomen—hindmost of three regions of an insect; contains much of the digestive, reproductive, and excretory systems

accession—an addition; in plant breeding, a plant with a distinct genotype used to develop a new cultivar

active ingredient—the portion of a pesticide that has a toxic or lethal effect on the target organism

acute toxicity—lethal effect of a single exposure (or a few exposures) to a compound at a relatively high rate; measured in LD_{50}

adsorb—collect in condensed form on the surface

adventitious—occurring accidentally, out of the ordinary course

aedeagal—pertaining to the aedeagus, or penis, of male insects

aeration—exposure to the air

aestivate—spend the summer in a dormant condition; opposed to *hibernate*

aggregated—occurring in "clumps" (some areas with high density, other areas with low density) instead of evenly distributed

alga—a one-celled to many-celled plant containing chlorophyll, having no true root, stem, or leaf, and found in damp places

alimentary—digestive

alkaline hydrolysis—relatively rapid breakdown of an insecticide (or other pesticide) in water with pH >7.0

antenna—one of a pair of segmented sensory organs, one on each side of the head

anterior—toward the front or head, as opposed to *posterior*

antibiosis—in biology, an association between two organisms in which one is adversely affected

apical—at or near the apex of a structure

apron—in golf, the fairway area in closest proximity to and in front of the putting green that adjoins the putting green collar and is mowed at the fairway height; sometimes called the approach

arcuate—arched, bowlike

arthropod—a member of a phylum of invertebrate animals having jointed legs and segmented bodies

articulation—a joint or the state of being jointed

aspirator—any apparatus for moving air, fluids, and so forth by suction

auricle—an earlike part or organ

axil—the upper angle between a leaf and the stem from which it grows

bacterium—a one-celled microorganism that has no chlorophyll and multiplies by simple division

balling—congregation of numerous male insects (e.g., Japanese beetles) attracted by pheromones emitted by virgin females

Berlese funnel—a combination of a heat source, usually an incandescent lamp, and an acute-angle funnel with a screen near the upper edge to hold foliage, thatch, or soil; apparatus forces living organisms into a collecting jar via heat

binomial—scientific name of a plant or animal, consisting of generic and species names

bioassay—a test used to determine the toxicity of a compound or infectivity of a parasitic organism by the use of a living organism

bionomics—branch of biology dealing with the adaptation of living things to their environment; ecology

biopesticide—a product that contains living organisms or by-products of living organisms and is used in the same manner as a traditional pesticide (e.g., repeated applications as needed using standard application equipment)

biotic—pertaining to or related to living organisms

biotype—a group of plants or animals with similar hereditary characteristics

bivoltine—having two generations in a year

black light—electromagnetic illumination peaking at 3,650 Å and attractive to night-flying insects; also causing fluorescence of certain compounds

blade—the flat portion of the leaf blade above the sheath

brachypterous—a short-winged adult, in which wings do not fully cover the abdomen

brachyptery—short-winged

braconid—a hymenopterous insect belonging to the family Braconidae, the members of which are parasitic on other insects

bract—in botany, a modified leaf, small and scalelike or large and brightly colored, growing at the base of a flower, for example, in flowering dogwood

brood—a generation of an insect that has a life cycle longer than a year and yet adults emerge annually (each brood refers to the year in which certain adults emerge)

caecal—of the caecum, a blind pouch or tubelike structure attached to the midgut

callow—teneral, or undeveloped and immature

caproic acid—a colorless, liquid fatty acid ($C_6H_{12}O_2$) found in butter and other animal fats

carbamate—any of a group of synthetic organic insecticides that are esters of *N*-methyl-carbamic acid (e.g., carbaryl, bendiocarb)

carina—an elevated ridge or keel

caudal—of or pertaining to the cauda, or the anal end of the insect body

caudomesal—directed toward the rear and toward the medial region, or middle of the body

cellulose—the chief substance composing the cell walls or woody part of plants

cephalothorax—the united head and thorax of arachnids or spiders and crustacea

chelicera—one of the pinching, pincer-like first pair of appendages of adult Chelicerata, such as ticks and mites

chitin—a colorless, nitrogenous polysaccharide secreted by the epidermis and applied to the hardened parts of an insect body

chlorinated cyclodiene—synthetic insecticides composed of highly chlorinated cyclic hydrocarbons (e.g., chlordane, dieldrin)

chlorophyll—the green, light-sensitive pigment of plants that in sunlight is capable of combining carbon dioxide and water to make carbohydrates

chlorotic—characterized by the fading of green color in plant leaves to light green or yellow

chorion—the outer covering of the insect egg

ciliate—fringed with a row of parallel hairs

claviform spot—spots clublike in form

claw (of leg)—a hollow, sharp, multicellular organ, generally paired, at the end of the insect leg

clitellum—in earthworms, the thickened, glandular, saddle-like portion of the body wall on (roughly) segments 31–37 and having a reproductive function

clone—an asexually produced progeny

clypeus—the part of the head of an insect which commonly bears the labrum on its anterior margin

coelomic cavity—the space between the viscera and the body wall

coition—sexual intercourse

collar—in botany, a narrow band marking the junction of the blade and leaf sheath; in golf, a narrow area adjoining the putting surface that is mowed at a height intermediate between that used for the fairway and the putting surface

complete metamorphosis—development of an insect through four distinct stages (egg, larva, pupa, adult); larva does not resemble adult; pupa serves as a transition ("resting") stage between larva and adult

compound eye—an aggregation of separate visual elements known as ommatidia

conifer—a cone-bearing tree or shrub, usually an evergreen

contact toxicity—toxicity of a compound when it is applied directly to the body rather than ingested

contiguous—touching

cornicle—in aphids, either of two honey tubes, the dorsal erect or semierect tubules in pairs that secrete a waxy liquid as a defense against enemies

costa—thickened anterior margin of any wing of insects

coxa—basal segment of the leg on insects or other arthropods

crochet—series of tiny hooks on the tip of the proleg of caterpillars

crown—compact series of nodes from which culms and roots arise; the area of a grass plant from which all growth is initiated

culm—stem of grass plant

cultivar—cultivated variety; in plants, one that differs from others within the same species and whose offspring retain those distinctive characteristics

cuticle—or cuticula, the outer covering of an insect formed by a noncellular layer of chitin

cuticula—see *cuticle*

cyst—a sac or vesicle (e.g., a protective sac for immature ground pearls)

dactyl—a finger or toe; a tarsal joint after the first one

degree-day—an estimate of heat energy; calculated by subtracting the base temperature for a given organism from average daily temperature

dermapteran—an insect belonging to the order Dermaptera, or earwigs

deutonymph—the second nymphal stage of Acari, which assumes the general nonsexual characteristics of the adult

diapause—period when physiological development of an insect (or other organism) ceases, usually during extended cold temperatures; similar in concept to "hibernation"

diffusion—spontaneous movement (usually scattering) of particles

diplopod—member of the class Diplopoda, known as millipedes

distal—near or toward the free end of any appendage

diurnal—active during daylight

diverticulum—an offshoot from a vessel or from the alimentary canal, usually blind or saclike

dormant—living in a state of reduced physiological activity

dorsal vessel—organ in insect that helps circulate hemolymph ("blood")

dorsomedian stripe—a line usually extending lengthwise along the middle of the dorsal surface of an insect; generally refers to lepidopteran larvae

dorsum—the upper surface, or back

eclosion—emergence of the adult insect from the pupa or of immature insect (nymph or larva) from egg

ecosystem—an ecological system; a functional system that includes the organisms of a natural community together with their environment

ectoparasite—parasitic organism living outside its host

ectothermic—"cold-blooded"; having no internal metabolic source of heat; body temperature usually mirrors ambient temperature

efficacy—effectiveness (of an insecticide)

elliptical—oblong-oval, the ends equally rounded

elytra—the anterior leathery or chitinous wings of beetles that serve as coverings to the hindwings

endoparasite—parasitic organism living inside its host

endophyte—any plant growing within another plant

entomogenous—growing in or on an insect, as do fungi, for example

entomopathogen—an agent (e.g., bacterium, nematode, fungus) that causes disease in insects

entomophagous—feeding on insects

entomophthorous—parasitic on insects; specifically, a fungus of the order Entomophthorales

enzyme—a complex organic substance produced by living cells and causing chemical changes in organic substances by catalytic action; indicated by the suffix *-ase*

epicranial suture—a Y-shaped suture (marking the junction of sclerites) on the dorsal surface of the head of a generalized insect

epidermis—the cellular layer of the skin underlying and secreting the cuticula

epimeron—the posterior of a thoracic pleuron or lateral region of a segment, usually small, narrow, or triangular

epipharynx—an organ, probably of taste, that is attached to the inner surface of the labrum

epithelium—primary animal tissue with cells usually close together; lines body cavities and ducts

epizootic—a disease temporarily prevalent among many animals

etiology—science or theory of the causes or origins of diseases

evagination—a turning inside out, with protrusion of an inner surface

excretion—act of eliminating a waste product

exoskeleton—entire body wall to the inner side, to which muscles are attached; the outside skeleton of insects

extravaginal growth—young vegetative stems that grow outside the basal leaf sheath by penetrating through the sheath

exuvia—the cast-off skin of larva or nymph at the time of molting

fairway—in golf, the area between the tee and putting green of variable lengths from about 100 to 500 m and mowed at heights of 1.3–3.0 cm

fallow ground—land plowed but not seeded for one or more growing seasons

fauna—animals of any given region or time

feather claw— distal portion of legs of eriophyid mites ending in rays and resembling a feather

fecundity—the quality or power of being prolific or fertile

femur—the thigh; in insects, usually the stoutest segment of the leg articulated proximally to the trochanter and distally to the tibia

filiform—threadlike; slender and of equal diameter

first generation—arises from eggs or offspring produced by spring-generation adults

flaccid—soft and limp; flabby

flagellum—that part of the antenna beyond the pedicel; a whip or whiplike process

foot-candle—illumination equal to 1 lumen per square foot

forage—a food or fodder plant

foregut—anterior section of the arthropod alimentary tract; includes esophagus and crop (where preliminary grinding of food occurs)

fossorial—formed for or with the habit of digging or burrowing

frass—solid larval insect excrement

fritz (Swedish)—empty kernel of grain as result of feeding by larvae of frit fly; source of common name

fungus—nucleated, usually filamentous spore-bearing organisms devoid of chlorophyll

fusiform—spindle shaped; broad at the middle and narrowing toward the ends

ganglion/ganglia—nerve center(s) composed of a cell mass and fibers and serving as a center from which impulses are transmitted

garrulous—talking much, noisy

gena—the cheek; the part of the head on each side below the eyes

genital sheath—portion of empty pupal case surrounding the terminal portion where the genital organs are located

genus—taxonomic category above species and below family

glabrous—smooth, hairless, and without punctures or structures

glutinous—gluey, slimy, viscid

gnathosoma—portion of Acari that resembles the head of generalized arthropods only in that the mouthparts are appended to it

gradual metamorphosis—development of an insect through three distinct stages (egg, nymph, adult); nymph resembles adult except development of wing pads is incomplete; for example, mole cricket, chinch bug, mealybug

granulated—formed into grains or crystals; mottled appearance

gravid—containing viable eggs

green—in golf, the putting green, the area of finest turf mowed at 4–8 mm that contains the cup and flag

gregarious—living in societies or communities but not social

grub—an insect larva; a term loosely applied but with specific reference to larvae of Coleoptera and Hymenoptera

gustatory—of or relating to the sense of taste

haltere—one of a pair of knobbed, movable filaments in Diptera that are situated on each side of the thorax and that represent the hindwings; sensors provide information on orientation of the insect

hamate—furnished with hooks

head capsule—the combined sclerites of the head, which form a hard compact case

hemelytron—the anterior wing in the Heteroptera, which has a thickened basal half and a membranous apical half

hemocoel—body cavity of insect

hemolymph—fluid that fills the insect body cavity and serves the dual purpose of vertebrate blood and lymph

hermaphrodite—an individual having characteristics of both sexes

hibernaculum—a larva's tent or a sheath, in which it hibernates

hibernate—to pass the winter in a dormant state

hindgut—posterior portion of the alimentary tract as far forward as the Malpighian tubes

holometabolous—having a complete transformation, with egg, larval, pupal, and adult stages distinctly separated

hymenopterous—of or relating to an insect in the order Hymenoptera, which includes ants, bees, and wasps, many of the latter being parasitic on other insects

hypha—one of the filaments that make up the mycelium of a fungus

hysterosoma—section of Acari body that occupies the third and fourth pair of legs and the posterior part of the abdomen

ichneumonid—a hymenopterous insect belonging to the family Ichneumonidae, the members of which are parasitic on other insects

idiosoma—the portion of Acari that assumes functions parallel to those of insects' abdomen, thorax, and portions of the head

incandescent (lamp)—a lamp in which the light is produced by a filament contained in a vacuum and heated to incandescence by an electric current

incisor—a cutting tooth

incomplete metamorphosis—development of an insect through three distinct stages (egg, naiad, adult); naiad does not resemble adult but is often aquatic; for example, dragonflies, mayflies

in copula—in copulation

indigenous—native to

insectivorous—feeding on insects

inseminate—fertilize with semen

instar—stage between molts

integument—outer covering, or cuticle, of the insect body

intercellular—between cells

internode—in botany, the section of a plant stem between two successive nodes or joints

intracellular—within cells

intravaginal growth—vegetative stems that grow upward within the enveloping basal leaf sheath

isotherm—a line on a map connecting two points on the earth's surface having the same mean temperature or the same temperature at a given time

labial palpus—a one- to four-jointed sensory appendage of the insect labium

labium—the lower lip; a compound structure that forms the floor of the mouth in mandibulate insects

labrum—the upper lip, which covers the base of the mandibles and forms the roof of the mouth in insects

lacinia—inner lobe of the first maxilla, articulated to the stipe and bearing brushes of hair or spines

lamellate—sheet or leaflike

lamellicorn—characteristic of a group of beetles that have their antennae terminating in lamellae, or thin platelike or leaflike processes

larva—an immature insect that undergoes complete metamorphosis and does not resemble the adult; for example, caterpillar, maggot, grub

larvaevorid—producing living young instead of eggs, as in some Diptera

leaf sheath—basal portion of a grass leaf that surrounds the stem

life cycle—the period between egg deposition and the attainment of sexual maturity as shown by egg laying

lignin—an organic substance forming the essential part of wood fiber

ligule—a thin outgrowth membrane attached to a leaf of grass at the point where the blade meets the leaf sheath

lodge—of crops, to fall or lie down; refers especially to hay or grain crops

lumen—unit of luminous flux equal to the light emitted in a unit solid angle by a uniform point source of one candle

lux—unit of illumination equal to 1 lumen per square meter

macroptery—long-winged

maggot—footless larva of Diptera

Malpighian tubules—the insect urinary system, composed of long, slender blind excretory tubes that vary in number and open into the hind intestine at its junction with the midintestine

mandibles—first pair of jaws in insects; stout and toothlike in chewing insects, needle-like in sucking insects, mouth hooks in muscid flies

maritime—on, near, or living near the sea

maxillae—second pair of "jaws" in a mandibulate insect; often adapted to manipulate food

maxillary palp—the pair of palps on the maxilla carried by the stipes on its outer end and sensory in function

meatus—a channel or duct

meconium—substance excreted by certain insects soon after their emergence from the chrysalis, or pupa

meridian—any lines of longitude running north and south through the poles on the globe or map

meristem (activity)—undifferentiated plant tissue consisting of cells actively growing and dividing, as at the lip of roots or stems

mesad—toward or in the direction of the median plane of a body

mesasoma—in scorpions, the broad seven-segmented section of the body immediately posterior to the last pair of legs

mesophyll—the leaf substance lying between the upper and lower epidermis; the parenchyma

mesothorax—the middle segment of the thorax; contains the middle pair of legs and, in winged adults, the first pair of wings

metamorphosis—a series of changes through which an arthropod passes in its development from the egg to the adult

metasoma—in scorpions, the narrow five-segmented posterior section of the body terminating in a sting

metasternum—underside of the metathorax

metathorax—the hindmost segment of the thorax; contains the third pair of legs and, in winged adults, the second pair of wings

micropyle—one of the minute openings in the insect egg through which spermatozoa enter in fertilization

microsporidium—any of a group of protozoa, some of which are pathogens to insects and other animals

midgut—true stomach

milorganite—activated sewage sludge that is steam sterilized and marketed by the City of Milwaukee, Wisconsin, as a turfgrass fertilizer

minim—something very small; in ants, the smallest and the first worker in a new colony

molt—to cast off or shed the outer skin and so forth at certain intervals prior to replacement of the cast-off parts by a new growth

multivoltine—having more than one generation in a year

muscid—a fly belonging to the superfamily Muscoidea

naiad—immature stage of an insect that goes through incomplete metamorphosis; often aquatic

nema—short for nematode

nematode—any of a class or phylum of elongated cylindrical worms that are parasitic in animals or plants or are free living in soil or water

nocturnal—active at night

node—in botany, a point of attachment for a leaf on a stem

nymph—young stage of insects with gradual metamorphosis that resembles the adult except for lack of wings; for example, chinch bugs, mole crickets

obligate (parasites)—a parasite living on one host exclusively

obtuse—not pointed; at an angle greater than a right angle; opposite of acute

ocellus/ocelli—simple eye(s) of insects, occurring singly or in small groups

olivaceous (dusting)—olive green; color of green olives

ommatidium—one of the visual elements that make up the compound eye of an arthropod

omnivorous—feeding generally on both animal and vegetable food

opisthosoma—in scorpions, the entire section of the body behind the legs and made of a broad, seven-segmented anterior portion and a narrow, five-segmented posterior section

orbicular—round and flat

orbicular spot—region of dark scales on front wing of certain moths, roughly circular and near leading edge of wing

organophosphate—synthetic insecticides that are esters of phosphoric or other phosphorus-derived acids (e.g., diazinon, chlorpyrifos)

oscillate—to move or travel back and forth between two points

ovarian yolk—concentration of yolk cells in the egg at the point where the cells of the developing embryo undergo cleavage

oversummer—to survive the summer (in a physiologically dormant state)

overwinter—to survive the winter (often in a physiologically dormant state)

overwintering stage—stage of development in which an insect survives the winter

oviparous—reproducing by eggs laid by the female

oviposit—deposit eggs

ovipositor—a tubular or valved structure by means of which eggs are deposited

ovoid—egg shaped

ovoviviparous—producing eggs that develop internally and hatch just before or just after the egg is extruded from the adult female's body

ovum/ova—egg(s)

palidium—in scarabaeid larva, a group of pali arranged in a single row or more and usually placed on the median of the venter of the lower anal lip

palpus—a process on a mouthpart of an arthropod that has a tactile or gustatory function

palus/pali—straight-pointed spine(s), a component of the palidium

parasite—any animal that lives in or on, or at the expense of, another

parasitic—of a parasite or parasitoid

parasitoid—an insect that parasitizes other insects (of a different species)

parasporal body—sporelike body apart from the spore proper in a bacterial spore

parenchyma—tissue consisting of large, thin-walled cells

parthenogenesis—reproduction by direct growth from egg cells without fertilization by sperm

pectinate—comblike, said especially of antennae with even processes like the teeth of a comb

pedicel (antenna)—a narrow basal part by which a larger part of an organ of an animal is attached; second joint of an antenna

pedipalp—one of the second pair of appendages in the cephalothorax, used in crushing prey

penultimate—next to the last

peripheral—relating to the outer margin

petiole—a stem or stalk

phenology—a branch of science dealing with the relations between climate and periodic biological phenomena

pheromone—any substance secreted by an animal that influences the behavior of other individuals of the same species

phloem—a complex, food-conducting vascular tissue in higher plants

photodegradation—breakdown of chemical bonds as a result of exposure to ultraviolet radiation in sunlight

photoperiod—length of time in a 24-hr period subject to sunlight

photosynthesis—a process by which green leaves manufacture carbohydrates using carbon dioxide, water, chlorophyll, and light

phylogenetic—of or related to phylogeny or based on natural evolutionary relationships

phylum—primary division of the animal kingdom; used in classification to indicate a series of related organisms

phytophagous—feeding on plants

phytotoxic—harmful to a plant; often said of a pesticide application that results in chlorotic spots or burns on foliage

piceous—pitchy black; black with a reddish tinge

piperonyl butoxide—a synthetic hydrocarbon compound used as a synergist in a mixture with insecticides to increase the effectiveness of the latter

pipet, pipette—a narrow glass tube into which liquid is drawn by suction and is retained by closing the upper end

pollination—the transfer of pollen from a stamen to a pistil; fertilization in flowering plants

polyembryony—the production of several embryos (genetically identical to each other) from a single egg

polyhedral—having many faces or sides

polyhedrosis—any of several viral diseases of insect larvae characterized by the breakdown of tissues and the presence of polyhedral granules

polymorph—in biology, one of several adult forms

polyphagous—eating many kinds of foods

polystand—a turfgrass community composed of plants of two or more cultivars and/or species

posterior—toward the rear, as opposed to *anterior*

postocular lobe—exoskeletal lobe behind the eyes

predator—any animal that preys on another

preoviposition—the period between emergence of adult female and initiation of oviposition

prepupa—a quiescent transitional period between the end of the larval period and the pupal period

primordial—of leaves, the earliest formed tissues

process—a prolongation of the surface or of an appendage, or any prominent body part not otherwise definable

proleg—fleshy leg (not a true leg) on abdomen of caterpillar

pronotum—the upper or dorsal surface of the prothorax

propodosoma—the anterior portion of the idiosoma, or body proper, of a mite, to which the legs are attached

prosoma—in scorpions, the anterior portion of the body bearing the eyes, pedipalps, and legs

prothoracic—of the prothorax

prothorax—anterior segment of thorax; contains first pair of legs

protonymph—the first nymphal stage of Acari to assume the general nonsexual characteristics of the adult

Protozoa—a phylum of the animal kingdom containing one-celled animals

protozoa—one-celled animals in general

proximal—part of an appendage nearest the body

pubescent—downy; clothed with soft, short, fine, closely set hairs

punctate—set with impressed points or punctures

pupa—the resting, inactive stage between the larva and the adult in all insects that undergo a complete metamorphosis

puparium—in Diptera, the thickened, hardened, barrel-like larval skin within which the pupa is formed

pygidium—the tergum (upper surface) of the last segment of the abdomen

pyrethrin—one of six separate compounds or a mixture of compounds extracted from certain chrysanthemum spp. flowers that have very rapid insecticidal action

raster—in scarabaeid larvae, a complex of definitely arranged bare places, hairs, and spines on the ventral surface of the last abdominal segment, in front of the anus

rectal sac—the enlarged anterior part of the rectum

recumbent—lying down; reclining

relative humidity—the ratio (expressed as a percentage) of the amount of water vapor actually present in the air to the greatest amount possible at the same temperature

reniform—kidney shaped

reniform spot—region of dark scales on the front wing of certain moths, roughly kidney shaped

resistance—a phenomenon in which populations of insects undergo genetic shift so the population is no longer sensitive to a particular insecticide

reticulation—formation or pattern; network

rhizome—a jointed underground stem that can produce roots and shoots at each node

rickettsia—rod-shaped nonfilterable microorganisms that cause various diseases

rosette—a cluster of leaves in crowded circles or spirals arising basally from a crown or apically from an axis, with greatly shortened internodes

rufous—pale red

runners—elongate growths produced by a plant; stolons

sagittal—longitudinal

saprophyte—any plant or animal living on dead or decaying vegetable matter

sarcophagid–flesh eating

scape—long basal joint in an antenna

scarab/scarabaeid—pertaining to members of the beetle family Scarabaeidae

sclerite–a hardened "plate" on the insect body wall; bounded by sutures

sclerotized—of an insect, hardened in definite areas by deposition or formation of substances other than chitin

scrobe—groove formed for the reception or concealment of an appendage, especially, in weevils, grooves at the sides of the rostrum to receive the scape of the antennae

septicemia—invasion of the blood by toxins produced by virulent microorganisms from a focus of infection (e.g., *Bacillus* bacteria)

septula—in scarabaeid grubs, the median narrow, bare region of the raster between the palidia

serosal cuticle—outer membrane

sessile—closely seated, without a stalk; in insects, immobile

seta/setae—slender hairlike appendage(s)

sheath (leaf)—basal tubular portion of a leaf surrounding the stem

sinuate—cut into sinuses; wavy; said especially of edges or margins

snout—the prolongation of the head of weevils at the end of which the mouthparts are located

sod—plugs, blocks, squares, or strips of turfgrass plus the adhering soil that are used for planting

spatulate—round and broad at the top, slender at the base

spermatheca—the sac or reservoir in the female that retains the sperm following copulation

sphecid—a hymenopterous insect belonging to the family Sphecidae, the members of which are parasitic on other insects

spinneret—an organ in the larva or adult that is used in spinning silk

spiracle—a breathing pore through which air enters the tracheae; in insects, located laterally on body segments

spittle—a frothy fluid secreted by insects; saliva

sporangium—a case within which the asexual spores are produced

sporulate—to undergo sporulation

sporulation—the formation of spores

spring generation—consists of the overwintering stage and all subsequent stages in that generation, culminating with emergence of adults

stadium—the interval between the molts of larvae; instar

sternum—the entire ventral division of any segment

stipe—the foot stalk of the maxilla, bearing the movable parts

stolon—a jointed, aboveground, creeping stem that can produce roots and shoots at each node and may originate extravaginally from the main stem

striate—marked with parallel, fine, impressed lines

stridulate—in insects, to make a shrill, creaking, grating, or hissing sound by rubbing two ridges or roughened surfaces against each other

stubble—the basal part of plants left after harvest

style—a short, slender, finger-like process; often projecting from last abdominal segment or from the apex of an antenna

stylet—a median dorsal element in the shaft of the ovipositor; one of the piercing structures in insects with piercing/sucking mouthparts (e.g., aphids)

sub-spiracular stripe—a colored or black lateral stripe on certain lepidopterous caterpillars beneath the spiracles

subterranean—existing under the surface of the earth

suture—an external groove in the body wall; a narrow membranous area between sclerites

sward—a carpet of grass or other ground cover; turf

synergist—in pesticide chemistry, a material used with an insecticide to enhance the activity of the insecticide; most often associated with synthetic pyrethroids

synthetic pyrethroid—class of insecticides based on structure of pyrethrin, a natural compound

tachinid—a dipteran insect belonging to the family Tachinidae, the members of which are parasitic on other insects

tarsus—the foot; the distal part of the insect leg that consists of one to five segments

tee—in golf, an area that is specially prepared from which the first stroke on each hole is made

teges—in scarabaeid larvae; a continuous dense or sparse patch of hooked or straight spines occupying almost the whole of the venter of the 10th abdominal segment when the palidium is absent

tegillum—in scarabaeid larvae, a paired patch of hooked or straight spines on each side of the palidia on the 10th abdominal segment

tegmen—the hardened leathery or horny forewing in Orthoptera

temporal—relating to time

teneral—condition of the adult shortly after emergence, when it is not entirely hardened or fully of mature color

tergum—the upper or dorsal surface of any body segment of an insect

terminal ampullae—in scarabaeid grubs, a pair of ovoidal structures; the posterior portion of the reproductive organs, found in the eighth abdominal sternum of females and ninth abdominal sternum of males

thatch—layer of plant litter from long-term accumulation of dead plant roots, crowns, rhizomes, and stolons between the zone of green vegetation and the soil surface

thermal unit—units relating to heat

thigmotactic—orientation in response to contact with some outside object or body

thorax—the second or intermediate region of the insect body, bearing the true legs and wings and composed of three segments (the prothorax, mesothorax, and metathorax)

threshold—the beginning point of something; in physiology, the point at which a stimulus is just strong enough to produce a response

tibia—in insects, the fourth division of the leg articulated at the proximal end to the femur and at the distal end to the tarsi

tiller—shoot, culm, or stalk arising from a crown bud; often a primary lateral stem

toxin—poisonous substance secreted by plants and animals

trachea—a spirally ringed internal elastic air tube in insects; an element of the respiratory system

tracheole—the smallest element of the insect respiratory system (i.e., a very small trachea)

transition zone—in turfgrass management, the area (roughly from Washington, D.C., west) in which both cool-season and warm-season grasses can grow

translocation—in botany, the movement of fluids through vascular tissue (phloem or xylem)

translucent—partially transparent, like frosted glass

transverse—lying across; set crosswise

trochanter—the second segment of an insect leg, between the coxa and femur

ultimate—last; terminal; final

univoltine—having one generation in a year

vapor pressure—the pressure of a vapor in equilibrium with its liquid or solid phase

vector—any organism that is the carrier of a disease-producing organism

venter—the undersurface of an animal

ventral—of the undersurface of an animal

vernation—in botany, the arrangement of leaves in a leaf bud

vertebrate—having a backbone or spinal column

vestigial—small or degenerate; the remains of a previously functional organ

violaceous—violet in color

virulence—quality of being poisonous; the relative infectiousness of a microorganism causing disease

virus—any of a large group of infectious agents 10–250 nm in diameter, composed of a protein sheath surrounding a nucleic acid core and capable of infecting animals, plants, and bacteria and totally dependent on living cells for reproduction

viviparous—bearing or bringing forth living young instead of laying eggs

volatilization—transformation from solid or liquid phase to gaseous phase; for example, evaporation of water

white grub—whitish, C-shaped larvae of insects belonging to the family Scarabaeidae

whorl—in botany, an arrangement of leaves or petals around a point on a stem

References

Adams, J. A. 1949a. The Oriental beetle as a turf pest associated with the Japanese beetle in New York. J. Econ. Entomol. 42: 366–371.

Addy, C. E., and J. D. Heyland. 1968. Canada goose management in eastern Canada and the Atlantic flyway, pp. 9–23. *In* R. L. Hine and C. Schoenfeld (eds.), Canada goose management. Dembar Educational Research Services, Madison, Wisc. 195pp.

Agriculture Canada. 1983. Japanese beetle/Scarabee japonais: Survey in Canada/Enquete au Canada. Biol. Programs Sect., Plant Health Div., Ottawa. 4 pp.

Ahmad, S., and C. R. Funk. 1983. Bluegrass billbug (Coleoptera: Curculionidae): Tolerance of ryegrass cultivars and selections. J. Econ. Entomol. 76: 414–416.

Ahmad, S., J. M. Johnson-Cicalese, W. K. Dickson, and C. R. Funk. 1986. Endophyte-enhanced resistance in perennial ryegrass to the bluegrass billbug, *Sphenophorus parvulus*. Entomol. Exp. Appl. 41: 3–10.

Ainslie, G. G. 1922. Webworms injurious to cereal and forage crops and their control. U.S. Dept. Agr. Farmers' Bull. No. 1258. 16 pp.

Ainslie, G. G. 1923a. Silver-striped webworm, *Crambus praefectellus*, Zincken. J. Agr. Res. 24: 415–426.

Ainslie, G. G. 1923b. Striped sod webworm, *Crambus mutabalis* Clemens. J. Agr. Res. 24: 399–414.

Ainslie, G. G. 1927. The larger sod webworm. U.S. Dept. Agr. Tech. Bull. No. 31. 17 pp.

Ainslie, G. G. 1930. The bluegrass webworm. U.S. Dept. Agr. Tech. Bull. No. 173. 25 pp.

Aldrich, J. M. 1920. European frit fly in North America. J. Agr. Res. 18: 451–473.

Alexander, C. P. 1919. The crane flies of New York. Pt. 1. Distribution and taxonomy of adult flies. Cornell Univ. Agr. Exp. Sta. Mem. 25: 766–993.

Allen, W. A., and R. L. Pienkowski. 1974. The biology and seasonal abundance of the frit fly, *Oscinella frit*, in reed canary grass in Virginia. Ann. Entomol. Soc. Amer. 67: 539–544.

Allen, W. A., and R. L. Pienkowski. 1975. Life tables for frit fly, *Oscinella frit*, in reed canary grass in Virginia. Ann. Entomol. Soc. Amer. 68: 1001–1007.

Allen, W. H. 1944. The Asiatic beetles in New Jersey. New Jersey Dept. Agr. Circ. No. 348. 18 pp.

Allsopp, P. G., M. G. Klein, and E. L. McCoy. 1992. Effect of soil moisture and soil texture on oviposition by Japanese beetle and rose chafer (Coleoptera: Scarabaeidae). J. Econ. Entomol. 85: 2194–2200.

Alm, S. R. 1996. The oriental beetle. TurfGrass Trends (July): 9–13.

Alm, S. R., M. G. Villani, T. Yeh, and R. Shutter. 1997. *Bacillus thuringiensis* serovar *japonensis* strain Buibui for control of Japanese and oriental beetle larvae control. Appl. Entomol. Zool. 32: 477–484.

Alm, S. R., T. Yeh, C. G. Dawson, and M. G. Klein. 1996. Evaluation of trapped beetle repellency, trap height, and string pheromone dispensers on Japanese beetle captures (Coleoptera: Scarabaeidae). Environ. Entomol. 25: 1274–1278.

Alm, S. R., T. Yeh, J. L. Hanula, and R. Georgis. 1992. Biological control of Japanese, oriental, and black turfgrass ataenius beetle (Coleoptera: Scarabaeidae) larvae with entomopathogenic nematodes (Nematoda: Steinernematidae, Heterorhabditidae). J. Econ. Entomol. 85: 1660–1665.

Anonymous. 1979. Scott's information manual for lawns. O. M. Scott and Sons, Marysville, Ohio. 98 pp.

Anonymous. 1983. Larra's theme: The undoing of mole crickets. Golf Course Management (April): 46–47.

Antonelli, A. L., and R. L. Campbell. 1984. The European crane fly: A lawn and pasture pest. Washington State Univ. Coop. Ext. Bull. No. 0856. 3 pp.

Archer, T. L., and G. J. Musick. 1976. Response of black cutworm larvae to light at several intensities. Ann. Entomol. Soc. Amer. 69: 476–478.

Arnett, R. H. 1985. American insects. Van Nostrand Reinhold, New York, N.Y. 850 pp.

Arnold, T. B., and D. A. Potter. 1987. Impact of a high-maintenance lawn-care program on nontarget invertebrates in Kentucky bluegrass turf. Environ. Entomol. 16: 100–105.

Baker, J. R. 1982. Insects and other pests associated with turf: Some important, common, and potential pests in the southeastern United States. North Carolina Agr. Ext. Service. AG-268. 108 pp.

Baker, P. B., R. H. Ratcliffe, and A. L. Steinhauer. 1981. Laboratory rearing of the hairy chinch bug. Environ. Entomol. 10: 226–229.

Balogh, J. C., and J. L. Anderson. 1992. Environmental impacts of turfgrass pesticides, pp. 221–353. In J. C. Balogh and W. J. Walker (eds.), Golf course management and construction. Lewis Publishers, Chelsea, Michigan. 951 pp.

Banerjee, A. C. 1967a. Flight activity of the sexes of crambid moths as indicated by light-trap catches. J. Econ. Entomol. 60: 383–390.

Banerjee, A. C. 1967b. Sod webworm parasites in Illinois. J. Econ. Entomol. 60: 1173–1174.

Banerjee, A. C. 1967c. Injury to grasses in lawns by *Acrolophus* sp. J. Econ. Entomol. 60: 1174.

Banerjee, A. C. 1968. Microsporidia diseases of sod webworms in bluegrass lawns. Ann. Entomol. Soc. Amer. 61: 544–545.

Banerjee, A. C. 1969a. Development of *Crambus trisectus* at controlled constant temperatures in the laboratory. J. Econ. Entomol. 62: 703–705.

Banerjee, A. C. 1969b. Sex attractants in sod webworms. J. Econ. Entomol. 62: 705–708.

Banerjee, A. C., and G. C. Decker. 1966. Studies on sod webworms, I. Emergence rhythm, mating, and oviposition behavior under natural conditions. J. Econ. Entomol. 59: 1237–1244.

Banks, N. 1904. A treatise on the acarina, or mites. Proc. U.S. Nat. Mus. Smithsonian Inst. 28: 1–114.

Barnes, H. F. 1937. Methods of investigating the bionomics of the common crane fly, *Tipula paludosa* Meigen, together with some results. Ann. Appl. Biol. 24: 356–368.

Baxendale, F. P. 1993. Billbugs: Characteristics and control. Golf Course Management (July): 115–119.

Baxendale, F. P. 1997. Insects and related pests of turfgrass, pp. 31–78. In F. P. Baxendale and R. G. Gaussoin (eds.), Integrated turfgrass management for the Northern Great Plains. Univ. of Nebraska, Lincoln, Neb. 236 pp.

Baxendale, F. P., J. M. Johnson-Cicalese, and T. P. Riordan. 1994. *Tridiscus sporoboli* and *Trionymus* sp. (Homoptera: Pseudococcidae): Potential new mealybug pests of buffalograss turf. J. Kans. Entomol. Soc. 67: 169–172.

Baxendale, F. P., T. P. Riordan, and T. M. Heng-Moss. 1997. Buffalograss chinch bugs. Neb. Facts (Univ. of Nebraska, Lincoln, Neb.). 2 pp.

Beard, J. B. 1973. Turfgrass science and culture. Prentice-Hall, Englewood Cliffs, N.J. 658 pp.

Beard, J. B. 1975. How to have a beautiful lawn: Easy steps in turfgrass establishment and care for aesthetic and recreational purposes. Intertec Publishing, Kansas City, Mo. 113 pp.

Beard, J. B. 1982. Turf management for golf courses. Pub. U.S. Golf Assoc. Burgess Publishing, Minneapolis, Minn. 642 pp.

Beard, J. B. 1984. Grasses for the transition zone. Ground Maintenance (Jan.): 60–62.

Beckhan, C. M., and M. Dupree. 1952. Attractants for the green June beetle, with notes on seasonal occurrence. J. Econ. Entomol. 45: 736–737.

Belluci, R. 1939. Respiratory metabolism of starved Japanese beetle larvae (*Popillia japonica* Newman) at different relative humidities. Physiol. Zool. 12: 50–56.

Bhatkar, A., W. H. Whitcomb, W. F. Buren, P. Callahan, and T. Carlysle. 1972. Confrontation behavior between *Lasius neoniger* (Hymenoptera: Formicidae) and the imported fire ant. Environ. Entomol. 1: 274–279.

Bianchi, F. A. 1935. Investigations on *Anomala orientalis* Waterhouse at Oahu Sugar Company, Ltd. Hawaii Plant Rec. 39: 234–255.

Bianchi, F. A. 1957. Notes and exhibitions. Proc. Hawaiian Entomol. Soc. 16: 184.

Blatchley, W. S. 1910. An illustrated descriptive catalog of the Coleoptera, or beetles (exclusive of the Rhyncophora), known to occur in Indiana. Nature Publishing, Indianapolis, Ind. 1,386 pp.

Blatchley, W. S. 1920. Orthoptera of northeastern America, with special reference to the faunas of Indiana and Florida. Nature Publishing, Indianapolis, Ind. 784 pp.

Blatchley, W. S. 1926. Heteroptera, or true bugs of eastern North America, with special reference to the faunas of Indiana and Florida. Nature Publishing, Indianapolis, Ind. 1,116 pp.

Blatchley, W. S., and C. W. Leng. 1916. Rhyncophora, or weevils, of North America. Nature Publishing, Indianapolis, Ind. 682 pp.

Boggess, E. K. 1983. Carnivores (meat-eating mammals), raccoons, pp. C-73–C-79. *In* R. M. Timm (ed.), Prevention and control of wildlife damage. Great Plains Agr. Council and Univ. Nebraska Coop. Ext. Service, Inst. Agr. and Nat. Resources, Lincoln, Neb. Looseleaf, 632 pp.

Bohart, R. M. 1940. Studies on the biology and control of sod webworms in California. J. Econ. Entomol. 33: 886–890.

Bohart, R. M. 1947. Sod webworms and other lawn pests in California. Hilgardia 17: 267–307.

Boratynski, K., E. Pancer-Koteja, and J. Koteja. 1982. The life history of *Lecanopsis formicarum* Newstead (Homoptera: Coccinea) [in English; Polish and Russian summaries]. Ann. Zool. 36: 517–536.

Borror, D. J., D. M. DeLong, and C. A. Triplehorn. 1981. An introduction to the study of insects. 5th edition. Saunders College Publishing, Dryden Press, Philadelphia, Penn. 827 pp.

Borror, D. J., C. A. Triplehorn, and N. F. Johnson. 1989. An introduction to the study of insects. 6th ed. Harcourt Brace College Publishers, New York, N.Y. 874 pp.

Boucias, D. 1984. Diseases, pp. 32–35. *In* T. J. Walker (ed.), Mole crickets in Florida. Florida Agr. Exp. Sta. Bull. 846.

Boving, A. G. 1939. Descriptions of the three larval instars of the Japanese beetle, *Popillia japonica* Newm. Proc. Washington Entomol. Soc. 41: 183–191.

Boving, A. G. 1942. A classification of larvae and adults of the genus *Phyllophaga* (Coleoptera: Scarabaeidae). Mem. Entomol. Soc. Washington No. 2. 96 pp.

Bowen, W. R., F. S. Morishita, and R. O. Oetting. 1980. Insects and related pests of turf, pp. 31–39. *In* W. R. Bowen, Comp., Turfgrass pests. Agr. Sci. Univ. California Pub. No. 4053. 53 pp.

Braman, S. K. 1995. Two-lined spittlebug, pp. 88–90. *In* Brandenburg, R. L., and M. G. Villani (eds.), Handbook of turfgrass insect pests. Entomological Society of America, Lanham, Md. 140 pp.

Braman, S. K., and W. G. Hudson. 1993. Patterns of flight activity of pest mole crickets in Georgia. Internatl. Turfgrass Soc. Res. J. 7: 382–384.

Brandenburg, R. L. 1995a. Ground pearls, pp. 62–63. *In* R. L. Brandenburg and M. G. Villani (eds.), Handbook of turfgrass insect pests. Entomological Society of America, Lanham, Md. 140 pp.

Brandenburg, R. L. 1995b. Intuitive forecasting of turfgrass insect pests. TurfGrass Trends (Nov.): 1–8.

Brandenburg, R. L. 1997. Managing mole crickets: Developing a strategy for success. TurfGrass Trends (Jan.): 1–8.

Brandenburg, R. L., and M. G. Villani, eds. 1995. Handbook of turfgrass insect pests. Entolomological Society of America, Lanham, Md. 140 pp.

Breen, J. P. 1992. Temperature and seasonal effects on expression of *Acremonium* endophyte-enhanced resistance to *Schizaphis graminum* (Homoptera: Aphididae). Environ. Entomol. 21: 68–74.

Breen, J. P. 1993. Enhanced resistance to fall armyworm (Lepidoptera: Noctuidae) in *Acremonium* endophyte-infected turfgrasses. J. Econ. Entomol. 86: 621–629.

Brindle, A. 1960. The larvae and pupae of British Tipulinae (Diptera: Tipulidae). Trans. Soc. Brit. Entomol. 14: 63–114.

Britton, W. E. 1925. A new pest of lawns. Connecticut Agr. Exp. Sta. Bull. Immed. Inform. 52: 25–28.

Brown, C. L., and W. K. Hock. 1989. The fate of pesticides in the environment and groundwater protection. Penn State Fact Sheet, Pennsylvania State Univ., State College, Penn. 6 pp.

Brussell, G. E., and R. L. Clark. 1968. An evaluation of Baygon for control of the hunting billbug in a zoysia grass lawn. J. Econ. Entomol. 61:1100.

Burrage, R. H., and G. G. Gyrisco. 1954. Distribution of third instar larvae of the European chafer and the efficiency of various sampling units for estimating their population. J. Econ. Entomol. 47: 1009–1014.

Burt, W. H., and R. P. Grossenheider. 1964. A field guide to the mammals. Houghton Mifflin, Boston, Mass. 284 pp.

Busching, M. K., and F. T. Turpin. 1976. Oviposition preferences of black cutworm moths among various crop plants, weeds, and plant debris. J. Econ. Entomol. 69: 587–590.

Busey, P., and B. J. Center. 1987. Southern chinchbug (Hemiptera: Heteroptera: Lygaeidae) overcomes resistance in St. Augustinegrass. J. Econ. Entomol. 80: 608–611.

Busey, P., and G. H. Snyder. 1993. Population outbreak of the southern chinch bug is regulated by fertilization. Internatl. Turfgrass Soc. Res. J. 7: 353–357.

Butler, G. D., Jr. 1963. The biology of the bermudagrass eriophyid mite. Arizona Agr. Exp. Sta. Rep. 219: 8–13.

Butt, B. A., and E. Cantu. 1962. Sex determination of lepidopterous pupae. U.S. Dept. Agr. ARS-33–75. 7 pp.

Butt, F. H. 1944. External morphology of *Amphimallon majalis* (Razoumowski) (Coleoptera, the European chafer). Cornell Univ. Agr. Exp. Sta. Mem. No. 266. 18 pp.; 13 plates.

Byers, G. W. 1961. The crane fly genus *Dolichopeza* in North America. Univ. Kansas Sci. Bull. 42: 665–924.

Byers, R. A. 1965. Biology and control of a spittlebug. *Prosapia bicincta* (Say), on coastal bermudagrass. Georgia Agr. Exp. Sta. Tech. Bull. n.s. 42. 26 pp.

Cameron, R. S. 1970. Control of a species of *Hyperodes*. New York Turfgrass Assoc. Bull. 86: 333–336.

Cameron, R. S., and N. E. Johnson. 1971a. Biology and control of turfgrass weevil, a species of *Hyperodes*. New York State Coll. Agr., Cornell Univ. Ext. Bull. No. 1226. 8 pp.

Cameron, R. S., and N. E. Johnson. 1971b. Biology of a species of *Hyperodes* (Coleoptera: Curculionidae), a pest of turfgrass. Search Agr. No. 1. 31 pp.

Cameron, R. S., and N. E. Johnson. 1971c. Chemical control of the "annual bluegrass weevil," *Hyperodes* sp. nr. *anthracinus*. J. Econ. Entomol. 64: 689–693.

Cameron, R. S., H. J. Kastl, and J. F. Cornman. 1968. *Hyperodes* weevil damages annual bluegrass. New York Turfgrass Assoc. Bull. No. 79. 2 pp.

Cartwright, O. L. 1974. *Ataenius, Aphotaenius,* and *Pseudataenius* of the United States and Canada (Coleoptera: Scarabaeidae: Aphodiinae). Smithsonian Contrib. Zool. 154: 1–106.

Case, R. M. 1983. Rodents (gnawing animals), pocket gophers, pp. B-13–B-26. *In* R. M. Timm (ed.), Prevention and control of wildlife damage. Great Plains Agr. Council and Univ. Nebraska Coop. Ext. Service, Inst. Agr. and Nat. Resources, Lincoln, Neb. Looseleaf, 632 pp.

Cazier, M. A. 1940. The species of *Polyphylla* in America, north of Mexico (Coleoptera: Scarabaeidae). Entomol. News 51: 134–139.

Chada, H. L. 1956. Biology of the winter grain mite and its control in small grains. J. Econ. Entomol. 49: 515–520.

Chada, H. L., and E. A. Wood. 1960. Biology and control of the rhodesgrass scale. U.S. Dept. Agr. Tech. Bull. No. 1221. 21 pp.

Chamberlin, T. R., C. L. Fluke, L. Seaton, and J. A. Callenbach. 1938. Population and host preference of June beetles. U.S. Dept. Agr. Bur. Entomol. and Plant Quarantine, 18 Suppl. to No. 4: 225–240.

Chambliss, C. E. 1895. The chinch bug, *Blissus leucopterus* (Say). Tennessee Agr. Exp. Sta. Bull. 8(4): 41–55.

Chapman, P. J., and S. E. Lienk. 1981. Flight periods of adults of cutworms, armyworms, loopers, and others (family Noctuidae) injurious to vegetable and field crops. Search Agr. No. 14. 43 pp.

Chittenden, F. H., and D. E. Fink. 1922. The green June beetle. U.S. Dept. Agr. Bull. No. 891. 52 pp.

Choban, R. G., and A. P. Gupta. 1972. Meiosis and early embryology of *Blissus leucopterus hirtus* Montandon (Heteroptera: Lygaeidae). Internatl. J. Insect Morph. Embryol. 1: 301–314.

Cobb, P. 1982. Mole crickets. Amer. Lawn Appl. (Sept./Oct.): 4–8.

Cobb, P. P. 1992. Controlling fall armyworms on lawns and turf. Alabama Cooperative Ext. Circular ANR-172. 4 pp.

Cobb, P. P., and K. R. Lewis. 1990. Searching for the best ways to control mole crickets. Golf Course Management (June): 26–28, 32, 34.

Cobb, P. P., and T. P. Mack. 1989. A rating system for evaluating tawny mole cricket, *Scapteriscus vicinus* Scudder, damage (Orthoptera: Gryllotalpidae). J. Entomol. Sci. 24: 142–144.

Cockfield, S. D., and D. A. Potter. 1983. Short-term effects of insecticidal applications on predaceous arthropods and oribatid mites in Kentucky bluegrass turf. Environ. Entomol. 12: 1260–1264.

Cockfield, S. D., and D. A. Potter. 1984. Predation on sod webworm (Lepidoptera: Pyralidae) eggs as affected by chlorpyrifos application to Kentucky bluegrass turf. J. Econ. Entomol. 77: 1542–1544.

Comstock, J. A. 1927. Butterflies of California. McBride Printing, Los Angeles, Cal. 334 pp.

Converse, J. 1982. Scott's guide to the identification of the turfgrass diseases and insects. O. M. Scott & Sons, Marysville, Ohio. 105 pp.

Coulson, J. C. 1962. The biology of *Tipula subnodicornis* Zetterstedt, with comparative observations on *Tipula paludosa* Meign. J. Animal Ecol. 31: 1–21.

Cowles, R. S., and M. G. Villani. 1994. Soil interactions with chemical insecticides and nematodes used for control of Japanese beetles larvae. J. Econ. Entomol. 87: 1014–1021.

Cowles, R. S., and M. G. Villani. 1996. Susceptibility of Japanese beetle, oriental beetle, and European chafer (Coleoptera: Scarabaeidae) to halofenozide, an insect growth regulator. J. Econ. Entomol. 89: 1556–1565.

Cranshaw, W. S., and M. G. Klein. 1994. Microbial control of insect pests of landscape plants, pp. 503–520. *In* A. R. Leslie (ed.), Handbook of integrated pest management for turf and ornamentals. CRC Press, Boca Raton, Fla. 560 pp.

Cranshaw, W. S., and C. R. Ward. 1996. Turfgrass insects in Colorado and northern New Mexico. Cooperative Ext. Bull., Colorado State Univ., Fort Collins, Colo. 38 pp.

Crawford, C. S. 1968. Oviposition rhythm in *Crambus teterrellus:* Temperature depression effects and apparent circadian periodicity. Ann. Entomol. Soc. Amer. 61: 1481–1486.

Crawford, C. S., and R. F. Harwood. 1964. Bionomics and control of insects affecting Washington grass seed fields. Washington Agr. Exp. Sta. Tech. Bull. No. 44. 25 pp.

Creighton, W. S. 1950. The ants of North America. Cosmos Press, Cambridge, Mass. 585 pp.; 57 plates.

Crocker, R. L. 1993. Chemical control of southern chinch bug in St. Augustinegrass. Internatl. Turfgrass Soc. Res. J. 7: 358–363.

Crocker, R. L., and J. B. Beard. 1982. Southern mole cricket moves further into Texas, pp. 58–61. *In* Texas Turfgrass Research, 1982. Texas Agr. Exp. Sta. Consolidated PR-4032–4055. 79 pp.

Crocker, R. L., and C. L. Simpson. 1981. Pesticide screening tests for the southern chinch bug. J. Econ. Entomol. 74: 730–731.

Crocker, R. L., and W. H. Whitcomb. 1980. Feeding niches of the big-eyed bugs *Geocoris bullatus, G. punctipes,* and *G. uliginosus* (Hemiptera: Lygaeidae: Geocorinae). Environ. Entomol. 9: 508–513.

Crocker, R. L., H. L. Cromroy, R. E. Woodruff, W. T. Nailon, Jr., and M. T. Longnecker. 1992. Incidence of *Caloglyphus phyllophaginus* (Acari: Acaridae) on adult *Phyllophaga* spp. and other Scarabaeidae (Coleoptera) in north central Texas. Ann. Entomol. Soc. Amer. 85: 462–468.

Crocker, R. L., R. M. Marengo-Lozada, J. A. Reinert, and W. H. Whitcomb. 1995a. Harvester ants, pp. 64–66. *In* Brandenburg, R. L., and M. G. Villani (eds.), Handbook of turfgrass insect pests. Entomological Society of America, Lanham, Md. 140 pp.

Crocker, R. L., W. L. Nailon, Jr., and J. A. Reinert. 1995b. May and June beetles, pp. 72–75. *In* Brandenburg, R. L., and M. G. Villani (eds.), Handbook of turfgrass insect pests. Entomol. Soc. of America, Lanham, Md. 140 pp.

Crocker, R. L., L. A. Rodriguez-del-Bosque, W. T. Nailor, Jr., and X. Wei. 1996. Flight periods in Texas of three parasites (Diptera: Pyrgotidae) of adult *Phyllophaga* spp. (Coleoptera: Scarabaeidae), and egg production by *Pyrgota undata*. Southwestern Entomol. 21: 317–324.

Crocker, R. L., C. L. Simpson, H. Painter, T. W. Fuchs, and R. E. Woodruff. 1982. White grub of southern masked chafer, *Cyclocephala immaculata*, found in Texas turfgrass, pp. 39–40. *In* Texas Turfgrass Research, 1982. Texas Agr. Exp. Sta. Consolidated PR-4032–4055. 79 pp.

Cromroy, H. L., and D. E. Short. 1981. Pests of three types of turfgrass in Florida. Amer. Lawn Appl. (Aug.): 32–37.

Crumb, S. E. 1929. Tobacco cutworms. U.S. Dept. Agr. Tech. Bull. No. 88. 179 pp.

Crumb, S. E. 1956. The larvae of the Phalaenidae. U.S. Dept. Agr. Tech. Bull. No. 1135. 356 pp.

Crutchfield, B. A., and D. A. Potter. 1995a. Tolerance of cool-season turfgrasses to feeding by Japanese beetle and southern masked chafer (Coleoptera: Scarabaeidae) grubs. J. Econ. Entomol. 88: 1380–1387.

Crutchfield, B. A., and D. A. Potter. 1995b. Damage relationships of Japanese beetle and southern masked chafer (Coleoptera: Scarabaeidae) grubs in cool-season turfgrasses. J. Econ. Entomol. 88: 1049–1056.

Crutchfield, B. A., and D. A. Potter. 1995c. Feeding by Japanese beetle and southern masked chafer grubs on lawn weeds. Crop Sci. 35: 1681–1684.

Curry, J. P. 1987. The invertebrate fauna of grassland and its influence on productivity. III. Effects on soil fertility and plant growth. Grass Forage Sci. 42: 325–341.

Dahlsson, S. O. 1974. Frit fly damage to turfgrass, pp. 418–420. *In* E. C. Roberts (ed.), Proc. 2d Int. Turfgrass Res. Conf. Amer. Soc. Agron., Madison, Wis. 602 pp.

Damback, C. A., and E. Good. 1943. Life history and habits of the cicada killer in Ohio. Ohio J. Sci. 43: 32–41.

Davidson, A. W., and D. A. Potter. 1995. Response of plant-feeding, predatory, and soil-inhabiting invertebrates to *Acremonium* endophyte and nitrogen fertilization in tall fescue turf. J. Econ. Entomol. 88: 367–379.

Davis, C. J. 1969. Notes on the grass webworm, *Herpetogramma licarsisalis* (Walker) (Lepidoptera: Pyraustidae), a new pest of turfgrass in Hawaii, and its enemies. Proc. Hawaiian Entomol. Soc. 20: 311–316.

Davis, J. J. 1919. Contributions to a knowledge of the natural enemies of *Phyllophaga*. Illinois Nat. Hist. Survey Bull. 13: 53–133.

Davis, J. J., and P. Luginbill. 1921. The green June beetle, or fig eater. North Carolina Agr. Exp. Sta. Bull. No. 242. 35 pp.

Davis, M. G., and D. R. Smitley. 1990a. Association of thatch with populations of hairy chinch bug (Hemiptera: Lygaeidae) in turf. J. Econ. Entomol. 83: 2370–2374.

Davis, M. G., and D. R. Smitley. 1990b. Relationship of hairy chinch bug (Hemiptera: Lygaeidae) presence and abundance to parameters of the turf environment. J. Econ. Entomol. 83: 2375–2379.

Davis, W. T. 1920. Mating habits of *Sphecius speciosus*, the cicada killing wasp. Bull. Brooklyn Entomol. Soc. 15:128–129.

Dean, H. A., and M. F. Schuster. 1958. Biological control of rhodesgrass scale in Texas. J. Econ. Entomol. 51: 363–366.

Dent, D. 1991. Insect pest management. CAB International, Wallingford, Oxon, UK. 604 pp.

Domek, J. M., and D. T. Johnson. 1988. Demonstration of semiochemically induced aggregation of the green June beetle, *Cotinis nitida* (L.) (Coleoptera: Scarabaeidae). Environ. Entomol. 17: 147–149.

Domek, J. M., and D. T. Johnson. 1991. Effect of food and mating on longevity and egg production in green June beetle (Coleoptera: Scarabaeidae). J. Entomol. Sci. 26: 345–349.

Dominick, C. B. 1964. Notes on the ecology and biology of the corn root webworm. J. Econ. Entomol. 57: 41–42.

Drees, B. M., R. W. Miller, S. B. Vinson, and R. Georgis. 1992. Susceptibility and behavioral response of red imported fire ant (Hymenoptera: Formicidae) to selected entomogenous nematodes (Rhabditida: Steinernematidae and Heterorhabditidae). J. Econ. Entomol. 85: 365–370.

Drees, B. M., C. L. Barr, S. B. Vinson, R. E. Gold, M. E. Merchant, and D. Kostraun. 1996. Managing red imported fire ants in urban areas. Texas Agr. Ext. Ser. B-6043. Texas A&M Univ. Sys., College Station, Tex. 18 pp.

Dudderar, C. R. 1983. Mole control—A problem for lawn applicators. Amer. Lawn Appl. (March/Apr.): 18–21.

Dudeck, A. E., J. B. Beard, J. A. Reinert, and S. I. Sifers. 1994. FloraTeX™ bermudagrass. Florida Agr. Exp. Sta. Bull. 891. 11 pp.

Duff, M. J. 1982. The mole crickets of Jekyll Island. Golf Course Management (May): 57–59.

Dunbar, D. M. 1971. Big-eyed bugs in Connecticut lawns. Connecticut Agr. Exp. Sta. Circ. No. 244. 6 pp.

Dunbar, D. M., and R. L. Beard. 1975. Present status of milky disease of Japanese and oriental beetles in Connecticut. J. Econ. Entomol. 68: 453–457.

Evans, D. L., and J. O. Schmidt (eds.) 1990. Insect defenses, adaptive mechanisms, and strategies of prey and predators. SUNY Press, New York, NY.

Facundo, H. T. 1997. Temporal and spatial distribution of the oriental beetle, *Anomala orientalis* (Coleoptera: Scarabaeidae) in a golf course environment. Ph.D. Thesis, Univ. of Rhode Island, Kingston, R.I.

Facundo, H. T., A. Zhang, P. S. Robbins, S. R. Alm, C. E. Linn, Jr., M. G. Villani, and W. L. Roelofs. 1994. Sex pheromone responses of the oriental beetle (Coleoptera: Scarabaeidae). Environ. Entomol. 23: 1508–1515.

Fagan, E. B., and L. C. Kuitert. 1969. Biology of the two-lined spittlebug, *Prosapia bicincta,* on Florida pastures (Homoptera: Cercopidae). Florida Entomol. 52:199–206.

Falk, J. H. 1982. Response of two turf insects, *Endria inimica* and *Oscinella frit,* to mowing. Environ. Entomol. 11: 29–31.

Felt, E. P. 1894. On certain grass-eating insects: A synopsis of the species of *Crambus* of the Ithaca fauna. Cornell Univ. Agr. Exp. Sta. Bull. 64: 47–102.

Ferron, P. 1978. Biological control of insect pests by entomogenous nematodes. Ann. Rev. Entomol. 23: 409–442.

Fleming, W. E. 1958. Biological control of the Japanese beetle, especially with entomogenous diseases. Proc. 10th Internatl Cong. 1956 (3):115–125.

Fleming, W. E. 1962. The Japanese beetle in the United States. U.S. Dept. Agr. Handbook No. 236. 30 pp.

Fleming, W. E. 1968. Biological control of the Japanese beetle. U.S. Dept. Agr. Tech. Bull. No. 1383. 78 pp.

Fleming, W. E. 1972. Biology of the Japanese beetle. U.S. Dept. Agr. Tech. Bull. No. 1449. 129 pp.

Fleming, W. E. 1976. Integrating control of the Japanese beetle—A historical review. U.S. Dept. Agr. Tech. Bull. No. 1545. 65 pp.

Fletcher, D. S. 1956. *Spodoptera mauritia* (Boisduval) and *S. triturata* (Walker), two distant species. Bull. Entomol. Res. 47: 215–217.

Forbes, W. T. M. 1954. Lepidoptera of New York and neighboring states. Noctuidae. Pt. 3. Cornell Univ. Agr. Exp. Sta. Mem. No. 329. 433 pp.

Forrest, T. G. 1986. Oviposition and maternal investment in mole crickets (Orthoptera: Gryllotalpidae): Effects of season, size, and senescence. Ann. Entomol. Soc. Amer. 79: 918–924.

Forschler, B. T., and W. A. Gardner. 1991. Parasitism of *Phyllophaga hirticula* (Coleoptera: Scarabaeidae) by *Heterorhabditis heliothidis* and *Steinernema carpocapsae.* J. Invertebr. Pathol. 58: 396–407.

Fox, D. J. J. 1957. Note on occurrence in Cape Breton Island of *Tipula paludosa* Meig. (Diptera: Tipulidae). Can. Entomol. 89: 288.

Frankie, G. W., H. A. Turney, and P. J. Hamman. 1973. White grubs in Texas turfgrass. Texas Agr. Ext. Service L-1131. 3 pp.

Fraser, M. L., and J. P. Breen. 1994. The role of endophytes in IPM for turf. pp 521–528. *In* A. R. Leslie (ed.), Handbook of integrated pest management for turf and ornamentals. CRC Press, Boca Raton, Fla. 560 pp.

French, J. C. 1964. Chinch bugs. Univ. Georgia Coll. Agr. Leaflet No. 20. 2 pp.

Friend, R. B. 1929. The Asiatic beetle in Connecticut. Connecticut Agr. Exp. Sta. Bull. 304: 585–664.

Funk, C. R., and R. H. Hurley. 1984. Seed facts update on perennial ryegrass. Lawn Care Ind. (Jan.): 38–38A.

Fushtey, S. G., and M. K. Sears. 1981. Turfgrass diseases and insect pests (descriptions, illustrations and controls). Min. Agric. Food Pub. No. 162. University of Guelph, Ontario. 32 pp.

Gambrell, F. L. 1943. Observations on the economic importance and control of the European chafer. New York State Agr. Exp. Sta. Bull. 703: 8–13.

Gambrell, F. L., S. C. Mendel, and E. H. Smith. 1942. A destructive European insect new to the United States. J. Econ. Entomol. 35: 289.

Gammon, E. T. 1961. The Japanese beetle in Sacramento. California Dept. Agr. Bull. 50: 221–235.

Garman, H. 1926. Two important enemies of bluegrass pastures. Kentucky Agr. Exp. Sta. Bull. 265: 29–47.

Gaussoin, R. 1995. Pesticide formulations. Golf Course Management (March): 49–51.

Gaylor, M. J., and G. W. Frankie. 1979. The relationship of rainfall to adult flight activity, and of soil moisture to oviposition behavior and egg and first instar survival in *Phyllophaga crinita*. Environ. Entomol. 8: 591–594.

Gelernter, W. 1996. Controlling the black turfgrass ataenius. Calif. Fairways (Jan./Feb.): 6–8.

Georgis, R., and R. Gaugler. 1991. Predictability in biological control using entomopathogenic nematodes. J. Econ. Entomol. 84: 713–720.

Gibb, T. J. 1995. Greenbug, pp. 60–61. *In* Brandenburg, R. L., and M. G. Villani (eds.), Handbook of turfgrass insect pests. Entomological Society of America, Lanham, Md. 140 pp.

Gibeault, V. A., K. Mueller, and J. Davidson. 1977. Dichondra. Div. Agr. Sci. Univ. California Leaflet No. 2983. 11 pp.

Gilbert, L. E., and L. W. Morrison. 1997. Patterns of host specificity in *Pseudacteon* parasitoid flies (Diptera: Phoridae) that attack *Solenopsis* fire ants (Hymenoptera: Formicidae). Environ. Entomol. 26: 1149–1154.

Glare, T. R., and A. M. Crawford. 1992. Viral diseases of scarabs, pp. 21–32. *In* T. A. Jackson and T. R. Glare (eds.), Use of pathogens in scarab pest management. Intercept Press, Andover, Hampshire, UK. 298 pp.

Glenister, C. S. 1994. Commercial biological controls for insect and mite pests of ornamentals, pp. 455–466. *In* A. R. Leslie (ed.), Handbook of integrated pest management for turf and ornamentals. CRC Press, Boca Raton, Fla. 560 pp.

Gordon, F. C., and D. A. Potter. 1985. Efficiency of Japanese beetle (Coleoptera: Scarabaeidae) traps in reducing defoliation of plants in the urban landscape and effect on larval density in turf. J. Econ. Entomol. 78: 774–778.

Grant, J. 1995. IPM of insects. TurfGrass Trends (Aug.): 3–7.

Grisham, J. 1994. Attack of the fire ant. BioScience 44(9): 587–590.

Gruttadaurio, J., E. E. Hardy, and A. S. Lieberman. 1978. Final report on an investigation of turfgrass land use acreages and selected maintenance expenditures across New York State. Dept. Floriculture and Ornamental Hort., New York State Coll. Agr. and Life Sci., Cornell Univ., Ithaca. 36 pp.

Guthrie, F. E., and G. C. Decker. 1954. The effect of humidity and other factors on the upper thermal death point of the chinch bug. J. Econ. Entomol. 47: 882–887.

Gyrisco, G. G., W. H. Whitcomb, R. H. Burrage, C. Logothetis, and H. H. Schwardt. 1954. Biology of European chafer, *Amphimallon majalis* Razoumowsky (Scarabaeidae). Cornell Univ. Agr. Exp. Sta. Mem. No. 328. 35 pp.

Hadley, C. H., and I. M. Hawley. 1934. General information about the Japanese beetle in the United States. U.S. Dept. Agr. Circ. No. 332. 22 pp.

Hallock, H. C. 1929. Known distribution and abundance of *Anomala orientalis* Waterhouse, *Aserica castanea* Arrow, and *Serica similis* Lewis in New York. J. Econ. Entomol. 22: 293–299.

Hallock, H. C. 1930. The Asiatic beetle, a serious pest in lawns. U.S. Dept. Agr. Circ. No. 117. 7 pp.

Hallock, H. C. 1933. Present status of two Asiatic beetles *(Anomola orientalis* and *Autoserica castanea)* in the United States. J. Econ. Entomol. 26: 80–85.

Hallock, H. C. 1935. Movements of larvae of the oriental beetle through soil. J. New York Entomol. Soc. 43: 413–425.

Hallock, H. C. 1936. Notes on biology and control of the Asiatic garden beetle. J. Econ. Entomol. 29: 348–356.

Hallock, H. C., and I. M. Hawley. 1936. Life history and control of the Asiatic garden beetle. Rev. ed. U.S. Dept. Agr. Circ. No. 246. 20 pp. [First edition, 1932, written by Hallock]

Hammond, G. H. 1940. White grubs and their control in eastern Canada. Dom. Can. Dept. Agr. Pub. No. 668. 18 pp.

Hanna, W. 1997. Genetic resistance to mole crickets in turf bermudagrass. TurfGrass Trends (Jan.): 17–18.

Hansen, J. D. 1987. Seasonal history of bluegrass billbug, *Sphenophorus parvulus* (Coleoptera: Curculionidae) in a range grass nursery. Environ. Entomol. 16: 752–756.

Hanson, A. A., F. V. Juska, and G. W. Burton. 1969. Species and varieties, pp. 370–409. *In* A. A. Hanson and F. V. Juska (eds.), Turfgrass science. Monogr. No. 14. Amer. Soc. Agr., Madison, Wisc. 715 pp.

Hanula, J. L. 1990. Epizootological investigations of the microsporidium *Ovavesicula popilliae* and bacterium *Bacillus popilliae* in field populations of the Japanese beetle (Coleoptera: Scarabaeidae). Environ. Entomol. 19: 1552–1557.

Hanula, J. L., and T. G. Andreadis. 1988. Parasitic microorganisms of Japanese beetle (Coleoptera: Scarabaeidae) and associated scarab larvae in Connecticut soils. Environ. Entomol. 17: 709–714.

Hanula, J. L., and T. G. Andreadis. 1992. Protozoan pathogens of scarabs, pp. 79–92. *In* T. A. Jackson and T. R. Glare (eds.), Use of pathogens in scarab pest management. Intercept Press, Andover, Hampshire, UK. 298 pp.

Harris, C. R., J. H. Mazurek, and G. V. White. 1962. The life history of the black cutworm, *Agrotis ipsilon* (Hufnagel), under controlled conditions. Can. Entomol. 94: 1183–1187.

Hawley, L. M. 1944. Notes on the biology of the Japanese beetle. U.S. Dept. Agr., Bur. Entomol. and Plant Quarantine E 615. 19 pp.

Hawthorne, D. W. 1983. Other mammals: Armadillos, pp. D-5–D-7. *In* R. M. Timm (ed.), Prevention and control of wildlife damage. Great Plains Agr. Council and Univ. Nebraska Coop. Ext. Service, Inst. Agr. and Nat. Resources, Lincoln, Neb. Looseleaf, 632 pp.

Haynes, K. F., and D. A. Potter. 1995. Chemically mediated sexual attraction of male *Cyclocephala lurida* (Coleoptera: Scarabaeidae) and other scarabaeid beetles to immature stages. Environ. Entomol. 24: 1303–1306.

Haynes, K. F., D. A. Potter, and J. T. Collins. 1992. Attraction of male beetles to grubs: Evidence for evolution of a sex pheromone from larval odor. J. Chem. Ecol. 18: 1117–1124.

Hayslip, N. C. 1943. Notes on biological studies of mole crickets at Plant City, Florida. Florida Entomol. 26: 33–46.

Hegner, R. W. 1942. College zoology. 5th ed. Macmillan, New York, N.Y. 817 pp.

Heinrichs, E. A. 1973. Bionomics and control of sod webworms. Bull. Entomol. Soc. Amer. 19: 89–97.

Heinrichs, E. H., and C. J. Southards. 1970. Susceptibility of the sod webworm *Pediasia trisecta* to biological control agents. Tennessee Farm Home Sci. (Jan., Feb., March): 30–32.

Heit, C. E., and H. K. Henry. 1940. Notes on the species of white grubs present in the Saratoga Forest Tree Nursery. J. Forest. 38: 944–948.

Heller, P. R. 1995. Asiatic garden beetle, pp. 26–28. *In* Brandenburg, R. L., and M. G. Villani (eds.), Handbook of turfgrass insect pests. Entomological Society of America, Lanham, Md. 140 pp.

Hellman, L. 1994. Turfgrass insect detection and sampling techniques, pp. 331–336. *In* A. R. Leslie (ed.), Handbook of integrated pest management for turf and ornamentals. CRC Press, Boca Raton, Fla.

Hellman, L. 1995. Green June beetle, pp. 57–59. *In* R. L. Brandenburg and M. G. Villani (eds.), Handbook of turfgrass insect pests. Entomological Society of America, Lanham, Md. 140 pp.

Henderson, F. R. 1983. Other mammals: Moles, pp. D-53–D-61. *In* R. M. Timm (ed.), Prevention and control of wildlife damage. Great Plains Agr. Council and Univ. Nebraska Coop. Ext. Service, Inst. Agr. and Nat. Resources, Lincoln, Neb. Looseleaf, 632 pp.

Heng-Moss, T. M., F. P. Baxendale, and T. P. Riordan. 1998. *Rhopus nigroclavatus* (Ashmead) and *Pseudaphycus* sp. (Hymenoptera: Encyrtidae): Two parasitoids of the buffalograss mealybugs, *Tridiscus sporoboli* (Cockerell) and *Trionymus* sp. J. Kansas Entomol. Soc. 71(1): 85–86.

Henry, H. K., and C. E. Heit. 1940. Flight records of *Phyllophaga* (Coleoptera: Scarabaeidae). Entomol. News 40: 279–282.

Hodges, R. W., et al. 1983. Check list of the Lepidoptera of America north of Mexico. E. W. Classey and Wedge Entomol. Research Found., London. 284 pp.

Hoffman, C. H. 1935. Biological notes on *Ataenius cognatus* (Lec.), a new pest of golf greens in Minnesota (Scarabaeidae—Coleoptera). J. Econ. Entomol. 28: 666–667.

Hoffmann, M. P., and A. C. Frodsham. 1993. Natural enemies of vegetable insect pests. Coop. Ext., Cornell Univ. Ithaca, N.Y. 64 pp.

Holldobler, B., and E. O. Wilson. 1990. The ants. Harvard University Press, Cambridge, Mass. 732 pp.

Hudson, W. G. 1987a. Variability in development of *Scapteriscus acletus* (Orthoptera: Gryllotalpidae). Florida Entomol. 70: 403–404.

Hudson, W. G. 1987b. Ontogeny of prey selection in *Sirthenea carinata*: generalist juveniles become specialist adults. Entomophaga 32: 399–406.

Hudson, W. G. 1988. Field sampling of mole crickets (Orthoptera: Gryllotalpidae: Scapteriscus): A comparison of techniques. Florida Entomol. 71: 214–216.

Hudson, W. G. 1989. Field sampling and population estimation of the tawny mole cricket (Orthoptera: Gryllotalpidae). Florida Entomol. 72: 337–343.

Hudson, W. G. 1994. Life cycles and population monitoring of pest mole crickets, pp. 345–349. *In* A. R. Leslie (ed.), Handbook of integrated pest management for turf and ornamentals. CRC Press, Boca Raton, Fla.

Hudson, W. G. 1995. Mole crickets, pp. 78–81. *In* R. L. Brandenburg and M. G. Villani (eds.), Handbook of turfgrass insect pests. Entomological Society of America, Lanham, Md. 140 pp.

Hudson, W. G., and K. B. Nguyen. 1989. Infection of *Scapteriscus vicinus* (Orthoptera: Gryllotalpidae) nymphs by *Neoaplectana* sp. (Rhabditida: Steinernematidae). Florida Entomol. 72: 383–384.

Hudson, W. G., and J. G. Saw. 1987. Spatial distribution of the tawny mole cricket, *Scapteriscus vicinus*. Entomol. Exp. Appl. 45: 99–104.

Hudson, W. G., J. H. Frank, and J. L. Castner. 1988. Biological control of *Scapteriscus* spp. mole crickets (Orthoptera: Gryllotalpidae) in Florida. Bull. Entomol. Soc. Amer. 34: 192–198.

Hudson, W. G., J. A. Reinert, and R. L. Crocker. 1995. Bermudagrass and related eriophyid mites, pp. 29–31. In R. L. Brandenburg and M. G. Villani (eds.), Handbook of turfgrass insect pests. Entomological Society of America, Lanham, Md. 140 pp.

Hunter, S. J. 1909. The greenbug and its enemies. Kansas Univ. Bull. No. 9. 163 pp.

Hurpin, B. 1953. Reconnaissance des sexes chez les larves de Coleopteres, Scarabaeidae. Bull. Soc. Entomol. France 58: 104–107.

Jackson, D. M., and R. L. Campbell. 1975. Biology of the European crane fly, *Tipula paludosa* Meigen, in western Washington (Tipulidae: Dipteral). Washington State Univ. Tech. Bull. No. 81. 23 pp.

Jackson, D. W., K. J. Vessels, and D. A. Potter. 1981. Resistance of selected cool and warm season turfgrasses to the greenbug *(Schizaphis graminum)*. HortScience 16: 558–559.

Jackson, T. A. 1984. Honey disease, an indicator of population decline in grass grub. Proc. N.Z. Weed Pest Control Conf. 37: 113–116.

Jackson, T. A., and T. R. Glare. 1992. Rickettsial diseases of scarabs, pp. 33–42. *In* T. A. Jackson and T. R. Glare (eds.), Use of pathogens in scarab pest management. Intercept Press, Andover, Hampshire, UK. 298 pp.

Jackson, T. A., J. F. Pearson, M. O'Callaghan, and H. K. Mahanty. 1992. Pathogen to product-development of *Serratia entomophila* (Enterobacteriaceae) as a commercial biological control agent for the New Zealand Grass Grub *(Costelytra zealandica)*, pp. 191–198. *In* T. A. Jackson and T. R. Glare (eds.), Use of pathogens in scarab pest management. Intercept Press, Andover, Hampshire, UK. 298 pp.

Janisch, E. 1941. Das temperatumptimum der Wiesenschnake *Tipula paludosa*. Mitt. Biol. Reichsanst 65: 38.

Jaynes, H. A., and T. R. Gardner. 1924. Selective parasitism by *Tiphia* sp. J. Econ. Entomol. 17: 366–369.

Jefferson, R. N., and C. O. Eades. 1952. Control of sod webworm in southern California. J. Econ. Entomol. 45: 114–118.

Jefferson, R. N., I. M. Hall, and F. S. Morishita. 1964. Control of lawn moths in southern California. J. Econ. Entomol. 57: 150–152.

Jeppson, L. R., H. H. Keifer, and E. W. Baker. 1975. Mites injurious to economic plants. Univ. California Press, Berkeley, Cal. 614 pp.; 74 plates.

Jepson, W. F., and A. J. Heard. 1959. The frit fly and allied stem boring Diptera in winter wheat and host grasses. Ann. Appl. Biol. 47: 114–130.

Jerath, M. L. 1960. Notes on larvae of nine genera of Aphodiinae in the United States (Coleoptera: Scarabaeidae). Proc. U.S. Nat. Mus. Smithsonian Inst. 111: 43–94.

Johnson, F. A. 1975. The bermudagrass mite *Eriophyes cynodoniensis* (Sayed) (Acari: Eriophyidae) in Florida, with reference to its injury, symptomology, ecology, and integrated control. Ph.D. Thesis. Univ. of Florida, Gainesville, Fla. 182 pp.

Johnson, J. P. 1941. *Cyclocephala (Ochrosidia) borealis* in Connecticut. J. Agr. Res. 62: 79–86.

Johnson, N. E., and R. S. Cameron. 1969. Phytophagous ground beetles. Ann. Entomol. Soc. Amer. 62: 909–914.

Johnson-Cicalese, J. M. 1997. Developing turfgrasses with enhanced insect resistance. TurfGrass Trends 6 (August): 1–6.

Johnson-Cicalese, J. M., and C. R. Funk. 1990. Additional host plants of four species of billbug found in New Jersey turfgrasses. J. Amer. Soc. Hort. Sci. 115: 608–611.

Johnson-Cicalese, J. M., and R. H. White. 1990. Effect of *Acremonium* endophytes on four species of billbug found on New Jersey turfgrasses. J. Amer. Soc. Hort. Sci. 115: 602–604.

Johnson-Cicalese, J. M., R. H. Hurley, G. W. Wolfe, and C. R. Funk. 1989. Developing turfgrasses with improved resistance to billbugs. 6th Internatl. Turfgrass Res. Conf. Proc. 107–110.

Johnson-Cicalese, J. M., G. W. Wolfe, and C. R. Funk. 1990. Biology, distribution, and taxonomy of billbug turf pests (Coleoptera: Curculionidae). Environ. Entomol. 19: 1037–1046.

Judge, F. D. 1972. Aspects of the biology of the gray garden slug *(Deroceras reticulatum* Muller*)*. Search Agr. 2: 1–18.

Kamm, J. A. 1969. Biology of the billbug *Sphenophorus venatus confluens,* a new pest of orchardgrass. J. Econ. Entomol. 62: 808–812.

Kamm, J. A. 1970. Effects of photoperiod and temperature on *Crambus trisectus* and *C. leachellus cypridalis* (Lepidoptera: Crambidae). Ann. Entomol. Soc. Amer. 63: 412–416.

Kamm, J. A. 1971. Environmental biology of a sod webworm *Crambus tutillus* (Lepidoptera, Crambinae). Entomol. Exp. Appl. 14: 30–38.

Kamm, J. A. 1973. Biotic factors that affect sod webworms in grass fields in Oregon. Environ. Entomol. 2: 94–96.

Kamm, J. A., and J. Capizzi. 1977. Control of grass seed insect pests, pp. 127–129. *In* Oregon insect control handbook. Oregon State Univ., Corvallis, Ore.

Kamm, J. A., and L. M. McDonough. 1979. Field tests with the sex pheromone of the cranberry girdler. Environ. Entomol. 8: 773–775.

Kamm, J. A., and L. M. McDonough. 1980. Synergism of the sex pheromone of the cranberry girdler. Environ. Entomol. 9: 795–797.

Kamm, J. A., P. D. Morgan, D. L. Overhulser, L. M. McDonough, M. Triebwasser, and L. N. Kline. 1983. Management practices for cranberry girdler (Lepidoptera: Pyralidae) in Douglas-fir nursery stock. J. Econ. Entomol. 76: 923–926.

Kawanishi, C. Y., C. M. Splittstoesser, H. Tashiro, and K. H. Steinkraus. 1974. *Ataenius spretulus*, a potentially important turf pest, and its associated milky disease bacterium. Environ. Entomol. 3: 177–180.

Kaya, H. K., and R. Gaugler. 1993. Entomopathogenic nematodes. Ann. Rev. Entomol. 38: 181–206.

Keifer, H. H., E. W. Baker, T. Kono, M. Delfinado, and W. E. Styer. 1982. An illustrated guide to plant abnormalities caused by eriophyid mites in North America. U.S. Dept. Agr. Handbook No. 573. 178 pp.

Kelsheimer, E. G. 1956. The hunting billbug, a serious pest of zoysia. Proc. Florida Hort. Soc. 69: 415–418.

Kelsheimer, E. G., and S. H. Kerr. 1957. Insects and other pests of lawns and turf. Univ. Florida Agr. Exp. Sta. Circ. S-96. 22 pp.

Kennedy, M. K. 1980. New webworm pests in Michigan lawns. Amer. Lawn Appl. (Nov./Dec.): 11–14.

Kennedy, M. K. 1981. Chinch bugs: Biology and control. Amer. Lawn Appl. (July/Aug.): 12–15.

Kerr, S. H. 1955. Life history of the tropical sod webworm *Pachyzancla phaeopteralis* Guenee. Florida Entomol. 38: 3–11.

Kerr, S. H. 1966. Biology of the lawn chinch bug, *Blissus insularis*. Florida Entomol. 49: 9–18.

Kerr, T. W. 1941. Control of white grubs in strawberries. Cornell Univ. Agr. Exp. Sta. Bull. No. 770. 40 pp.

Kindler, S. D., and E. J. Kinbacher. 1975. Differential reaction of Kentucky bluegrass cultivars to the bluegrass billbug, *Sphenophorus parvulus* Gyllenhal. Crop Sci. 15: 873–874.

Kindler, S. D., and E. J. Kinbacher. 1982. Sampling for eggs, larvae, and adults of the bluegrass billbug. Crop Sci. 22: 677–678.

Kindler, S. D., and S. M. Spomer. 1986. Observations on the biology of the bluegrass billbug, *Sphenophorus parvulus* Gyllenhal (Coleoptera: Curculionidae), in an eastern Nebraska sod field. J. Kansas Entomol. Soc. 59: 26–31.

Kindler, S. D., R. Staples, S. M. Spomer, and O. Adeniji. 1983. Resistance of bluegrass cultivars to biotypes C and E greenbug (Homoptera: Aphididae). J. Econ. Entomol. 76: 1103–1105.

King, G. B. 1901. The Coccidae of British North America. Can. Entomol. 33: 193–200.

Kissinger, D. G. 1964. Curculionidae of America north of Mexico: A key to the genera. Taxonomic Publishing, South Lancaster, Penn. 143 pp.

Klein, M. G. 1981. Mass trapping for suppression of Japanese beetles, pp. 183–190. *In* E. R. Mitchell (ed.), Management of insect pests with semiochemicals. Plenum Publishing, New York, N.Y. 514 pp.

Klein, M. G. 1995. Microbial control of turfgrass insects, pp. 95–106. *In* R. L. Brandenburg and M. G. Villani (eds.), Handbook of turfgrass insect pests. Entomological Society of America, Lanham, Md. 140 pp.

Klein, M. G., and R. Georgis. 1993. Persistence of control of Japanese beetle (Coleoptera: Scarabaeidae) larvae with steinernematid and heterorhabditid nematodes. J. Econ. Entomol. 85: 727–730.

Klots, A. B. 1951. A field guide to the butterflies of North America east of the Great Plains. Riverside Press, Cambridge, Mass. 349 pp.

Knight, J. E. 1983. Carnivores (meat-eating mammals): Skunks, pp. C-81–C-86. *In* R. M. Timm (ed.), Prevention and control of wildlife damage. Great Plains Agr. Council and Univ. Nebraska Coop. Ext. Service, Inst. Agr. and Nat. Resources, Lincoln, Neb. Looseleaf, 632 pp.

Knowles, B. H., and J. A. Dow. 1993. The crystal endotozins of *Bacillus thuringiensis*: Models for their mechanism of action on the insect gut. BioEssays 15(7): 469–475.

Koehler, P. G., and D. E. Short. 1976. Control of mole crickets in pasture grass. J. Econ. Entomol. 69: 229–232.

Kouskolekas, C. A., and R. L. Self. 1974. Biology and control of the ground pearl in relation to turfgrass infestation, pp. 421–423. *In* E. C. Roberts (ed.), Proc. 2d Int. Turfgrass Res. Conf. Amer. Soc. Agron., Madison, Wisc. 602 pp.

Krantz, G. W. 1957. Winter grain mite *(Penthaleus major)*. Plant Pest Control Div., U.S. Dept. Agr., Coop. Econ. Insect Rep. 7: 302.

Krantz, G. W. 1978. A manual of acarology. 2d ed. Oregon State Univ. Book Stores, Corvallis, Ore. 509 pp.

Krueger, S., M. G. Villani, J. P. Nyrop, and D. W. Roberts. 1991. Effect of soil environment on the efficacy of fungal pathogens against scarab grubs in laboratory bioassays. J. Biol. Control 1: 203–209.

Ladd, T. L., Jr. 1970. Sex attraction in the Japanese beetle. J. Econ. Entomol. 63: 905–908.

Ladd, T. L., Jr., M. G. Klein, and J. H. Tumlinson. 1981. Phenethyl proprionate + eugenol + geraniol (3:7:3) and Japonilure: A highly effective joint lure for Japanese beetles. J. Econ. Entomol. 74: 665–667.

Laigo, F. M., and M. Tamashiro. 1966. Virus and insect parasite interaction in the lawn armyworm, *Spodoptera mauritia acronyctoides* (Guenee). Proc. Hawaiian Entomol. Soc. 19: 233–237.

Lambert, B., and M. Peferoen. 1992. Insecticidal promise of *Bacillus thuringiensis*. BioScience 42(2): 112–122.

LaPlante, A. A., Jr. 1966a. How to control the lawn armyworm. Univ. Hawaii Coop. Ext. Entomol. Notes No. 1. 2 pp.

LaPlante, A. A., Jr. 1966b. How to control the hunting billbug. Univ. Hawaii Coop. Ext. Entomol. Notes No. 2. 2 pp.

Latch, C. C. M. 1976. Studies on the susceptibility of *Oryctes rhinoceros* to some entomogenous fungi. Entomophaga 21: 31–38.

Laughlin, R. 1958. Desiccation of eggs of the crane fly *Tipula oleracea*. Nature 182: 613.

Lawrence, K. O. 1982. A linear pitfall trap for mole cricket and other soil arthropods. Florida Entomol. Sci. Notes 65: 376–377.

Leal, W. S. 1993. (Z)- and (E)-tetradec-7-en-2-one, a new type of sex pheromone from the oriental beetle. Naturwissenschaften 80: 86–87.

Leonard, D. E. 1966. Biosystemics of the *leucopterus* complex of the genus *Blissus* (Heteroptera: Lygaeidae). Connecticut Agr. Exp. Sta. Bull. 677: 1–47.

Leonard, D. E. 1968. A revision of the genus *Blissus* (Heteroptera: Lygaeidae) in eastern North America. Ann. Entomol. Soc. Amer. 61: 239–250.

Liu, H. J., and F. L. McEwen. 1979. The use of temperature accumulations and sequential sampling in predicting damaging populations of *Blissus leucopterus hirtus*. Environ. Entomol. 8: 512–515.

Lofgren, C. S., W. A. Banks, and B. M. Glancey. 1975. Biology and control of imported fire ants. Ann. Rev. Entomol. 20: 1–30.

Loughrin, J. H., D. A. Potter, and T. R. Hamilton-Kemp. 1995. Volatile compounds induced by herbivory act as aggregation kairomones for the Japanese beetle (*Popillia japonica* Newman). J. Chem. Ecol. 21: 1457–1467.

Loughrin, J. H., D. A. Potter, T. R. Hamilton-Kemp, and M. E. Byers. 1997. Diurnal emissions of volatile compounds by Japanese beetle damage grape leaves. Phytochemistry 45: 919–923.

Ludwig, D. 1932. The effect of temperature on the growth curves of the Japanese beetle (*Popillia japonica* Newman). Physiol. Zool. 5: 431–447.

Luginbill, P. 1922. Bionomics of the chinch bug. U.S. Dept. Agr. Bull. No. 1016. 14 pp.

Luginbill, P. 1928. The fall armyworm. U.S. Dept. Agr. Tech. Bull. No. 34. 92 pp.

Luginbill, P. 1938. Control of common white grubs in cereal and forage crops. U.S. Dept. Agr. Farmers' Bull. No. 1798. 19 pp.

Luginbill, P., and H. R. Painter. 1953. May beetles of the United States and Canada. U.S. Dept. Agr. Tech. Bull. No. 1060. 102 pp.; 78 plates.

Maddock, D. R., and C. F. Fehn. 1958. Human ear invasion by adult scarabaeid beetles. J. Econ. Entomol. 51: 546–547.

Maercks, H. 1939. Die-Wiesenschnaken und ihre Bekampfung. Kranke Pflanze 16: 107–110.

Mahr, D. L., and R. Kachadoorian. 1984. Turfgrass disorder: Bluegrass billbug. Univ. Wisconsin Urban Phytonarian Ser. A 3234. 2 pp.

Mailloux, G., and H. T. Streu. 1979. A sampling technique for estimating hairy chinch bug *(Blissus leucopterus hirtus* Montandon, Hemiptera: Lygaeidae) populations and other arthropods from turfgrass. Ann. Entomol. Soc. Quebec 24: 139–143.

Mailloux, G., and H. T. Streu. 1981. Population biology of the hairy chinch bug *(Blissus leucopterus hirtus,* Montandon: Hemiptera: Lygaeidae). Ann. Entomol. Soc. Quebec 26: 51–90.

Mailloux, G., and H. T. Streu. 1982. Bionomics of the larger sod webworm, *Pediasia trisecta* (Walker) (Lepidoptera: Pyralidae: Crambinae. Ann. Entomol. Soc. Quebec 27: 68–74.

Malcolm, D. R. 1955. Biology and control of the timothy mite, *Paratetranychus pratensis* (Banks). Washington Agr. Exp. Sta. Tech. Bull. No. 17. 35 pp.

Martignoni, M. E., and P. J. Iwai. 1986. A catalogue of viral diseases of insects, mites, and ticks. U.S. Dept. Agr. For. Serv. Gen. Tech. Rep. PNQ-195.

Matheny, E. L., Jr. 1971. Seasonal abundance, distribution, and egg studies of sod webworm moths (Lepidoptera: Pyralidae: Crambinae) in Tennessee. Ph.D. Thesis. Univ. of Tennessee, Knoxville, Tenn. 102 pp.

Matheny, E. L., Jr., and E. A. Heinrichs. 1972. Chorion characteristics of sod webworm eggs. Ann. Entomol. Soc. Amer. 65: 238–246.

Matheny, E. L., Jr., and R. L. Kepner. 1980. Maxillae of the mole crickets, *Scapteriscus acletus* Rehn and Hebard, and *S. vicinus* Scudder (Orthoptera: Gryllotalpidae): A new means of identification. Florida Entomol. 63: 512–514.

Matheson, R. 1951. Entomology for introductory courses, 2d ed. Comstock Publishing Co., Cornell Univ. Press, Ithaca, N.Y. 630 pp.

Mathias, J. K., R. H. Ratcliffe, and J. L. Hellman. 1990. Association of an endophyte fungus in perennial ryegrass and resistance to the hairy chinch bug (Hemiptera: Lygaeidae). J. Econ. Entomol. 83: 1640–1646.

Maxwell, K. E., and G. F. McLeod. 1936. Experimental studies of the hairy chinch bug. J. Econ. Entomol. 29: 339–343.

McCarty, L. B., and M. L. Elliott. 1994. Pest management strategies for golf courses, pp. 193–202. *In* A. R. Leslie (ed.), Handbook of integrated pest management strategies for turf and ornamentals. Lewis Publishers, Boca Raton, Fla. 660 pp.

McCrea, R. J. 1972. The dichondra flea beetle (Genus *Chaetocnema*) in southern California (Coleoptera: Chrysomelidae). M.S. Thesis. California State Coll., Long Beach, Cal. 83 pp.

McDonough, L. M., and J. A. Kamm. 1979. Sex pheromone of the cranberry girdler, *Chrysoteuchia topiaria* (Zellar) (Lepidoptera: Pyralidae). J. Chem. Ecol. 5: 211–219.

McDonough, L. M., J. A. Kamm, D. A. George, C. L. Smithhisler, and S. Voerman. 1982. Sex attractant for the western lawn moth *Tehama bonifatella* Hulst. Environ. Entomol. 11: 711–714.

McGregor, R. A. 1976. Florida turfgrass survey, 1974. Florida Crop and Livestock Rep. Service. 33 pp.

Merchant, M. E., and R. L. Crocker. 1995. White grubs in Texas turfgrass. Texas Agr. Ext. Serv. L-1131, 6 pp.

Metcalf, C. L., W. P. Flint, and R. L. Metcalf. 1962. Destructive and useful insects: Their habits and control. 4th ed. McGraw-Hill, New York, N.Y. 1,087 pp.

Milner, R. J. 1992. The selection of strains of *Metarrhizium anisopliae* for control of Australian sugar cane white grubs, pp. 209–216. *In* T. A. Jackson and T. R. Glare (eds.), Use of pathogens in scarab pest management. Intercept Press, Andover, Hampshire, UK. 298 pp.

Mitchell, W. C., and C. L. Murdoch. 1974. Insecticides and their application frequency for control of turf insects in Hawaii. Down to Earth 30: 17–23.

Morishita, F. S., W. Humphrey, L. C. Johnston, and R. F. Jefferson. 1971. Control of billbugs on turf. California Turfgrass Culture 21: 13–14.

Morrill, W. L., and E. F. Suber. 1976. Biology and control of *Sphenophorus coesifrons* (Gyllenhal) (Coleoptera: Curculionidae) in bahiagrass. J. Georgia Entomol. Soc. 11: 283–288.

Morrison, W. P., B. C. Pass, and C. S. Crawford. 1972. Effect of humidity on eggs of two populations of the bluegrass webworm. Environ. Entomol. 1: 218–221.

Muma, M. H. 1944. The attraction of *Cotinus nitida* by caproic acid. J. Econ. Entomol. 37: 855–856.

Murdoch, C. L., and H. Tashiro. 1976. Host preference of the grass webworm, *Herpetogramma licarsisalis,* to warm season turfgrasses. Environ. Entomol. 5: 1068–1070.

Murphy, J. A., S. Sun, and L. L. Betts. 1993. Endophyte-enhanced resistance to billbug (Coleoptera: Curculionidae), sod webworm (Lepidoptera: Pyralidae), and white grub (Coleoptera: Scarabaeidae) in tall fescue. Environ. Entomol. 22: 699–703.

Neiswander, C. R. 1938. The annual white grub, *Ochrosidia vinosa* Burm., in Ohio lawns. J. Econ. Entomol. 31: 340–344.

Neiswander, C. R. 1963. The distribution and abundance of May beetles in Ohio. Ohio Agr. Exp. Sta. Res. Bull. No. 951. 35 pp.

Nguyen, K. B., and G. C. Smart, Jr. 1991. Pathogenicity of *Steinernema scapterisci* to selected invertebrates. J. Nemat. 23: 7–11.

Nickle, D. A., and J. L. Castner. 1984. Introduced species of mole crickets in the United States, Puerto Rico, and the Virgin Islands (Orthoptera: Gryllotalpidae). Ann. Entomol. Soc. Amer. 77: 450–465.

Nickle, D. A., and W. Frank. 1988. Pest mole crickets, *Scapteriscus acletus* (Orthoptera: Gryllotalpidae), reported established in Arizona. Florida Entomol. 71: 90–91.

Niemczyk, H. D. 1976. A new grub problem in golf course turf. Golf Course Superintendent (Mar.): 26–29.

Niemczyk, H. D. 1977. Thatch: A barrier to control of soil-inhabiting insects pests of turf. Weeds Trees Turf (Feb.): 16–19.

Niemczyk, H. D. 1978. The winter grain mite: Winter pest of turf. Weeds Trees Turf (Feb.): 22–23.

Niemczyk, H. D. 1980a. Insects and their control. Lawn Care Ind. (Mar.): 34–46.

Niemczyk, H. D. 1980b. New evidence indicates greenbug overwinters in North. Weeds Trees Turf (June): 64–65.

Niemczyk, H. D. 1980c. Proper pesticide application for effective greenbug control. Lawn Care Ind. (Oct.): 16–19.

Niemczyk, H. D. 1981. Destructive turf insects. HDN Books, Wooster, Ohio. 48 pp.

Niemczyk, H. D. 1982. Chinch bug and bluegrass billbug control with spring application of chlorpyrifos, pp. 85–89. *In* H. D. Niemczyk and B. J. Joyner (eds.), Advances in turfgrass entomology. Hammer Graphics, Piqua, Ohio. 150 pp.

Niemczyk, H. D. 1983. The bluegrass billbug: A frequently misdiagnosed pest of turfgrass. Amer. Lawn Appl. (May/June): 4–7.

Niemczyk, H. D., and R. A. Chapman. 1987. Evidence of enhanced degradation of isofenphos in turfgrass thatch and soil. J. Econ. Entomol. 80: 880–882.

Niemczyk, H. D., and D. M. Dunbar. 1976. Field observations, chemical control, and contact toxicity experiments on *Ataenius spretulus,* a grub pest of turf grass. J. Econ. Entomol. 69: 345–348.

Niemczyk, H. D., and C. Frost. 1978. Insecticide resistance found in Ohio bluegrass billbugs. Ohio Rep. 63: 22–23.

Niemczyk, H. D., and H. R. Krueger. 1987. Persistence and mobility of isazofos in turfgrass thatch and soil. J. Econ. Entomol. 80: 950–952.

Niemczyk, H. D., and J. R. Moser. 1982. Greenbug occurrence and control on turfgrasses in Ohio, pp. 105–111. *In* H. D. Niemczyk and B. G. Joyner (eds.), Advances in turfgrass entomology. Hammer Graphics, Piqua, Ohio. 150 pp.

Niemczyk, H. D., and K. T. Power. 1982. Greenbug buildup in Ohio is linked to overwintering. Weeds Trees Turf (June): 36.

Niemczyk, H. D., and G. S. Wegner. 1979. Life history and control of the black turfgrass ataenius. Ohio Rep. 64: 85–88.

Niemczyk, H. D., and G. S. Wegner. 1982. Life history and control of the black turfgrass ataenius (Coleoptera-Scarabaeidae), pp. 113–117. *In* H. D. Niemczyk and B. G. Joyner (eds.), Advances in turfgrass entomology. Hammer Graphics, Piqua, Ohio. 150 pp.

Niemczyk, H. D., Z. Filary, and H. Krueger. 1988. Movement of insecticide residues in turfgrass thatch and soil. Golf Course Management (Feb.): 22, 26.

Niemczyk, H. D., R. A. J. Taylor, M. P. Tolley, and K. T. Power. 1992. Physiological time-driven model for predicting first generation of the hairy chinch bug (Hemiptera: Lygaeidae) on turfgrass in Ohio. J. Econ. Entomol. 85: 821−829.

Nyrop, J. P., M. G. Villani, and J. Grant. 1995. A control decision rule for European chafer grubs infesting residential lawns. Environ. Entomol. 24: 521−528.

O'Brien, J. M. 1983. Rodents (gnawing mammals): Voles, pp. B-147−B-152. *In* R. M. Timm (ed.), Prevention and control of wildlife damage. Great Plains Agr. Council and Univ. Nebraska Coop. Ext. Service, Inst. Agr. and Nat. Resources, Lincoln, Neb. Looseleaf, 632 pp.

Oi, D. H., R. M. Pereira, J. L. Stimac, and L. A. Wood. 1994. Field applications of *Beauveria bassiana* for control of the red imported fire ant (Hymenoptera: Formicidae). J. Econ. Entomol. 87: 623−630.

Okamura, G. T. 1959. Illustrated key to the lepidopterous larvae attacking lawns in California. California Dept. Agr. Bull. 48: 15−21.

Oliver, A. D. 1982a. The red imported fire ant as a lawn insect problem. Amer. Lawn Appl. (May/June): 4−8.

Oliver, A. D. 1982b. The fall armyworm as an annual pest. Amer. Lawn Appl. (Nov./Dec.): 18−22.

Oliver, A. D. 1984. The hunting billbug—One among the complex of turfgrass insect and pathogen problems. Amer. Lawn Appl. (March/Apr.): 24−27.

Oliver, A. D., and J. B. Chapin. 1981. Biology and illustrated key for the identification of twenty species of economically important noctuid pests. Louisiana Agr. Exp. Sta. Bull. No. 733. 26 pp.

Oliver, A. D., and K. N. Komblas. 1981. Southern chinch bug in Louisiana. Amer. Lawn Appl. (July/Aug.): 26−31.

Opler, P. A., and G. O. Krizek. 1984. Butterflies east of the Great Plains. John Hopkins Univ. Press, Baltimore, Md. 274 pp.

Oschmann, M. 1979. Violet trays for recording the population dynamics of the frit fly *(Oscinella frit* L.). Archiv Phytopathol. Pflanzenschutz 15: 197−203.

Palmer, R. S. 1962. Handbook of North American birds. Yale University Press, New Haven, Conn.

Paris, O. H. 1963. The ecology of *Armadillidum vulgare* (Isopoda: Oniscoidea) in California grassland: Food, enemies and weather. Ecol. Monogr. 33: 1−22.

Parkman, J. P., and J. H. Frank. 1992. Infection of sound-trapped mole crickets, *Scapteriscus* spp., by *Steinernema scapterisci*. Florida Entomol. 75: 163−165.

Parkman, J. P., J. H. Frank, K. B. Nguyen, and G. C. Smart, Jr. 1994. Inoculative release of *Steinernema scapterisci* (Rhabditida: Steinernematidae) to suppress pest mole crickets (Orthoptera: Gryllotalpidae) on golf courses. Environ. Entomol. 23: 1331−1337.

Parkman, J. P., J. H. Frank, T. J. Walker, and D. J. Schuster. 1996. Classical biological control of *Scapteriscus* spp. (Orthoptera: Gryllotalpidae) in Florida. Environ. Entomol. 25: 1415−1420.

Parkman, J. P., W. G. Hudson, J. H. Frank, K. B. Nguyen, and G. C. Smart, Jr. 1993. Establishment and persistence of *Steinernema scapterisci* (Rhabditida: Steinernematidae) in field populations of *Scapteriscus* spp. mole crickets (Orthoptera: Gryllotalpidae). J. Entomol. Sci. 28: 182−190.

Pemberton, C. E. 1955. Notes and exhibitions. Proc. Hawaiian Entomol. Soc. 15: 373.

Pendland, J. C., and D. G. Boucias. 1987. The hyphomycete *Sorosporella-Syngliocladium* from mole cricket, *Scapteriscus vicinus*. Mycopathologia 99: 25−30.

Peterson, R. T. 1980. A field guide to the birds. 4th ed. Houghton Mifflin, Boston, Mass. 384 pp.

Polivka, J. B. 1960a. Effect of lime applications to soil on Japanese beetle larval population. J. Econ. Entomol. 53: 476−477.

Polivka, J. B. 1960b. Grub population in turf varies with pH levels in Ohio soils. J. Econ. Entomol. 53: 860−863.

Polivka, J. B. 1963. Control of hairy chinch bug, *Blissus leucopterus hirtus* Mont., in Ohio. Ohio Agr. Exp. Sta. Res. Circ. No. 122. 8 pp.

Poprawski, T. J. 1994. Insect parasites and predators of *Phyllophaga anxia* (LeConte) (Col., Scarabaeidae) in Quebec, Canada. J. Appl. Entomol. 117: 1−9.

Poprawski, T. J., and W. N. Yule. 1990a. Bacterial pathogens of *Phyllophaga* spp. (Col., Scarabaeidae) in southern Quebec, Canada. J. Appl. Entomol. 109: 414−422.

Poprawski, T. J., and W. N. Yule. 1990b. A new small iridescent virus from grubs of *Phyllophaga anxia* (LeConte) (Col., Scarabaeidae). J. Appl. Entomol. 110: 63–67.

Poprawski, T. J., and W. N. Yule. 1991a. Incidence of fungi in natural populations of *Phyllophaga* spp. and susceptibility of *Phyllophaga anxia* (LeConte) (Col., Scarabaeidae) to *Beauveria bassiana* and *Metarrhizium anisopliae* (Deuteromycotina). J. Appl. Entomol. 112: 359–365.

Poprawski, T. J., and W. N. Yule. 1991b. *Chroniodiplogaster aerivora* (Cobb) (Rhabditida: Diplogasteridae), a natural enemy of white grubs, *Phyllophaga* Harris (Coleoptera: Scarabaeidae). Biocontrol Sci. Technol. 1: 311–321.

Poprawski, T. J., and W. N. Yule. 1992a. *Actinocephalus* sp., a protozoan (Eugregarinidia: Actinocephalidae) associated with June beetles, *Phyllophaga* Harris. (Coleoptera: Scarabaeidae). Can. Entomol. 124: 391–396.

Poprawski, T. J., and W. N. Yule. 1992b. Acari associated with *Phyllophaga anxia* (LeConte) (Coleoptera: Scarabaeidae) in southern Quebec and eastern Ontario. Can. Entomol. 124: 397–403.

Porter, K. B., G. L. Peterson, and O. Vise. 1982. A new greenbug biotype. Crop Sci. 22: 847–850.

Potter, D. A. 1980. Flight activity and sex attraction of northern and southern masked chafers in Kentucky turfgrass. Ann. Entomol. Soc. Amer. 73: 414–417.

Potter, D. A. 1981a. Biology and management of masked chafer bugs. Amer. Lawn Appl. (July/Aug.): 2–6.

Potter, D. A. 1981b. Seasonal emergence and flight of northern and southern masked chafers in relation to air and soil temperature and rainfall patterns. Environ. Entomol. 10: 793–797.

Potter, D. A. 1982a. Greenbugs on turfgrass: An informative update. Amer. Lawn Appl. (March/Apr.): 20–25.

Potter, D. A. 1982b. Influence of feeding by grubs of the southern masked chafer on quality and yield of Kentucky bluegrass. J. Econ. Entomol. 75: 21–24.

Potter, D. A. 1983. Effect of soil moisture on oviposition, water absorption, and survival of southern masked chafer (Coleoptera: Scarabaeidae) eggs. Environ. Entomol. 12: 1223–1227.

Potter, D. A. 1995a. Masked chafers, pp. 70–72. *In* Brandenburg, R. L., and M. G. Villani (eds.), Handbook of turfgrass insect pests. Entomological Society of America, Lanham, Md. 140 pp.

Potter, D. A. 1995b. Beneficial and innocuous invertebrates in turf, pp. 101–104. *In* R. L. Brandenburg and M. G. Villani (eds.), Handbook of turfgrass insect pests. Entomological Society of America, Lanham, Md. 140 pp.

Potter, D. A. 1998. Destructive turfgrass insects: Biology, diagnosis, and control. Ann Arbor Press, Inc., Chelsea, Mich. 344 pp.

Potter, D. A., and F. C. Gordon. 1984. Susceptibility of *Cyclocephala immaculata* (Coleoptera: Scarabaeidae) eggs and immatures to heat and drought in turfgrass. Environ. Entomol. 13: 794–799.

Potter, D. A., and K. F. Haynes. 1993. Field-testing pheromone traps for predicting masked chafer (Coleoptera: Scarabaeidae) grub density in golf course turf and home lawns. J. Entomol. Sci. 28: 205–212.

Potter, D. A., M. C. Buxton, C. T. Redmond, C. G. Patterson, and A. J. Powell. 1990. Toxicity of pesticides to earthworms (Oligochaeta: Lumbricidae) and effect on thatch degradation in Kentucky bluegrass turf. J. Econ. Entomol. 83: 2362–2369.

Potter, D. A., C. G. Patterson, and C. T. Redmond. 1992. Influence of turfgrass species and tall fescue endophyte on feeding ecology of Japanese beetle and southern masked chafer grubs (Coleoptera: Scarabaeidae). J. Econ. Entomol. 85: 900–909.

Potter, D. A., A. J. Powell, P. G. Spicer, and D. W. Williams. 1996. Cultural practices affect root-feeding white grubs (Coleoptera: Scarabaeidae) in turfgrass. J. Econ. Entomol. 89: 156–164.

Potter, M. F. 1995. Biting and stinging pests, pp. 91–94. *In* R. L. Brandenburg and M. G. Villani (eds.), Handbook of turfgrass insect pests. Entomological Society of America, Lanham, Md. 140 pp.

Randell, R., J. D. Butler, and T. D. Hughes. 1972. The effect of pesticides on thatch accumulation and earthworm populations in Kentucky bluegrass turf. HortScience 7: 64–65.

Ratcliffe, B. C. 1991. The scarabs of Nebraska. Bull. Univ. Neb. State Mus. Vol. 12, 333 pp.

Ratcliffe, R. H. 1982. Evaluation of cool-season turfgrasses for resistance to the hairy chinch bug, pp.

13–18. *In* H. D. Niemczyk and B. J. Joyner (eds.), Advances in turfgrass entomology. Hammer Graphics, Piqua, Ohio. 150 pp.

Ratcliffe, R. H., and J. J. Murray. 1983. Selection for greenbug (Homoptera: Aphididae) resistance in Kentucky bluegrass cultivars. J. Econ. Entomol. 76: 1221–1224.

Redmond, C. T., and D. A. Potter. 1995. Lack of efficacy of in vivo- and putatively in vitro-produced *Bacillus popilliae* against field populations of Japanese beetle (Coleoptera: Scarabaeidae) grubs in Kentucky. J. Econ. Entomol. 88: 846–854.

Reese, J. C., L. M. English, T. R. Yonke, and M. L. Fairchild. 1972. A method for rearing black cutworms. J. Econ. Entomol. 65: 1047–1050.

Regnier, R. 1939. Contribution a l'etude des hannetons: Un grand ennemi des gazons: *Amphimallon majalis* Razoumowsky. Ann. Epiphyt. Phytogenet. 5: 257–265.

Regniere, J., R. L. Rabb, and R. E. Stinner. 1981. *Bacillus popilliae:* Effect of soil moisture and texture on the survival and development of eggs and first instar grubs. Environ. Entomol. 10: 654–660.

Reid, W. 1983. European chafer/Hanneton europeen: Situation in Canada/situation au Canada. Biol. Programs Sect., Plant Health Div., Ottawa. 5 pp.

Reinert, J. A. 1972. New distribution and host record for the parasitoid *Eumicrosoma benefica.* Florida Entomol. 55: 143–144.

Reinert, J. A. 1973. Sod webworm control in Florida turfgrass. Florida Entomol. 56: 333–337.

Reinert, J. A. 1974. Tropical sod webworm and southern chinch bug control in Florida. Florida Entomol. 57: 275–280.

Reinert, J. A. 1975. Life history of the striped grassworm, *Mocis latipes.* Ann. Entomol. Soc. Amer. 68: 201–204.

Reinert, J. A. 1978. Natural enemy complex of the southern chinch bug in Florida. Ann. Entomol. Soc. Amer. 71: 728–731.

Reinert, J. A. 1982a. The bermudagrass stunt mite. U.S. Golf Assoc. Green Sec. Rec. 20: 9–12.

Reinert, J. A. 1982b. Insecticide resistance in epigeal insect pests of turfgrass: 1, A review, pp. 71–75. *In* H. D. Niemczyk and B. J. Joyner (eds.), Advances in turfgrass entomology. Hammer Graphics, Piqua, Ohio. 150 pp.

Reinert, J. A. 1982c. Southern chinch bug resistance to insecticides: A method for quick diagnosis of chlorpyrifos (OP) resistance and alternate controls. Florida Turfgrass Proc. 30: 64–78.

Reinert, J. A. 1983a. Field experiments for insecticidal control of sod webworms (Lepidoptera: Pyralidae) in Florida turfgrass. J. Econ. Entomol. 76: 150–153.

Reinert, J. A. 1983b. Foraging sites of the southern mole cricket, *Scapteriscus acletus* (Orthoptera: Gryllotalpidae). Proc. Florida State Hort. Soc. 96: 149–151.

Reinert, J. A. 1983c. The bermudagrass stunt mite. Florida Green (Spring): 34–38.

Reinert, J. A., and P. Busey. 1983. Resistance of bermudagrass selections to the tropical sod webworm (Lepidoptera: Pyralidae). Environ. Entomol. 12: 1844–1845.

Reinert, J. A., and H. L. Cromroy. 1981. Bermudagrass stunt mite and its control in Florida. Proc. Florida State Hort. Soc. 94: 124–126.

Reinert, J. A., and S. H. Kerr. 1973. Bionomics and control of lawn chinch bugs. Bull. Entomol. Soc. Amer. 19: 91–92.

Reinert, J. A., and K. M. Portier. 1983. Distribution and characterization of organophosphate-resistant southern chinch bugs (Heteroptera: Lygaeidae) in Florida. J. Econ. Entomol. 76: 1187–1190.

Reinert, J. A., and D. E. Short. 1981. Turf devastation in the Southeast. Golf Course Management (Apr.): 22–28.

Reinert, J. A., B. D. Bruton, and R. W. Toler. 1980. Resistance of St. Augustinegrass to southern chinch bug and St. Augustine decline strain of Panicum mosaic virus. J. Econ. Entomol. 73: 602–604.

Reinert, J. A., A. E. Dudeck, and G. H. Snyder. 1978. Resistance in bermudagrass to the bermudagrass mite. Environ. Entomol. 7: 885–888.

Reinert, J. A., P. R. Heller, and R. L. Crocker. 1995. Chinch bugs, pp. 38–42. *In* R. L. Brandenburg and M. G. Villani (eds.), Handbook of turfgrass insect pests. Entomological Society of America, Lanham, Md. 140 pp.

Reinhard, H. J. 1940. The life history of *Phyllophaga lanceolata* (Say) and *Phyllophaga crinita* Burmeister. J. Econ. Entomol. 33: 572–578.

Rennie, J. 1917. On the biology and economic significance of *Tipula paludosa* Meign. Pt. 2. Hatching, growth, and habits of larva. Ann. Appl. Biol. 3: 116–137.

Rex, E. G. 1931. Facts pertaining to the Japanese beetle. New Jersey Dept. Agr. Circ. No. 180. 31 pp.

Rings, R. W. 1977. An illustrated field key to common cutworms, armyworms, and looper moths in north central states. Ohio Agr. Res. Dev. Center Res. Circ. No. 227. 60 pp.

Rings, R. W., and G. J. Musick. 1976. A pictorial field key to the armyworms and cutworms attacking corn in the north central states. Ohio Agr. Res. Dev. Center Res. Circ. No. 221. 36 pp.

Rings, R. W., F. J. Arnold, A. J. Keaster, and G. J. Musick. 1974a. A worldwide annotated bibliography of the black cutworm *Agrotis ipsilon* (Hufnagel). Ohio Agr. Res. Dev. Center Res. Circ. No. 198. 106 pp.

Rings, R. W., B. A. Baughman, and F. J. Arnold. 1974b. An annotated bibliography of the bronzed cutworm. Ohio Agr. Res. Dev. Center Res. Circ. No. 200. 36 pp.

Rings, R. W., B. A. Johnson, and F. J. Arnold. 1976. A worldwide, annotated bibliography of the variegated cutworm, *Peridroma saucia* Hubner. Ohio Agr. Res. and Dev. Center Res. Circ. No. 219. 126 pp.

Ritcher, P. O. 1940. Kentucky white grubs. Kentucky Agr. Exp. Sta. Bull. 401: 71–157.

Ritcher, P. O. 1949. May beetles and their control in the inner bluegrass region of Kentucky. Kentucky Agr. Exp. Sta. Bull. No. 542. 12 pp.

Ritcher, P. O. 1966. White grubs and their allies: A study of North American scarabaeid larvae. Oregon State Univ. Press, Corvallis, Ore. 219 pp.

Rivers, R. L., K. S. Pike, and Z. B. Mayo. 1977. Influence of insecticides and corn tillage systems on larval control of *Phyllophaga anxia*. J. Econ. Entomol. 70: 794–796.

Roberts, E. C., and B. C. Roberts. 1988. Lawn and sports turf benefits. The Lawn Institute, Pleasant Hill, Tenn.

Robinson, W. H., and M. P. Tolley. 1982. Sod webworms associated with turfgrass in Virginia. Amer. Lawn Appl. (July/Aug.): 22–25.

Romoser, W. S., and J. G. Stoffolano, Jr. 1998. The science of entomology. 5th ed. WCB/McGraw-Hill, Boston, Mass. 605 pp.

Roselle, R. E. 1975. Bluegrass billbug. Univ. Nebraska Coop. Ext. Nebguide G75–236. 2 pp.

Rothwell, N. L. 1997. Effects of golf course maintenance and cultural practices on *Ataenius spretulus* Haldeman and *Aphodius granarius* (L.) on a perennial ryegrass golf course in Michigan. M.S. Thesis. Michigan State Univ., East Lansing, Mich.

Rothwell, N. L., and D. Smitley. 1999. Impact of golf course mowing practices on *Ataenius spretulus* Haldeman (Coleoptera: Scarabaeidae) and its natural enemies. Environ Entomol. 28: 358–366.

Rowe, W. J., II, and D. A. Potter. 1996. Vertical stratification of feeding by Japanese beetle within linden tree canopies: selective foraging or height per se? Oecologia 108: 459–466.

Sargent, S. 1982. The sod webworm, turfgrass pest. Amer. Lawn Appl. (Mar./Apr.): 4–6.

Satterthwait, A. F. 1931. *Anaphoidea calendrae* Gahan, a mymarid parasite of eggs of weevils of the genus *Calendra*. J. New York Entomol. Soc. 39: 171–190.

Satterthwait, A. F. 1932. How to control billbugs destructive to cereals and forage crops. U.S. Dept. Agr. Farmers' Bull. No. 1003. 22 pp.

Satterthwait, A. F. 1933. Larval instars and feeding of the black cutworm, *Agrotis ipsilon* Rott. J. Agr. Res. 46: 517–530.

Saxena, P. N., and H. L. Chada. 1971. The greenbug *Schizaphis graminum*: Mouth parts and feeding habits. Ann. Entomol. Soc. Amer. 64: 897–904.

Schread, J. C. 1964. Insect pests of Connecticut lawns. Connecticut Agr. Exp. Sta. Circ. No. 212. 10 pp.

Schread, J. C. 1970a. Chinch bug control. Connecticut Agr. Exp. Sta. Circ. No. 233. 6 pp.

Schread, J. C. 1970b. The annual bluegrass weevil. Connecticut Agr. Exp. Sta. Circ. No. 234. 6 pp.

Schuder, D. L. 1964. The control of sod webworms *(Crambus* spp.*)* in Indiana. Proc. Indiana Acad. Sci. 72: 164–166.

Schumann, G. L., P. J. Vittum, M. L. Elliott, and P. P. Cobb. 1997. IPM handbook for golf courses. Ann Arbor Press, Chelsea, Mich. 264 pp.

Schurr, K. M., and R. W. Rings. 1964. Uniform terminology for generations of multivoltine insects. Bull. Entomol. Soc. Amer. 10: 89–91.

Sears, M. K. 1978. Hairy chinch bugs in lawns. Ontario (Canada) Ministry of Agri. and Food Factsheet AGDEX 626. 2 pp.

Sears, M. K. 1979. Damage to golf course fairways by *Aphodius granarius* (L.) (Coleoptera: Scarabaeidae). Proc. Entomol. Soc. Ontario. 109: 48.

Sears, M. K., and K. Maitland. 1983. Biology of the turfgrass scale insect, *Lecanopsis formicarum*. 1983 Turfgrass Res. Ann. Rep. Univ. of Guelph, Ontario. 43 pp.

Sears, M. K., and K. Maitland. 1984. Biology of the turfgrass scale. 1984 Turfgrass research annual report. Univ. of Guelph, Ontario. 38 pp.

Sellke, K. 1937. Beobachtungen uber die Bekampfung von Wiesenschnakenlarven (*Tipula pallidosa* Meig und *Tipula czizeki* de J.). Z. Angew. Entomol. 24: 277–284.

Selvan, S., R. Gaugler, and J. Cambell. 1993. Efficacy of entomopathogenic nematode strains against Japanese beetle larvae. J. Econ. Entomol. 86: 353–359.

Semlitsch, R. D. 1986. Life history of the northern mole cricket, *Neocurtilla hexadactyla* (Orthoptera: Gryllotalpidae), utilizing Carolina-bay habitats. Ann. Entomol. Soc. Amer. 79: 256–261.

Shapiro, I. 1975. Courtship and mating behavior of the fiery skipper, *Hylephelia phyleus* (Hesperiidae). J. Res. Lepid. 14: 125–141.

Shaw, M. W., P. Blasdale, and R. M. Allan. 1974. A comparison between the ODCB technique and heat extraction of soil cores for estimating leatherjacket populations. Plant Pathol. 23: 60–66.

Shearman, R. C., D. M. Bishop, D. H. Steinegger, and A. H. Bruneau. 1983. Kentucky bluegrass cultivar and blend response to bluegrass billbug. HortScience 18: 441–442.

Shetlar, D. J. 1991. Billbugs in turfgrass. Ohio Home, Yard, and Garden Facts, Ohio Coop. Ext. HYG-2502-91. 4 pp.

Shetlar, D. J. 1995a. Turfgrass insect and mite management, pp. 171–344. *In* T. L. Watschke, P. H. Dernoeden, and D. J. Shetlar (eds.), Managing turfgrass pests. Lewis Publishers, Ann Arbor, Mich. 361 pp.

Shetlar, D. J. 1995b. Cutworms, pp. 46–48. *In* R. L. Brandenburg and M. G. Villani (eds.), Handbook of turfgrass insect pests. Entomological Society of America, Lanham, Md. 140 pp.

Shetlar, D. J. 1995c. Managing black cutworms without pesticides. Golf Course Management (April): 49–51.

Shetlar, D. J., P. R. Heller, and P. D. Irish. 1983. Turfgrass insect and mite manual with an index of registered materials. Pennsylvania Turfgrass Council, Bellefonte, Penn. 63 pp.

Shetlar, D. J., R. E. Sulman, and R. Georgis. 1988. Irrigation and the use of entomogenous nematodes *Neoaplectana spp.* and *Heterorhabditis heliothidis* (Rhabditida: Steinernematidae and Heterorhabditidae) for control of Japanese beetle (Coleaoptera: Scarabaeidae) larvae in turfgrass. J. Econ. Entomol. 81: 1318–1322.

Shorey, H. H., R. H. Burrage, and G. G. Gyrisco. 1960. The relationship between environmental factors and the density of European chafer (*Amphimallon majalis*) larvae in permanent pasture sod. Ecology 41: 253–258.

Short, D. 1973. Field evaluation of insecticides for controlling mole crickets in turf. Down to Earth 29: 3–5.

Short, D. E., and P. G. Koehler. 1979. A sampling technique for mole crickets and other pests in turfgrass and pasture. Florida Entomol. 62: 282–283.

Short, D. E., and J. A. Reinert. 1982. Biology and control of mole crickets in Florida, pp. 119–124. *In* H. D. Niemczyk and B. G. Joyner (eds.), Advances in turfgrass entomology. Hammer Graphics, Piqua, Ohio. 150 pp.

Shurtleff, M. C., and R. Randell. 1974. How to control lawn diseases and pests. Intertec Publishing, Kansas City, Mo. 97 pp.

Smiley, R. W. 1983. Compendium of turfgrass diseases. Amer. Phytopathol. Soc. St. Paul, Minn. 102 pp.

Smith, L. B., and C. H. Hadley. 1926. The Japanese beetle. U.S. Dept. Agr. Circ. No. 363. 67 pp.

Smitley, D. R. 1995. European chafer, pp. 50–52. *In* R. L. Brandenburg and M. G. Villani (eds.), Handbook of turfgrass insect pests. Entomological Society of America, Lanham, Md. 140 pp.

Smitley, D. R., T. W. Davis, and N. L. Rothwell. 1998. Spatial distribution of *Ataenius spretulus* Haldeman and *Aphdius granaruis* (Coleoptera: Scarabaeidae), and predaceous insects across golf course fairways and roughs. Environ. Entomol. 27: 1336–1349.

Snodgrass, R. E. 1935. Principles of insect morphology. McGraw-Hill, New York, N.Y. 667 pp.

Snow, J. W., and P. S. Callahan. 1968. Biological and morphological studies of the granulate cutworm, *Feltia subterranea* (F.), in Georgia and Louisiana. Univ. Georgia Coll. Agr. Exp. Sta. Res. Bull. No. 42. 23 pp.

Sorenson, K. A., and H. E. Thompson. 1971. Mating behavior of the buffalograss webworm, *Surattha identella*. J. Kansas Entomol. Soc. 44: 329–331.

Southwood, T. R. E., W. F. Jepson, and H. F. Van Emden. 1961. Studies on the behavior of *Oscinella frit* L. (Diptera) adults of the panicle generation. Entomol. Exp. Appl. 4: 196–210.

Sparks, B. 1995. Red imported fire ant, pp. 84–85. *In* R. L. Brandenburg and M. G. Villani (eds.), Handbook of turfgrass insect pests. Entomological Society of America, Lanham, Md. 140 pp.

Splittstoesser, C., and H. Tashiro. 1977. Three milky disease bacilli from a scarabaeid, *Ataenius spretulus*. J. Invertebr. Pathol. 30: 436–438.

Splittstoesser, C. M., C. Y. Kawanishi, and H. Tashiro. 1978. Infection of the European chafer, *Amphimallon majalis*, by *Bacillus popilliae:* Light and electron microscope observations. J. Invertebr. Pathol. 31: 84–90.

Splittstoesser, C. M., H. Tashiro, S. L. Lin, K. H. Steinkraus, and B. J. Fiori. 1973. Histopathology of the European chafer, *Amphimallon majalis*, infected with *Bacillus popilliae*. J. Invertebr. Pathol. 22: 161–167.

Stahnke, G. K., S. E. Brauen, A. L. Antonelli, and R. L. Goss. 1993. Alternatives for European crane fly control in turfgrass. Internatl. Turfgrass Soc. Res. J. 7: 375–381.

Stimac, J. L., R. M. Pereira, S. B. Alves, and L. A. Wood. 1993. *Beauveria bassiana* (Balsamo) Vuillemin (Deuteromycetes) applied to laboratory colonies of *Solenopsis invicta* Buren (Hymenoptera: Formicidae) in soil. J. Econ. Entomol. 86: 348–352.

Stockton, W. D. 1956. A review of the Nearctic species of the genus *Hyperodes* Jekel (Coleoptera: Curculionidae). Ph.D. Thesis. Cornell Univ., Ithaca, N.Y. 122 pp.

Stockton, W. D. 1963. New species of *Hyperodes* Jekel and a key to the Nearctic species of the genus (Coleoptera: Curculionidae). Bull. So. California Acad. Sci. 62: 140–149.

Street, J. R., R. Randell, and G. Clayton. 1978. Greenbug damage found on Kentucky bluegrass. Weeds Trees Turf (Oct.): 26.

Streu, H. T. 1981. Winter grain mite. Amer. Lawn Appl. (March/Apr.): 28–31.

Streu, H. T., and J. B. Gingrich. 1972. Seasonal activity of the winter grain mite in turfgrass in New Jersey. J. Econ. Entomol. 65: 427–430.

Streu, H. T., and H. D. Niemczyk. 1982. Pest status and control of winter grain mite, pp. 101–104. *In* H. D. Niemczyk and B. J. Joyner (eds.), Advances in turfgrass entomology. Hammer Graphics, Piqua, Ohio. 150 pp.

Streu, H. T., and L. M. Vasvary. 1966. 1966 report of turfgrass research, pp. 78–82. Rutgers University, New Brunswick.

Stringfellow, T. L. 1969. Turfgrass insect research in Florida, pp. 19–33. *In* H. T. Streu and R. T. Bangs (eds.), Proceedings of Scott's turfgrass research conference. O. M. Scott & Sons, Marysville, Ohio. 89 pp.

Strobel, J. 1971. Turfgrass. Proc. Florida Turfgrass Management Conf. 19: 19–28.

Sutherland, D. W. S. [Chairman]. 1978. Common names of insects and related organisms (1978 revision): An update. Bull. Entomol. Soc. Amer. 24: 408.

Swezey, O. H. 1946. Insects of Guam, Il. Bernice P. Bishop Mus. Bull. 189: 184.

Tanada, Y. 1955. Notes and exhibitions. Proc. Hawaiian Entomol. Soc. 15: 384.

Tanada, Y., and J. W. Beardsley. 1957. Probable origin and dissemination of a polyhedrosis virus of an armyworm in Hawaii. J. Econ. Entomol. 50: 118–120.

Tanada, Y., and J. W. Beardsley. 1958. A biological study of the lawn armyworm, *Spodoptera mauritia* (Boisduval), in Hawaii (Lepidoptera:Phalaenidae). Proc. Hawaiian Entomol. Soc. 16: 411–436.

Tanada, Y., and H. K. Kaya. 1993. Insect pathology. Academic Press, San Diego, Cal. 666 pp.

Tashiro, H. 1976a. Biology of the grass webworm, *Herpetogramma licarsisalis* (Lepidoptera: Pyraustidae), in Hawaii. Ann. Entomol. Soc. Amer. 69: 797–803.

Tashiro, H. 1976b. Hyperodes weevil, a serious menace to *P. annua* in Northeast. Golf Superintendent (Mar.): 34–37.

Tashiro, H. 1977. Colonization of the grass webworm, *Herpetogramma licarsisalis*, and its adaptability for laboratory tests. Proc. Hawaiian Entomol. Soc. 22: 533–539.

Tashiro, H. 1983. Insecticides: Properties and insects. Proc. 37th Ann. New York State Turfgrass Conf. 7: 77–81.

Tashiro, H., and W. E. Fleming. 1954. A trap for European chafer surveys. J. Econ. Entomol. 47: 618–623.

Tashiro, H., and F. L. Gambrell. 1963. Correlation of European chafer development with the flowering period of common plants. Ann. Entomol. Soc. Amer. 56: 239–243.

Tashiro, H., and W. C. Mitchell. 1985. Biology of the fiery skipper, *Hylephila phyleus* (Lepidoptera: Hesperiidae), a turfgrass pest in Hawaii. Proc. Hawaiian Entomol. Soc. 25: 131–138.

Tashiro, H., and W. Neuhauser. 1973. Chlordane-resistant Japanese beetle in New York. Search Agr. No. 3. 6 pp.

Tashiro, H., and K. E. Personius. 1970. Current status of the bluegrass billbug and its control in western New York home lawns. J. Econ. Entomol. 63: 23–29.

Tashiro, H., and R. W. Straub. 1973. Progress in the control of turfgrass weevil, a species of *Hyperodes*. Down to Earth 29: 8–10.

Tashiro, H., S. I. Gertler, M. Beroza, and N. Green. 1964. Butyl sorbate as an attractant for the European chafer. J. Econ. Entomol. 57: 230–233.

Tashiro, H., G. G. Gyrisco, F. L. Gambrell, B. J. Fiori, and H. Breitfeld. 1969. Biology of the European chafer *Amphimallon majalis* (Coleoptera Scarabaeidae) in northeastern United States. New York State Agr. Exp. Sta. Bull. No. 828. 71 pp.

Tashiro, H., J. G. Hartsock, and G. G. Rohwer. 1967. Development of blacklight traps for European chafer surveys. U.S. Dept. Agr. Tech. Bull. No. 1366. 52 pp.

Tashiro, H., C. L. Murdoch, and W. C. Mitchell. 1983. Development of a survey technique for larvae of the grass webworm and other lepidopterous species in turfgrass. Environ. Entomol. 12: 1428–1432.

Terry, L. A., D. A. Potter, and P. G. Spicer. 1993. Insecticides affect predatory arthropods and predation on Japanese beetle (Coleoptera: Scarabaeidae) eggs and fall armyworm (Lepidoptera: Noctuidae) pupae in turfgrass. J. Econ. Entomol. 86: 871–878.

Thorvilson, H. G., S. A. Philips, Jr., A. A. Sorenson, and M. R. Trostle. 1987. The straw itch mite, *Pyemotes tritici* (Acari: Pyemotidae) as a biological control agent of red imported fire ants, *Solenopsis invicta* (Hymenoptera: Formicidae). Florida Entomologist 70(1): 440–444.

Timm, R. M., ed. 1983. Prevention and control of wildlife damage. Great Plains Agr. Council and Univ. Nebraska Coop. Ext. Service, Inst. Agr. and Natur. Resources, Lincoln, Neb. Looseleaf, 632 pp.

Tolley, M. P. 1983. Resting site preferences for sod webworm moths. Amer. Lawn Appl. (Nov./Dec.): 27–28.

Tolley, M. P. 1995. Frit fly, pp. 55–57. *In* R. L. Brandenburg and M. G. Villani (eds.), Handbook of turfgrass insect pests. Entomological Society of America, Lanham, Md. 140 pp.

Tolley, M. P., and H. D. Niemczyk. 1988a. Seasonal abundance, oviposition activity, and degree-day prediction of adult frit fly (Diptera: Chloropidae) occurrence on turfgrass in Ohio. Environ. Entomol. 17: 855–862.

Tolley, M. P., and H. D. Niemczyk. 1988b. Upper and lower threshold temperatures and degree-day estimates for development of the frit fly (Diptera: Chloropidae) at eight constant temperatures. J. Econ. Entomol. 81: 1346–1351.

Tolley, M. P., and W. H. Robinson. 1986. Seasonal abundance and degree-day prediction of sod webworm (Lepidoptera: Pyralidae) adult emergence in Virginia. J. Econ. Entomol. 79: 400–404.

Townsend, M. L., D. C. Steinkraus, and D. T. Johnson. 1994. Mortality response of green June beetle (Coleoptera: Scarabaeidae) to four species of entomopathogenic nematodes. J. Entomol. Sci. 29: 268–275.

Traniello, J. F. A., and S. C. Levings. 1986. Intra- and intercolony patterns of nest dispersion in the ant, *Lasius neoniger*: Correlations with territoriality and foraging ecology. Oecologia 69: 413–419.

Troester, S. J., W. G. Ruesink, and R. W. Rings. 1982. A model of black cutworm *(Agrotis ipsilon)* development: description, uses, and implications. Illinois Agr. Exp. Sta. Bull. No. 774. 33 pp.

Tumlinson, J. H., M. G. Klein, R. E. Doolittle, T. L. Ladd, and A. T. Proveaux. 1977. Identification of the female Japanese beetle sex pheromone: Inhibition of male response by an enantiomer. Science 197: 789–792.

Turgeon, A. J. 1991. Turfgrass management. Prentice-Hall, Inc. (division of Simon & Schuster), Englewood Cliffs, N.J. 418 pp.

Tuttle, D. M., and G. D. Butler, Jr. 1961. A new eriophyid mite infesting bermudagrass. J. Econ. Entomol. 54: 836–838.

Ulagaraj, S. M. 1975. Mole crickets: Ecology, behavior, and dispersal flight (Orthoptera: Gryllotalpidae: *Scapteriscus*). Environ. Entomol. 4: 265–273.

Ulagaraj, S. M. 1976. Sound production in mole crickets (Orthoptera: Gryllotalpidae: *Scapteriscus*). Ann. Entomol. Soc. Amer. 69: 299–306.

Ulagaraj, S. M., and T. J. Walker. 1973. Phonotaxis of crickets in flight: Attraction of male and female crickets to male calling sounds. Science 182:1278–1279.

U.S. Department of Agriculture. 1951–1980. Cooperative economic insect report (1951–1975) and cooperative plant pest report (1976–1980).

U.S. Department of Agriculture. 1972. Japanese beetle quarantines. U.S. Dept. Agr., Animal and Plant Health Serv., Plant Protection and Quarantine Programs, and Can. Dept. Agr., cooperating with affected states. Cooperative Economic Insect Report 22: between 216–217.

U.S. Department of Commerce. 1981. Brief history of measurement systems with a chart of the modernized metric system. National Bur. Standards Spec. Pub. No. 304A. 4 pp.

Vance, A. M., and B. A. App. 1971. Lawn insects: How to control them. U.S. Dept. Agr. Home Garden Bull. No. 53. 23 pp.

Van Zwaluwenburg, R. H. 1918. The changa, or West Indian mole cricket. Porto Rico Agr. Exp. Sta. Bull. No. 23. 28 pp.

Vaurie, P. 1951. Revision of the genus *Calendra* (formerly *Sphenophorus)* in the United States and Mexico (Coleoptera: Curculionidae). Bull. Amer. Mus. Nat. Hist. 98: 33–186.

Vickery, R. A. 1929. Studies on the fall armyworm in the Gulf Coast district of Texas. U.S. Dept. Agr. Tech. Bull. No. 138. 64 pp.

Villani, M. G. 1994. Predicting grub damage in turf. TurfGrass Trends (Aug.): 1–7.

Villani, M. G. 1995. Relationships among soil insects, soil insecticides, and soil physical properties. TurfGrass Trends (Sept.): 11–17.

Villani, M. G. 1996. Multiple considerations in turfgrass and landscape pest management. TurfGrass Trends (May): 1–7.

Villani, M. G., and R. J. Wright. 1988a. Use of radiography in behavioral studies of turfgrass-infesting scarab grub species (Coleoptera: Scarabaeidae). Bull. Entomol. Soc. Amer. 34: 132–144.

Villani, M. G., and R. J. Wright. 1988b. Entomogenous nematodes as biological control agents of European chafer and Japanese beetle (Coleoptera: Scarabaeidae) larvae in turf. J. Econ. Entomol. 81: 484–487.

Villani, M. G., J. P. Nyrop, and S. R. Krueger. 1992. A case study of the impact of the soil environment on insect/pathogen interactions: Scarabs in turfgrass, pp. 111–126. *In* T. A. Jackson and T. R. Glare (eds.), Use of pathogens in scarab pest management. Intercept Press, Andover, Hampshire, UK. 298 pp.

Villani, M. G., R. J. Wright, and P. B. Baker. 1988. Differential susceptibility of Japanese beetle, oriental beetle, and European chafer (Coleoptera: Scarabaeidae) larvae to five soil insecticides. J. Econ. Entomol. 81: 785–788.

Villani, M. G., L. L. Allee, A. Diaz, and P. R. Robbins. 1999. Adaptive strategies of edaphic arthropods. Annu. Rev. Entomol. 44: 233–256.

Vinson, S. B., and L. Greenberg. 1986. The biology, physiology, and ecology of imported fire ants, pp. 193–225. In S. B. Vinson (ed.), Economic impact and control of social insects. Greenwood, Westport, Conn.

Vinson, S. B., and A. A. Sorensen. 1986. Imported fire ants: Life history and impact. Texas A&M Univ., College Station, and Texas Dept. of Agr., Austin, Tex.

Vittum, P. J. 1979. A taxonomic study of two species of *Hyperodes* (Coleoptera: Curculionidae) damaging turfgrass in New York State. M.S. Thesis. Cornell Univ., Ithaca, N.Y. 50 pp.

Vittum, P. J. 1980. The biology and ecology of the annual bluegrass weevil, *Hyperodes* sp. near *anthracinus* (Dietz) (Coleoptera: Curculionidae). Ph.D. Thesis. Cornell Univ., Ithaca, N.Y. 117 pp.

Vittum, P. J. 1984. Effect of lime applications on Japanese beetle (Coleoptera: Scarabaeidae) grub populations in Massachusetts soils. J. Econ. Entomol. 77: 687–690.

Vittum, P. J. 1986. Biology of the Japanese beetle (Coleoptera: Scarabaeidae) in eastern Massachusetts. J. Econ. Entomol. 79: 387–391.

Vittum, P. J. 1994. Enhanced efficacy of isazophos against Japanese beetle (Coleoptera: Scarabaeidae) grubs using sub-surface placement technology. J. Econ. Entomol. 87: 162–167.

Vittum, P. J. 1995a. Annual bluegrass weevil, pp. 21–23. In R. L. Brandenburg and M. G. Villani (eds.), Handbook of turfgrass insect pests. Entomological Society of America, Lanham, Md. 140 pp.

Vittum, P. J. 1995b. Black turfgrass ataenius, pp. 35–37. In R. L. Brandenburg and M. G. Villani (eds.), Handbook of turfgrass insect pests. Entomological Society of America, Lanham, Md. 140 pp.

Vittum, P. J. 1995c. Japanese beetle, pp. 66–69. In R. L. Brandenburg and M. G. Villani (eds.), Handbook of turfgrass insect pests. Entomological Society of America, Lanham, Md. 140 pp.

Vittum, P. J. 1997a. Chemical classes of turfgrass insecticides. TurfGrass Trends (Jan.): 9–16.

Vittum, P. J. 1997b. Insecticide formulations. TurfGrass Trends (Apr.): 1–8.

Vittum, P. J. 1997c. How insecticides work. TurfGrass Trends (July): 9–12.

Vittum, P. J. 1997d. Insecticides and environmental issues. TurfGrass Trends (Aug.): 7–15.

Vittum, P. J. 1997e. Insect monitoring techniques and setting thresholds. TurfGrass Trends (Dec.): 8–15.

Vittum, P. J. 1997f. Strategies for insect control. TurfGrass Trends (Nov.): 9–15.

Vittum, P. J., and M. R. McNeill. 1999. Suitability of *Listronotus maculicollis* (Coleoptera: Curculionidae) as a host for *Microctonus hyperodae* (Hymenoptera: Braconidae). J. Econ. Entomol. 92: In press.

Vittum, P. J., and H. Tashiro. 1980. Effect of soil pH on survival of Japanese beetle and European chafer larvae. J. Econ. Entomol. 73: 577–579.

Vittum, P. J., and H. Tashiro. 1987. Seasonal activity of *Listronotus maculicollis* (Coleoptera: Scarabaeidae) on annual bluegrass. J. Econ. Entomol. 80: 773–778.

Wadley, F. M. 1931. Ecology of *Toxoptera graminum*, especially as to factors affecting importance in the northern United States. Ann. Entomol. Soc. Amer. 24: 325–395.

Walkden, H. H. 1950. Cutworms, armyworms, and related species attacking cereal and forage crops in the Central Great Plains. U.S. Dept. Agr. Circ. No. 849. 52 pp.

Walker, T. J. 1982. Sound traps for sampling mole cricket flights (Orthoptera: Gryllotalpidae: *Scapteriscus*). Florida Entomol. 65: 105–110.

Walker, T. J. (ed.). 1984. Mole crickets in Florida. Univ. Florida Inst. Food Agr. Sci. Bull. 846.

Walker, T. J., and D. A. Nickle. 1981. Introduction and spread of pest mole crickets: *Scapteriscus vicinus* and S. *aletus* reexamined. Ann. Entomol. Soc. Amer. 74: 158–163.

Walker, T. J., J. A. Reinert, and D. J. Schuster. 1983. Geographical variation in flights of mole crickets, *Scapteriscus* spp. (Orthoptera: Gryllotalpidae). Ann. Entomol. Soc. Amer. 76: 507–517.

Walton, W. R. 1921. The green-bug, or spring grain aphis: How to prevent its periodical outbreaks. U.S. Dept. Agr. Farmers' Bull. No. 1217. 11 pp.

Walton, W. R. 1929. Cutworms on golf green. U.S. Golf Assoc. Green Sec. Bull. 9: 156–157.

Watson, J. R., H. E. Kaerwere, and D. P. Martin. 1992. The turfgrass industry, pp. 29–88. In D. V. Waddington, R. N. Carrow, and R. C. Shearman (eds.), Turfgrass, Agron. Monogr. No. 32. American Society of Agronomy, Madison, Wis. 805 pp.

Weaver, J. E., and J. D. Hacker. 1978. Bionomical observations and control of *Ataenius spretulus* in West Virginia. West Virginia Univ. Agr. Forest. Exp. Sta. Current Rep. No. 72. 16 pp.

Weaver, J. E., and R. A. Sommers. 1969. Life history and habits of the short-tailed cricket, *Anurogryllus muticus*, in central Louisiana. Ann. Entomol. Soc. Amer. 62: 337–342.

Webster, F. M., and W. J. Phillips. 1912. The spring grain aphid, or "greenbug." U.S. Bur. Entomol. Bull. No. 110. 153 pp.

Wegner, G. S., and H. D. Niemczyk. 1981. Bionomics and phenology of *Ataenius spretulus*. Ann. Entomol. Soc. Amer. 74: 374–384.

Weinzierl, R., and T. Henn. 1994a. Beneficial insects and mites, pp 443–454. *In* A. R. Leslie (ed.), Handbook of integrated pest management for turf and ornamentals. CRC Press, Boca Raton, Fla. 560 pp.

Weinzierl, R., and T. Henn. 1994b. Botanical insecticides and insecticidal soaps, pp 541–556. *In* A. R. Leslie (ed.), Handbook of integrated pest management for turf and ornamentals. CRC Press, Boca Raton, Fla. 560 pp.

Werner, F. G. [Chairman]. 1982. Common names of insects and related organisms. Entomological Society of America, Lanham, Md. 132 pp.

Wessel, R. D., and J. B. Polivka. 1952. Soil pH in relation to Japanese beetle populations. J. Econ. Entomol. 45: 733–735.

Wheeler, E. H. 1946. The pathogenicity of *Bacillus lentimorbus* Dutky and strains of *Bacillus popilliae* Dutky to larvae of *Amphimallon majalis* (Razoumowski), Scarabaeidae. Ph.D. Thesis. Cornell Univ., Ithaca, N.Y. 83 pp.

White, R. T. 1947. Milky disease infecting *Cyclocephala* larvae in the field. J. Econ. Entomol. 40: 912–914.

Wildermuth, V. L. 1916. California green lacewing fly. J. Agr. Res. 6: 517.

Wilkinson, A. T. S. 1969. Leatherjackets—A new pest in British Columbia. Canada Agr. (Summer). 2 pp.

Wilkinson, A. T. S., and H. S. Gerber. 1972. Description, life history, and control of leatherjackets. Brit. Columbia Dept. Agr. Pub. No. 72–5. 2 pp.

Wilkinson, A. T. S., and H. R. MacCarthy. 1967. The marsh crane fly, *Tipula paludosa* Mg., a new pest in British Columbia (Diptera: Tipulidae). J. Entomol Soc. Brit. Columbia 64: 29–34.

Williams, D. F., and S. D. Porter. 1996. Update on USDA-ARS biological control studies on fire ants, pp. 143–144. *In* Proc. 1996 Imported Fire Ant Research Conf., New Orleans, La.

Williamson, R. C., and D. A. Potter. 1997a. Does aerification promote black cutworm infestations? Golf Course Management (Jan.): 54–57.

Williamson, R. C., and D. A. Potter. 1997b. Oviposition of black cutworm (Lepidoptera: Noctuidae) on creeping bentgrass putting greens and removal of eggs by mowing. J. Econ. Entomol. 90: 590–594.

Williamson, R. C., and D. A. Potter. 1997c. Nocturnal activity and movement of black cutworms (Lepidoptera: Noctuidae) and response to cultural manipulations on golf course putting greens. J. Econ. Entomol. 90: 1283–1289.

Williamson, R. C., and D. A. Potter. 1997d. Turfgrass species and endophyte effects on survival, development, and feeding preference of black cutworms (Lepidoptera: Noctuidae). J. Econ. Entomol. 90: 1290–1299.

Williamson, R. C., and D. J. Shetlar. 1995. Oviposition, egg location, and diel periodicity of feeding by black cutworm (Lepidoptera: Noctuidae) on bentgrass maintained at golf course cutting heights. J. Econ. Entomol. 88: 1292–1295.

Wills, W., L. Lengel, Jr., and G. Whitmyre. 1969. False parasitism by the Asiatic garden beetle *Maladera castanea* (Arrow) (Coleoptera: Melolonthidae). Entomol. Soc. Pennsylvania, Melsheimer Entomol. Ser. No. 3. 3 pp.

Wilson, E. O. 1955. A monographic revision of the ant genus *Lasius*. Bull. Mus. Comparative Zool. Harvard 113: 1–201.

Wilson, J. W. 1932. Coleoptera and Diptera collected from a New Jersey sheep pasture. J. New York Entomol. Soc. 40: 77–93.

Wolcott, G. N. 1941. The establishment in Puerto Rico of *Larra americana* Saussure. J. Econ. Entomol. 34: 53–56.

Woodruff, R. E. 1966. The hunting billbug, *Sphenophorus venatus vestitus* Chittenden, in Florida (Coleoptera: Curculionidae). Florida Dept. Agr. Entomol. Circ. No. 45. 2 pp.

Woodruff, R. E. 1973. Arthropods of Florida and neighboring land areas, Vol. 8. The scarab beetles of Florida. Florida Dept. Agr. Consumers Service Bur. Entomol. Contrib. No. 260. 220 pp.

Woodruff, R. E., and B. M. Beck. 1989. The scarab beetles of Florida (Coleoptera: Scarabaeidae). Part II. The May or June beetles (genus *Phyllophaga*). Arthropods of Florida and neighboring land areas. 13: 1–225. Florida Dept. Agr. and Consumer Serv., Div. Plant Ind., Gainesville, Fla.

Wright, R. J., M. G. Villani, and F. Agudelo-Silva. 1988. Steinernematid and heterorhabditid nematodes for control of larval European chafers and Japanese beetles (Coleoptera: Scarabaeidae) in potted yews. J. Econ. Entomol. 81(1): 152–157.

Wylie, W. D. 1944. *Crambus haytiellus* (Zincken) as a pest of carpet grass. Florida Entomol. 27: 5–9.

Yeager, L. E. 1949. Forest leaf chafers and white grubs lamellicornia, pp. 161–186. *In* F. C. Craighead (ed.), Insect enemies of eastern forests. U.S. Dept. Agr. Misc. Pub. No. 657. 679 pp.

Yeh, T., and S. R. Alm. 1995. *Steinernema glaseri* (Nematoda: Steinernematidae) for biological control of Japanese and oriental beetle larvae (Coleoptera: Scarabaeidae). J. Econ. Entomol. 88: 1251–1255.

Zhang, A., H. T. Facundo, P. S. Robbins, C. E. Linn, Jr., J. L. Hanula, M. G. Villani, and W. L. Roelofs. 1994. Identification and synthesis of the female sex pheromone of the oriental beetle, *Anomala orientalis* (Coleoptera: Scarabaeidae). J. Chem. Ecol. 20: 2415–2427.

Zhang, A., P. S. Robbins, W. S. Leal, C. E. Linn, Jr., M. G. Villani, and W. L. Roelofs. 1997. Essential amino acid methyl esters: Major sex pheromone components of the cranberry white grub, *Phyllophaga anxia* (Coleoptera: Scarabaeidae). J. Chem. Ecol. 23: 231–245.

Index

Page numbers in *italics,* refer to figures and tables. "Pl" indicates the color plates, found between pages 140 and 141.

Acari. *See* Chigger mite; Mites; Ticks
Aceria neocynodonis. See Bermudagrass mite
Acremonium. See Endophytes
Acrididae, 271–272, Pl. 53
Acrolophidae, 87, 277, Pl. 55
Acrolophus spp. *See* Burrowing sod webworms
Actinocephalus spp., 194
Adelina, 337
Aedes. See Mosquitoes
Aestivation
    bronzed cutworm, 118
    winter grain mite, 38
Agriphila, (webworms). *See also* Sod webworms
    ruricolella, 88, 95
    vulgivagella, 88
Agropyron smithii. See Wheatgrass
Agrostis spp. *See* Bentgrass
    alba, 276
Agrotis ipsilon. See Black cutworm
Alfalfa (*Medicago sativa*), 127
Alfalfa cutworm. *See* Variegated cutworm
Amber disease, 335
Amara spp., 61, 62
Amblyseius spp., 128. *See also* Predatory mites
Amphimallon majalis. See European chafer
Amphypyrinae, 106, 121
Anagyrus antoninae, 84
Anaphoidea calendrae, 225
Anatomy
    grass, 1–2
        phloem, 72
        rhizome, 2, *2*
        stolon, 2, *2*
        vernation (rolled, folded), 2–3, *2*
    insect, external anatomy
        abdomen, 12, *13*
        antenna, 12, *13, 134,*
            filiform, 113, Pl. 23
            lamellate, *132,* 133
            pectinate, 113, Pl. 23

chitin, 12
elytra, 132
exoskeleton, 12
eyes, 12, *13*
head capsule, 12
mouthparts, 17–19, *18, 19*
raster of scarabaeid grubs. *See* Scarabaeidae: larvae
spiracles, *13,* 15, *134*
stridulating organ, 45, Pl. 9
tarsal claws, 175, *176*
thorax: pro, meso, meta, 12, *134*
tibia, *13*
    dactyl differences, 46, *46*
    fossorial, 132–133, *132*
trachea, *13*
wing, *13,* 109
    costa, 119
    orbicular spot, 108, *109*
    reniform spot, 108, *109*
insect, internal anatomy
    central nervous system, 14, *14*
    circulatory system, *14,* 15
    digestive system, 14–15, *14*
    respiratory system (spiracle, trachea), *13,* 15
    salivary system, *14, 18*
mite. *See* Mites: anatomy
Andrenidae, 289
Annelida, 283
Annual bluegrass (*Poa annua,* and *P. annua reptans,* turf-grass)
    adaptation, 5–6
    characteristics, 6, *9,* Pl. 1, Pl. 3
    insect pests
        annual bluegrass weevil, 232
        black turfgrass ataenius, 142
        greenbug, 72
Annual bluegrass weevil (*Listronotus maculicollis*), 230–242
    history and distribution, 231, *232*
    host plants and damage, 232–233, Pl. 49

Annual bluegrass weevil (*continued*)
  importance, 231
  life stages
    adult, 233, *234*, Pl. 48
      extraction from hibernation media, 239
      sex differentiation, 233–234, Pl. 48
    egg, 234, *234*, Pl. 48
    larva, 234, *234*, 236, Pl. 48
    prepupa and pupa, *234*, 236, Pl. 48
  natural enemies, 241–242
  phenology of populations, 236
  populations
    in fairways and tees, *237*
    in hibernation sites, 239–240
  sampling techniques
    for fairway and tee populations, *237*
    for hibernating populations, 239
  seasonal cycle, 236–237, *237*
    degree-day relations, *238*
    hibernation (overwintering), 237
    insecticide timing, *344*, 346
  taxonomy, 230
Annual ryegrass or Italian ryegrass (*Lolium
  multiflorum*), 4, *5*, 6
Anomala
  (*Pachystethus*) *lucicola*, 212
  orientalis. See Oriental beetle
Anopheles. See Mosquitoes
Antonina graminis. See Rhodesgrass mealybug
Ants
  history and distribution, 256
  host plants and damage, 256–257
  importance, 255
  life stages, 257, *260, 263,* Pl. 52
  seasonal cycle, 257–258
  taxonomy, 255
  turf-infesting species
    black imported fire ant (*Solenopsis richteri*), 255
    cornfield ant (*Lasius alienus*), 255
    fire ant (*Solenopsis geminata*), 255
    harvester ant (*Pogonomyrmex* spp.)
      history and distribution, 261–262, *262*
      host plants and damage, 262, Pl. 52
      life stages, 262, *263*
      natural enemies, 264
      seasonal cycle, 263–264
    mound-building ant (*Formica exsectoides*), 255
    pavement ant (*Tetramorium caespitum*), 255
    predatory ants, 329
    red ant (*Formica pallidefulva*), 255
    red imported fire ant (*Solenopsis invicta*)
      history and distribution, 258, *259*
      host plants and damage, 258, Pl. 52
      life stages, *260*
      natural enemies, 260–261
      seasonal cycle, 259–260
    southern fire ant (*Solenopsis xyloni*), 255
    turfgrass ant, *Lasius neoniger*
      history and distribution, 264
      host plants and damage, 264–265, Pl. 52
      seasonal cycle, 265
Anurogryllus muticus. See Short-tailed cricket
Apanteles marginiventris, 124

Apanteles spp., 100, 112, 128
Aphididae, 71
Aphidius testaceipes, 78
Aphidoletes spp., 79
Aphodinae, 141
Aphodius granarius, A. pardalis, 149–151
  history and distribution, 149, *150*
  host plants and damage, 149
  life stages, 149–151
  natural enemies, 151
  seasonal cycle, 151
  taxonomy, 149
Aphodius, v. *Ataenius*, distinguishing characteristics.
  See Ataenius, v. *Aphodius*, distinguishing characteristics
Aplomyia confusionis, 100
Arachnida, 285, 286, 287
Araneae, 285
Argasidae (soft ticks), 286
Argiope aurantia, 82
Armadillo, nine-banded (*Dasypus novemcinctus
  taxanus*)
  damage to turfgrass by, 298
  distribution, 298
  as predator of turfgrass insects, 298
Armyworm (*Pseudoletia unipuncta*)
  distribution, 107, 116
  habitat, 106
  host plants and damage, 107, 116
  importance, 107
  life stages
    adult, 108–109, 116, *117*
    egg, 109, *117*
    larva, 110, 116, *117*
  natural enemies, 112–113
  sampling techniques, 112
  seasonal cycle, 110
    and timing of insecticide applications, *344*, 347
  taxonomy, 106
Armyworms, en masse movement, 106, 117
Arthropoda, phylum, 12
Aserica castanea. See Asiatic garden beetle
Asiatic beetle. *See* Oriental beetle
Asiatic garden beetle (*Maladera castanea*), 167–172
  history and distribution, 167
  host plants and damage, 168, Pl. 27
  human ear invasion, 167
  importance, 167
  life stages, 169, *169*, Pl. 36
    distinctive characters of larvae, 169, Pl. 28,
      Pl. 36
  natural enemies, 172
  sampling for populations, 319
  seasonal cycle, *170*, 171
  taxonomy, 167
Aspergillus flavus, 260
Aspirator (insect-collecting device), 311, *311*
Ataenius
  cognatus, 141
  spretulus. See Black turfgrass ataenius
Ataenius, v. *Aphodius*, distinguishing characteristics
  adult, 149, *150*
  larva, 150, *150*

Autocerica castanea. See Asiatic garden beetle
Azadirachta indica, 339

Bacillus lentimorbus. See Milky disease
Bacillus popilliae. See Milky disease
Bacillus thuringiensis, 112, 206, 214, 333. *See also* Natural enemies of turfgrass pests: pathogens: bacteria
Bacteremia, 333
Baculovirus, 337
Bahiagrass (*Paspalum notatum,* turfgrass)
    adaptation, 5
    characteristics, 8, *9,* Pl. 2, Pl. 4
    insect pests
        chinch bug, southern, 64
        mole crickets, 42
        striped grassworm, 278
        tropical sod webworm, 103
        two-lined spittlebug, 80
    use, 8
Balaustium spp., 39
Banks grass mite (*Oligonychus pratensis*), 40−41
Barley (*Hordeum vulgare*), 36, 78
Beauveria bassiana, (fungal pathogen), 333, Pl. 57
    dependency on weather, 62
    host insects
        billbug, bluegrass, 225
        billbug, hunting, 228
        European chafer, 183
        chinch bugs, 61, 62, 67, *67,* Pl. 57
        cranberry girdler, 100
        May and June beetles, 193
        mole cricket, 51
        red imported fire ant, 260
        sod webworms, 100
    See also Natural enemies of turfgrass pests: fungi
Beauveria brongniartii, 183
Bembicinae, 266
Bentgrass (*Agrostis* spp., turfgrass)
    adaptation, 4, 7
    characteristics, 7, *9,* Pl. 1, Pl. 3
    insect and mite pests
        annual bluegrass weevil, 231
        black cutworm, 113
        black turfgrass ataenius, 142
        fiery skipper, 124
        frit fly, 250
        hairy chinch bug, 56
        Japanese beetle, 198
        masked chafer, 158−159
        turfgrass scale, 275
        variegated cutworm, 115
        webworms, 90
        winter grain mite, 36
        yellow-striped armyworm, 120
    species
        colonial (*A. tenius*), 4, 5, 7
        creeping (*A. palustris*), 4, 5, 7
        velvet (*A. canina*), 5, 7
    uses, 7
Berlese funnel. *See* Sampling devices for turfgrass insects
Bermudagrass (*Cynodon* spp., turfgrass)
    adaptation, 4, 7

characteristics, 7−8, *9,* Pl. 2, Pl. 4
insect pests
    bermudagrass scale, 275
    billbug, hunting, 226
    billbug, Phoenician, 27, 229
    chinch bug, southern, 24, 64
    dichondra flea beetle, 216
    fall armyworm, 118
    fiery skipper, 124
    Fissicrambus haytiellus, 90
    grass webworm, 101
    ground pearl, 26, 85
    lawn armyworm, 121
    mole crickets, 26, 44
    Phyllophaga crinita, 186
    Rhodesgrass mealybug, 23, 82
    tropical sod webworm, 103
    two-line spittlebug, 24, 80
    yellow-striped armyworm, 120
mite pests
    Banks grass mite, 23, 40
    bermudagrass mite, 23, 30
species
    African (*C. transvaalensis*), 7, 8
    common (*C. dactylon*), 7, 8
    Magennis (*C. magenisii*), 7, 8
    Tifway hybrids (*C. dactylon,* X *C. transvaalensis*), 7, 8
uses, 7−8
Bermudagrass mite (*Eriophes cynodoniensis*)
    history and distribution, 31
    host plants and damage, 23, 31
    importance, 30
    life stages
        adult, 32, *32,* Pl. 7
        egg and nymph, 32, Pl. 7
    natural enemies, 33
    resistant turfgrass accessions and cultivars, 33
    sampling techniques, 33
    seasonal cycle, 32−33
    taxonomy, 30
Bermudagrass scale (*Odonaspis ruthae*), 275
Bermudagrass stunt mite. *See* Bermudagrass mite
Big-eyed bug (*Geocoris bullatus, G. uligenosus*)
    characteristics, 68, 329, Pl. 60
    as predator of hairy and southern chinch bugs, 61, 63, *67,* 68, Pl. 60
Billbugs. *See* Bluegrass billbug; Denver billbug; Hunting billbug; Phoenician billbug
Birds
    as insect predators. *See* Natural enemies of turfgrass pests: birds
    as turfgrass pests. *See* Crow, American; Grackle, common; Starling, European
Bird's-foot trefoil (*Lotus corniculatus*), 60
Biting midges (*Culicoides, Leptoconops*), 289
Blackbirds, as insect predators, 100, 157, 225, 242
Black cutworm (*Agrotis ipsilon*)
    history and distribution, 107, *107,* 113
    host plants and damage, 107, 113
    importance, 107
    life stages
        adult, 108−109, *114,* Pl. 23
        sex differentiation, 113, Pl. 23

Black cutworm (*continued*)
    egg, 109, *114*, Pl. 23
    larva and pupa, 109–110, *114*, Pl. 23
    number of generations, 113
    rearing, 111
    seasonal cycle, 110, 113–115
    taxonomy and synonyms, 106
Black fairway beetle. *See* Black turfgrass ataenius
Black-light trap, Pl. 71
    to attract annual bluegrass weevil, *239*, 240
    to attract black turfgrass ataenius, 145
    to attract moths of sod webworms, armyworms, cut-
      worms, 96, 110, *111*
    to attract scarabaeid beetles, 163, 171, 180, 192
    See also Sampling devices for turfgrass insects
Black locust (*Robinia pseudoacacia*), 146, *179*, 345, Pl.
    32
Black turfgrass ataenius (*Ataenius spretulus*)
    history and distribution, 141–142
    host plants and damage, 142, Pl. 31
    importance, 141
    life stages (adult, egg, larva, pupa), 143, *143*, Pl. 31
      differences between *Ataenius,* and *Aphodius,* 149,
        150
    natural enemies
      avian predators, 148
      invertebrate predators, 148
      milky disease, 148, Pl. 57
    sampling techniques for, 147
    seasonal life cycle, 144–145, *144*
      degree-day accumulations and development, 146,
        345
      on fairways, *144*, 145
      influence of latitude on generations, 144
      overwintering, 145
      phenology, 146
      and timing of insecticide application, 146, 345–
        346
    taxonomy, 141
    threshold populations, 147
Black widow spider. *See* Spiders
Blissinae, 54
Blissus
    insularis. See Chinch bug, southern
    leucopterus hirtus. See Chinch bug, hairy
    leucopterus leucopterus. See Chinch bugs
    leucopterus occiduus. See Chinch bug, buffalograss
Blue disease, 336
Bluegrass (*Poa* spp., turfgrass)
    adaptations, 6
    characteristics, 6, *9*, Pl. 1, Pl. 3
    insect pests. *See* Annual bluegrass; Kentucky blue-
      grass: Insect pests
    mite pests. *See* Kentucky bluegrass: Mite pests
    species
      annual (*P. annua*), 6, *9*, 232, Pl. 1, Pl. 3
      Canada (*P. compressa*), 72, 232
      Kentucky (*P. pratensis*), 6, *9*, 232, Pl. 1, Pl. 3
      rough (*P. trivialis*), 6, 232
    use, 6
    See also Annual bluegrass; Kentucky bluegrass
Bluegrass billbug (*Sphenophorus parvulus*)
    history and distribution, 220, *221*

host plants and damage, 220–221, Pl. 47
importance, 219
life stages
    adult, 221, *222*, Pl. 46
      sex differentiation, 222, Pl. 46
      species differentiation, 221, Pl. 47
    egg, 222, *222*, Pl. 46
    larva, 222, *222*, Pl. 46
      sawdust-like frass, 221, 223, 224
    pupa, 222, *222*, Pl. 46
natural enemies
    Beauveria bassiana, 225
    insect parasitoids, 225
    Steinernema carpocapsae, 225
    vertebrate predators, 225
and resistant turfgrass cultivars
    tall fescue, 225
    Kentucky bluegrass, 224–225
    perennial ryegrass, 225
sampling techniques for, 224
seasonal cycle, 222–223
    degree-day accumulations, 224
    and timing of insecticide application, 346
taxonomy, 219
Bluegrass webworm. *See Parapediasia: teterrella*
Blue oat mite. *See* Winter grain mite
Botanicals, 339
Bottlebrush-buckeye (*Aesculus parviflora*), 198
Brachyptery. *See* Chinch bugs
Bronzed cutworm (*Nephelodes minians*)
    history and distribution, 107, 117
    host plants and damage, 107, 118
    importance, 107
    life stages
      adult, 108–109, 118, Pl. 22
      egg, 109, 118
      larva, 109–110, 118, Pl. 22
    natural enemies, 112
    seasonal cycle and generations, 110, 118
    taxonomy, 107
Brown recluse spider. *See* Spiders
Bryobia praetiosa. See Clover mite
Bt. *See Bacillus thuringiensis*
Buffalograss (*Buchloe dactyloides*, turfgrass)
    adaptations, 5, 7
    characteristics, 5, 7, *9*
    insect pests
      buffalograss chinch bug, 69
      buffalograss mealybug, 273
    mite pest
      buffalograss mite, 34
Buffalograss chinch bug. *See* Chinch bug,
    buffalograss
Buffalograss mealy bug, 273, Pl. 15
Buffalograss mite (*Eriophyes slyhuisi*), 34
Buffalograss webworm. *See Surattha identella*
Burrowing sod webworms (*Acrolophus* spp.)
    burrowing sod webworm (*A. popeanellus*), 277,
      Pl. 55
    cigarette-paper sod webworm (*A. plumifrontellus*),
      277, Pl. 55
Butterfly. *See* Fiery skipper
Butyl sorbate (insect attractant), 180

Cages, for insect rearing, 128, *128*
Calendra. See Sphenophorus
Caloglyphus phyllophaginus, 194
Canada bluegrass. *See* Bluegrass: Canada
Cannibalism, by mole crickets, 49
Caproic acid (insect attractant), 156
Carbamate insecticides. *See* Insecticides
Carnivora, 297–298
Castorbean (*Ricinus communis*), 198
Catalpa, common (*Catalpa bignonioides*), 177, *179*
Centipedegrass (*Eremochloa ophuiroides,* turfgrass)
 adaptation, 8
 characteristics, 8, *9,* Pl. 2, Pl. 4
 insect pests
  grass webworm, 101
  ground pearl, 85
  hunting billbug, 226
  Rhodesgrass mealybug, 82
  southern chinch bug, 64
  tropical sod webworm, 103
  two-lined spittlebug, 80
 use, 8
Centrifugation, for chinch bug sampling, 62
Centruroides sculpturatus. See Scorpions
Cercis canadensis. See Redbud, phenology of
Cercopidae, 79
Cetoniinae, 152
Chaetocnema spp. *See* Dichondra flea beetle
Changa. *See* Mole crickets
Chewings fescue. *See* Fescue
Chigger mite (*Trombicula irritans*), 287
Chinch bug, buffalograss (*Blissus occiduus*), 68–70
 history and distribution, *55,* 68
 host plants and damage, 69, Pl. 12
 life stages, 69, Pl. 12
 sampling techniques, 69
 seasonal cycle, 69
Chinch bug, hairy (*Blissus leucopterus hirtus*), 54–63
 diapause, 61
 history and distribution, *55,* 56
 host plants and damage, 56, Pl. 12
 importance, 55
 laboratory rearing, 61
 life stages
  adult, 56, *57,* Pl. 11
  egg, 57, *57,* Pl. 11
  nymph, 57–58, *57,* Pl. 11
 mortality factors, 61
 natural enemies, 62–63
 and resistant turfgrass cultivars, 62
 sampling techniques for
  Berlese funnel, 62
  centrifugation, 62
  flotation, 61–62, Pl. 70
  irritants, 315
  pitfall traps, 319, *320*
 seasonal cycle, 58–59, *58*
  degree-day accumulations and development, 60–61
  generations, 58
  phenology of oviposition, 59–60
  and timing of insecticide applications, 61, 345
 taxonomy, 54
 weather's influence on winter mortality, 61

Chinch bug, southern (*Blissus insularis*), 63–68
 history and distribution, *55,* 64
 host plants and damage, 64
  influence of fertilizer on, 64
  influence of moisture on, 65
  influence of thatch on, 64
 importance, 63
 life stages, *57,* 65, Pl. 12
 natural enemies, 67–68
 resistance to insecticides, 63
 and resistant turfgrass accessions and cultivars, 66–67
 seasonal cycle and generations, 65–66
 taxonomy, 54
 threshold populations, 66
Chinch bugs
 brachyptery, 55, 69, Pl. 11
 leucopterus, complex, 54
 macroptery in, 55
 nymphal separations, 57–58
 sampling techniques for, 61–62, 315, 319, Pl. 70
 sex differentiation, 57, Pl. 11
 species differentiation, 54–55
 See also Chinch bug, buffalograss; Chinch bug, hairy; Chinch bug, southern
Chinese lawngrass. *See* Centipedegrass
Chipmunk, eastern (*Tamias striatus*), 295
 as predator on green June beetle, 157
Chloropidae, 249
Chrysomelidae, 215
Chrysanthemum cinerariaefolium, 339
Chrysopa californica, 39
Chrysopa plorabunda, 79
Chrysoteuchia topiaria, (cranberry girdler, subterranean webworm)
 feeding habits, 93
 history and distribution, 88–89
 host plants and damage, 90–91
 life stages, 91–93, *92,* Pl. 17
 natural enemies, 100
 number of generations, 93, *94*
 oviposition, 96–97
 seasonal cycle, 93–95, *94,* 96–97
 sex pheromone, 96
Cicada, Cicadidae, 272, Pl. 53
 as prey for cicada killer, 267, Pl. 63
 seasonal cycles, 272
Cicada killer (*Specius speciosus*), 266–268
 damage, 267, Pl. 63
 distribution, 266
 importance, 266
 life stages, 267, Pl. 63
 natural enemies, 268
 seasonal cycle, 267–268
 sting, 267
 taxonomy, 266
Cigarette-paper sod webworm (*Acrolophus plumifrontellus*), 277, Pl. 55
Cirphis unipuncta. See Armyworm
Click beetle. *See* Wireworms
Clover, 232
 red (*Trifolium pratense*), 127
 white (*Trifolium repens*), 60, 113

Clover mite (*Bryobia praetiosa*), 39–40, *40*, Pl. 8
Coccidae, 275, 276
Coccinella 9-notata, 79
Coleoptera
  as predators. *See* Natural enemies of turfgrass pests:
    predators
  turfgrass-infesting families
    Chrysomelidae, 215
    Curculionidae, 219, 230, 280
    Elateridae, 279
    See also Scarabaeidae
Collection of insects, for identification, 310–312
  aerial and sweeping net, 311
  aspirator, 311
  hand, 311
  transportation of live insects, 312
Common cutworm. *See* Variegated cutworm
Conifers, 88
Control
  biological, 305–306, 326–340. *See also* Natural
    enemies of turfgrass pests
  cultural practices, 302–304
    fertilization, 303
    irrigation, 303
    mowing, 302–303
    soil conditions, 304
    thatch management, 303
  resistant cultivars
    endophytes, 305
    resistant parentage, 305
  timing of insecticide applications, 343–347. *See also*
    Insecticides
Convolvulaceae, 215
Cool-season grasses, 5–7, *9*, Pl. 1, Pl. 3
Cordyceps ravenelii, 193
Corn root webworm. *See Crambus: caliginosellus*
Corvidae, 292
Cotinis nitida. See Green June beetle
Cottony grass scale (*Eriopeltis festucae*), 276, Pl. 54
Couchgrass. *See* Bermudagrass
Coxiella popilliae. See Blue disease
Crab spider, 330
Crambinae, 87, 88, 101
Crambus, (webworms)
  agitatillus, 88
  caliginosellus, 87, 88
    distribution, 89, *89*
    fecundity, 97
    host plants and damage, 90, 96
    importance, 87–88
    life stages, 92, *92*, 97, Pl. 19
    seasonal cycle and generations, 93, *94*
  laqueatellus, 88, Pl. 17
    fecundity, 97
  leachellus, 88
    diapause, 98
  luteolellus, 88
  perlellus, 88
  praefectellus
    burrows and cocoons, Pl. 18, Pl. 19
    distribution, 89, *89*
    life stages, 92, Pl. 19
    seasonal cycle, 93

  sperryellus, Pl. 17
    distribution, *89,* 90
    life stages, 92, 98
    seasonal cycle, 93, *94, 95*
    taxonomy, 87
  tutillus, 88, 98
Cranberry (host of *Chrysoteuchia topiaria*), 88, 90
Cranberry girdler. *See Chrysoteuchia topiaria*
Creeping bentgrass. *See* Bentgrass
Cricetidae, 295
Crow, American (*Corvus brachyrhynchos*), 292, Pl. 66
  as insect predator, 148, 166, 209, 214, 225
Ctenocephalides canis, C. felis. See Fleas
Culex. See Mosquitoes
Culicidae, 289
Culicoides. See Biting midges
Curculionidae. *See* Annual bluegrass weevil; Bluegrass
  billbug; Denver billbug; Hunting billbug; Phoenician
  billbug; Vegetable weevil
Curled dock (*Rumex crispus*), 113
Cutworms, 106–118
  history and distribution, 107, *107*
  host plants and damage, 107
  importance, 107
  life stages, 108–110, *114*, Pl. 23
  sampling techniques
    for adults, 112
    for larvae, 112, 315
  seasonal cycle, 110, *111*
    and timing of insecticide applications, 345
  species, 106
  synonyms, 106
  taxonomy, 106
  See also names of individual species
Cyclocephala
  borealis. See Masked chafer, northern
  lurida. See Masked chafer, southern
  pasadenae. See Masked chafer
Cyclodiene-resistant insects. *See* Insecticides
Cylindrorhininae, 230
Cynodon spp. *See* Bermudagrass

Dandelion (*Taraxacum officinale*), 232
Dark sword grass moth. *See* Black cutworm
Dasypodidae, 298
Degree-day accumulations, effect on development
  of annual bluegrass weevils, 236, *238*
  of black turfgrass ataenius, 146
  of bluegrass billbug, 224
  of frit fly, 254
  of hairy chinch bug, 60–61
  of masked chafers, 163
Deltacephalus sonorus (leafhopper), 274
Denver billbug (*Sphenophorus cicatristriatus*)
  distribution, 229, *229*
  host plants and damage, 229
  life stages, 221–222, *222*, 229
  seasonal cycle, 229
  taxonomy, 219
Dermaptera. *See* Earwig
Detection and diagnosis of insect damage, 310, 312–319
  active sampling, 313–318
    flotation, 314–315, Pl. 70

Detection and diagnosis of insect damage (*continued*)
  hibernating quarters, 314
  inspection of food plants, 313–314
  irritating drench, 315–316, Pl. 70
  rating system, 316, Pl. 70
  soil sample, 316–317, Pl. 71
  vacuum device, 317
  early symptoms, 313
  passive sampling, traps, 318–320
    black-light, 318–319, Pl. 71
    pheromone, 319, Pl. 70
    pitfall, 319–320, *320*, Pl. 70
  relating symptom to cause, 310
    presence of insect, 309–310
    presence of insect frass, 310
    soil examination, 310, 316–317
    value of early detection, 310
Detergent for insect drench, 91, 112, 315–316
DeVac (insect-collecting suction device), 254
Diapause, 27
  breaking of, 61
  influence of photoperiod and temperature on, 98
  summer aestivation, 38–40
  winter diapause, 61, 98
Diaspididae, 275
Dichlorobenzene (insect irritant), 249
Dichondra (*Dichondra* spp., lawn plant)
  characteristics, 10, Pl. 5, Pl. 45
  as host to slugs and snails, 284
  and insect pests, 113, 216–217, 274, 276, 278, 280
  as lawns, 10, Pl. 5, Pl. 45
Dichondra flea beetle (*Chaetocnema repens*), 215–218
  history and distribution, 215–216
  host plants and damage, 216, Pl. 45
  importance, 215
  life stages, 217, *217*, Pl. 45
  rearing, 218
  related species, 215
  seasonal cycle, 217–218
  taxonomy, 215
Diogmites discolor, 195
Diplopoda, 284, Pl. 66
Diptera, 243, 249, 281, Pl. 51
Dormyrmex pyramicus, 264
Douglas fir (*Pseudotsuga menziesii*), 90
Draeculacephala minerva (leafhopper), 273–274
Drought
  dormancy of turf during, 8, 10, Pl. 6
  insect damage during, 56, 90–91, 221
    difficulty of diagnosis, 56, 91, 221
Dynastinae, 158

Earthworms, 283–284, Pl. 65
Earwigs, 288, Pl. 59
Eastern yellow jacket (*Vespula maculifrons*), 268–269, Pl. 64
Economic threshold, 341–342. *See also* economic thresholds under individual species
Egg, egg pod, of insects, 101, 161, 272
  chorion, 101, 161
  oversummering
    of clover mite, 39, Pl. 8
    of winter grain mite, 38, Pl. 8

Elateridae. *See* Wireworm
Endophytes, 337
  Epichtoe typhinum, 337
  Neotyphodium coenophialum
    and resistance to armyworms, 112
    and resistance to bluegrass billbugs, 225
    and resistance to greenbug, 78
    and resistance to sod webworms, 98–99
  Neotyphodium lolii
    and resistance to bluegrass billbug, 225
    and resistance to greenbug, 78
    and resistance to hairy chinch bug, 62
    and resistance to sod webworm, 98
Endotoxin, 334
Entomophthora, 51, 82
Epichtoe typhinum, 337
Enzymes, toxic salivary
  of chinch bugs, 24
  of greenbugs, 72
Erax aestuans, 100
Eremochloa ophuiroides. *See* Centipedegrass
Eriopeltis festucae. *See* Cottony grass scale
Eriophyes cynodoniensis. *See* Bermudagrass mite
Eriophyes slyhuisi. *See* Buffalograss mite
Eriophyes zoysiae. *See* Zoysiagrass mite
Eriophyidae, Eriophyoidea, 14, 30
Eucelatoria armigera, 102
Eumicrosoma benefica, 61, 67
European chafer (*Rhizotrogus* [*Amphimallon*] *majalis*, 172–183
  differentiation from May or June beetles, 175, *176*
  history and distribution, 173, *174*
  host plants and damage, 173–174, Pl. 27
  importance, 172–173
  life stages, 175–176, Pl. 37
    adult, 175, *175*, 177–180, Pl. 38
      feeding, 173
      response to chemicals, colors, and radiation, 180
    egg, 175, *175*
      oviposition sequence, Pl. 29
    larva, 175–176, *175*
      larva-to-adult sequence, Pl. 30
    prepupa and pupa, 175, *176*, Pl. 37
  natural enemies, 182–183
    milky disease, 182
    miscellaneous microorganisms, 183
    parasitoids and predators, 183
  sampling techniques to determine population size
    larvae, 316–317
    observing swarming adults, 314
    trapping adults with black-light traps, 180, 318
  seasonal cycle, 177–182
    influence of temperature on development, 181
    phenology, 177, *179*
    population trends, 180–181
    winter mortality, 181
  sex differentiation
    adult, 176, Pl. 37
    larva, *137*, 176–177
    pupa, *137*, 177
  taxonomy, 172

European crane fly (*Tipula paludosa*), 243–249
history and distribution, 243, *244*
host plants and damage, 244, Pl. 51
importance, 243
life stages (adult, egg, larva, pupa), 245, *246*, Pl. 51
moisture requirements, 248
mortality factors, 248
natural enemies, 249
sampling techniques for, 249
seasonal cycle, 245–248, *247*
and timing of insecticide application, 345
taxonomy, 243
threshold populations, 249
European earwig (*Forficula auricularia*), 288, Pl. 59
European starling. *See* Starling, European
Excoprosopa fasciata, 195
Exomala orientalis. *See* Oriental beetle

Fall armyworm (*Spodoptera* [*Laphygma*] *frugiperda*)
distribution, *108*, 118
host plants and damage, 118–119
importance, 107
life stages, 119, *119*, Pl. 24
seasonal cycle, 119–120
taxonomy, 106
Feltia subterranea. *See* Granulate cutworm
Fescue (*Festuca* spp., turfgrass)
adaptation, 5
characteristics, 6–7, *9*, Pl. 1, Pl. 3
insect pests
bluegrass billbug, 220
cutworms, 107
fall armyworm, 118
greenbug, 72
hairy chinch bug, 56
Japanese beetle, 198
masked chafer, 158–159
sod webworms, 90
mite pests
winter grain mite, 36
species
chewings (*F. rubra commutata*), 5, 6
red (*F. rubra*), 5, 6
tall (*F. arundinacea*), 5, 6
uses, 6
Fiddle back. *See* Spiders: brown recluse
Fiery skipper (*Hylephila phyleus*), 124–128
history and distribution, 124
host plants and damage, 124–125
importance, 124
life stages, 125–127, *126*, Pl. 26
natural enemies, 128
rearing, 128
seasonal cycle, 127
taxonomy, 124
Fig eater. *See* Green June beetle
Fir (*Abies*), 90
Fire ants. *See* Ants
Fissicrambus, (webworms)
haytiellus, *88*
damage, 90
mutabilis, Pl. 17
abundance, 96

damage, 90
distribution, 89
life stages and size, *92*
Fleahopper. *See* Leafbugs, fleahoppers
Fleas
cat flea (*Ctenocepholides felis*), 289, Pl. 64
dog flea (*Ctenocepholides canis*), 289
Flotation (for insect survey). *See* Sampling devices, for
turfgrass insects
Flowering dogwood (*Cornus florida*), phenology, 236,
346, Pl. 50
Fluorescent lamps. *See* Sampling devices for turfgrass
insects: black-light traps
Forficula auricularia. *See* European earwig
Formica exsectoides, *F. pallidefulva*. *See* Ants
Formicidae, Formicinae, 255
Forsythia (*Forsythia* spp.), phenology of, 236, 346, Pl. 50
Frit fly (*Oscinella frit*), 249–254
attraction to white objects, 251, 253
fritz and derivation of name, 250
history and distribution, 250
host plants and damage, 250
importance, 249–250
life stages, 251–252, *251*, Pl. 51
natural enemies, 254
sampling techniques, 254
seasonal cycle, 252–254, *252*
and timing of insecticide application, 345
taxonomy, 249
R,Z-Furanone (Japonilure), 203
Fusarium spp., 193–194

Gastropoda, 284
Geocoris bullatus, 63, 67–68
Geocoris uligenosus, 63, 67–68
Geomyidae, 295
Geranium (*Pelargonium* spp.), 198
Golden digger wasp. *See* Cicada killer
Golf
balls (white) as insect attractants, 251
cup cutter as sampling tool, 316
designated areas affected by turf insects
aprons and collars, 231–232, 249
fairways, 36, 125, 141, 231–232, 295
greens, 36, 113, 141, 231–232, 249, 295
roughs, 295
tees, 141, 231–232
insect pests exclusive to golf courses
annual bluegrass weevil, 25, 232
black turfgrass ataenius, 26, 142
Gophers. *See* Pocket gophers
Grackle, common (*Quiscalus guiscala*), 292, Pl. 66
as insect predator, 53, 166, 208, 214
Graminae, 1, 72, 87
Granulate cutworm (*Feltia subterranea*), 278, Pl. 55
Grass. *See* Anatomy: grass; Turf, turfgrass
Grasshoppers, 271–272, Pl. 53
Grass webworm (*Herpetogramma licarsisalis*), 100–103
history and distribution, 100
host plants and damage, 101, Pl. 21
importance, 100
life stages, 101, Pl. 20, Pl. 21
natural enemies, 102

Grass webworm (*continued*)
  rearing, 102
  sampling techniques for, 102, 315
  seasonal cycle, 101–102
  taxonomy, 100
Greasy cutworm. *See* Black cutworm
Greenbug (*Schizaphis graminum*), 71–79
  biotypes, 73
  history and distribution, 72, *72*
  host plants and damage, 72, Pl. 13
    leaf blade populations, 73
  importance, 71
  life stages, 73–74
    adult forms, 73–74, *74*, Pl. 13
    egg, 74, *74*, Pl. 13
    nymph, 74, *74*, Pl. 13
  natural enemies, 78–79
  resistance
    of grass species and cultivars
      antibiosis, 78
      endophytes, 78
    to insecticides, 71
  seasonal cycle, 74–77
    development of winged forms, 77
    north and south of 35th latitude, 75–76, *76*
    and timing of insecticide application, 345
  taxonomy, 71
Green June beetle (*Cotinis nitida*), 152–157
  history and distribution, 152, *153*
  host plants and damage, 153, Pl. 33
  importance, 152
  life stages, 153–154
    adult, 153–154, *154*, Pl. 33, Pl. 34
      epimeron (mesothorax), Pl. 34
    egg, 154, *154*, Pl. 33, Pl. 34
    larva, 154, *154*, Pl. 33
      dorsal locomotion, 156, Pl. 33
    prepupa and pupa, 154, *154*, Pl. 34
  natural enemies, 157
  sampling for populations, 313
  seasonal cycle, 155–157, *155*
  taxonomy, 152
Ground pearls (*Margarodes meriodionalis, Eumarga-rodes laingi*), 84–86
  history and distribution, 84, *85*
  importance, 84
  host plants and damage, 85, Pl. 16
  life stages, 85, *86*, Pl. 16
  seasonal cycle, 85–86
  taxonomy, 84
Growing degree-day accumulation. *See* Degree-day accumulations, effect on development
Gryllidae, 271
Gryllotalpidae, Gryllotalpinae, *Gryllotalpa* spp., 42

Hairy chinch bug. *See* Chinch bug, hairy
Halticinae, 215
Harpalus erraticus, 183
Harpalus pennsylvanicus, 183
Hederinae, 106
Hemiptera
  Heteroptera. *See* Big-eyed bug; Chinch bugs; Tarnished plant bug

Homoptera. *See* Bermudagrass scale; Buffalograss mealybug; Cicada, Cicadidae; Cottony grass scale; Greenbug; Ground pearls; Leafhoppers; Rhodesgrass mealybug; Turfgrass scale; Two-lined spittlebug
Herpetogramma
  licarsisalis. See Grass webworm
  phaeopteralis. See Tropical sod webworm
Hesperiidae, Hesperiinae, 124
Heterorhabditis bacteriophora, 157, 183, 207, 214, 228, 336, Pl. 58. *See also* Natural enemies of turfgrass pests: nematodes
Hibernation sites
  of annual bluegrass weevil, 237, 239
  of black turfgrass ataenius, 145
  of bluegrass billbug, 223
  of hairy chinch bug, 59
Hippodamia convergens, 79
Horogenes spp., 105
Horse chestnut (*Aesculus hippocastanium*), 146, *179*, 345, Pl. 32
Humans, arthropods harmful or annoying to
  harmful turfgrass arthropods
    ants, 255–265
    bees, wasps, yellow jackets, 266–269
    centipedes, chiggers, scorpions, ticks, 285–287
    midges, mosquitoes, 289
    spiders, 285–286
  nuisance turfgrass arthropods
    insects, 288, 289
    mites, 286
Hunting billbug (*Sphenophorus venatus vestitus*), 226–228
  history and distribution, 226, 227
  host plants and damage, 226–227
  importance, 226
  life stages
    adult, 227, Pl. 47
      sex differentiation, 227
      species differentiation, 219, 227, Pl. 47
    egg, 227
    larva, 227
  natural enemies, 228
  seasonal cycle, 228
  taxonomy, 219, 226
Hydrogen cyanide, as exudate of millipedes, 285
Hylephila phyleus. See Fiery skipper
Hymenoptera. *See* Ants; Cicada killer; Wild bees; Yellow jackets
Hyperodes. See Annual bluegrass weevil; *Listronotus maculicollis*

Icteridae, 292
Identification of insects, 309–312
IGR. *See* Insect Growth Regulator
Imported fire ants. *See* Ants: turf-infesting species
Insect Growth Regulator, 338–339, Pl. 58
Insecticides, 347–359
  adsorption, 348, 356–357
  classes, 350–352
    carbamates, 351
      bendiocarb (Turcam), 342, *343*, 348, *349*, *350*, 351, 353
      carbaryl (Sevin), 36, *37*, *349*, *350*, 351, 353

Insecticides (*continued*)
    chlorinated cyclodienes, chlorinated hydrocarbons, 350–351
      chlordane, 63, 88, 342, 348, 352
      dieldrin, 88, 342, 352
    chloronicotinyls, 351–352
      imidacloprid (Merit), *349, 350*
    organophosphates, 351
      acephate, *349, 350,* 358
      chlorpyrifos (Dursban), 63, 71, 342, *343,* 347, *349–350,* 351–353
      diazinon, 63, 71, *343, 349, 350,* 352–353
      ethoprop (Mocap), 342, *349, 350,* 353
      fonofos, *349, 350*
      isazophos (Triumph), 342, 353, 358
      isofenphos (Oftanol), 342, *349, 350,* 351, 353
      malathion, 71
      trichlorfon (Dylox, Proxol), 342, *343,* 348, *349, 350,* 351, 353, 358
    phenylpyrazoles, 352
      fipronil, *349,* 352
    synthetic pyrethroids, 351
      bifenthrin, *349, 350*
      cyfluthrin, *349, 350*
      lambda-cyhalothrin, *349, 350*
      fluvalinate, *349, 350*
    degradation, 357–358
    effect on nontarget organisms, 353
      sod webworms, 88
      winter grain mite, 36, *37*
    formulations, 353–355
    influence of thatch on efficacy, 342–343, *343*
    movement (transfer), 357
    nomenclature, 347–348
    persistence, 348–349, *349*
    precautions in using, 358–359
    resistance, 352
      resistant turfgrass insects
        bluegrass billbug, 220
        greenbug, 71
        sod webworm, 88
        southern chinch bug, 63
    timing of application
      daily, 347
      seasonal, 343–347, *344*
    toxicity, *350,* 355–356
    trade names, 347–348
    vapor pressure, 348, *349*
    water solubilities, 348, *349*
Insectivora, 293–294
Insect life stages and development. *See* Life stages and development, insect
Invertebrate parasitoids and predators of turfgrass insects. *See* Natural enemies of turfgrass pests
Irritants, to force insects to surface, 315–316, Pl. 70
    detergents, 50, 315
    orthodichlorobenzene, 249
    pyrethrins (with or without piperonyl butoxide), 50, 315–316
Istocheta aldrichi, 208
Italian ryegrass. *See* Annual ryegrass
Ixodidae (hard ticks), 286–287, Pl. 64

Japanese beetle (*Popillia japonica*), 196–209
    "balling" of adults, 203, Pl. 41
    California infestations and eradication, 197
    history and distribution, 196–197, *197*
    host plants and damage, 198–199, Pl. 27, Pl. 42
      adults, 198, Pl. 42
      larvae, 198–199, Pl. 27
    importance, 196
    life stages, 199–200, *200,* Pl. 41, Pl. 42
      adult, 199, *200,* Pl. 41, Pl. 42
        sex differentiation, 199, Pl. 41
      egg, 199, *200,* Pl. 41
      larva, 199–200, *200,* Pl. 41
        larva-to-adult sequence, Pl. 43
        vertical distribution in soil, 204, *205*
      prepupa and pupa, 200, *200*
    lures for adults, 203
    natural enemies, 206–209. *See also* Milky disease: Japanese beetle; *Istocheta aldrichii; Bacillus thuringiensis*
    sampling for populations, 313, 316
    seasonal cycle, 200–206, *201*
      influence of altitude and latitude, 202, *202*
      winter mortality, 204
    sex pheromone, 203
    taxonomy, 196
Japanese lawngrass. *See* Zoysiagrass
Japonilure. *See R,Z,* -Furanone
Java citronella oil-eugenol (insect attractant), 180
June beetles (agricultural pests), 183. *See also* May or June beetles

Kentucky bluegrass (*Poa pratense,* turfgrass)
    characteristics, 5, 6, *9,* Pl. 1, Pl. 3
    insect pests
      annual bluegrass weevil, 232
      Aphodius granarius, 149
      black turfgrass ataenius, 142
      bluegrass billbug, 220
      bronzed cutworm, 118
      frit fly, 250
      greenbug, 72
      hairy chinch bug, 56
      Japanese beetle, 198
      masked chafers, 158–159
      webworms, 90
    mite pests
      Banks grass mite, 40
      clover mite, 39
      winter grain mite, 36
    resistant or tolerant cultivars
      to bluegrass billbug, 224–225
      to greenbug, 78
      to hairy chinch bug, 62
      to sod webworms, 99
    uses, 5, 6
Kikuyugrass (*Pennisetum clandestinum,* pasture and turfgrass), 101
Killing and preservation of insects, 311–312
    labeling, 312
Korean velvet lawngrass. *See* Zoysiagrass

Lachnosterna. See May or June beetle
Labidura riparia, 68
Lamellicorn beetles (leaf chafers, scavengers), 129
Lantana (*Lantana camara*), 125
Laphygma. See Fall armyworm
Larra bicolor, 52
Larger sod webworm. *See Pediasia trisecta*
Lasiochilus pallidulus, 68
Lasius alienus. See Ants
Lasius neoniger. See Ants
Latrodectus mactans, (black widow spider). *See* Spiders
Lawn
   for bowling green, 7, 125
   definition, 1
   residential, 1, 73, 115, 124, 220, 226, 244, 256, 275, 295
   for tennis court, 7
   See also Turf, turfgrass
Lawn armyworm (*Spodoptera mauritia*), 121–123
   history and distribution, 121
   host plants and damage, 121, Pl. 25
   importance, 121
   life stages, 121–122, Pl. 25
   natural enemies, 123
   seasonal cycle, 122–123
   taxonomy, 121
Lawn chinch bug. *See* Chinch bug, southern
Lawn grass. *See* Zoysiagrass
Lawn moth. *See* Sod webworms, temperate region
Leach's crambus. *See Crambus: leachellus*
Leafbugs, fleahoppers (*Spanogonicus* spp.), 274–275
Leafhoppers (*Endria* spp. and others), 273–274, Pl. 53
Leaf- and stem-infesting insects, 23
Leatherjacket. *See* European crane fly
Leconopsis formicarum. See Turfgrass scale
Lepidoptera. *See* Armyworm; Burrowing sod webworms; Cutworms; Fall armyworm; Fiery skipper; Lucerne moth; Sod webworms, temperate region; Striped grassworm
Leptoconops, (biting midge), 289
Life stages and development, insect, 15–17, *15, 16*
   adult, 15
      teneral, 177, 237, *237,* Pls. 30, 31, 43, 46, 48
   cocoon, 93
   cyst (ground pearl nymph), 85
   diapause, 27, 61, 98
   egg, 15
   generation vs. brood: distinguishing characteristics, 16–17
   hibernaculum, 93, 100
   instar, 15–16
   larva, 15
   nymph, 15
   oviparity, 17
   parthenogenesis, 16
   polyembryony, 16
   prepupa, 135
   pupa, 15
   viviparity, 16
   voltinism, 16

Lights, insects attracted to, 167, 171, 180, 192. *See also* Black-light traps
Listroderes costirostris obliquus. See Vegetable weevil
Listronotus maculicollis. See also Annual bluegrass weevil
   differentiation from *Sphenophorus,* 221–222
Lolium, endophytic fungus. *See* Endophytes: *Neotyphodium lolii* .
Lotus corniculatus. See Bird's-foot trefoil
Loxosceles reclusa. See Spiders: brown recluse
Lucerne moth (*Nomophila noctuella*), 276, Pl. 55
Lycosidae. *See* wolf spider
Lydina polidoides, 100
Lygaeidae, 54
Lygus lineolaris. See Tarnished plant bug
Lysiphlebus testaceipes, 78

Macrocentrus crambivorus, 100
Macroptery. *See* Chinch bugs
Magennis bermudagrass. *See* Bermudagrass
Magicicada septendecim. See Periodical cicada
Magnifying lens, 310–311
Maladera castanea. See Asiatic garden beetle
Malt extract, as insect attractant, 156
Mammals, 293–298
   as insect predators. *See* Armadillo; Chipmunk; Moles; Raccoon; Shrew; Skunks
   as turfgrass pests (vegetarians). *See* Pocket Gophers; Voles
Manilagrass lawngrass. *See* Zoysiagrass
March flies, 281–282
Margarodes meriodionalis. See Ground pearls
Margarodidae, 84
Masked chafer (*Cyclocephala pasadenae*), 158
Masked chafer, northern (*Cyclocephala borealis*) and southern (*C. immaculata-lurida*), 158–166
   history and distribution, 158, *159*
   host plants and damage, 158–159, Pl. 27
   importance, 158
   influence of rainfall on adult flights, *162,* 163
   life stages, 159–160
      adult, 159–160, *160,* Pl. 35
         sex differentiation, 160, *160*
      egg, larva, prepupa, pupa, 160, *160,* Pl. 35
   natural enemies, 166
   response to black light, 163
   sampling for populations
      adults, 318
      larvae, 316–317
   seasonal cycle, 161–165, *162*
      degree-day accumulation, 163
   sex pheromones, 163
   species differences
      in adults, 160, Pl. 35
      in nightly flights, 163, *164*
      in seasonal occurrence, 161, *162,* 163
   taxonomy, 158
   threshold populations, 165
May or June beetles (*Phyllophaga* spp.), 183–195
   history and distribution, 185–186, *184*
   host plants and damage, 186
   importance, 184

May and June beetles (*continued*)
  life stages, *187*
    adults, 186–187, Pl. 39
    eggs, 187
    larvae, 187–188
      rastral patterns, *188*, Pls. 28, 40
    pupae, 188
  major species, *184*, 184
  natural enemies, 193–195
  sampling for populations
    soil sampling, 316–317
    adult trapping, 192, 318–319
  seasonal cycles, 190–191
  taxonomy, 183
Megilla maculata, 79
Melolonthinae, 129, 167
Metamorphosis, insect, 15–16, *15, 16*
Metarhizium anisopliae, 333, Pl. 57
  Host arthropods
    European chafer, 183
    green June beetles, 157
    Japanese beetles, 207
    May and June beetles, 193
    mole cricket, 51
    red imported fire ant, 260
  See also Natural enemies of turfgrass pests:
    pathogens, fungi
Meteorus laphygmae, 102
Mice, meadow. *See* Voles
Microcrambus elegans, (pretty crambus), 88, Pl. 17
  adult resting, 95
  fecundity, 97
  generations, *94*
Micrococcus nigrofaciens, 157
Microctonus aethiopoides, 242
Microctonus hyperodae, 242
Midges. *See* Biting midges
Milky disease, 334, Pl. 57
  host grubs
    Aphodius spp., 151, Pl. 57
    black turfgrass ataenius, 148, Pl. 57
    European chafer, 182, Pl. 57
    Japanese beetle, 206, Pl. 57
    May or June beetle, 193
    Oriental beetle, 214
    southern masked chafer, 166
  site of infection in host, 334–335
  See also Natural enemies of turfgrass pests:
    pathogens, bacteria
Milky spore disease. *See* Milky disease
Millipedes, 284–285, Pl. 65
Mites
  anatomy, *13,* 13–14, 30
    Chelicerae, *13,* 13–14
    gnathosoma, hysterosoma, idiosoma, propodoso-
      ma, *13,* 13–14
  life stages: larva, deutonymph, protonymph, 16
  sampling technique for, 318
Mocis latipes. See Striped grassworm
Mole crickets (*Scapteriscus* spp. and *Neocurtilla* spp.),
  42–53
  history and distribution, *43,* 43–44
  host plants and damage, 44–45, Pl. 10

  importance, 42
  life stages, 45–46, *45,* Pl. 9
    adult, 45–46, *45,* Pl. 9
    egg, *45,* 46
    nymph, *45,* 46
  natural enemies, 51–53
    microorganisms, 51–52, Pl. 70
    parasitoids and predators, 52–53
  sampling and trapping techniques for, 50–51, Pl. 70
  seasonal cycle, *47,* 47–50
    and timing of insecticide application, *344,* 345
  sex differentiation, 45, Pl. 9
  sound traps for, 48–49
  species differentiation, 45–46, *46*
  taxonomy, 42
  threshold populations, 51
  turf-infesting species
    northern (*Neocurtilla hexadactyla*), 270
    short-winged (*Scapteriscus abbreviatus*), 42
    southern (*Scapteriscus acletus*), 42–53, Pl. 9
    tawny (*Scapteriscus vicinus*), 42–53, Pl. 9
Moles, 293–294
  common species
    California (*Scapanus latimanus*), 294
    eastern (*Scalopus aquaticus*), 294, Pl. 67
    hairy-tailed (*Parascalops breweri*), 294
    star-nose (*Condylura cristata*), 294, Pl. 67
  control, 294
  damage to turfgrass by, 293
  as insect predators, 293–294
Mollusca, 284
Mosquitoes (*Aedes, Anopheles, Culex*), 289
Mouthparts, insect
  Chewing, 17, *18, 19*
  Sucking, 17–18, *18, 19*
  Tearing-rasping, 19
  See also Anatomy: insect, external anatomy
Mustelidae. *See* Skunks
Myiophasia metallica, 225
Myrmicinae. *See* Ants
Myzine quinquecincta, 195

Narrow-leaf plantain. *See* Plantain, narrow-leaf plantain
Natural enemies of turfgrass pests
  parasitoids, 330–332, Pls. 62, 63
    parasitoid families
      Aphidiidae (aphidiid wasps), 331, Pl. 62
      Braconidae (braconid wasps), 100, 102, 105,
        128, 242, 331–332, Pl. 63
      Ichneumonidae (ichneumonid wasps), 102, 128,
        332, Pl. 63
      Phoridae (humped-backed flies), 261, 331, Pl. 62
      Pyrgotidae (light flies), 194, 331, Pl. 62
      Scelionidae (scelionid wasps), 62–63, 67–68,
        331
      Scoliidae (scoliid wasps), 157, 214, 332, Pl. 63
      Sphecidae (digger wasps), 52, 332, Pl. 63
      Tachinidae (tachinid flies), 52, 100, 102, 115,
        195, 208, 225, 249, 331, Pl. 62
      Tiphiidae (tiphiid wasps), 166, 172, 195, 208,
        331, Pl. 62
      Trichogrammatidae (trichogrammatid wasps),
        102, 332, Pl. 63

Natural enemies of turfgrass pests (*continued*)
  pathogens, 332–337, Pls. 57–58
    bacteria, 112, 148, 151, 157, 166, 182, 193, 206, 207, 214, 260, 333–335, Pl. 57. *See also Bacillus thuringiensis;* Milky disease
    fungi, 51, 62, 67, 157, 100, 183, 193–194, 207, 225, 228, 260, 333, Pl. 57. *See also Beauveria bassiana; Metarhizium anisopliae*
    microsporidia, 99, 123, 207, 337
    nematodes, 52, 112, 157, 183, 194, 206, 214, 225, 228, 241, 249, 261, 335–336, Pl. 58. *See also Heterorhabditis bacteriophora: Steinernema*
    protozoans, 194, 207, 214, 337
    rickettsias, 172, 182, 207, 336
    viruses, 193, 337
  predators, 327–330
    invertebrate, 327–330, Pls. 59–61
      Acari (predatory mites), 33, 39, 100, 194, 261, 330, Pl. 61
      Araneae (spiders), 53, 67–68, 285, 330, Pl. 61
      Asilidae (robber flies), 100, 195, 329, Pl. 61
      Bombyliidae (bee flies), 195, 329, Pl. 61
      Carabidae (ground beetles), 100, 148, 183, 288, 328, Pl. 59
      Chilopoda (centipedes), 285, 330, Pl. 61
      Chrysopidae (green lacewings) 39, 79, 328, Pl. 59
      Cicindellidae (tiger beetles), 329, Pl. 60
      Coccinellidae (ladybird beetles), 79, 328, Pl. 60
      Dermaptera (earwigs), 100, 288, 328, Pl. 59
      Formicidae (ants), 67–68, 100, 208, 264, 329–330
      Histeridae (hister beetles), 148, 328, Pl. 59
      Lygaeidae (big-eyed bugs), 63, 67–68, 329, Pl. 60
      Reduviidae (assassin bugs), 67–68, 329, Pl. 60
      Sarcophagidae (flesh flies), 157, 329, Pl. 61
      Staphylinidae (rove beetles), 100, 148, 288, 328, Pl. 59
      Syrphidae (flower flies), 79, 329
    vertebrate, Pls. 66–69
      birds, 53, 100, 103, 148, 157, 166, 172, 195, 264, 208–209, 214, 225, 291–293, Pl. 66. *See also* Crow, American; Grackle, common; Starling, European
      mammals, 53, 79, 157, 166, 195, 208, 214, 225, 293–294, 295–298, Pls. 67–69. *See also* Armadillo; Chipmunks, Moles; Raccoon; Skunks
      reptiles, 225, 264
      toads, 53, 103, 225, 242, 264, Pl. 21
Neocunoxoides andrei, 33
Neotyphodium. See Endophytes
Nephelodes minian. See Bronzed cutworm
Nephila clavipes, 82
Nets, for insect collection, *311*
nicotine, 340
Nitrogen
  for turfgrass tolerance to insect activity, 64, 303–304
Noctuidae, 106. *See also* Armyworm; Cutworms; Fall armyworm; Granulate cutworm; Striped grassworm
Nomophila noctuella. See Lucerne moth
Northern masked chafer. *See* Masked chafers, northern
"No-see-ums." *See* Biting midges
Nosema spp., 99

Notophallus dorsalis. See Winter grain mite
Nursery stock injured by scarabaeid grubs, 90, 173, 174, 199, 210

Oats (*Avena sativa*), 36, 72, 90
Ochrosidia villosa. See Masked chafers, northern
Odonaspis ruthae. See Bermudagrass scale
Oligochaeta, 283
Oligonychus "rickiella" pratensis. See Banks grass mite
Orchardgrass (*Dactylus glomeratus*, pasturegrass), 226
Organophosphate insecticides. *See* Insecticides
Orgilus detectiformis, 100
Ormia depleta, 52
Oriental beetle (*Exomala orientalis*), 209–214
  history and damage, 209, *210*
  host plants and damage, 210, Pl. 27
  importance, 209
  life stages, 210–212, *211*, Pl. 44
  mixed populations with Japanese beetles, 209
  natural enemies, 214
  seasonal cycle, 212–214
  sex differentiation, 210, Pl. 44
  taxonomy, 209
Oriental garden beetle. *See* Asiatic garden beetle
Orthoptera. *See* Grasshoppers; Mole crickets
Oscinella frit. See Frit fly
Ovavesicula popilliae, 207

Pachyzancla spp. *See* Grass webworm; Tropical sod webworm
Painted leafhopper (*Endria inimica*), 327
Paneled crambus. *See Crambus: laqueatellus*
Parapediasia, (sod webworms)
  decorella, 88
  teterrella, (bluegrass webworm), Pl. 17
    damage, 90–91, Pl. 18
    history and distribution, 89–90
    fecundity, 97
    importance, 87
    natural enemies, 99
    oviposition, 96
    seasonal cycle
      generations, 93, *94, 95*
    sex pheromone, 97
    size, 92
    taxonomy, 87
Parthenogenesis. *See* Life stages and development, insect
Paspalum, (turfgrass)
  notatum. See Bahiagrass
  vaginaturm, ("seashore" paspalum), 80, 315
Passeriformis. *See* Crow, American; Grackle, common; Starling, European
Pathogens of insects. *See* Natural enemies of turfgrass pests
Pea mite. *See* Winter grain mite
Pearly underwing moth. *See* Variegated cutworm
Pediasia trisecta, (larger sod webworm)
  history and distribution, 89
  fecundity, 97
  host plants, damage, and importance, 89
  life stages
    adult, 91, Pl. 17

Pediasia trisecta (*continued*)
  egg, 91–92
  larva, 93
  size, 91
  natural enemies, 99
  seasonal cycle, 93, *94*
    diapause, 93
  taxonomy, 87
Penicillium spp., 194
Pennisetum clandestinum. See Kikuyugrass
Penthalus major. See Winter grain mite
Perennial ryegrass (*Lolium perenne*, turfgrass)
  adaptation, 5, 6, *9*, Pl. 1, Pl. 3
  characteristics, 6
  endophytic fungi association. *See* Endophytes
  insect and mite pests
    annual bluegrass weevil, 232
    black turfgrass ataenius, 142
    bluegrass billbug, 220
    fall armyworm 118
    greenbug, 72
    hairy chinch bug, 56
    Japanese beetle, 198
    masked chafer, 158–159
    webworms, 90
    winter grain mite, 39
  resistant cultivars
    to bluegrass billbug, 225
    to hairy chinch bug, 62
    to sod webworms, 98–99
  uses, 6
Peridroma saucia (*margaritosa*). *See* Variegated
  cutworm
Periodical cicadas (*Magicicada* spp., *M. septendecim*),
  272, Pl. 53
pH. *See* Soil
Pheidole tysoni, 100
Phenethyl lures, for Japanese beetle, 203
Phenology (insect-plant relationships)
  of annual bluegrass weevil, 236, Pl. 50
  of black turfgrass ataenius, 146
  of European chafer, 177, *179*, Pl.38
  of hairy chinch bug, 60
Pheromone, sex
  in management programs, 338
  of green June beetles, 156
  of Japanese beetle, 202–203
  of masked chafers, 163
  of May and June beetles, 192
  of oriental beetle, 213
  of sod webworms, 96
Pheropsophu spp., 53
Phoenician billbug (*Sphenophorus phoeniciensis*),
  229–230
  distribution, *229*
  seasonal cycle, 229
  taxonomy, 219
Phyllopertha horticola, (scarabaeid beetle), 212
Phyllophaga spp. *See* May or June beetle
Phytoseiulus. See Predatory mites
Pillbugs, 284, Pl. 65
Piperonyl butoxide (synergist with insect irritants), 339,
  374

Plantain, narrow-leaf plantain (*Plantago lanceolata*), 168,
  232
Plant resistance
  endophytic ryegrasses and fescues, 6, 62, 78, 112,
    224–225, 305, 337
  research techniques
  resistant cultivars
    cool-season grasses, 62, 78, 224–225, 322
    warm-season grasses, 33, 66–67, 105
Poa spp. *See* Bluegrass
Poaceae, (formerly *Graminae*), 1
Pocket gophers, 295–297, *296*, Pl. 68
  common species and geographic ranges, 296–297
    northern (*Thomomys talpoides*), 296
    plains (*Geomys bursarius*), 297
    southeastern (*Geomys pinetis*), 297
    valley (*Thomomys bottae*), 296
  control, 297
Pogonomyrmex. See Ants: harvester ant
Polyphylla
  description, 280
  habits and life history, 280
  hosts and damage, 279
  importance, 279
  species: *comes, decimlineata, variolosa*, 279
Popillia japonica. See Japanese beetle
Predatory mites, 330, Pl. 61
Preoviposition insect control. *See* Control
Preservation of turfgrass arthropods. *See* Killing and
  preservation of insects
Pretty crambus. *See Microcrambus elegans*
Procyonidae. *See* Raccoon
Prosapia bicincta. See Two-lined spittlebug
Psara spp. *See* Grass webworm; Tropical sod webworm
Pseudacteon spp., 261
Pseudalatia unipuncta. See Armyworm
Pseudaphycus spp., 273
Pseudococcidae. *See* Rhodesgrass mealybug
"Punkies." *See* Biting midges
Pyemotes tritici, 261
Pyralidae, 87
pyrethrin, 50, 102, 224, 280, 339, 351, 375. *See also*
  Irritants
pyrethrum, 90, 339, 351
Pyrgota undata, 194
Pyrustinae, 100

Quarantine regulations, 168, 180, 196

Raccoon (*Procyon lotor*), 297–298, Pl. 69
Rainfall, influence of
  on Asiatic garden beetle survival, 172
  on frit fly egg survival, 253
  on insect activity, 322–323
  on Japanese beetle oviposition, 203
  on masked chafers oviposition, 184
  on May beetle flights, 191
  on mole crickets flights, 48
  on turf regeneration, 10, 303, Pl. 5
Rasters. *See* Scarabaeidae
Rearing insects
  techniques
    for dichondra flea beetle, 218

Rearing insects (*continued*)
  for fiery skipper, 128
  for grass webworm, 102
  hor hairy cinch bug, 61
Redbud (*Cercis canadensis*), phenology of, 236
Red fescue. *See* Fescue
Red-legged earth mite. *See* Winter grain mite
Red top (*Agrostis alba*), 276
Reed canary grass (*Phalaris arundinaceae*, forage grass), 250
Reniform of moths' wings, 108, *109*
Resistance, insecticidal. *See* Insecticides
Rhizotrogus majalis. See European chafer
Rhodesgrass (*Chloris gayana*, pasturegrass), 82
Rhodesgrass mealybug (*Antonina graminis*), 82–84
  history and distribution, 82, *83*
  host plants and damage, 82–83
  importance, 82
  life stages, 83–84, Pl. 15
  natural enemies, 84
  seasonal cycle, 84
  taxonomy, 82
Rhodesgrass scale. *See* Rhodesgrass mealybug
Rhopus nigroclavatus, 273
Rhynchophorinae, 219, 226
Rice (*Oryza sativa*), 72
Rosaceae, 198
Rose of sharon (*Hibiscus syriacus*), phenology of, 146, Pl. 32
Rough bluegrass. *See* Bluegrass
Rutelinae, 196, 209
Ruth's scale. *See* Bermudagrass scale
Rye (*Secale cereale*), 36, 72
Ryegrasses (*Lolium* spp., turfgrasses)
  annual (*L. multiflorum*). *See* Annual ryegrass or Italian ryegrass
  perennial (*L. perenne*). *See* Perennial ryegrass
Rhyncophorinae, 219

Salivary system, insect. *See* Anatomy: salivary system
Sampling devices for turfgrass insects
  berlese funnel, 317, *317*
  black-light traps, 145, 318, Pl. 71
  (golf) cup cutter, 316, Pl. 71
  DeVac suction device, 254, 317
  enclosure for irritant application, 315
  insect net, 311, *311*
  pheromone or floral traps, 319, Pl. 70
  pitfall traps: conventional and linear, 319–320, *320*, *321*
  sod cutter, 317, *317*
  sound trap, 50–51
  submergence and floatation cylinders, 314–315, Pl. 70

Sand hornet. *See* Cicada killer
Sarcophaga spp., 157
Scapteriscus spp. *See* Mole crickets
Scarabaeidae (*See also names of individual species*)
  host plants and damage
    adult feeding, 130–131, Pl. 42
    larval feeding, 131–132
      progression of damage, 131–132, Pl. 27
      season of damage, 131

importance, 129–130
  destructive stages, 130
  grub complex, 129
life stages, 132–135, 138
  adults, 132–133
    external features, 134
    feeding habits, 138–139
    oviposition sequence, Pl. 29
  eggs, 133
  larvae, 133–135
    external features, 134–135
    digestive system, 135
    larva-to-adult sequence, 140, Pl. 30
    rastral patterns, 135, *136*, Pl. 28
    size differences, *133*
    vertical movement, 139
  prepupae and pupae, 135, 138
sampling for populations, 313, 316–317
seasonal cycles
  and timing of insecticide application, *344*
sexual differentiation of stages, *137*, 138
taxonomy and nomenclature, 129
Schizaphis graminum. See Greenbug
Sciuridae. *See* Chipmunk
Scolia dubia, 157
Scolia manilae, 214
Scorpiones, 286
Scorpions, 286
Season of turfgrass insect damage, 27–29, *28*
Septicemia, 333
Serratia. See Natural enemies of turfgrass pests: pathogens, bacteria
Serratia entomophila, 207
Seventeen-year locust. *See* Periodical cicadas
Sex pheromone. *See* Pheromone, sex
Shiny leaf chafer (subfamily Rutelinae), 196
Short-tailed cricket (*Anurogryllus muticus*), 271
Short-winged mole cricket. *See* Mole crickets
Shrew, short-tailed (*Blarina brevicauda*), 294
  as predator of turfgrass insects, 294
Silver-barred crambus. *See* Crambus: sperryellus
Silver-striped webworm. *See* Crambus: praefectellus
Siphona geniculata, 249
Skipper. *See* Fiery skipper
Skunks
  common species and geographic ranges
    hog-nose (*Conepatus leuconotus*), 297
    hooded (*Mephitis macroura*), 297
    spotted (*Spilogale putorius*), 297
    striped (*Mephitis mephitis*), 297, Pl. 69
  damage to turfgrass by, 297, Pl. 69
  as predator of turfgrass insects, 297
Slugs, 284, Pl. 65
Snails, 284
Snout moths. *See* Sod webworms, temperate region
Sod cutter, 317, *317*
Sod webworms, temperate region, 87–100
  common names, 88
  history and distribution, 88–90, *89*
  host plants and damage, 90–91, Pl. 18
  importance, 87–88
  description and activity of life stages
    adults, 91, *92*, 95–97, Pl. 19

Sod webworms (*continued*)
  eggs, 91, *92*, Pl. 19
  larvae, 92–93, *92*, 97–98, Pl. 18, Pl. 19
    size comparisons, 92
  pupae and cocoons, 91, *92*, 98
  major species, 88
  natural enemies, 99–100
  population thresholds, 99
  and resistant turfgrass cultivars, 98–99
  sampling for populations
    of adults with black-light traps, 318–319, Pl. 17
    of larvae with Berlese funnel, 317–318, *318*
    of larvae with irritants, 315–316
  seasonal cycle
    diapause, 98
    generations in, 93–95, *94*
    and timing of insecticide application, *344*
  sex pheromones, 96
  summer drought, 88, 91
  taxonomy, 87
  See also names of individual species
Soil
  castings, 283, Pl. 65
  influence of compaction on insects, 304
  influence of moisture on insects, 130, 139, 164–165, 181, 303
  influence of pH on insects, 304
  influence of temperature on insects, 165, 174, 181
  influence of texture on insects, 304
  inhabitants, 25–27
  sampling for insects in, 316
    using cup cutter, 316
    using sod cutter, 317, *317*
Soil-thatch interface, as insect feeding zone, 25–27
Solenopsis diplorhoptrum, 208
Solenopsis geminata, 208
Solenopsis molesta, 264
Sorghum (*Sorghum vulgare*), 56
Soricidae, 294
Sorosporella uvella, 51
Southern chinch bug. *See* Chinch bug, southern
Southern masked chafer. *See* Masked chafer, southern
Sowbugs, 284, Pl. 65
Spanogonicus albofasciata. *See* White-marked leafhopper
Species differentiation
  Annual bluegrass weevil vs. bluegrass billbug, 221–222
  Aphodius, vs. *Ataenius*, species, 149, 150, *150*
  Billbug species, 219, 229, Pl. 47
  European chafer vs. May-June beetles, 175–176, *176*
  masked chafers, northern vs. southern, 160, *160*
  mole crickets, southern vs. tawny, 45–46, Pl. 9
Specius speciosa. *See* Cicada killer
Sphenophorus, (*Calendra;* billbugs), 219–230
  differentiation from annual bluegrass weevil, 221–222
  species
    cicatristriatus, 219
    parvulus, 219–225
    phoeneciensis, 219, 229
    venatus vestitus, 219, 226
  species differentiation, 219, 229, Pl. 47

  See also Bluegrass billbug; Denver billbug; Hunting billbug; Phoenician billbug
Spiders
  harmful to humans
    black widow (*Latrodectus mactans*), 285
    brown recluse (*Loxosceles reclusa*), 285
  as insect predators, 53 285, 330, Pl. 61
Spodoptera
  frugiperda. *See* Fall armyworm
  mauritia. *See* Lawn armyworm
Spring grain aphis. *See* Greenbug
Starling, European (*Sturnus vulgaris*), 291, 330, Pl.4
  description and range, 291
  as indicator of turfgrass insects, 291
  as predator of turfgrass insects, 291
St. Augustinegrass (*Stenotaphrum secundatum,* turf grass)
  adaptation, 7, 8
  arthropod pests
    Banks grass mite, 40
    fiery skipper, 124
    grass webworm, 101
    ground pearl, 85, 82
    Rhodesgrass mealybug, 82
    southern chinch bug, 64
    tropical sod webworm, 103
  characteristics, 7, 8, *9*, Pl
  resistant cultivars and accessions, 305
  use, 8
Steinernema
  carpocapsae, 112, 147, 225, 228, 241, 249, 336
  feltiae, 157, 214, 249, 336
  glaseri, 157, 183, 206, 214, 225, 336
  riobravis, 112, 336
  scapterisci, 52, 336
Stem-thatch interface, as insect feeding zone, 23–25
Stenaptinus, 53
Steneotarsonemus spirifex, 33
Stenotaphrum secundatum. *See* St. Augustinegrass
Striped grassworm (*Mocis latipes*), 88, 278–279, Pl. 55
Striped sod webworm. *See Fissicrambus: mutabilis*
Stomatomia floridensis, 100
Sturnidae, 291
Subterranean webworm. *See Chrysoteuchia topiaria*
Surattha identella, (Buffalograss webworm), 88
Survey techniques. *See* Sampling devices for turfgrass insects

Tall fescue. *See* Fescue
Talpidae, 293
Tarnished plant bug (*Lygus lineolaris*), 274
Tehama bonifatella (western lawn moth)
  distribution, 90
  habits of larva, 98
  seasonal cycle and generations, 93–95, *94*
  sex pheromone, 96
  taxonomy, 87, Pl. 17
Telenomus nawai, 124
Temperature. *See* Degree-day accumulations
Terpinyl acetate. *See* Malt extract
Tetramorium caespitum. *See* Ants: pavement ant
Tetranychidae, 39–40

Thatch
    influence on insecticide efficacy, 342–343
    influence on insect populations, 303
    and stem inhabitants, 23–25
    and root inhabitants, 25–27
Thatch-soil interface. *See* Soil-thatch interface
Thatch-stem interface. *See* Stem-thatch interface
Thelohania spp., 99
Thermal units. *See* Degree-day accumulations
Ticks, 286–287, Pl. 64
Tiphia popillivora, (Japanese tiphia), 208
Tiphia spp., 166, 195
Tiphia vernalis, (spring or Korean tiphia), 208
Tipula paludosa. *See* European crane fly
Tipulidae, Tipulinae, 243
Toxemia, 333
Toxicity, measurements of. *See* Insecticides: toxicity
Toxoptera. *See* Greenbug
Transition belt of cool- and warm-season grasses, 5
Trapping for insects. *See* Sampling devices for turfgrass
    insects
Trichogramma minutum, 124
Trichogramma semifumatum, 102
Trombicula irritans. *See* Chigger mite
Tropical sod webworm (*Herpetogramma phaeop-*
    toralis), 103–105
    history and distribution, 103
    host plants and damage, 103
    importance, 103
    life stages, 104
    natural enemy, 105
    and resistant turfgrasses, 105
    seasonal cycle, 104–105
    sex differentiation, 104
    taxonomy, 103
Turf, turfgrass
    adaptation zones. *See* Zones
    anatomy. *See* Anatomy: grass
    appearance, rooting habits, stem and leaf character-
        istics
        cool-season grasses, 9, Pl. 1, Pl. 3
        warm-season grasses, 9, Pl. 2, Pl. 4
    climatic determinations of cool-season and warm-
        season grasses, 4–5
    cutting height and insect damage, 302–303
    definitions, 1–2, 2
    dormancy, 8, 10, Pl. 6
    fertilization influence on insect damage, 303–304,
        323
    major grasses
        cool-season, 5–7, Pl. 1
        warm-season, 7–8, Pl. 2
    vigor influence of damage threshold, 322, 323
    water management, 303, 322
    See also names of individual turfgrasses
Turfgrass scale (*Lecanopsis formicarum*), 275–276,
    Pl. 54
Turfgrass weevil. *See* Annual bluegrass weevil
Two-lined spittlebug (*Prosapia bicincta*) 78–82
    history and distribution, 80, 80, Pl. 14
    host plants and damage, 80, Pl. 14
    life stages, 80–81, 81, Pl. 14
    natural enemies, 82

seasonal cycle, 81
taxonomy, 79

Unarmed rustic. *See* Variegated cutworm
Urola nivalis, (webworm), 88

Vagabond crambus. *See Agriphila: vulgivagella*
Vanhoutte spirea (*Spiraea vanhouttei*), 131, 177, *179*, Pl.
    38
Variegated cutworm (*Peridroma saucia*)
    description, 115, Pl. 22
    history and distribution, 115
    host plants and damage, 115
    seasonal cycle, 115–116
Vegetable weevil (*Listronotus costirostris obliquus*),
    280, Pl. 56
Vertebrate predators of turfgrass insects. *See* Natural
    enemies of turfgrass pests
Vespidae, 266
Vespula maculifrons. *See* Yellow jacket
Voles, 295, Pl. 67
Voltinism. *See* Life stages and development

Warm-season grasses. *See* Turf, turfgrass: major grasses
Webworms. *See* Sod webworms
Western lawn moth. *See Tehama bonifatella*
Wheat (*Triticum sativum*), 36, 56
Wheatgrass (*Agropyron smithii*, turfgrass), 4
White grubs
    annual, 129
    common terms, 129, 183
    See also names of individual species
White-marked fleahopper (*Spanogonicus albofasciatus*),
    274–275
Wild bees, 289–290
Winter grain mite (*Penthalus major*)
    history and distribution, 35–36, *35*
    host plants and damage, 36, Pl. 8
    importance, 35
    life stages, 36–38, Pl. 8
    natural enemies, 39
    seasonal cycle, 37, 38
    taxonomy and synonyms, 35
Wireworms, 279, Pl. 56
Wolf spider, 330

Xenarthra. *See* Armadillos
Xylocoris vicarius, 68

Yellow crambus. *See Crambus: luteolellus*
Yellow jackets, 268–269, Pl. 64
Yellow rocket mustard (*Barbarea vulgaris*), 113

Zones, turfgrass adaptation, 3–5
    cool, arid-semiarid, *3*, 4
    cool, humid, *3*, 4
    warm, arid-semiarid, *3*, 4
    warm, humid, *3*, 4
Zoysiagrass (*Zoysia* spp., turfgrass), 7–8
    adaptation, 8
    characteristics, 8, *9*, Pl. 2, Pl. 14
    insect pests
        hairy chinch bug, 56

Zoysiagrass (*continued*)
    Phoenician billbug, 229
    southern chinch bug, 63
    ground pearls, 85
    lawn armyworm, 121
    tropical sod webworms, 103

    species
      japonica, (Japanese lawngrass), 7
      matrella, (Manilagrass lawngrass), 7
      tenufolia, (Korean velvet lawngrass), 7
    use, 8
Zoysiagrass mite (*Eriophyes zoysiae*), 34